Dairy Microbiology and Biochemistry

Recent Developments

Dairy Microbiology and Biochemistry

Recent Developments

Editors

Barbaros H. Özer
Ankara University
Faculty of Agriculture
Department of Dairy Technology
Ankara, Turkey

Gülsün Akdemir-Evrendilek
Abant Izzet Baysal University
Faculty of Engineering and Architecture
Department of Food Engineering
Golkoy, Bolu, Turkey

CRC Press
Taylor & Francis Group
Boca Raton London New York

CRC Press is an imprint of the
Taylor & Francis Group, an **informa** business
A SCIENCE PUBLISHERS BOOK

CRC Press
Taylor & Francis Group
6000 Broken Sound Parkway NW, Suite 300
Boca Raton, FL 33487-2742

International Standard Book Number: 978-1-4822-3502-9 (Hardback)

Library of Congress Cataloging-in-Publication Data

Dairy microbiology and biochemistry : recent developments / editors: Barbaros H. Özer, Gülsün Akdemir-Evrendilek.
pages cm
Includes bibliographical references and index.
ISBN 978-1-4822-3502-9 (hardcover : alk. paper) 1. Dairy microbiology. 2. Dairy products industry. I. ?zer, Barbaros. II. Akdemir-Evrendilek, Gulsun.

QR121.D283 2014
338.1'762142--dc23

2014013169

Visit the Taylor & Francis Web site at
http://www.taylorandfrancis.com

and the CRC Press Web site at
http://www.crcpress.com

Science Publishers Web site at
http://www.scipub.net

Preface

Dairy industry is the largest sector of the food industry in many countries. Much scientific research has been dedicated to the development of milk and dairy products, especially dairy microbiology and biochemistry. The amount and variety of dairy products are also increasing, and the technology is improving. This has brought about changes both in the range of, and handling of milk products which provided the motivation to produce this book. The basic philosophy behind this book is to provide readers with the newest scientific and legislative information regarding milk and dairy products with specific emphasis on food safety.

The book contains 16 chapters written by distinguished authors in their own field. Chapter 1 deals with the microbiology of raw milk and role of milking practices including animal health and welfare, and post-milking treatments to milk on the microbiological quality of raw milk. Chapters 2 and 9 give an overview on dairy starter technology and probiotic dairy products' technology, respectively. Chapters 3 and 4 provide brief information about the genetic properties of lactic acid bacteria that are widely employed in the manufacture of dairy products and biopreservation by lactic acid bacteria, respectively. Chapters 5, 6, 7, 8, 10 and 11 are dedicated to the microbiology of dairy products including cheese, fermented milks, powdered and concentrated milks, ice cream, etc. Functional ingredients that are indigenously present in milk and milk products and/or are formed during and/or after processing of milk are discussed in Chapter 12 in detail. The demand for minimally processed foods has been increasing all over the world. This trend has also affected dairy industry. Chapter 13 covers this aspect. It provides an overview of the non-thermal technologies used in the production of dairy products with specific emphasis on microbial safety of the end products. Chapters 14 and 15 are dedicated to the microbial safety systems for dairy processing and rapid detection of pathogenic microorganisms in dairy products, respectively. The last chapter compares the current regulations in microbial quality control of milk and dairy products that are in effect in various countries including EU, Russia and Japan.

The book is primarily intended for use by those who are involved in dairy research and processing in academia and industry as well as undergraduate, graduate students in dairy science and technology. It is not an easy task to bring individual chapters together to produce a book, but the quality of the contributions has made the editorial function a pleasure, and our sincere gratitude is extended to all those concerned.

Barbaros H. Özer
Gülsün Akdemir-Evrendilek

Contents

Preface v

List of Contributors ix

1. **Microbiology of Raw Milk** 1
Golfo Moatsou and *Ekaterini Moschopoulou*

2. **Dairy Starter Cultures** 39
Zeynep Ustunol

3. **Recent Advances in Genetics of Lactic Acid Bacteria** 68
Nefise Akçelik, Ömer Şimşek and *Mustafa Akçelik*

4. **Biopreservation by Lactic Acid Bacteria** 86
Per E.J. Saris

5. **Microbiology of Processed Liquid Milk** 95
Ebru Şenel and *Ayşe Gürsoy*

6. **Cheese Microbiology** 113
Manuela Pintado, Adriano Gomes da Cruz and *Patricia B. Zacarchenco Rodrigues de Sá*

7. **Primary Biochemical Events During Cheese Ripening** 134
A.A. Hayaloglu and *P.L.H. McSweeney*

8. **Microbiology and Biochemistry of Yogurt and Other Fermented Milk Products** 167
Barbaros Özer

9. **Development of Fermented Milk Products Containing Probiotics** 214
Claude P. Champagne

10. **Microbiology of Cream, Butter, Ice Cream and Related Products** 245
Hamid Ghoddusi and *Barbaros Özer*

11. **Microbiology of Evaporated, Condensed and Powdered Milk** 271
Ayse Demet Karaman and *Valente B. Alvarez*

12. **Functional Dairy Ingredients** 288
Ana Raquel Madureira, Ana Gomes and *Manuela Pintado*

13. **Non-Thermal Processing of Milk and Milk Products for Microbial Safety** 322
Gulsun Akdemir Evrendilek

14. **Microbiological Safety Systems for Dairy Processing** 356
Theo Varzakas

15. **Strategies for Rapid Detection of Milk-borne Pathogens** 390
Keith A. Lampel

16. **Current Regulations in Microbiological Control of Milk and Dairy Products** 404
Theo Varzakas, Ilya Vladimirovich Nikolaev and *Olga Vladimirovna Koroleva*

Index 439

Color Plate Section 447

List of Contributors

Prof. Dr. Mustafa Akçelik
Department of Biology, Faculty of Science, Ankara University, 06100, Tandogan, Ankara, Turkey.
Phone: + 90 (312) 212 6720
Fax: + 90 (312) 223 2395
Email: akcelik@science.ankara.edu.tr

Dr. Nefise Akçelik
Instititute of Biotechnology, Ankara University, 06100, Tandogan, Ankara, Turkey.
Phone: +90 (312) 212 6720
Fax: +90 (312) 223 2395
Email: nakcelik@ankara.edu.tr

Prof. Dr. Valente B. Alvarez
The Ohio State University, Gould Food Industries Center, Dept. Food Science and Technology, 2015 Fyffe Road, Columbus, OH, USA, 43210.
Phone: +1 (614) 688 4961
Fax: +1 (614) 688 5459
Email: alvarez.23@osu.edu

Dr. Claude P. Champagne
Food R & D Centre, Agriculture and Agri-Food Canada, 3600 Casavant, St. Hyacinthe, QC, J2S 8E3, Canada.
Phone: +1 (450) 768 3238
Fax: +1 (450) 773 8461
Email: Claude.Champagne@agr.gc.ca

Prof. Dr. Adriano Gomes da Cruz
Instituto Federal de Educação, Ciencia e Technologia do Rio de Jenerşo (IFRJ), Departmento de Alimentos , Rua Senador Furtado, 171, Maracanã, Rio de Jenerio, Brazil.
Phone: +55 (21) 2566 7000
Fax: +55 (21) 2566 7000
Email: food@globo.com

Dr. Hamid Ghoddusi
London Metropolitan University, School of Human Sciences, Microbiology Research Unit, London, UK.
Phone: +44 (20) 7133 4196
Fax: +44 (20) 7133 4682
Email: h.ghoddusi@londonmet.ac.uk

Dr. Ana Gomes
Centro de Biotecnologia e Química Fina, Escola Superior de Biotecnologia, Universidade Católica do Porto, Rua Dr. António Bernardino de Almeida 4200-072 Porto, Portugal.
Phone: +351 (22) 558 0084
Fax: +351 (22) 509 0351
Email: amgomes@porto.ucp.pt

Dr. Ayşe Gürsoy
Ankara University Faculty of Agriculture Department of Dairy Technology, Ankara, Turkey.
Phone: +90 (312) 596 1353
Fax: +90 (312) 318 2219
Email: gursoy@agri.ankara.edu.tr

Dr. A. A. Hayaloglu
Department of Food Engineering, Inonu University, 44280 Malatya, Turkey.
Phone: +90 (422) 377 4792
Fax: +90 (422) 411 0046
Email: adnan.hayaloglu@inonu.edu.tr

Dr. Ayse Demet Karaman
Adnan Menderes University, Faculty of Agriculture, Dept. Dairy Technology, Aydin, Turkey 09100.
Phone: +90 (256) 772 70 22
Fax: +90 (256) 773 72 33
Email: demet.karaman@adu.edu.tr

Dr. Keith A. Lampel
Food and Drug Administration, Center for Food Safety and Applied Nutrition, 5100 Paint Branch Parkway, College Park, Maryland, USA.
Phone: +1 (240) 402 2007
Fax: +1 (240) 402 2599
Email: Keith.lampel@fda.hhs.gov

Dr. Ana Raquel Madureira
Centro de Biotecnologia e Química Fina, Escola Superior de Biotecnologia, Universidade Católica do Porto, Rua Dr. António Bernardino de Almeida 4200-072 Porto, Portugal.
Phone: +351 225 5880044
Fax: +351 22 509 0351
Email: rmadureira@porto.ucp.pt

Prof. Dr. P.L.H. McSweeney
Department of Food and Nutritional Sciences, University College, Cork, Ireland.
Phone: +353 (21) 490 2011
Fax: +353 (21) 427 6398
Email: p.mcsweeney@ucc.ie

Dr. Golfo Moatsou
Laboratory of Dairy Research, Department of Food Science and Human Nutrition, Agricultural University of Athens, Iera Odos 75, Athens 118 55, Greece.
Phone: +30 (210) 529 4630/680
Fax: +30 (210) 5294672
Email: mg@aua.gr

Dr. Ekaterini Moschopoulou
Laboratory of Dairy Research, Department of Food Science and Human Nutrition, Agricultural University of Athens, Iera Odos, Athens, Grece.
Phone: +30 (210) 529 4680
Fax: +30 (210) 529 4672
Email: catmos@aua.gr

Dr. Manuela Pintado
Centro de Biotecnologia e Química Fina, Escola Superior de Biotecnologia, Universidade Católica do Porto, Rua Dr. António Bernardino de Almeida 4200-072 Porto, Portugal.
Phone: +351 (22) 558 0097
Fax: +351 (22) 509 0351
Email: mpintado@porto.ucp.pt

Dr. Patricia B. Zacarchenco Rodrigues de Sá
Instituto de Technologia de Alimentos (ITAL), Centro de Tecnologia de Laticínos (TECNOLAT) Avenida Brasil, São Paulo, Brazil.
Phone: +55 (19) 374 31860
Fax: +55 (19) 374 31862
Email: pblumer@ital.sp.gov.br

Prof. Dr. Per E.J. Saris
Department of Food and Environmental Sciences, University of Helsinki, P.O. Box 56, FI-00014, Finland.
Phone: + 358 9 19159369
Mobile: + 358 50 5203696
Fax: + 358 9 19159322
Email: per.saris@helsinki.fi

Dr. Ebru Şenel
Ankara University Faculty of Agriculture Department of Dairy Technology, Ankara, Turkey.
Phone: +90 (312) 596 1300
Fax: +90 (312) 318 2219
Email: senel@agri.ankara.edu.tr

Dr. Ömer Şimşek
Department of Food Engineering, Faculty of Engineering, Pamukkale University, 20700, Kinikli, Denizli, Turkey.
Phone: + 90 (258) 296 3015
Fax: + 90 (258) 296 3262
Email: omers@pau.edu.tr

Prof. Dr. Zeynep Ustunol
Michigan State University Department of Food Science and Human Nutrition, 2105 S. Anthony Hall, 474 S. Shaw Lane, E. Lansing, MI 48824, USA.
Phone: +1 (517) 355 7713/184
Fax: +1 (517) 353 1676
Email: ustunol@anr. msu.edu

Dr. Theo Varzakas
Technological Educational Institute of Peloponnese, School of Agricultural Technology, Food Technology and Nutrition, Department of Food Technology, Antikalamos, 24100, Kalamata, Greece.
Phone: +30 (272) 104 5279
Fax: +30 (272) 104 5234
Email: theovarzakas@yahoo.gr

Dr. Nikolaev Ilya Vladimirovich
Federal State Budget Research Institution of Science, A.N. Bach Institute of Biochemistry of Russian Academy of Sciences, Moscow, Russia.
Phone:+7 (495) 954 4477
Fax:+7 (495) 954 2732
Email: ilya_mbf@yahoo.com

Prof. Koroleva Olga Vladimirovna
Federal State Budget Research Institution of Science, A.N. Bach Institute of Biochemistry of Russian Academy of Sciences, Moscow, Russia.
Phone: +7 (495) 952 8799
Fax: +7 (495) 954 2732
Email: koroleva@inbi.ras.ru

CHAPTER 1

Microbiology of Raw Milk

*Golfo Moatsou** and *Ekaterini Moschopoulou*

1.1 Microorganisms associated with raw milk

Milk is an ideal medium for microbial growth because of its high water content, neutral pH and biochemical composition. Therefore, raw milk may contain various kinds of microorganisms with variable characteristics in respect to classification, morphology and physiology. Very important for the quality of raw milk and dairy products are bacteria that predominate among all kinds of milk microorganisms. Bacteria in raw milk can be spoilage or pathogenic with mesophilic, psychrophilic or thermophilic behavior. In brief, bacterial growth is divided into four phases: i.e., lag, exponential or log, stationary and dying-off phases (Walstra et al. 2006). Multiplication of bacteria shows a geometric progression and the bacterial growth during the log phase is described by the generation time (g) that is the time needed for a full cell division. Generation time in raw milk depends mainly on species or strains of bacteria as well as temperature, pH, level of oxygen, inhibitors and nutrients. Thus, the profile of initial microflora and the handling of raw milk regarding hygienic and temperature conditions are the determinative factors for raw milk quality before processing. Raw milk microflora is of critical importance for consumers' safety and quality and shelf-life of dairy products. Raw milk microflora could be grouped as indigenous or contaminants and also as spoilage or pathogenic microorganisms.

1.1.1 Indigenous microflora

Normally the udder of a healthy animal is habited by bacteria that belong to genera *Streptococcus*, *Staphylococcus* and *Micrococcus* which account for >50% of overall raw milk flora, followed by *Corynebacterium*,

*Corresponding author

Escherichia coli and others (ICMSF 1998). Microbial counts of aseptically drawn milk is <100 cfu ml^{-1} (Walstra et al. 2006), but in practice they usually range from <1000 cfu ml^{-1} to 20,000 cfu ml^{-1} (Chambers 2002).

1.1.2 Contaminant microorganisms

After secretion, the initial microbial load of raw milk changes because microorganisms from different sources, i.e., environment, infected udder, milk equipment, etc., enter the milk. The contaminant microorganisms, which belong to different genera, are distributed as follows: *Lactobacillus, Corynebacterium, Microbacterium, Pseudomonas, Escherichia, Alcaligenes, Acinetobacter, Bacillus, Clostridium,* yeasts and moulds at levels of <10%, *Lactococcus* and *Streptococcus* at varying levels from 0 to 50% and *Micrococcus* and *Staphylococcus* at levels varying from 30 to 99% (Chambers 2002, Frank and Hassan 2002).

In general, both indigenous and contaminant microflora of raw milk are classified into two categories: (a) the spoilage and (b) the pathogenic microorganisms, which can cause dairy animal or human diseases. From technological point of view, spoilage microorganisms present in milk are grouped as: (i) psychrotrophs which can grow during milk storage at 6°C or less, (ii) coliforms and other Gram-negative bacteria which are associated with poor hygienic production, (iii) thermoduric bacteria which can survive pasteurization conditions, (iv) sporeforming bacteria which produce heat-resistant spores, (v) lactic acid bacteria of which some are beneficial as they are used as starters in fermented dairy products, and (vi) yeasts and moulds. Table 1.1 presents the spoilage microorganisms associated with raw milk, their common sources and spoilage potential.

1.2 Types of spoilage microorganisms present in raw milk

1.2.1 Psychrotrophic bacteria

The most effective measure for controlling the microbiological quality of raw milk is immediate cooling after milking and storage at low temperatures, i.e., at 2–7°C. Cooling decreases effectively the growth rate of the main spoilage microorganisms, but it also changes the microbiological profile of raw milk dramatically. Under these conditions, psychrotrophic bacteria predominate the milk microflora depending on the duration of the refrigerated storage. The most common psychrotrophic bacteria in raw milk are listed in Table 1.2. As defined by Jay (2000), psychrotrophs are microorganisms that can grow at temperatures between 0 and 7°C and produce visible colonies or turbidity within 7–10 days. There are particular mechanisms of the psychrotrophic cells that are related to their ability to grow at low

Table 1.1 Predominant spoilage microorganisms associated with raw milk.

Microorganism	Common sources	Spoilage
Sporeforming bacteria		
Bacillus cereus[a,c]	Feed, dung, soil, dust	Sweet curdling, bitty cream in pasteurized milk and cream
Bacillus subtilis[b]	Feed, dung, soil, dust	Spoil sterilized cream
Geobacillus stearothermophilus[b]	Feed, soil	Spoil evaporated milk
Clostridium tyrobutyricum[b]	Soil, silage, dung	Late blowing in cheese
Coliforms		
Escherichia coli[a,c]	Feces, milking utensils, contaminated water	Spoil milk and cheese
Klebsiella aerogenes	Feces, milking utensils, contaminated water	Spoil milk
Lactic acid bacteria		
Lactobacillus spp.	Milking utensils, parlor	Sour milk
Lactococcus spp.	Milking utensils, parlor	Sour milk
Leuconostoc spp.	Milking utensils, parlor	Sour milk
Streptococcus thermophilus[b]	Milking utensils, parlor	Sour milk
Enterococcus spp.	Feces	Spoil milk
Psychrotrophs		
Pseudomonas spp., *Achromobacter* spp., *Aeromonas* spp., *Alcaligenes* spp.,[b] *Chromobacterium* spp., *Flavobacterium* spp.	Milking utensils, cold-stored milk	Protein and fat hydrolysis in cold-stored milk
Thermoduring bacteria		
Micrococcus spp., *Microbacterium* spp.	Milking utensils	Growth on pasteurized products
Yeasts	Dust, milking utensils	Spoil cheese, butter, sweetened condensed milk
Molds	Dust, dirty surfaces, feed	Spoil yogurt, cheese, butter, sweetened condensed milk

[a]Pathogen; [b]Thermoduric; [c]Some strains are psychrotrophic
Data compiled from: Chambers (2002), Frank and Hassan (2002) and Walstra et al. (2006).

temperatures. At low temperatures, they increase the synthesis of neutral lipids and phospholipids with an elevated proportion of unsaturated fatty acids, regulating thus their lipid fluidity. Growth of psychrotrophs decreases more slowly than that of mesophiles when temperature is decreased, since the former sustain enzymatic activities at low temperatures. Furthermore, psychrotrophic membranes transport solutes efficiently at low temperatures opposite to mesophiles. This behavior is due to the greater fluidity of their membrane resulted from its lipid synthesis in combination to the activity of their transport proteases under these conditions (Jay 2000).

One of the major characteristics of psychrotrophic bacteria in milk is their ability to produce extracellular enzymes that can attack milk constituents. Bacteria of this group are in general inactivated by heat treatments applied to market milk products; however, their enzymes may be extremely heat-

Table 1.2 Most common psychrotrophic bacteria in raw milk.

Groups	Pathogens[a]	Gram staining
Pseudomonas		–
Flavobacterium		–
Alcaligenes		–
Enterobacteriaceae	*Escherichia coli* O157:H7, *Yersinia enterocolitica*	–
Acinetobacter		–
Aeromonas	*Aeromonas hydrophila*	–
Bacillus	Strains of *B. cereus*	+
Clostridium		+
Arthrobacter[b]		+
Streptococcus[b]		+
Corynebacteria[b]		+
Lactobacillus		+
Micrococcus[b]		+
Others	*Listeria monocytogenes*	+

[a] Occurrence of pathogens or sporeformers within these groups
[b] Thermoduric strains surviving pasteurization are included in these groups.
Data compiled from: Champagne et al. (1994), Sørhaug and Stepaniak (1997), Stepaniak (2003).

resistant remaining active in the products, decreasing thus their shelf-life. In fact, an initial count as little as 10^2 cfu ml^{-1} can spoil raw milk during cold storage within five days (Frank and Hassan 2002). Development of bitter taste and off-flavors or gelation in UHT products have been also related to these enzymes. The heat-stable proteinases from *Pseudomonas* strains affect the plasmin system in milk by releasing plasmin and plasminogen from the casein micelles into the whey fraction or can stimulate plasminogen activators (Fajardo-Lira and Nielsen 1998, Frohbieter et al. 2005). Moreover, some thermoduric bacteria that can survive pasteurization and are able to grow during refrigerated storage of dairy products belong to this group. Finally, the existence of pathogens and sporeformers in this group stresses the significance of psychrotrophs for both raw and processed milks. In the present section, aspects of psychrotrophic microflora in relation only to raw milk will be presented.

Psychrotrophic bacteria found in raw milk belong mainly to Gram-negative genera *Pseudomonas, Enterobacter, Flavobacterium, Klebsiella, Aeromonas, Acinetobacter, Alcaligenes, Achromobacter* and *Serratia* (Table 1.2). Among them, *Pseudomonas fluorescens, P. putida, P. fragi, P. putrefaciens* and *P. lundensis* are the most commonly isolated species, whereas other species and genera have been reported as minor groups (Cousin 1982, Ternström et al. 1993, Jayarao and Wang 1999). Gram-negative microflora is estimated to account for >90% of the total psychrotrophic microflora of raw milk (Muir

1996a). *Pseudomonas* spp. and *Enterobacteriaceae* are the most abundant microorganisms in cold stored raw milk; the former account for up to 95% of the isolates, whereas the latter vary from 3 to about 15% (Griffiths et al. 1987, Champagne et al. 1994, Stepaniak 2003). Finally, apart from Gram-negative bacteria, Gram-positive bacteria such as *Bacillus*, *Clostridium*, *Corynebacterium*, *Microbacterium*, *Micrococcus*, *Streptococcus* and *Lactobacillus* are also psychrotrophs of raw milk (Sørhaug and Stepaniak 1997). Gram-positive bacteria are a small part of psychrotrophic microflora of milk, being ≤14% of the isolates obtained from cold stored raw milks (Champaigne et al. 1994, Stepaniak 2003). The sporeforming *Bacillus* and *Clostridium* species, and thermoduric *Micrococcus* spp., *Corynebacterium* spp. and *Streptococcus* spp. are the most important members of Gram-positive psychrotrophs in raw milk. A study has indicated that many novel species and genera are yet to be defined since about 20% of the psychrotrophic isolates from raw milk from four farms, analyzed over a 10-month period, were considered as novel species (Hantsis-Zacharov and Halpern 2007).

Both initial level of bacterial counts and storage temperature affect psychrotrophic counts and, therefore, the storage life of raw milk under refrigerated conditions. At 6–8°C, the generation times of psychrotrophs vary substantially among genera, species and strains, i.e., from 4 to 12 hr. According to the findings of Griffiths et al. (1987), in farm bulk tank milk, the time taken for psychrotrophs count to increase from the initial count of 2.6×10^2 cfu ml^{-1} to 10^6 cfu ml^{-1} was 2.9 days at 6°C and five days at 2°C. Storage of raw milk at 2°C can result in a 1.8-fold increase in storage life compared to storage at 6°C. The relation between initial microflora and time taken for psychrotrophs count to reach the critical level of 10^6 cfu ml^{-1} is linear but this correlation weakens with the decrease of storage temperature from 6 to 2°C, indicating that only a part of these bacteria are able to grow well at 2°C (Griffiths et al. 1987). Also, Griffiths and Phillips (1988) showed that the lag phase duration depends strongly on storage temperature. Forty eight hours of cold storage is in general considered as a critical time interval for the psychrotrophic growth. However, Lafarge et al. (2004) found that the counts of psychrotrophs increased markedly at 4°C within 24 hr. Therefore, the shelf-life of raw milk at >4°C cannot be really controlled and deep cooling to 2°C is suggested for a 48 hr increase of raw milk's shelf-life (Muir 1996a,b, Walstra et al. 2006). Finally, it has to be taken into account that psychrotrophic bacteria of the bulk milk that enter a dairy silo after a 48 hr stay in farm storage tanks can be in the exponential growth phase, therefore silo milk is more susceptible to spoilage (Muir 1996a).

Ma et al. (2003) reported that low (3.1×10^4 cells ml^{-1}) and high (1.1×10^6 cells ml^{-1}) somatic cell counts had no effect on microbial growth during cold storage of fresh raw milk. Although, milks had standard plate counts about one log cycle higher than psychrotrophic bacterial counts, i.e., 10^4 and

10^3 cfu ml^{-1}, respectively; after seven days of storage at 4°C, the counts of both groups were similar, i.e., close to 10^6 cfu ml^{-1}. A remarkable increase of proteolysis was observed at 4°C between 7 and 14 days of storage, whereas, lipolysis in terms of meq FFA g^{-1} of milk was significant after 14 days at 4°C. Ercolini et al. (2009) reported that various strains of *Pseudomonas* spp. isolated from raw milk exhibited full growth in synthetic growth media at 20°C within 1–2 days and at 30°C within 1–7 days. The growth was substantially suppressed at 7°C, since 5–17 days were needed for full growth. Furthermore, the proteolytic activity was affected by storage temperature and in some cases it was not apparent at 7°C.

Hantsis-Zacharov and Halpern (2007) reported that an average percentage of psychrotrophic bacteria, which are able to grow at both 7°C and 30°C in the total mesophilic bacteria in raw milk, was 14.7%. The dominant genera were *Pseudomonas* and *Acinetobacter* of the *Gammaproteobacteria* class and a seasonal effect was observed, i.e., *Gammaproteobacteria* were predominant in spring and winter, *Bacilli* in summer and *Actinobacteria* in autumn. Although all four classes were observed in the milk obtained from different dairies, each dairy had a unique "bacterial profile". In 72% of 75 milk samples taken from bulk milks in Denmark, Gram-negative, oxidase-positive bacteria were found, whereas in 28% of milk samples psychrotrophic bacteria were dominant, with *P. fluorescens* and *Pseudomonas* spp. being the common species (Holm et al. 2004). Similar microbial profile but much higher counts have been also reported for raw cold stored ovine milk (Sanjuan et al. 2003, de Garnica et al. 2011).

Rasolofo et al. (2010) demonstrated that the class of *Bacilli* accounted for about two-third of the 16S RNA clones isolated from raw milk stored at 4°C for three days, followed by *Staphylococcus* spp. (one-third) and *Clostridium* spp. Under these conditions, *Pseudomonas* spp. accounted only for 2.4% of the isolates, whereas it increased noticeably after three days at 8°C. However, after seven days of storage at 4°C, this genus was by far the most abundant one accounting for 94.2% of total 16S rDNA clones. The shares of *Bacilli* (3.7%) and *Staphylococcus* (1.6%) were fairly small. However, limitations of culture-independent techniques must be taken into account in the assessment of raw milk microflora during storage, i.e., differentiation between DNA of live and dead cells, nucleic acid extraction, the target region and the molecular methods employed (Quigley et al. 2011).

Phillips and Griffiths (1987) estimated the temperature-growth parameters of psychrotrophs in dairy products. The apparent activation energy (μ) and conceptual minimum temperature (T_o) for growth of psychrotrophs were from 10.9 to 26.1 kcal mol^{-1} and 260.2 to 270.1°K, respectively. These values varied depending on the *Pseudomonas* strain and the dairy products used as growth medium. *Pseudomonas* spp. and in particular *P. fluorescens* is the most important psychrotroph for the spoilage of raw or pasteurized milk, although it accounts for <10% of the initial milk microflora. This Gram-negative genus includes aerobic, motile, catalase- and oxidase-positive rods with largely non-fermentative metabolism. However, it is strongly lipolytic and proteolytic capable of hydrolyzing all available casein fractions. It includes species with the shortest generation times at 0–7°C and the lowest theoretical minimum growth temperatures, i.e., –10°C (Sørhaug and Stepaniak 1997, Jay 2000, McPhee and Griffiths 2002).

P. fluorescens is found in soil and water but a major source for milk is the milking utensils, the storage tanks and the transport equipment. The surfaces of milk handling equipment must be cleaned and disinfected properly to avoid the development of bacterial biofilms. It has been proven that these surface associated three-dimensional groups of bacterial colonies grown on milk residues are held together by glycocalyx produced during bacterial metabolism, and can survive after several cleaning processes. Psychrotrophic *Pseudomonas* spp., *E. coli* strains, heat-resistant streptococci and *Bacillus* spores have a major role in the formation of biofilms in raw milk handling equipment (Kulozik 2002). This is particularly important for *P. fluorescens*, *P. fragi* and *P. lundensis* which are the most common spoilage psychrotrophs of cold stored raw milk. According to McPhee and Griffiths (2002), the most possible sources of contamination of raw milk by *Pseudomonas* spp. are milk pipelines, agitators, dipsticks, outlet plugs and cocks of the milk storage tanks as well as air separator, milk-meter, milk sieve and the suction hose in milk tanker. Adequate cleaning-in-place (CIP) procedures for raw milk handling equipment must be implemented considering also the hardness of utilized water in relation to mineral deposits on the surfaces (Chambers 2002, Walstra et al. 2006, McPhee and Griffiths 2002).

The psychrotrophic microflora of raw milk comprises some important pathogens including Gram-negative *Aeromonas hydrophila* and *Yersinia enterocolitica*, Gram-positive *Listeria monocytogenes* and toxin-producing *Bacillus cereus* strains. If the latter exceeds 1×10^7 cfu ml^{-1}, it can produce two types of toxins: heat-labile diarrheagenic and heat-stable emetic toxins.

As mentioned earlier, the major characteristic of psychrotrophs is the production of extracellular enzymes which play active role in degradation of certain milk compounds resulting in flavor defects. The critical counts in regard to the production of these enzymes and the appearance of flavor defects in the products are about 10^6 cfu ml^{-1} for *Pseudomonas* spp. and *Bacillus* spp. (Stepaniak 2003). These enzymes are usually produced from *P. fluorescens* in the late log/early stationary growth phase when the cell count is high, i.e., $\geq10^6$ cfu ml^{-1} (Dunstall et al. 2005).

Extracellular thermostable proteinases, lipases, phospholipases, exopeptidases and glycosidases are produced by psychrotrophic bacteria in milk. They are secreted at the end of the stationary phase of the growth and they can accumulate in tanks and pipelines not properly cleaned. In addition, psychrotrophs also produce intracellular and cell-bound peptidases and esterases. The extracellular enzymes of psychrotrophs belong to two different groups (Sørhaug and Stepaniak 1997):

- Proteinases, lipases and phospholipases that survive pasteurization and UHT treatment but are not active above the temperatures of 50–60°C. They have temperature optima at 30–45°C, require low activation energy and thus are more active at 4–7°C than enzymes of mesophiles. Usually deviations of the inactivation kinetics from Arrhenius plot are observed. Of particular importance is the sensitivity of most of the proteinases and lipases from *Pseudomonas* spp. to heat treatment at low temperatures, i.e., 50–60°C, opposite to their notorious heat-resistance at temperatures >100°C.
- Thermoenzymes from some thermophilic bacteria that are stable at 60–80°C or higher; the inactivation kinetics of these enzymes do not deviate markedly from the Arrhenius plot.

Due to their technological significance, the enzymes of psychrotrophs have been studied extensively for both their specificities and heat tolerance. The findings of these studies have been presented in several papers, which are the source of below-presented information (Law 1979, Cousin 1982, Christiansson 2002, McPhee and Griffiths 2002, Stepaniak 2003, Champagne et al. 1994, Sørhaug and Stepaniak 1997, Dunstall et al. 2005, de Jonghe et al. 2010).

- At low temperatures, proteinases, lipases and phospholipases from *Pseudomonas* spp. are synthesized at the end of log phase of growth,
- Low temperatures may induce the production of *Pseudomonas* proteinases, which are a diverse group of metallo-enzymes containing one zinc atom, up to 16 calcium atoms per molecule and have milk clotting activity. They also hydrolyze all milk caseins but not whey proteins. Mostly, a single proteinase is secreted by a particular strain, although two or three types can also be produced by particular strains,
- In general, *Pseudomonas* produces only one type of lipase that can hydrolyze many natural oils and synthetic triglycerides. Also, phospholipase (lecithinase) produced by *Pseudomonas* can hydrolyze the milk fat globule membrane that leaves the milk fat unprotected against lipase action,
- Extracellular proteolytic activity, which is heat-stable at 140°C for 5 s has been observed in *Bacillus, Enterobacter, Serratia, Alcaligenes, Flavobacterium* and *Achromobacter* strains or species. Similarly, lipase secreted by *Bacillus, Enterobacter, Serratia, Citrobacter, Moraxella* and

Achromobacter can withstand high heat treatments. Finally, heat-stable phospholipase are also produced by *Bacillus*, *Flavobacterium*, *Alcaligenes* and *Aeromonas* species or strains,

- Proteinases of Gram-positive and Gram-negative psychrotrophs exhibit different specificities against individual caseins. At first, their activities result in the development of off-flavor in dairy products, whereas gelation and sweet curdling are observed after the advancement of proteolysis,
- Various psychrotrophic *Bacillus* spp. induce flavor defects in milk stored at 7°C when their counts exceed 10^7 cfu ml^{-1},
- Lipolytic activity of thermoduric psychrotrophs (e.g., *Corynebacterium*, *Micrococcus*) results often in rancid and fruity off-flavors in dairy products,
- Phospholipases, proteinases and glycosidases from *Pseudomonas*, *Citrobacter* and *Enterobacter* can damage milk fat globule membrane by synergistic action,
- *B. cereus* is a very important psychrotolerant for the dairy industry in relation to its ability to produce endospores that survive heat treatments. Vegetative cells of the most strains produce proteases causing sweet curdling, when their counts are >10^6 cfu ml^{-1}. They also produce lipase and phospholipase acting against phospholipids causing fat accumulation defect in cream, also called *"bitty cream"* defect. On contrary, Gram-negative psychrotrophs do usually not cause bitty cream defect in dairy products. The majority of *B. cereus* strains do not grow on lactose but can ferment other carbohydrates, e.g., glucose, fructose, trehalose, N-acetyl glucosamine and mannose,
- Some strains of *Bacillus* spp., e.g., strains of *B. amyloliquefaciens*, *B. claussi*, *B. subtilis* and *Paenibacillus polymyxa* (formerly *B. polymyxa*) (an aerobic sporeformer) are able to reduce nitrate to nitrite, which is very important for controlling the growth of *Clostridium* in cheese, and
- Recently, lecithinase (phospholipase) activity has been found in *P. polymyxa*, which can also produce gas from lactose fermentation.

1.2.2 Thermoduric bacteria

Under pasteurization conditions, i.e., heat treatment at 63°C for 30 min or equivalent, non-sporeforming pathogens, yeasts and moulds, Gram-negative and many Gram-positive bacteria are destroyed. However, thermodurics and thermophiles can survive under these conditions and along with sporeformers can decrease the shelf-life of dairy products kept under non-refrigerated conditions. Thermoduric bacteria can survive at high temperatures but do not necessarily grow at these temperatures opposite to thermophiles that require high temperatures for their growth (Jay 2000). They do not form spores and they can be a very important spoilage factor for pasteurized dairy products provided the psychrotrophic count of raw milk and recontamination have been efficiently controlled. Thermoduric bacteria are very important for cheese because cheese-making conditions are favorable for their growth. In general, a laboratory pasteurization count exceeding 500 cfu thermoduric bacteria per ml indicates major thermoduric problem in the raw milk production chain (Hayes and Boor 2001). Thermoduric species of raw milk include *Microbacterium* spp. (e.g., *M. lacticum*), *Micrococcus* spp., spores of *Bacillus* and *Clostridium*, *Streptococcus* (e.g., *S. thermophilus*), *Corynebacterium* spp., *Enterococcus* spp. (e.g., *E. faecium*) and *Lactobacillus* spp. (Walstra et al. 2006).

The sources of thermodurics in raw milk are infected udder and outside udder and teats, as well as soil, water and milking machines. They can grow fast along with lactic acid bacteria, when raw milk is kept under non-refrigerated conditions. A healthy udder is also a source of thermodurics in raw milk since the predominant types of bacteria inside a healthy udder include *Micrococcus* (mesophilic, aerobic, Gram-positive cocci), *Streptococcus* (mesophilic, facultative anaerobic, Gram-positive cocci) and *Corynebacterium* (non-sporeforming, mesophilic, facultative anaerobic and Gram-positive irregular rods). *Enterococcus* spp. come mainly from animal environment. They are mesophilic, facultative anaerobes and Gram-positive cocci which are used as indicators of sanitation (Frank and Hassan 2002, Ray 2004).

In general, thermodurics dominate sections in milk production chain, where other bacteria do not survive due to high temperatures, e.g., during the high temperature applied for cleaning the milking units or regeneration section of the pasteurizer. $D_{63°C}$-values for *Enterococcus* spp. in skim milk range from 2.6 to 10.3, *E. faecium* being the most thermoduric with a $D_{84°C}$-value of 2.5 to 7.5 min. *S. thermophilus*, *E. durans* and *E. faecalis* may colonize in the regeneration section (Walstra et al. 2006).

1.2.3 Sporeforming bacteria

These microorganisms belong mainly to the genera of *Bacillus*, *Clostridium* and *Geobacillus*. They are Gram-positive, aerobic or facultative anaerobic, except for *Clostridium* spp. that are strictly anaerobic. The latter grow in cheese rather than in milk. *C. tyrobutyricum* causes late blowing in hard-type cheeses with high pH and low salt, fermenting lactic acid to produce butyric acid, CO_2 and H_2. *C. sporogenes* and *C. butyricum* are also involved in cheese defects such as putrid spots in the Swiss cheese. Although *C. perfringens* has not been widely associated with milk-based powdered products, owing to its survival under extreme conditions, it may pose a potential health risk in milk powder (Frank and Hassan 2002, Burgess et al. 2010).

The main aspects of bacterial sporulation and germination are presented by Ray (2004). In brief, the spores of bacterial cells are inside the cell, i.e., one endospore per cell, opposite to yeast and moulds spores. Bacterial spores are spheroid or oval with terminal, central or off-center position. Their formation is triggered by the environmental conditions, i.e., reduction of nutrients (mainly carbon, nitrogen and phosphorus) and changes in optimum pH and temperature conditions. Bacterial sporulation takes place only at the end of completion of DNA replication. The bacterial spores are metabolically inactive or dormant but they can emerge as one vegetative cell per spore under favorable conditions. Some spores of *Bacillus* and *Clostridium* may need a fairly long time before germination and they are called superdormant spores. The life cycle of a sporeforming bacterium has a vegetative cycle (by binary fission) and a spore cycle. The latter goes through several stages, which are genetically controlled and affected by environmental parameters and biochemical processes. Spores can be activated before germination by sub-lethal heat treatments, radiation, high pressure, sonication and extreme pH. Recently, Burgess et al. (2010) reviewed the findings about spore formation and germination of sporeforming bacteria important for dairy science and technology. Spore formation within a dairy environment may be related to magnesium, calcium and potassium compounds that are very important for spore structure as well as for the activation of the spore formation process. Activation, germination and outgrowth are the three steps needed for the change from a spore to a vegetative cell. Heat treatment, application of chemicals and decrease of pH to 2–3 can activate spores. Heat activation is species-specific. Superdormant spores may require higher activation temperature. Germination may be triggered by nutrients such as L-alanine or by high pressure as well as salts or lysozyme. The most common aerobic sporeforming bacteria occurring in raw milk with some examples of their heat resistance are presented in Table 1.3.

Endo-sporeforming bacteria like *Bacillus* spp. and *Paenibacillus lactis* isolated from raw milk are present in the environment of dairy farms and dairy factories (Scheldeman et al. 2005, Huck et al. 2008). *B. cereus*, *B. subtilis*, *B. licheniformis*, *B. sporothermodurans* and *Geobacillus stearothermophilus* (formerly *B. stearothermophilus*) are the most common *Bacillus* species found in raw milk (Table 1.1). They form heat- and chemical-resistant spores that cause defects in heat-treated dairy products; thus, determining their shelf-life. *B. cereus* causes sweet curdling of pasteurized milk and fat aggregation in cream (bitty cream). *B. subtilis* and *B. licheniformis* are more heat-resistant than *B. cereus* and spoil both sterilized and UHT milks. The thermophilic *G. stearothermophilus* is the most heat-resistant of this group causing flat sour spoilage and sweet curdling defects in evaporated milks. The vegetative

Table 1.3 Aerobic sporeforming bacteria occurring in raw milk.

Type	**Genera**	**Usual counts in raw milk**	**Most common species**	**Inactivation of endospores (indicative D-values)**[d]
Psychrotolerant	*Bacillus* *Paenibacillus*	0.003–3.5 spores ml^{-1}	*B. cereus, B. circulans, P. polymyxa, B. mycoides, B. thuringiensis*	$D_{100°C} \approx 0.3–10$ min and $D_{121°C} \approx 0.04$ min for *B. cereus*
Mesophilic	*Bacillus*	0–965 cfu ml^{-1}	*B. licheniformis, B. pumilus, B. subtilis*	$D_{100°C} \approx 20–103$ min for *B. licheniformis;* $D_{121°C} \approx 0.03–0.5$ min for *B. subtilis*
Facultative thermophiles or thermotolerant[a]	*Bacillus*	-	*B. licheniformis, B. coagulans, B. pumilus, B. sporothermodurans, B. subtilis*	$D_{100°C} \approx 14–160$ min and $D_{121°C} \approx 2–3.5$ min for *B. sporothermodurans*
Obligate thermophiles[a]	*Geobacillus, Anoxybacillus*	<10 cfu ml^{-1}	*Geobacillus stearothermophilus,*[b,c] *Anoxybacillus flavithermus*[b]	$D_{100°C} \approx 3000$ min and $D_{121°C} \approx 2.5–4$ min for *G. stearothermophilus*

[a] Consist the group of thermophilic sporeforming bacteria
[b] Formerly classified in the genus *Bacillus*
[c] Identified previously as *B. stearothermophilus*
[d] Time needed to reduce the number of spores by a factor of 10

Data compiled from: McKinnon and Pettipher (1983), Phillips and Griffiths (1986), Griffiths and Phillips (1990), Sutherland and Murdoch (1994), Sørhaug and Stepaniak (1997), Christiansson (2002), McGuiggan et al. (2002), Stepaniak (2003), Scheldeman et al. (2006), Walstra et al. (2006), Huck et al. (2008), Burgess et al. (2010).

cells of sporeforming bacteria are not very resistant to heat similarly to the spores of moulds and yeasts (Ternström et al. 1993, Scheldeman et al. 2005, Walstra et al. 2006).

Thermophilic sporeforming bacilli are potent spoilage factors for the dairy products and are usually enumerated by means of aerobic plate count incubated at 55°C for several days. This group of bacteria is divided into: i. *Obligate thermophiles*: growing only at high optimum temperatures (55–60°C) including *Anoxybacillus flavithermus* and *Geobacillus* spp., and ii. *Facultative thermophiles* or *thermotolerant bacilli*: growing at both mesophilic and thermophilic temperatures (~30–55°C), depending on the strain. Strains of *Bacillus licheniformis, B. coagulans, B. pumilus, B. sporothermodurans* and *B. subtilis* belong to the latter group (Burgess et al. 2010).

The counts of thermophilic sporeforming bacteria are usually very low in raw milk, i.e., less than 10 cfu ml^{-1} but they are present at high numbers in filter cloth, green crop and fodder. The significance of this group for heat-treated products is due to their ability to produce heat-resistant spores (HRS). HRS occurring in raw milk may survive during/after UHT treatment or industrial sterilization having $D_{100°C}$-values from 14.1 to 32 min. Although heat processed dairy products are kept at low temperatures that do not favor the growth of thermophilic bacilli, *G. stearothermophilus*, and most strains of *B. licheniformis, B. subtilis* and *B. coagulans* are generally associated with particular defects in UHT and canned milk, and cream due to their ability to produce acid- and heat-stable proteases and lipases. These defects are not usually apparent in pasteurized milk because storage at low temperatures for a shorter period (usually up to seven days) is not favorable for the germination of the spores (Kalogridou-Vassiliadou 1992, Chen et al. 2004). Therefore, the presence of this sporeforming group in dairy products like heat-treated milks and especially in dairy-based powders is usually not related to raw milk but to the processing conditions and the biofilm formation on the processing equipment (Scott et al. 2007, Burgess et al. 2010).

Of particular importance are the psychrotolerant sporeforming bacteria because their spores survive pasteurization, germinate and multiply under refrigerated conditions. The psychrotrophic spores coming from raw milk are more important for the keeping quality of dairy products than post-pasteurization contamination unlikely to the spores of thermophiles (Sørhaug and Stepaniak 1997, Champagne et al. 1994). The term "psychrotolerant endo-sporeforming spoilage bacteria" has been used by Huck et al. (2008) for members of *Bacillus* spp. and *Paenibacillus* spp. The authors observed that a great part of these genera (i.e., *B. licheniformis, B. pumilus, B. subtilis, B. weihenstephanensis, P. amylolyticus* and *Paenibacillus* spp.) existed in both dairy farms' and dairy plants' environments, although their distribution was different. In particular, they reported that *Bacillus* spp.

accounted for 87% of the isolates from the milk in the farms, 8.8% in the raw milk tank trucks, 48.2% in the dairy silos and 23% in the pasteurized milk. An average incidence of psychrotrophic *Bacillus* spores of 32.4% and 44% was reported for bulk tank and creamery silo raw milk samples which were analyzed monthly over a three-year period, respectively. The same group of researchers estimated that the spores of these bacteria were present in 58% of raw milk from farm bulk milks taken in the period from May to June. After storage at 6°C for 7 d, 10% of the milk samples was found to contain *Bacillus* spp. at >10^5 cfu ml^{-1} levels (Phillips and Griffiths 1986, Griffiths and Phillips 1990). Strains of *B. cereus* were by far the most common psychrotrophic *Bacillus* spp. isolated from milk production chain, followed by strains of *B. circulans* and *Paenibacillus polymyxa*. At 6°C, generation times and lag times of these strains varied between 7 and 23 hr, and 3 and 276 hr, respectively (Griffiths and Phillips 1990, Sutherland and Murdoch 1994). Finally, it has been shown that housing and feeding strategies, i.e., conventional vs. organic farm, affect the counts and diversity of aerobic spores during late summer/autumn and winter periods. In particular, higher numbers of thermotolerant organisms and lower numbers of *B. cereus* were found in milk produced in conventional farms compared to milk obtained from organic farms (Coorevits et al. 2008).

The germination of mesophilic strains peaks at 15 and 30°C (Griffiths and Phillips 1990, Christiansson 2002, Stepaniak 2003, Burgess et al. 2010). The maximum germination activity of spores of *Bacillus* spp. is at 15°C with a possible second maximum peak at 5°C. Temperatures higher than 72°C applied in HTST pasteurization may induce germination of these spores. Activation of thermophilic spores at 110°C before germination was observed (Stepaniak et al. 2003).

Among psychrotrophic bacilli, *B. cereus* has been studied extensively due to its technological significance. The physiology and incidence of *B. cereus* has been reviewed by Christiansson (2002). In brief, its optimum growth temperature is 30–37°C with an upper limit between 37 and 48°C. The minimum growth temperature is in general within 5–6°C but a few strains can grow at 4°C. Minimum and maximum growth pHs are 4.3 and 9.3, respectively. Its sporulation needs 16–24 hr and occurs at the late logarithmic and early stationary phase of growth. When sufficient levels of nutrients are available, *B. cereus* do not form endospores under refrigerated conditions. Germination depends on temperature and it is stimulated by HTST pasteurization or by pasteurization in the range from 72 to 85°C. However, during cold storage of pasteurized milk, a lag phase of several days is observed. Nevertheless, *B. cereus* vegetative cells are killed by pasteurization and UHT also ensures the inactivation of its endospores.

The highest levels of psychrotrophic spores are observed in the mid-summer and early autumn months apparently due to the contamination of cow's teats with soil on pasturing (Phillips and Griffiths 1986, Sutherland and Murdoch 1994, Slaghuis et al. 1997, Christiansson 2002, McGuiggan et al. 2002). Furthermore, the number of psychrotrophic spores is highly variable between different dairy plants and manufacturing days (Stepaniak 2003). The heat resistance of the spores is influenced by sporulation conditions, the physiological state of the microorganism, the composition of the heating medium and strategies used for their recovery and enumeration (Scheldeman et al. 2006).

The counts of psychrotrophic *Bacillus* spores recovered from raw milk in creamery silos or from the balance tank close to pasteurizer ranged from 0.02 to 3.5 spores ml^{-1} (McKinnon and Pettipher 1983, McGuiggan et al. 2002). The mean spore counts of mesophilic *Bacillus* spp. from raw milk samples obtained from different points of milk production chain (milking machine, bulk tank, tanker, dairy silo and after pasteurization) ranged from 0 to 965 spores ml^{-1}, in which *B. licheniformis*, *B. pumilus* and *B. subtilis* were predominant. Also, the mesophilic spore counts were the highest in winter periods and lowest in summer, attributed mainly to the contact of udder surfaces with contaminated winter bedding (Phillips and Griffiths 1986, Sutherland and Murdoch 1994).

McGuiggan et al. (2002) found that the counts of recovered mesophilic *Bacillus* spores, ranged from 1.4×10^1 to 2.4×10^5 spores ml^{-1} in milk samples obtained from the balance tank throughout a year, the highest counts being observed in the mid to late summer. The respective numbers of thermophilic *Bacillus* spores were found from 0.08 to 54 spores ml^{-1}. Both mesophilic and thermophilic *Bacillus* spores were significantly correlated with somatic cell count (SCC) of the samples, the former positively and the latter negatively. McGuiggan et al. (2002) reported also correlations between the recovery of various *Bacillus* spores and free amino acids and metal ions concentrations in milk.

In conclusion, at a storage temperature lower than 6°C, *B. cereus* does not grow. *B. circulans* is the major spoilage factor of milk stored under these conditions. The possible spoilage factors of heat-treated milk at 100°C and kept at relatively high temperatures are *B. licheniformis* and *B. subtilis* (Walstra et al. 2006). Finally, the toxinogenic effect of most aerobic sporeformers related to the consumption of milk is not common since sweet curdling or bitty cream defects in the products made them unacceptable by the consumers.

1.2.4 Coliforms

The genera *Escherichia, Enterobacter, Klebsiella, Proteus, Serratia, Hafnia* and *Citrobacter* are grouped as coliforms. They are originated from the digestive tract of milking animal and their presence in raw milk is usually associated with the unhygienic conditions of the production line, although they can rapidly build up in biofilms on milking equipment (Chambers 2002). These microorganisms utilize proteins and lactose, and are able to produce CO_2, causing defects in cheese like early blowing of hard cheeses and poor structure of soft cheeses (Frank and Hassan 2002, Walstra et al. 2006). They are Gram-negative asporogenous rods that can grow aerobically or facultative anaerobically at 37°C, some of them are psychrotrophic and all are sensitive to pasteurization. Coliform counts in bulk tank raw milk vary considerably, i.e., from 0 to 4.7 $\log_{10}$ cfu ml^{-1} (Jayarao and Wang 1999).

1.2.5 Lactic acid bacteria

Lactic acid bacteria (LAB) originate from the gastrointestinal tract (GIT) of milking animals soon after the birth. They produce mainly lactic acid from lactose, causing souring of milk. Many of them are exploited by the dairy industry in making starter cultures for fermented products including yogurt, cheese and butter. The spoilage LAB belong mainly to the genera *Lactobacillus*, *Lactococcus*, *Propionibacterium*, *Leuconostoc* and *Enterococcus* and are usually heterofermentative causing off-flavors and texture defects in cheeses. *Lactobacillus* genus is a heterogeneous microbial group containing about 135 species and 27 sub-species whose classification is constantly being changed (Bernardeau et al. 2008). *Lb. brevis* and *Lb. casei* subsp. *pseudoplantarum* cause open texture in Cheddar cheese due to the production of gas. Lactobacilli are also responsible for forming white insoluble crystals of calcium lactate in ripened hard cheeses, sulphite or phenolic-like flavors and pink discoloration (Frank and Hassan 2002). Mesophilic and some thermophilic LAB are killed by low pasteurization, i.e., at 72°C for 15 s (Walstra et al. 2006).

The biodiversity of LAB in milk depends on the kind of milk and other external parameters during milking. For example, LAB flora of raw ewes' milk was dominated by enterococci (~40%), lactococci (14–20%), leuconostocs (8–18%) and lactobacilli (10–30%) (Medina et al. 2001, Samelis et al. 2009), while raw goat's milk was dominated by lactobacilli (Colombo et al. 2010). Owing to its biodiversity, raw milk is an excellent source of technologically interesting strains of LAB employed as starters in traditional cheeses that are made especially from ewes' or goats' milk.

1.2.6 Yeasts and molds

Yeasts and molds originate usually from contaminated environment of the dairy farm or processing plant. They cause defects in cheese and yogurt. The most common yeasts found in milk are *Debaryomyces hansenii*, *Kluyveromyces marxianus* var. *marxianus*, *K. marxianus* var. *lactis*, *Saccharomyces cerevisiae*, *Candida lacticondensi*, *C. famata*, *C. versatilis*, *C. lusitaniae* and *Yarrowia lipolytica* (formerly *Candida lipolytica*) whereas molds belong to the genera *Rhizomucor*, *Rhizopus* and *Aspergillus* (Frank and Hassan 2002). The geometric mean of yeasts in bulk tank milks with microbial counts of $>3\times10^4$ cfu ml^{-1} was estimated to be 5.2×10^3 (Holm et al. 2004). It has been suggested that the ability of several yeasts to grow under refrigerated conditions in combination to the ability of some strains to produce proteinase and phospholipase may affect the quality of raw milk stored under low temperatures (Roostita and Fleet 1996, Melville et al. 2011).

1.3 Important pathogenic microorganisms present in raw milk

Raw milk can be a source of food-borne human diseases caused by pathogens. Their prevalence, like other non-pathogenic microorganisms, is affected by numerous factors such as farm size, number of animals, hygiene, farm management practices, geographical location and season (Oliver et al. 2005). Pathogenic microorganisms can be transferred to raw milk either from animals, i.e., zoonotic pathogens or from contaminated environment, i.e., exogenous pathogens. Most of the pathogenic microorganisms in milk can cause the three types of microbial food-borne diseases: (a) milk-borne infection, (b) milk-borne intoxication and (c) milk-borne toxicoinfection (Ray 2004). Table 1.4 gives an overview of pathogenic microorganisms present in raw milk while the most significant ones are discussed below.

1.3.1 Salmonella spp.

Salmonellae, a member of the family of *Enterobacteriaceae*, are natural inhabitants of the GIT of animals. They are Gram-negative, non-sporulating, facultative anaerobic rods. *Salmonella* spp. are mesophilic with optimum growth temperature of 35–37°C, but can grow at the temperature range of 5–46°C. They are sensitive to pasteurization, to low pH, e.g., pH <4.5, and do not multiply at a_w <0.94, especially when combined with pH <5.5 (Ray 2004). Systematic surveys in the USA showed that 0.2–8.9% of the isolates obtained from bulk tank milk or from in-line milk filters were *Salmonella* spp. (Oliver et al. 2005, Jayarao et al. 2006). *Salmonella* spp. were detected in 15% of raw bovine colostrum samples from dairy herds in Pennsylvania (Houser et al. 2008). More recently, about 28% of the dairy operations inspected, where raw milks were sampled from different parts of the dairy operations, were found to be contaminated with *S. enterica*, being more pronounced in milk filters (van Kessel et al. 2011). The in-line filter testing has been proposed as a more sensitive measure of *Salmonella* spp. in raw milk than

Table 1.4 Groups of microbial pathogens possibly occurred in raw milk.

Organism	Disease	Main source of contamination
Enterobacteriaceae		
Escherichia coli, including 0157:H7	Gastroenteritis, hemolytic uremic syndrome, thrombotic thrombocytogenic purpua (TTP)	Feces, water, biofilms on the milking equipment
Salmonella	Gastroenteritis, typhoid fever	Feces, water
Shigella	Gastroenteritis	Feces, water
Yersinia enterocolitica[a]	Gastroenteritis	Feces, water
Other Gram-negative bacteria		
Aeromonas hydrophila[a]	Gastroenteritis	Cold-stored milk
Brucella spp.	Brucellosis (Bang's disease)	Sick animal
Campylobacter jejuni	Gastroenteritis	Feces
Pseudomonas aeruginosa[a]	Gastroenteritis	Cold-stored milk
Gram-positive spore-formers		
Bacillus cereus[b]	Intestinal intoxication	Soil, biofilms
Bacillus anthracis	Anthrax	Soil
Clostridium perfringens	Gastroenteritis	Silage feed, air
Clostridium botulinum	Botulism	Silage feed, air
Gram-positive cocci		
Staphylococcus aureus	Emetic intoxication	Mastitic udder
Streptococcus agalactiae	Sore throat	
Streptococcus pyogenes	Scarlet fever, sore throat	
Streptococcus zooepidemicus	Pharyngitis, nephritic sequelae	
Miscellaneous Gram-positive bacteria		
Corynebacterium spp.	Diphtheria	
Listeria monocytogenes[a]	Listeriosis	Dairy farm environment
Mycobacterium bovis	Tuberculosis	Sick animal
Mycobacterium tuberculosis	Tuberculosis	Sick animal
Mycobacterium paratuberculosis	Johne's disease (only for ruminants)	Feces
Spirochetes		
Leptospira interrogans	Leptospirosis	Sick animal
Rickettsia		
Coxiella burnetii	Q fever	Sick animal
Viruses		
Enterovirus, including polioviruses, rotoviruses, *Coxsackie* viruses	Enteric infection	Feces, water
FMD virus	Foot-and-mouth disease	Sick animal
Hepatitis virus	Infectious hepatitis	Feces
Fungi		
Molds	Mycotoxicosis	Air, feed
Protozoa		
Entamoeba histolytica	Amoebiasis	Water
Cryptosporidium muris	Cryptosporidiosis	Water
Cryptosporidium parvum	Cryptosporidiosis	Sick animal, feces
Giardia lamblia	Giardiasis	Water
Toxoplasma gondii	Toxoplasmosis	Water

[a]Psychrotrophic; [b]Some strains are psychrotrophic. **Data compiled from:** ICMSF (1998), Hayes and Boor (2001), Walstra et al. (2006), Jooste and Anelich (2008).

testing the milk alone (van Kessel et al. 2008). The gastroenteric form of non-typhoid salmonellosis is frequently linked to the consumption of raw milk. Worldwide outbreaks of salmonellosis have been reported elsewhere in detail (de Buyser et al. 2001, Mazurek et al. 2004, Oliver et al. 2009).

1.3.2 *Escherichia coli*

Escherichia coli belongs to the family of *Enterobacteriaceae* and has been recognized as the most important indicator of fecal contamination of water and raw food products. *E. coli* strains are responsible for three main clinical syndromes, namely the enteric and diarrheal diseases, urinary tract infections and sepsis/meningitis, and thus they have grouped into enteroaggregative *E. coli* (EAggEC), enteroinvasive *E. coli* (EIEC), enteropathogenic *E. coli* (EPEC) and enterotoxigenic *E. coli* (ETEC). Strains of the latter group produce cytotoxins called verotoxins (VTEC) or shiga-toxins (STEC) and colonize in the intestinal track of healthy animals. *E. coli* serotype O157:H7 (EHEC) is the most studied strain of *E. coli*, followed by 026 and 0111 serotypes belong to STEC group (also called enterohemorrhagic *E. coli*) (Fremaux et al. 2008, Jooste and Anelich 2008). Most of *E. coli* strains are not heat-resistant and are readily destroyed by the pasteurization process; however, EHEC is an acid-resistant strain and thus can grow on acidified milk products such as yogurt and fresh acid cheese (Lekkas et al. 2006). Cattle feces are considered as a major reservoir of EHEC (Weimer 2001) and hence surveys on its isolation from the bulk tank milk are being continuously carried out worldwide.

EHEC was recently detected in 1.1% of bulk tank milk samples and in 6.3% of milk filters samples from dairies in the USA (van Kessel et al. 2011) as well as in the milk filters of 12% of the examined Irish farms (Murphy et al. 2005). It was also detected in raw ewes' milk in Greece and Spain (Dontorou et al. 2003, Caro et al. 2006) and in goats' milk in Bergamo region of Italy (Foschino et al. 2002). Moreover, after an extensive survey in Switzerland, 16.3% of goats' milk and 12.7% of ewes' milk samples were found PCR-positive for this microorganism (Muehlherr et al. 2003). Isolation rates for STEC were reported to be ranging from 0.8 to 3.2% with no presence of EHEC (Oliver et al. 2005, Jayarao et al. 2006, Cobbold et al. 2008). Prevalence (0.6%) of 026VTEC was reported in raw water buffalo milk in Italy (Lorusso et al. 2009). Recent outbreaks of milk-borne diseases caused by EHEC and associated with consumption of raw milk in the USA, have been reviewed by Oliver et al. (2009).

1.3.3 *Yersinia enterocolitica*

Yersinia enterocolitica belongs to the family of *Enterobacteriaceae*. This pathogen causes acute gastroenteritis, enterocolitis and mesenteric lymphadenitis as well as various extra-intestinal disorders (Jooste and Anelich 2008). It is a psychrotrophic microorganism and thus is highly susceptible to pasteurization. Raw milk often contains *Y. enterocolitica,* although dairy cattle are not considered reservoirs of this pathogen. It is thought to contaminate raw milk through contacting with animal feces or polluted water supplies. Prevalence of *Y. enterocolitica* in bulk tank milk was reported at rates varying from 1.2 to 6.1% (Oliver et al. 2009).

1.3.4 *Listeria* spp.

Listeria spp., which belongs to the family of *Listeriaceae,* is commonly found in the dairy farm environment and thus raw milk can be contaminated through various sources (Fox et al. 2009, Schoder et al. 2011). *Listeria monocytogenes* causes human listeriosis, a serious invasive disease that causes abortion in pregnant women and meningitis, encephalitis and septicemia in neonates and immunocompromised adults. Among 13 known serotypes of *L. monocytogenes,* the 4b, 1/2 and 1/2b are the most common ones that account for 89–96% of human listeriosis (Jooste and Anelich 2008).

L. monocytogenes is frequently isolated from raw farm milk or bulk tank milk samples throughout the world, since it can grow on steel and rubber surfaces and at refrigeration temperatures. Recent studies have shown that this microorganism takes part in biofilm formation in the milking parlor equipment, thus milk is contaminated when it passes through the pipeline system into the tank (Latorre et al. 2010, van Kessel et al. 2011). Contamination of bulk tank milk by *L. monocytogenes* was reported to range from 1.0 to 12.6% of the raw milk samples tested in the USA dairy farms (Oliver et al. 2005) and from 0.3 to 7.5% in other countries (Yoshida et al. 1998, Uraz and Yücel 1999, Baek et al. 2000, Hamdi et al. 2007, Moshtaghi and Mohammadpour 2007). *Listeria* spp. in bovine milk were reported to be higher in spring (14.3%) than in autumn (4.8%) (Yoshida et al. 1998), but the opposite results were noted in raw caprine milk (Gaya et al. 1996). Reported outbreaks of milk-borne listeriosis result mainly from the consumption of soft cheese made from raw milk rather than the consumption of raw milk (de Buyser 2001, Swaminathan and Gerner-Smidt 2007, Oliver et al. 2009).

1.3.5 *Staphylococcus* spp.

Staphylococcus aureus belongs to the family of *Micrococcaceae* and is the most significant causative agent of mastitis in dairy cows. The enterotoxigenic strains produce enterotoxins that are classified according to serotypes into A-H groups and toxic shock syndrome toxin (TSST) (Oliver et al. 2005). Staphylococcal enterotoxin is heat-resistant and survives pasteurization.

Enterotoxigenic *S. aureus*, that is coagulase-positive, was isolated from bulk tank milk samples from the USA dairies at levels varying from 27.4 to 37%, whereas colostrum was implicated in higher isolation rates, i.e., 42% (Oliver et al. 2009). Prevalence of *S. aureus* in raw bovine and caprine milk samples from Norwegian dairies was reported at rates from 47 to 75% and from 2 to 96%, respectively (Jørgensen et al. 2005, Jakobsen et al. 2011). Coagulase-negative *Staphylococcus* spp. were isolated at levels of >74% and the dominant species reported were *S. chromogenes, S. hyicus* and *S. epidermis* (Oliver et al. 2009). *S. aureus* outbreaks were reported to be more related to the consumption of raw milk cheeses than the consumption of raw milk (de Buyser et al. 2001, Rosengren et al. 2010).

1.3.6 Campylobacter spp.

Campylobacter jejuni belongs to the family of *Campylobacteriaceae* and is the most important causative agent of abortion in cattle and sheep and an etiological agent of human gastroenteritis (Jooste and Anelich 2008). It is widely isolated from feces of infected cattle as well as from raw milk. *C. jejuni* is acid- and heat-sensitive, and thus is killed by pasteurization. Prevalence of *C. jejuni* in bulk tank milk was reported to range from 0.4 to 12.3% (Whyte et al. 2004, Oliver et al. 2005). Outbreaks of campylobacteriosis due to consumption of raw milk were reported in the USA (Oliver et al. 2009), Netherlands (Heuvelink et al. 2009) and Hungary (Kálmán et al. 2000) during the last decade.

1.3.7 Mycobacterium avium subsp. *paratuberculosis*

Mycobacterium avium subsp. *paratuberculosis* (MAP) is a member of the family of *Mycobacteriaceae* and is the causative agent of paratuberculosis which is a zoonotic disease also known as John's disease. MAP has also been linked to Crohn's disease (CD), a chronic human disease of the terminal ileum, although the relationship between MAP and CD is yet to be established (Waddell et al. 2008).

This microorganism can be present in raw milk through feces of infected animal at numbers varying from 250 cfu ml^{-1} to more than 10^4 cfu ml^{-1} (Skovgaard 2007). In raw milk, MAP was found at ≤65 cfu ml^{-1} (Boulais et al. 2011). It is a short, thin, Gram-positive, acid-fast rod which is considered to survive under HTST pasteurization (Weimer 2001). It was shown that MAP was completely inactivated by HTST pasteurization or by combination of

HTST and homogenization in about 97% of the milk samples tested (Grant et al. 2005). Numerous studies concerning the prevalence of MAP in raw milk and milk products as well as its sensitivity towards heat treatments have been recently reviewed by Gill et al. (2011).

1.3.8 Others

Mycobacteriun tuberculosis is the causative agent of tuberculosis, one of the most pervasive and destructive human and animal disease in the past. It is the most heat-resistant non-sporeforming Gram-positive pathogen that is killed during pasteurization of milk at 72°C for 15 s. *Brucella abortus* and *B. melitensis* are the members of the family of *Brucellaceae* and are responsible for the most prevalent bacterial zoonosis called brucellosis that still exists, especially in developing countries. They are killed easily by low pasteurization. *Coxiella burnetii* of *Coxiellaceae* family is responsible for Q fever. Although, *C. burnetii* is relatively heat-resistant, it is killed by regular pasteurization treatment (Walstra et al. 2006). Despite Q fever being a neglected zoonosis for many years, *C. burnetti* is still present in raw milk worldwide. Recently this microorganism has been detected in about 42% of the commercial raw milk samples analyzed in the USA (Loftis et al. 2010).

1.4 Sources of contamination of milk

Milk after secretion from udder is immediately contaminated through various sources including the udder, environment and different milking practices. The hygienic control of these contaminating sources is crucial for microbiological quality of raw milk.

1.4.1 Udder hygiene-mastitis

The udder hygiene affects the microbial load of bulk tank milk, since poor teat cleanliness is associated with high bacterial counts. The numbers of bacteria in milk immediately after milking of healthy cow under hygienic conditions is about ≤10.000 cfu ml^{-1} (Walstra et al. 2006). However in practice, raw milk has much higher bacterial counts due to probably poor hygienic conditions of milking. It has been calculated that the amount of transmitted dirt attached to the exterior of teat to milk is on average 59 mg l^{-1} (Vissers et al. 2007a). The total bacteria counts of milk are positively correlated with the amount of soiling on the teats prior to udder preparation for milking, but coliforms are negatively associated with clipping udder hair (Elmoslemany et al. 2010). On the other hand, it seems that the teat surface is the main source of beneficial lactobacilli and propionibacteria (Vacheyrou et al. 2011).

In the case of inflammatory disease of udder (mastitis), the most common causative microorganisms such as *S. aureus*, *S. agalactiae*, *S. dysagalactiae*, *S. uberis*, *E. coli*, *C. freundii*, *Enterobacter* spp., *Klebsiella* spp. and *Actinomyces pyogenes* pass into milk together with somatic cells at numbers varying with the stage of mastitis. *L. monocytogenes*, coagulase-negative *Staphylococcus* spp., *S. pyogenes*, *P. aeruginosa*, *C. bovis*, *M. bovis*, *B. cereus*, *B. abortus* and *C. burnetii* have also been associated with mastitis (Weimer 2001). Mastitis causes reduced milk yield and thus economic loss to the dairy industry. It is estimated that each doubling of somatic cell count above 5×10^4 cells ml^{-1} reduces milk yield by 0.4 kg d^{-1} in primiparous cows and 0.6 kg d^{-1} in multiparous cows (Hortet and Seegers 1998). Individual healthy cows have milk SCC with <50,000 cells ml^{-1} but bulk tank milk contains usually SCC of >200,000 cells ml^{-1} (Barbano et al. 2006). In the case of sub-clinical mastitis, both udder and milk appear normal and the disease can only be detected by analyzing the milk. Mastitis milk has high SSC, ca. $>4\times10^5$ cells ml^{-1}, and different biochemical composition than milk of healthy animals, e.g., increased pH and chlorine ions content and decreased lactose content (Ogola et al. 2007). In the case of clinical mastitis, changes in both udder, i.e., swelling and milk are easily detected macroscopically by the milker. In a recent study in France, the distribution of pathogenic microorganisms responsible for clinical mastitis was found to be *Streptococcus uberis* (22.1%), *E. coli* (16%) and coagulase-positive *Staphylococci* (15.8%), whereas for sub-clinical mastitis was found as follows: coagulase-positive *Staphylococcus* spp. (30.2%), coagulase-negative *Staphylococcus* spp. and *S. dysagalactiae* (9.3%) (Botrel et al. 2010). A positive correlation of thermophilic aerobic bacteria and yeasts and moulds in the cases of infectious mastitis and mastitis caused by *Staphylococcus* spp. has been also observed, but it seems that the microbiological quality of raw milk is more affected by the lack of udder hygiene and environmental contaminations than the mastitis frequency in dairy herds (Souto et al. 2008).

Consequently, the strategies for mastitis control must include improved hygiene in the farm environment regarding particularly the health and cleanliness of teats. For example, the efficient pre-milking teat preparation, i.e., teat wash with water or NaOCl and drying can reduce the total microbial counts up to 50% during winter housing (Chambers 2002).

1.4.2 Environment

The environment (air, soil, water, animal feces and feed) of dairy farm is the main contributor of contaminants and/or pathogenic microorganisms even in farms with milking room. In an extensive study, environmental samples (teat surface, air before and/or after work, air of milking room, hay of the day, hay of the month and settled dust) collected from 16 farms in France showed that fungi in the environment contained *Aspergillus* spp., *Penicillium* spp., *Eurotium* spp. and white yeasts, while bacteria were *Lactobacillus delbrueckii* spp., *Lb. paracasei*, *P. freudenreichii*, *Acinetobacter* spp., *Corynebacterium* spp., *Staphylococcus* spp. and *Streptococcus* spp. Most of these microorganisms were also found in raw milk (Vacheyrou et al. 2011). Other studies showed that there was a correlation between the level of hygienic standards of the farm and the occurrence of *L. monocytogenes* (Fox et al. 2009), as well as between housing and bedding of the cow and the prevalence of *S. aureus* or *Streptococcus* spp. (Ferguson et al. 2007). Soil, forage, grass silage, maize silage and dry hay are sources of sporeforming microorganisms. Hay dust is a reservoir of *B. subtilis* (Walstra et al. 2006). Feed, especially silage, is considered to be the main source of contamination of raw milk with *Bacillus* spp. spores (Te Giffel et al. 2002, Walstra et al. 2006, Vissers et al. 2007b) or clostridia (Julien et al. 2008). To minimize the transmission of *Bacillus* spp. or butyric acid bacteria spores to milk via the route of feed and to ensure a concentration in farm tank of less than 3 $\log_{10}$ spores l^{-1}, the initial contamination level in the feed should be lower than 3 $\log_{10}$ spores g^{-1} (Vissers et al. 2007b,c). Also, it has been calculated that *B. cereus* spores in farm tank milk during the grazing period could be up to 3.5 $\log_{10}$ spores l^{-1} and during the housing period up to 2 $\log_{10}$ spores l^{-1}. Moreover, it has been estimated that if all farms minimize the soil contamination of teats and if teat cleaning is optimized, a 99% reduction of *B. cereus* spores could be achieved during grazing period (Vissers et al. 2007d).

The water used in dairy farm is also of high importance since it is associated with poor bacteriological quality of milk (Perkins et al. 2009). Water used in primary milk production must be of good quality, i.e., free of pathogens and fecal contamination. Psychrotrophic Gram-negative rods like *Pseudomonas*, *Achromobacter*, *Flavobacterium* and *Alcaligenes* spp. often exist in contaminated water (Walstra et al. 2006). High alkalinity of water is associated with high bacterial counts in bulk tank milk, while high temperature of water and use of softeners are associated with lower bacterial counts (Elmoslemany et al. 2009a).

Cattle feces are considered to be a big reservoir of spoilage or pathogenic microorganisms, but it seems that they are not so important source of contamination of raw milk with enterococci, lactobacilli or coliforms (Kagkli et al. 2007). In contrast, feces is the main source of bifidobacteria and for this reason *Bifidobacterium pseudolongum* has been proposed to be used as fecal indicator rather than *E. coli* to assess the hygienic quality of raw milk (Beerens et al. 2000, Delcenserie et al. 2011).

Weather temperature is also a factor determining the microbial counts of raw milk. Seasonal data showed that the lowest bacterial counts tended to occur in winter, while counts of coliforms and somatic cells seemed to be always high in summer (Elmoslemany et al. 2009b, 2010). Potentially toxic *B. cereus sensu lato* (the species group comprising *B. anthracis*, *B. thuringiensis* and *B. cereus*) was isolated from raw milk at higher levels during spring and summer months (Bartoszewicz et al. 2008).

1.4.3 Milking practices

Hand milking is of high risk for contaminating milk because milker may transfer the pathogens to milk, while milking machines, pipelines and milk filters act as reservoirs of high heat-resistant sporeforming microorganisms (Scheldeman et al. 2005). Also, different hygienic milking practices can affect the balance between bacterial populations. Raw milks produced under high level of hygiene contain more Gram-positive non-lactic acid bacteria (*Corynebacteriaceae* and *Micrococcaceae*), while those milks produced under less intensive hygienic practices are dominated by Gram-negative bacteria and *Lactococcus lactis* or *Brevibacterium lines* and *Leuconostoc mesenteroides* (Verdier-Metz et al. 2009). Poorly cleaned and disinfected milking equipment have residual milk which often contains bacterial load of ca. 10^9 cfu ml^{-1} and 1 ml of this residual milk is able to increase the bacterial load of 100 liters of the newly collected milk to 10,000 cfu ml^{-1} (Walstra et al. 2006). Biofilm formation on the surfaces of milking utensils has detrimental effect and must be controlled (Kulozik 2002). Consequently, CIP protocol including pre-rinsing with water, circulation of alkaline/acidic and disinfectant solutions and finally rinsing with water is required for milking as well as for storage units of raw milk (Chambers 2002, Walstra et al. 2006). In general, the Good Manufacturing Practices (GMP) must be applied to the dairy farms in order to produce high quality milk. For more information about microbial safety systems of milk and milk products, readers may refer to Chapter 14.

1.5 Microbial standards for raw milk

According to European legislation (European Union 2006), raw milk must come from animals free of brucellosis and tuberculosis, which do not show any symptoms of infections, diseases or they do not suffer from any infection of the genital tract with enteritis, enteritis with diarrhea and fever or an inflammation of an udder that can affect human health through milk consumption. Raw cow's milk collected from production holdings within the European Community (EC) must be produced according to the hygienic requirements described in EC Regulation 1662/2006 (European Union 2006) and must meet a set of microbial criteria, i.e., the two-months rolling geometric average of microbes (with at least two samples per month) grown on Plate Count Agar at 30°C should be ≤100,000 cfu ml^{-1} and the two-months geometric mean of SCC should be ≤400,000 ml^{-1}. However, immediately before processing into dairy products, raw cow's milk may have a plate count of ≤300,000 cfu ml^{-1}. According to the same EC regulation, raw milk from other species should have a 2-month rolling geometric average with at least two samples per month less than ≤1,500,000 plate count at 30°C. Moreover, if a dairy product is to be produced without any heat treatment then the raw milk from other species should have plate count of ≤500,000 cfu ml^{-1}.

In the USA, cow's and goat's milks must not exceed the SCC limit of 400,000 ml^{-1} and 1,000,000 ml^{-1}, respectively (Grade 'A' Pasteurized Milk Ordinance 2009). Raw cow's milk produced in Canada must have aerobic mesophilic bacteria count of ≤50,000 cfu ml^{-1} and SCC of ≤1,000,000 ml^{-1} (National Dairy Code of Canada 2011). Raw milk for direct human consumption as drinking milk is not generally encouraged because even the most appropriate hygienic procedures do not always ensure the absence of pathogens. For this reason, when raw milk is sold for human consumption, the words 'raw milk' must be clearly shown on the label of packaging (European Union 2006). In the Standard 1.6.1 of Australia and New Zealand, microbiological criteria for unpasteurized milk for retail use are set for *Campylobacter*, *E. coli*, *L. monocytogenes*, *Salmonella* and standard plate count (SPC). In particular, *Campylobacter*, *L. monocytogenes* and *Salmonella* must not be detected in 25 ml; SPC (30°C for 72 hr) >2.5×10^{5}, coliforms >10^{3} and *E. coli* >9 cfu ml^{-1} in one sample would cause the lot to be rejected (Food Standards Code 2001).

1.6 Improving microbial quality of raw milk during storage

1.6.1 Anti-microbial factors of raw milk

Raw milk contains indigenous anti-microbial factors that belong to the fraction of minor serum proteins such as the enzymes lactoperoxidase and lysozyme, the iron binding protein lactoferrin and immunoglobulins (antibodies). All these proteins can act against a broad spectrum of microorganisms and viruses; thus they are potential inhibitors of microbial growth in raw milk during storage.

Lactoperoxidase-thiocyanate-hydrogen peroxide (LPO) system

Lactoperoxidase (LPO, EC 1.11.1.7) is one of the most important indigenous milk enzymes in terms of both concentration and function. LPO is mainly responsible for the anti-microbial properties of raw milk in the presence of sufficient quantities of thiocyanate ion (SCN^-) and hydrogen peroxide

(H_2O_2). Its concentration in bovine milk is high (i.e., 10–30 μg ml^{-1}) compared to other indigenous enzymes. LPO is a 78 kDa glycoprotein containing one heme group and ~10% carbohydrate with a very ordered monomeric structure stabilized by eight disulphide bonds and one calcium ion. It catalyzes the reaction of $H_2O_2 + 2HA \rightarrow 2H_2O + 2A$, where HA is an oxidizable substrate. In this respect, the oxidation of SCN^- from animal feed to $OSCN^-$ (hypothiocyanite anion) is catalyzed by LPO. The $OSCN^-$ inhibits bacteria, fungi and viruses. The function of LPO-SCN^--H_2O_2 anti-microbial system in raw milk is controlled by the variable concentration of its two latter components. In particular, the SCN^- concentration in milk varies with breed, species, udder health and feeding type. Normally, H_2O_2 is not a raw milk constituent but it comes from aerobic metabolism of lactobacilli and lactococci or from polymorphonuclear leukocytes during phagocytosis. It may also be added to milk or may come from added H_2O_2-generating systems, e.g., glucose oxidase (Pruitt and Kamau 1994, Kussendrager and van Hooijdonk 2000, Seifu et al. 2005, Moatsou 2010).

The bacteriostatic and/or bactericidal activities of the LPO system have been reviewed by Shakeel-Ur-Rehman and Farkye (2002) and Seifu et al. (2005). Gram-negative, catalase-positive bacteria such as *Pseudomonas*, coliforms, *Salmonella* spp. and *Shigella* spp. are inhibited, or killed if H_2O_2 is added exogenously. Gram-positive, catalase-negative bacteria including streptococci and lactobacilli are inhibited but not killed by the LPO system. The $OSCN^-$ is bactericidal for enteric pathogens, i.e., *E. coli* strains, by damaging the inner membrane. The LPO system also inhibits EHEC (Enterohaemorrhagic *E. coli*), shows bactericidal activity against *C. jejuni* and inhibits substantially vegetative cells of various *B. cereus* strains. Acid production and oxygen uptake in strains of *S. mutans*, *S. salivarius*, *S. sanguis* and *S. mitis* are also inhibited by this anti-microbial system.

Activation of the LPO system in raw milk through the addition of SCN^- and H_2O_2 is a successful mean of preserving raw milk and inhibiting pathogens during storage and transportation in the lack of refrigeration and thermization (Codex Alimentarius 1991). This practice is known as "cold pasteurization". There are guidelines for the preservation of raw milk by the LPO system (IDF 1988, Codex Alimentarius 1991). The suggested practical application is the addition of 14 ml NaSCN and 30 mg H_2O_2 L^{-1} of milk, within 2–3 hr from the time of milking. The inhibitory effect of treatment is inversely affected by the temperature of raw milk, e.g., 7–8 hr or 24–26 hr for storage at 30 or 15°C, respectively.

Lysozyme

Another indigenous milk enzyme with potential anti-microbial properties is lysozyme (EC 3.2.1.17), also called muramidase (peptidoglycan-Nacetylmuranoylhydrolase), that cleaves the peptidoglycan in the bacterial cell-wall. It is an anti-bacterial agent present in many body fluids and in the milk of many mammalian species causing lysis of many types of bacteria. Its concentration in bovine milk is <0.03 mg ml^{-1}, increasing with increase in somatic cell count. It acts mostly against Gram-positive and a few Gram-negative bacteria; however, its effect on raw milk's shelf-life is not significant due to its low concentration (Farkye 2002, Walstra et al. 2006).

Lactoferrin

Lactoferrin (LF) is a non-hemic 80 kDa iron binding glycoprotein of the transferrin family, which is a very small fraction of milk serum proteins, i.e., 0.1–0.4 mg ml^{-1}. LF is a multifunctional protein that exhibits anti-microbial, anti-viral, anti-protozoan and anti-oxidant activities along with immunomodulation, modulation of cell growth and binding or inhibition of bioactive compounds (Wakabayashi et al. 2006, García-Montoya et al. 2012). There are contradictory reports about factors affecting LF concentration in bovine milk; nevertheless, effects of daily milk production, parity, stage of lactation and somatic cell count on the LF concentration in milk have been reported (Tsuji et al. 1990, Cheng et al. 2008).

The molecular structure of LF, i.e., a single polypeptide chain of about 700 amino acids folded in two symmetrical lobes, is the reason for its anti-microbial properties that are expressed through various mechanisms. Iron sequestration, membrane destabilization, targeting of bacterial virulence mechanisms and host invasion strategies are the modes of LF's anti-microbial action (Jenssen and Hancock 2009, García-Montoya et al. 2012). It exhibits strong anti-bacterial activity against a broad spectrum of Gram-positive and Gram-negative bacteria, especially against pathogens. For example, iron sequestering action of LF inhibits the growth of *G. stearothermophilus* and *S. aureus*, by depriving them of this nutrient. Altering bacterial virulence by LF is the major mechanism against *L. monocytogenes* and an iron-independent interaction of LF with bacterial cell acts against *S.* Typhimurium. Cation chelation, damage of the bacterial membrane or altering of the outer membrane permeability by LF show inhibitory effect on *E. coli* (Jenssen and Hancock 2009, García-Montoya et al. 2012). However, its role in the extension of shelf-life of raw bovine milk is negligible due to its low concentrations (Walstra et al. 2006).

Immunoglobulins

Immunoglobulins (Igs) are antibodies distinguished in several classes. They are heat-sensitive glycoproteins with heterogeneous composition and IgG, IgA and IgM classes occur in milk serum. IgG predominates Igs in bovine milk (roughly 2/3 of total Igs). While Igs occupy 70–80% of the total protein in the bovine colostrums, their proportion in mature milk decreases to 1–2%. Concentrations of Igs in milk and colostrum differ from those in blood and are highly variable affected by species, breed, stage of lactation and health status of milking animals. They have many functions, i.e., augment leucocytes activities, activate complement-mediated bacteriolytic reactions, prevent the adhesion of microbes to surfaces, inhibit bacterial metabolism, agglutinate bacteria and neutralize toxins and viruses (Marnilla and Korhonen 2002).

A description of Igs inhibitory action in raw milk is given by Walstra et al. (2006). In brief, IgG with a molar mass of 150,000 Da can act against many antigens and may inhibit bacterial growth, whereas the role of IgA (Mw of 385,000 Da) in milk is not clear. IgM (agglutinin) with a molar mass of 900,000 Da may be an antibody against polysaccharide groups of bacterial cell wall and can flocculate "particles" including bacteria and viruses (agglutination). In the IgM class, cryoglobulins are included, which can precipitate at low temperatures agglutinating other particles. Also, lactenins L_1 and L_3, which are inhibitors of Gram-positive bacteria, are in the IgM class.

1.6.2 Refrigeration during collection and storage of raw milk

Apart from milking procedures, the pre-storage conditions and transportation also affect the raw milk quality. According to European legislation the raw milk on the farm must be held at ≤8°C immediately after milking or at ≤6°C if it is not collected every day (European Union 2006). Raw milk on most dairy farms is usually stored in a bulk tank but in the case of very small enterprises it is still collected and stored in cans (churns). Depending on the farm location, refrigerated milk is collected every day or every two or three days by tankers and is transported to the dairy factory for processing. Milk tankers are insulated to ensure that raw milk arrives at its destination at ≤5°C; although, in practice this temperature limit is extended up to 6–9°C or >10°C in the case of churns (Walstra et al. 2006). Cold storage inhibits the growth of mesophilic microorganisms and especially those producing heat-resistant toxins, e.g., *S. aureus* associated to raw milk until the time of processing. The generation time of the most important bacterial groups of milk at 30°C is around 30 min, whereas it ranges between 4 and >20 hr at 5°C (Chambers 2002, Walstra et al. 2006). *Bacillus* spp. spores cannot germinate in refrigerating bulk tank milk and the kind of the present genera and species will depend on those that are present in raw milk initially. Consequently, the dominant flora of refrigerated milk for 2–3 days is psychrotrophs with *Pseudomonas* spp. and especially

P. fluorescens being the most frequently present. After 4–5 days of storage substantial growth of psychrotrophs occurs and if their total numbers are $>5\times10^5$ cfu ml^{-1} there is a risk to have produced their heat-stable enzymes (Walstra et al. 2006).

1.6.3 Thermal treatments during storage of raw milk

In order to reduce the counts of microorganisms and avoid in particular the growth of psychrotrophs, a mild heat treatment with minimum heat damage of milk components should be applied to milk as soon as possible after its arrival at the dairy. This treatment is known as "thermization". Thermization usually allows milk to be stored 2–3 days after milking without deterioration. Typical temperature-time combinations for thermization are 68°C for 10 s or 65°C for 20 s. This heat treatment reduces the numbers of microorganisms by a factor of 3 $\log_{10}$ or 4 $\log_{10}$, while keeping alkaline phosphatase active (Stepaniak and Rukke 2003, de Jong 2008). However, the final level of microorganisms will depend on their initial counts as well as on the thermization temperature. It was reported that thermization at 60°C for 30 s had significantly inactivated Gram-negative bacteria, i.e., *Listeria* spp., *Enterobacteria* and coagulase-positive *Staphylococci*, but moderate effect against Gram-positive bacteria was reported under the same conditions. Thermization at 67°C for 30 s is much more effective to all bacterial groups. In particular, thermization at 60–63°C causes a 4 log reduction of *Salmonella* spp. and a 2 log reduction of *L. monocytogenes* (Stepaniak and Rukke 2003, Samelis et al. 2009). Also extracellular enzymes of psychrotrophs can be inactivated by low temperature thermization. For example, thermization of raw milk at 50–60°C and 60–70°C leads to inactivation of proteinases and lipases from *Pseudomonas* spp., respectively. Interestingly, thermization at 65°C for 10 s may stimulate the germination of *Bacillus cereus* spores; nevertheless vegetative cells that may appear after 6 hr at 10°C would be destroyed by pasteurization (Stepaniak and Rukke 2003).

1.6.4 Non-thermal treatments

Two centrifugation techniques, the clarification and bactofugation, are used for 'cleaning' milk. Both techniques operate to separate bacterial cells and milk components based on the differences in relative densities. Clarification removes mainly foreign particles, i.e., dirt, hairs, etc., whereas bactofugation removes bacterial spores. Bactofugation takes place at 60–65°C and can remove up to 99% of the spores (Hayes and Boor 2001, Walstra et al. 2006).

Microfiltration, a membrane processing technique using 0.1–10 μm pore-size membranes, improves microbial quality of raw milk by an average decimal reduction (DR) of 3.5, which means that if initial skim milk contains 20,000 cfu ml^{-1}, the microfiltered milk will contain less than 10 cfu ml^{-1}. For bacterial spores, the DR is higher than 4.5 (Saboya and Maubois 2000).

Finally, alternative novel methods to heat treatment have been widely investigated during the last two decades. For example, storage of raw milk under low carbon dioxide (CO_2) pressures, i.e., 68 to 689 kPa, reduces bacterial growth rates without causing precipitation of milk proteins. The combination of low CO_2 pressure and refrigeration preserves raw milk for longer time. If 1500 ppm CO_2 is added to high quality raw milk, it can be stored at 4°C for 14 days with low proteolysis and lipolysis and with a standard plate count of $\leq 3\times10^5$ cfu ml^{-1} (Ma et al. 2003, Rajagopal et al. 2005). This treatment is more effective on Gram-negative bacteria than on Gram-positives and spores (Singh et al. 2012). Other novel non-thermal methods that may be employed in keeping bacteriological quality of raw milk are discussed in Chapter 13.

1.8 Conclusion

The fast growth of various kinds of microorganisms is the major drawback of raw milk in terms of both consumers' safety and keeping quality of dairy products. The profile of microorganisms in raw milk is configured by the status of animal health, the milking environment, the milking practices and the storage strategies. The counts of pathogens, psychrotrophs, thermodurics and sporeformers must be controlled or eliminated during the production and pre-processing storage period of raw milk. Factors affecting their growth and proliferation have been constantly studied. The outcome of these studies is the implementation of hygienic and technological practices during raw milk collection and storage to control efficiently its microflora.

References

Baek, S.Y., S.Y. Lim, D.H. Lee, K.H. Min and C.M. Kim. 2000. Incidence and characterization of *Listeria monocytogenes* from domestic and imported foods in Korea. J. Food Protect. 63: 186–189.

Barbano, D.M., Y. Ma and M.V. Santos. 2006. Influence of raw milk quality on fluid milk shelf life. J. Dairy Sci. 89Suppl. 1: E15–19.

Bartoszewicz, M., B.M. Hansen and I. Swiecicka. 2008. The members of the *Bacillus cereus* group are commonly present contaminants of fresh and heat-treated milk. Food Microbiol. 25: 588–596.

Beerens, H., B. Hass Brac de la Perriere and F. Gavini. 2000. Evaluation of the hygienic quality of raw milk based on the presence of bifidobacteria: The cow as a source of faecal contamination. Int. J. Food Microbiol. 54: 163–169.

Bernardeau, M., J.P. Vernoux, S. Henri-Dubernet and M. Guéguen. 2008. Safety assessment of dairy microorganisms: The *Lactobacillus* genus. Int. J. Food Microbiol. 126: 278–285.

Botrel, M.A., M. Haenni, E. Morignat, P. Sulpice, J.Y. Madec and D. Calavas. 2010. Distribution and antimicrobial resistance of clinical and subclinical mastitis pathogens in Dairy Cows in Rhône-Alpes, France. Foodborne Pathog. Dis. 7: 479–487.

Boulais, C., R. Wacker, J.C. Augustin, M.H. Ben Cheikh and F. Peladan. 2011. Modeling the occurrence of *Mycobacterium avium* subsp. *paratuberculosis* in bulk raw milk and the impact of management options for exposure mitigation. J. Food Protect. 74: 1126–1136.

Burgess, S.A., D. Lindsay and S.H. Flint. 2010. Thermophilic bacilli and their importance in dairy processing. Int. J. Food Microbiol. 144: 215–225.

Caro, I., V.M. Fernández-Barata, A. Alonso-Llamazares and M.R. García-Armesto. 2006. Detection, occurrence and characterization of *Escherichia coli* O157:H7 from raw ewe's milk in Spain. J. Food Protect. 69: 920–924.

Chambers, J.V. 2002. The microbiology of raw milk. pp. 39–85. *In:* R.K. Robinson [ed.]. Dairy Microbiology Handbook, 3rd edition. John Wiley and Sons Inc, Ames, Iowa, USA.

Champagne, C.P., R.R. Laing, D. Roy and A.A. Mafu. 1994. Psychrotrophs in dairy products: Their effects and their control. Critical Rev. Food Sci. Nutr. 34: 1–30.

Chen, L., T. Coolbear and R.M. Daniel. 2004. Characteristics of proteinases and lipases produced by seven *Bacillus* spp. isolated from milk powder production lines. Int. Dairy J. 14: 495–504.

Cheng, J.B., J.Q. Wang, D.P. Bu, G.L. Liu, C.G. Zhang, H.Y. Wei, L.Y. Zhou and J.Z. Wang. 2008. Factors affecting the lactoferrin concentration in bovine milk. J. Dairy Sci. 91: 970–976.

Christiansson, A. 2002. *Bacillus cereus*. pp. 123–128. *In:* H. Roginski, J.W. Fuquay and P.F. Fox [eds.]. Encyclopedia of Dairy Science. Elsevier, London, UK.

Cobbold, R.N., M.A. Davis, D.H. Rice, M. Szymanski, P.I. Tarr, T.E. Besser and D.D. Hancock. 2008. Associations between bovine, human, and raw milk and beef isolates of non-O157 Shiga toxigenic *Escherichia coli* within a restricted geographic area of the United States. J. Food Protect. 71: 1023–1027.

Codex Alimentarius. 1991. Guidelines for the preservation of raw milk by use of lactoperoxidase system. Standard CAC/GL 13. Available at http://www.codexalimentarius.org/standards/list-of-standards.

Colombo, E., L. Franzetti, M. Frusca and M. Scarpellini. 2010. Phenotypic and genotypic characterization of lactic acid bacteria isolated from artisanal Italian goat cheese. J. Food Protect. 73: 657–662.

Coorevits, A., V. de Jonge, J. Vandroemme, R. Reekamns, J. Heyrman, W. Messens, P. de Vos and M. Heyndrickx. 2008. Comparative analysis of the diversity of aerobic sporeforming bacteria in raw milk from organic or conventional farms. Sys. Appl. Microbiol. 31: 126–140.

Cousin, M.A. 1982. Presence and activity of psychrotrophic microorganisms in milk and dairy products: A review. J. Food Protect. 45: 172–207.

de Buyser, M.L., B. Dufour, M. Maire and V. Lafarge. 2001. Implication of milk and milk products in food-borne diseases in France and in different industrialised countries. Int. J. Food Microbiol. 67: 1–17.

de Garnica, M.L., J.A. Santos and C. Gonzalo. 2011. Influence of storage and preservation on microbiological quality of silo ovine milk. J. Dairy Sci. 94: 1922–1927.

de Jong, P. 2008. Thermal processing of milk. pp. 1–34. *In:* T.J. Britz and R.K. Robinson [eds.]. Advance Dairy Science and Technology. Blackwell Publishing, Oxford, UK.

de Jonghe, V., A. Coorevits, J. de Block, E. van Coillie, K. Grijspeerdt, L. Herman, P. de Vos and M. Heynrickx. 2010. Toxinogenic and spoilage potential of aerobic spore-formers isolated from raw milk. Int. J. Food Microbiol. 136: 318–325.

Delcenserie, V., F. Gavini, B. China and G. Daube. 2011. *Bifidobacterium pseudolongum* are efficient indicators of animal fecal contamination in raw milk cheese industry. BMC Microbiol. 11: Art 178. Available at: http://www.biomedcentral.com/content/pdf/1471-2180-11-178.pdf.

Dontorou, C., C. Papadopoulou, G. Filioussis, V. Economou, I. Apostolou, G. Zakkas, A. Salamoura, A. Kansouzidou and S. Levidiotou. 2003. Isolation of *Escherichia coli* O157:H7 from foods in Greece. Int. J. Food Microbiol. 82: 273–279.

Dunstall, G., M.T. Rowe, G.B. Wisdom and D. Kilpatrick. 2005. Effect of chorus sensing agents on the growth kinetics of *Pseudomonas* spp. of raw milk origin. J. Dairy Res. 72: 276–280.

Elmoslemany, A.M., G.P. Keefe, I.R. Dohoo and B.M. Jayarao. 2009a. Risk factors for bacteriological quality of bulk tank milk in Prince Edward Island dairy herds. Part 1: Overall risk factors. J. Dairy Sci. 92: 2634–2643.

Elmoslemany, A.M., G.P. Keefe, I.R. Dohoo and R.T. Dingwell. 2009b. Microbiological quality of bulk tank raw milk in Prince Edward Island dairy herds. J. Dairy Sci. 92: 4239–4248.

Elmoslemany, A.M., G.P. Keefe, I.R. Dohoo, J.J. Wichtel, H. Stryhn and R.T. Dingwell. 2010. The association between bulk tank milk analysis for raw milk quality and on-farm management practices. Prev. Veter. Medic. 95: 32–40.

Ercolini, D., F. Russo, I. Ferraino and F. Villani. 2009. Molecular identification of mesophilic and psychrotrophic bacteria from raw cow's milk. Food Microbiol. 26: 228–231.

European Union. 2006. Commission Regulation (EC) No 1662/2006 of 6 November 2006 amending Regulation (EC) No 853/2004 of the European Parliament and of the Council laying down specific hygiene rules for food of animal origin. Official Journal of the European Union L320, 18.11.2006, 1–10.

Fajardo-Lira, C.E. and S.S. Nielsen. 1998. Effect of psychrotrophic micro-organisms on the plasmin system in milk. J. Dairy Sci. 81: 901–908.

Farkye, N.Y. 2002. Enzymes indigenous to milk: Other enzymes. pp. 943–948. *In:* H. Roginski, J.W. Fuquay and P.F. Fox [eds.]. Encyclopedia of Dairy Science. Elsevier, London, UK.

Ferguson, J.D., G. Azzaro, M. Gambina and G. Licitra. 2007. Prevalence of mastitis pathogens in Ragusa, Sicily, from 2000 to 2006. J. Dairy Sci. 90: 5798–5813.

Food Standards Code. 2001. Standard 1.6.1. of Australia and New Zealand. Microbiological limits for food with additional guideline criteria. http://www.comlaw.gov.au/Details/F2012C00862/Supporting%20Material/Text.

Foschino, R., A. Invernizzi, R. Barucco and K. Stradiotto. 2002. Microbial composition, including the incidence of pathogens, of goat milk from the Bergamo region of Italy during a lactation year. J. Dairy Res. 69: 213–225.

Fox, E., T. O'Mahony, M. Clancy, R. Dempsey, M. O'Brien and K. Jordan. 2009. *Listeria monocytogenes* in the Irish dairy farm environment. J. Food Protect. 72: 1450–1456.

Frank, J.F. and A.N. Hassan. 2002. Microorganisms associated with milk. pp. 1786–1796. *In:* H. Roginski, J.W. Fuquay and P.F. Fox [eds.]. Encyclopedia of Dairy Sciences. Elsevier, London, UK.

Fremaux, B., C. Prigent-Combaret and C. Vernozy-Rozand. 2008. Long-term survival of Shiga toxin-producing *Escherichia coli* in cattle effluents and environment: An updated review. Vet. Microbiol. 132: 1–18.

Frohbieter, K.A., B. Ismail, S.S. Nielsen and K.D. Hayes. 2005. Effects of *Pseudomonas fluorescens* M3/6 bacterial protease on plasmin system and plasmin activation. J. Dairy Sci. 88: 3392–3401.

García-Montoya, I.A., T.S. Cendón, S. Arévalo-Gallegos and Q. Rascón. 2012. Lactoferrin a multiple bioactive protein: An overview. Biochim. Biophy. Acta 1820: 226–236.

Gaya, P., C. Saralegui, M. Medina and M. Nuñez. 1996. Occurrence of *Listeria monocytogenes* and other *Listeria* spp. in raw caprine milk. J. Dairy Sci. 79: 1936–1941.

Gill, C.O., L. Saucier and W.J. Meadus. 2011. *Mycobacterium avium* subsp. *paratuberculosis* in dairy products, meat, and drinking water. J. Food Protect. 74: 480–499.

Grade "A" Pasteurized Milk Ordinance. 2009. Revision [online]. http://www.fda.gov.

Grant, I.R., A.G. Williams, M.T. Rowe and D.D. Muir. 2005. Efficacy of various pasteurization time-temperature conditions in combination with homogenization on inactivation of *Mycobacterium avium* subsp. *paratuberculosis* in milk. Appl. Environ. Microbiol. 71: 2853–2861.

Griffiths, M.W., J.D. Phillips and D.D. Muir. 1987. Effect of low-temperature storage on the bacteriological quality of raw milk. Food Microbiol. 4: 285–291.

Griffiths, M.W. and J.D. Phillips. 1988. Modeling the relation between bacterial growth and storage temperature in pasteurized milks of varying hygienic quality. J. Soc. Dairy Technol. 41: 96–102.

Griffiths, M.W. and J.D. Phillips. 1990. Incidence, source and some properties of psychrotrophic *Bacillus* spp. found in raw and pasteurized milk. J. Soc. Dairy Technol. 43: 62–66.

Hamdi, T.M., M. Naïm, P. Martin and C. Jacquet. 2007. Identification and molecular characterization of *Listeria monocytogenes* isolated in raw milk in the region of Algiers (Algeria). Int. J. Food Microbiol. 116: 190–193.

Hantsis-Zacharov, E. and M. Halpern. 2007. Culturable psychrotrophic bacterial communities in raw milk and their proteolytic and lipolytic traits. Appl. Environ. Microbiol. 73: 7162–7168.

Hayes, M.C. and K. Boor. 2001. Raw milk and fluid milk products. pp. 59–76. *In:* E.H. Marth and J.L. Steele [eds.]. Applied Dairy Microbiology. Marcel Dekker Inc., New York, USA.

Heuvelink, A.E., C. van Heerwaarden, A. Zwartkruis-Nahuis, J.J.H.C. Tilburg, M.H. Bos, F.G.C. Heilmann, A. Hofhuis, T. Hoekstra and E. de Boer. 2009. Two outbreaks of campylobacteriosis associated with the consumption of raw cows' milk. Int. J. Food Microbiol. 134: 70–74.

Holm, C., L. Jepsen, M. Larsen and L. Jespersen. 2004. Predominant microflora of downgraded Danish bulk milk tank. J. Dairy Sci. 87: 1151–1157.

Hortet, P. and H. Seegers. 1998. Calculated milk production losses associated with elevated somatic cell counts in dairy cows: Review and critical discussion. Vet. Res. (Paris). 29: 497–510.

Houser, B.A., S.C. Donaldson, S.I. Kehoe, A.J. Heinrichs and B.M. Jayarao. 2008. A survey of bacteriological quality and the occurrence of *Salmonella* in raw bovine colostrums. Foodborne Path. Dis. 5: 853–858.

Huck, J.R., M. Sonnen and K.J. Boor. 2008. Tracking heat-resistant, cold-thriving fluid milk spoilage bacteria from farm packaged product. J. Dairy Sci. 91: 1218–1228.

IDF. 1988. Code of practice for the presentation of raw milk by lactoperoxidase system. IDF Bulletin No: 224. International Dairy Federation, Brussels, Belgium.

ICMSF (International Commission on Microbiological Specifications for Foods). 1998. Milk and dairy products. pp. 521–576. *In:* Microorganisms in Foods 6: Microbial Ecology of Food Commodities. Blackie Academic & Professional, London, UK.

Jakobsen, R.A., R. Heggebø, E.B. Sunde and M. Skjervheim. 2011. *Staphylococcus aureus* and *Listeria monocytogenes* in Norwegian raw milk cheese production. Food Microbiol. 28: 492–496.

Jay, J.M. 2000. Modern Food Microbiology. 6th ed. Aspen Publishers Inc., New York, USA.

Jayarao, B.M. and L. Wang. 1999. A study on the prevalence of gram-negative bacteria in bulk tank milk. J. Dairy Sci. 82: 2620–2624.

Jayarao, B.M., S.C. Donaldson, B.A. Straley, A.A. Sawant, N.V. Hegde and J.L. Brown. 2006. A survey of foodborne pathogens in bulk tank milk and raw milk consumption among farm families in Pennsylvania. J. Dairy Sci. 89: 2451–2458.

Jenssen, H. and R.E.W. Hancock. 2009. Antimicrobial properties of lactoferrin. Biochimie 91: 19–29.

Jooste, P.J. and L.E.C.M. Anelich. 2008. Safety and quality of dairy products. pp. 153–182. *In:* T.J. Britz and R.K. Robinson [eds.]. Advance Dairy Science and Technology. Blackwell Publishing, Oxford, UK.

Jørgensen, H.J., T. Mork, H.R. Høgåsen and L.M. Rørvik. 2005. Enterotoxigenic *Staphylococcus aureus* in bulk milk in Norway. J. Appl. Microbiol. 99: 158–166.

Julien, M.C., P. Dion, C. Lafrenière, H. Antoun and P. Drouin. 2008. Sources of clostridia in raw milk on farms. Appl. Environ. Microbiol. 74: 6348–6357.

Kagkli, D.M., M. Vancanneyt, P. Vandamme, C. Hill and T.M. Cogan. 2007. Contamination of milk by enterococci and coliforms from bovine faeces. J. Appl. Microbiol. 103: 1393–1405.

Kálmán, M., E. Szöllsi, B. Czermann, M. Zimányi, S. Szekeres and M. Kálmán. 2000. Milkborne *Campylobacter* infection in Hungary. J. Food Protect. 63: 1426–1429.

Kalogridou-Vassiliadou, D. 1992. Biochemical activities of *Bacillus* species isolated from sour evaporated milk. J. Dairy Sci. 75: 2681–2686.

Kulozik, U. 2002. Biofilm formations. pp. 152–156. *In:* H. Roginski, J.W. Fuquay and P.F. Fox [eds.]. Encyclopedia of Dairy Sciences. Elsevier, London, UK.

Kussendrager, K.D. and A.C.M. van Hooijdonk. 2000. Lactoperoxidase: Physicochemical properties, occurrence, mechanism of action and applications. British J. Nutr. 84: 519–525.

Lafarge, V., J.C. Ogier, V. Girard, V. Maladen, J.C. Lereau, A. Gruss and A. Delacroix-Busset. 2004. Raw milk bacterial population shifts attributable to refrigeration. Appl. Environ. Microbiol. 70: 5644–5650.

Latorre, A.A., J.S. van Kessel, J.S. Karns, M.J. Zurakowski, A.K. Pradhan, K.J. Boor, B.M. Jayarao, B.A. Houser, C.S. Daugherty and Y.H. Schukken. 2010. Biofilm in milking equipment on a dairy farm as a potential source of bulk tank milk contamination with *Listeria monocytogenes*. J. Dairy Sci. 93: 2792–2802.

Law, B.A. 1979. Reviews on the progress of dairy science: Enzymes of psychrotrophic bacteria and their effects on milk and milk products. J. Dairy Res. 46: 573–581.

Lekkas, C., A. Kakouri, E. Paleologos, L.P. Voutsinas, M.G. Kontominas and J. Samelis. 2006. Survival of *Escherichia coli* O157:H7 in Galotyri cheese stored at 4 and 12°C. Food Microbiol. 23: 268–276.

Loftis, A.D., R.A. Priestley and R.F. Massung. 2010. Detection of *Coxiella burnetii* in commercially available raw milk from the United States. Foodborne Path. Dis. 7: 1453–1456.

Lorusso, V., A. Dambrosio, N.C. Quaglia, A. Parisi, G. La Salandra, G. Lucifora, G. Mula, S. Virgilio, L. Carosielli, A. Rella, M. Dario and G. Normanno. 2009. Verocytotoxin-producing *Escherichia coli* 026 in raw water buffalo (bubalus bubalis) milk products in Italian J. Food Protect. 72: 1705–1708.

Ma, Y., D.M. Barbano and M. Santos. 2003. Effect of CO_2 addition to raw milk on proteolysis and lipolysis at 4°C. J. Dairy Sci. 86: 1616–1631.

Marnilla, P. and H. Korhonen. 2002. Immunoglobulins. pp. 1950–1956. *In:* H. Roginski, J.W. Fuquay and P.F. Fox [eds.]. Encyclopedia of Dairy Sciences. Elsevier, London, UK.

Mazurek, J., E. Salehi, D. Propes, J. Holt, T. Bannerman, L.M. Nicholson, M. Bundesen, R. Duffy, and R.L. Moolenaar. 2004. A multistate outbreak of *Salmonella enterica* serotype *typhimurium* infection linked to raw milk consumption—Ohio, 2003. J. Food Protect. 67: 2165–2170.

McGuiggan, J.T.M., D.R. McCleery, D.R. Hannan and A. Gilmour. 2002. Aerobic sporeforming bacteria in bulk raw milk: Factors influencing the numbers of psychrotrophic, mesophilic and thermophilic *Bacillus* spores. Int. J. Dairy Technol. 55: 100–107.

McKinnon, C.H. and G.L. Pettipher. 1983. A survey of sources of heat-resistant bacteria in milk with particular reference to psychrotrophic sporeforming bacteria. J. Dairy Res. 50: 163–170.

McPhee, J.D. and M.W. Griffiths. 2002. *Pseudomonas* spp. pp. 2340–2345. *In:* H. Roginski, J.W. Fuquay and P.F. Fox [eds.]. Encyclopedia of Dairy Sciences. Elsevier, London, UK.

Medina, R., M. Katz, S. Gonzalez and G. Oliver. 2001. Characterization of the lactic acid bacteria in ewe's milk and cheese from northwest Argentina. J. Food Protect. 64: 559–563.

Melville, P.A., N.R. Benites, M. Ruz-Peres and E. Yokoya. 2011. Proteinase and phospholipase activities and development at different temperatures of yeasts isolated from bovine milk. J. Dairy Res. 78: 385–390.

Moatsou, G. 2010. Indigenous enzymatic activities in ovine and caprine milks. Int. J. Dairy Technol. 1: 16–31.

Moshtaghi, H. and A.A. Mohamadpour. 2007. Incidence of *Listeria* spp. in raw milk in Shahrekord, Iran. Foodborne Path. Dis. 4: 107–110.

Muehlherr, J.E., C. Zweifel, S. Corti, J.E. Blanco and R. Stephan. 2003. Microbiological quality of raw goat's and ewe's bulk-tank milk in Switzerland. J. Dairy Sci. 86: 3849–3856.

Muir, D.D. 1996a. The shelf-life of dairy products: 1. Factors influencing raw milk and fresh products. J. Soc. Dairy Technol. 49: 24–32.

Muir, D.D. 1996b. The shelf-life of dairy products: 2. Raw milk and fresh products. J. Soc. Dairy Technol. 49: 44–48.

Murphy, B.P., M. Murphy, J.F. Buckley, D. Gilroy, M.T. Rowe, D. McCleery and S. Fanning. 2005. In-line milk filter analysis: *Escherichia coli* O157 surveillance of milk production holdings. Int. J. Hygiene Environ. Health 208: 407–413.

National Dairy Code of Canada. 2011. National Dairy Code-1997, revised 2010. Production and processing requirements, 5th ed. (part 1). 20 p. http://www.dairyinfo.gc.ca/pdf/Dairy%20Code %20Revised_May%202011_NDC %20Part% 20I%20_final2_e.pdf.

Ogola, H., A. Shitandi and J. Nanua. 2007. Effect of mastitis on raw milk compositional quality. J. Vet. Sci. 8: 237–242.

Oliver, S.P., B.M. Jayarao and R.A. Almeida. 2005. Foodborne pathogens in milk and the dairy farm environment: Food safety and public health implications. Foodborne Path. Dis. 2: 115–129.

Oliver, S.P., K.J. Boor, S.C. Murphy and S.E. Murinda. 2009. Food safety hazards associated with consumption of raw milk. Foodborne Path. Dis. 6: 793–806.

Perkins, N.R., D.F. Kelton, K.J. Hand, G. MacNaughton, O. Berke and K.E. Leslie. 2009. An analysis of the relationship between bulk tank milk quality and wash water quality on dairy farms in Ontario, Canada. J. Dairy Sci. 92: 3714–3722.

Phillips, J.D. and M.W. Griffiths. 1986. Factors contributing to the seasonal variation of *Bacillus* spp. in pasteurized dairy products. J. Appl. Bacteriol. 61: 275–280.

Phillips, J.D. and M.W. Griffiths. 1987. The relation between temperature and growth of bacteria in dairy products. Food Microbiol. 4: 173–185.

Pruitt, K.M. and D.N. Kamau. 1994. Quantitative analysis of bovine lactoperoxidase system on bacterial growth and survival. pp. 73–87. *In:* Proceedings of the International Dairy Federation Seminar on *Indigenous antimicrobial agents of milk—recent developments.* Uppsala, Sweden 31 Aug–1 Sept. Special Issue 9404. International Dairy Federation, Brussels, Belgium.

Quigley, L., O. O'Sullivan, T.P. Beresford, R.P. Ross, G.F. Fitzgerald and P.D. Cotter. 2011. Molecular approaches to analyzing the microbial composition of raw milk and raw milk cheese. Int. J. Food Microbiol. 150: 81–94.

Rajagopal, M., B.G. Werner and J.H. Hotchkiss. 2005. Low pressure CO_2 storage of raw milk: Microbiological effects. J. Dairy Sci. 88: 3130–3138.

Rasolofo, E.A., D. St-Gelais, G. LaPointe and D. Roy. 2010. Molecular analysis of bacterial population structure and dynamics during cold storage of untreated and treated milk. Int. J. Food Microbiol. 138: 108–118.

Ray, B. 2004. Fundamental Food Microbiology. 3rd ed. CRC Press, New York, USA.

Roostita, R. and G.H. Fleet. 1996. Growth of yeasts in milk and associated changes to milk composition. Int. J. Food Microbiol. 31: 205–219.

Rosengren, Å., A. Fabricius, B. Guss, S. Sylvén and R. Lindqvist. 2010. Occurrence of foodborne pathogens and characterization of *Staphylococcus aureus* in cheese produced on farm-dairies. Int. J. Food Microbiol. 144: 263–269.

Saboya, L.V. and J.L. Maubois. 2000. Current development of microfiltration technology in the dairy industry. Lait 80: 541–553.

Samelis, J., A. Lianou, A. Kakouri, C. Delbès, I. Rogelj, B.M. Bojana and M.C. Montel. 2009. Changes in the microbial composition of raw milk induced by thermization treatments applied prior to traditional Greek hard cheese processing. J. Food Protect. 72: 783–790.

Sanjuan, S., J. Rua and M.R. Garcio-Armesto. 2003. Microbial flora of technological interest in raw ovine milk during 6°C storage. Int. J. Dairy Technol. 56: 143–148.

Scheldeman, P., A. Pil, L. Herman, P. de Vos and M. Heyndrickx. 2005. Incidence and diversity of potentially highly heat-resistant spores isolated at dairy farms. Appl. Environ. Microbiol. 71: 1480–1494.

Scheldeman, P., L. Herman, S. Foster and M. Heyndrickx. 2006. *Bacillus sporothermodurans* and other highly heat-resistant spore formers in milk. J. Appl. Microbiol. 101: 542–555.

Schoder, D., D. Melzner, A. Schmalwieser, A. Zangana, P. Winter and M. Wagner. 2011. Important vectors for *Listeria monocytogenes* transmission at farm dairies manufacturing fresh sheep and goat cheese from raw milk. J. Food Protect. 74: 919–924.

Scott, S.A., J.D. Brooks, J. Rakonjac, K.M.R. Walker and S.H. Flint. 2007. The formation of thermophilic spores during the manufacture of whole milk powder. Int. J. Dairy Technol. 60: 109–117.

Seifu, E., E.M. Buys and E.F. Donkin. 2005. Significance of the lactoperoxidase system in dairy industry and its potential application: A review. Trends Food Sci. Technol. 16: 137–154.

Shakeel-ur-Rehman and N.Y. Farkye. 2002. Lactoperoxidase. pp. 938–941. *In:* H. Roginski, J.W. Fuquay and P.F. Fox [eds.]. Encyclopedia of Dairy Sciences. Elsevier, London, UK.

Singh, P., A.A. Wani, A.A. Karim and H.C. Langowski. 2012. The use of carbon dioxide in the processing and packaging of milk and dairy products: A review. Int. J. Dairy Technol. 65: 161–177.

Skovgaard, N. 2007. New trends in emerging pathogens. Int. J. Food Microbiol. 120: 217–224.

Slaghuis, B.A., M.C. TeGiffel, R.R. Beumer and A. Geert. 1997. Effect of pasturing on the incidence of *Bacillus cereus* spores in raw milk. Int. Dairy J. 7: 201–205.

Sørhaug, T. and L. Stepaniak. 1997. Psychrotrophs and their enzymes in milk and dairy products: Quality aspects. Trends Food Sci. Technol. 8: 35–41.

Souto, L.I.M., C.Y. Minagawa, E.O. Telles, M.A. Garbuglio, M. Amaku, R.A. Dias, S.T. Sakata and N.R. Benites. 2008. Relationship between occurrence of mastitis pathogens in dairy cattle herds and raw-milk indicators of hygienic-sanitary quality. J. Dairy Res. 75: 121–127.

Stepaniak, L. and E.O. Rukke. 2003. Thermization of milk. pp. 2619–2623. *In:* H. Roginski, J.W. Fuquay and P.F. Fox [eds.]. Encyclopedia of Dairy Sciences. Elsevier, London, UK.

Stepaniak, L. 2003. Bacteria other than *Pseudomonas* spp. pp. 2345–2351. *In:* H. Roginski, J.W. Fuquay and P.F. Fox [eds.]. Encyclopedia of Dairy Sciences. Elsevier, London, UK.

Sutherland, A.D. and R. Murdoch. 1994. Seasonal occurrence of psychrotrophic *Bacillus* species in raw milk and studies on the interactions of *Bacillus* spp. Int. J. Food Microbiol. 21: 279–292.

Swaminathan, B. and P. Gerner-Smidt. 2007. The epidemiology of human listeriosis. Microbes Infect. 9: 1236–1243.

Te Giffel, M.C., A. Wagendorp, A. Herrewegh and F. Driehuis. 2002. Bacterial spores in silage and raw milk. Antonie van Leeuwen. Int. J. Gen. Mol. Microbiol. 81: 625–630.

Ternström, A., A.M. Lindberg and G. Molin. 1993. Classification on the spoilage flora of raw and pasteurized bovine milk, with special reference to *Pseudomonas* and *Bacillus*. J. Appl. Bacteriol. 75: 25–34.

Tsuji, S.Y., F. Hirata, F. Mukai and S. Ohtagaki. 1990. Comparison of lactoferrin content in colostrums between different cattle breeds. J. Dairy Sci. 73: 125–128.

Uraz, G. and N. Yücel. 1999. The isolation of certain pathogen microorganisms from raw milk. Central Eur. J. Pub. Health 7: 145–148.

Vacheyrou, M., A.C. Normand, P. Guyot, C. Cassagne, R. Piarroux and Y. Bouton. 2011. Cultivable microbial communities in raw cow milk and potential transfers from stables of sixteen French farms. Int. J. Food Microbiol. 146: 253–262.

van Kessel, J.S., J.S. Karns, D.R. Wolfgang, E. Hovingh, B.M. Jayarao, C.P. van Tassell and Y.H. Schukken. 2008. Environmental sampling to predict fecal prevalence of *Salmonella* in an intensively monitored dairy herd. J. Food Protect. 71: 1967–1973.

van Kessel, J.S., J.S. Karns, J.E. Lombard and C.A. Kopral. 2011. Prevalence of *Salmonella enterica, Listeria monocytogenes* and *Escherichia coli* virulence factors in bulk tank milk and in-line filters from U.S. dairies. J. Food Protect. 74: 759–768.

Verdier-Metz, I., V. Michel, C. Delbès and M.C. Montel. 2009. Do milking practices influence the bacterial diversity of raw milk? Food Microbiol. 26: 305–310.

Vissers, M.M.M., F. Driehuis, M.C. TeGiffel, P. de Jong and J.M.G. Lankveld. 2007a. Quantification of the transmission of microorganisms to milk via dirt attached to the exterior of teats. J. Dairy Sci. 90: 3579–3582.

Vissers, M.M.M., M.C. TeGiffel, F. Driehuis, P. de Jong and J.M.G. Lankveld. 2007b. Minimizing the level of *Bacillus cereus* spores in farm tank milk. J. Dairy Sci. 90: 3286–3293.

Vissers, M.M.M., F. Driehuis, M.C. TeGiffel, P. de Jong and J.M.G. Lankveld. 2007c. Minimizing the level of butyric acid bacteria spores in farm tank milk. J. Dairy Sci. 90: 3278–3285.

Vissers, M.M.M., M.C. TeGiffel, F. Driehuis, P. de Jong and J.M.G. Lankveld. 2007d. Predictive modeling of *Bacillus cereus* spores in farm tank milk during grazing and housing periods. J. Dairy Sci. 90: 281–292.

Waddell, L.A., A. Rajić, J. Sargeant, J. Harris, R. Amezcua, L. Downey, S. Read and S.A. McEwen. 2008. The zoonotic potential of *Mycobacterium avium* spp. *paratuberculosis*: A systematic review. Can. J. Pub. Health 99: 145–155.

Wakabayashi, H., K. Yamauchi and M. Takase. 2006. Lactoferrin research, technology and applications. Int. Dairy J. 16: 1241–1251.

Walstra, P., J.T.M. Wouters and T.J. Geurts. 2006. Dairy Science and Technology. 2nd ed. CRC Press, New York, USA.

Weimer, P.J. 2001. Microbiology of the dairy animal. pp. 1–58. *In:* E.H. Marth and J.L. Steele [eds.]. Applied Dairy Microbiology. Marcel Dekker Inc., New York, USA.

Whyte, P., K. McGill, D. Cowley, R.H. Madden, L. Moran, P. Scates, C. Carroll, A. O'Leary, S. Fanning, J.D. Collins, E. McNamara, J.E. Moore and M. Cormican. 2004. Occurrence of *Campylobacter* in retail foods in Ireland. Int. J. Food Microbiol. 95: 111–118.

Yoshida, T., M. Sato and K. Hirai. 1998. Prevalence of *Listeria* species in raw milk from farm bulk tanks in Nagano prefecture. J. Vet. Med. Sci. 60: 311–314.

CHAPTER 2

Dairy Starter Cultures

Zeynep Ustunol

2.1 Introduction

Starter cultures are selected microorganisms that are deliberately added to milk to initiate and carry out desired fermentation under controlled conditions in the production of fermented dairy products. Currently, cultured dairy products are driving the growth of dairy foods consumptions (IDFA 2011). Starter cultures play an essential role in the manufacturing, flavor and texture development of fermented dairy foods; therefore, they are of great industrial significance. Furthermore, data accumulating on the potential health benefits of ingesting these organisms have created additional interest in starter bacteria. Traditionally, starter cultures were developed through screening procedures and selection by trial and error. Today, starter cultures for fermented foods are developed mainly by design. Advances made in genetics and molecular biology have provided additional opportunities for studying the genomics of these economically significant organisms. This has allowed engineering of cultures that focuses on rational improvement of the industrially useful strains.

2.2 A brief historic overview

A starter culture is a microbial preparation that is added intentionally to achieve desirable changes under controlled conditions of fermentation. These preparations include lactic acid bacteria, propionibacteria, as well as surface ripening bacteria, yeasts and molds. Most starter cultures used today have their origins with lactic acid bacteria. Lactic acid bacteria are widespread in nature. They have been used in the production of fermented dairy foods for over 4000 years. Chr. Hansen Laboratory in 1878 was the first to introduce commercial lactic starter cultures for the manufacture of cheese and cultured butter in North America. Initially, these were liquid cultures in glass bottles that required a few transfers prior to their use for the manufacture of cheese or cultured butter. They were produced in Denmark and shipped to America together with milk coagulants, and were distributed from New York City. In 1880 Chr. Hansen's operations moved to Little Falls, NY. At the time, Little Falls was the center of cheese production in the U.S.

Initially, cultures were collected over the years from factories that had excellent and uniform success of making raw milk cheese. Early commercial starter cultures consisted of a mixture of lactococci and leuconostocs (flavor enhancing) organisms. Eventually, they were purified. They were grown in pasteurized non-fat skim milk and maintained as stable multiple-mixed strain cultures (Cogan et al. 2007). In the late 1930's first commercial 'Dry Lactic Starters' were made available. The first freeze-dried starters became available in the mid-1950s. In 1965, the first cryogenically frozen mother culture for inoculating bulk starter was made available commercially. In 1975, Marschall Dairy Laboratory introduced the first highly concentrated cryogenically frozen starter cultures for the direct setting of cultured milk for cheese, cultured buttermilk, sour cream, etc. Chr. Hansen's followed with their competitive direct to the vat set (DVS) cultures in 1978 (Cogan et al. 2007).

Initially external and later internal pH control systems were used to further grow starter cultures to higher number. In the external pH control system, pH was monitored and lactic acid produced by lactic acid bacteria was continuously neutralized by the introduction of ammonium hydroxide to maintain a specified pH. This prevented acid injury to the organisms and allowed them to grow to higher cell numbers. This method was used initially to produce bulk set cultures in the early 1960's. Internal pH control systems achieve this with the use of buffer salts incorporated into the medium that maintains the pH at the specified target. The concentrated, active cultures were further concentrated by centrifugation. Stability and subsequent high activity of the starter culture was obtained by cryogenically freezing them in liquid nitrogen (Hoeier et al. 1999, Porubcan and Sellars 1979). All types of starter cultures were produced by these described procedures. Today all commercial companies that sell starter cultures to the dairy industry use these basic processes (or some modifications depending on the organism). Today, highly stable and active concentrated cultures for direct inoculations, or freeze-dried cultures are readily available to the industry. These cultures have eliminated the need for culture preparation which often required several days of transfers to obtain the desired levels of activity.

2.3 Classification of starter cultures

Traditionally, fermentation in dairy foods was by natural wild-type lactic acid bacteria that were found in raw milk. Although there are 12 genera of lactic acid bacteria, today, primarily four genera *Lactococcus*, *Leuconostoc*, *Streptococcus* and *Lactobacillus* are used as dairy starter cultures. More recently, *Bifidobacterium* genera have also been added to the list due to the perceived health benefits of these organisms, thus their inclusion in fermented dairy foods. Inclusion of enterococci in this group has been controversial since these are fecal organisms, and have been recognized as opportunistic pathogens (Franz et al. 1999). However, they are sometimes found in undefined mixed strain cultures and contribute to ripening and flavor development in cheese (Broome et al. 1990, McSweeney et al. 1993, Crow et al. 2001, Dudley and Steele 2005). All lactic acid bacteria used as starter cultures in the dairy industry are Gram-positive, catalase-negative, non-motile, non-sporeformers, cocci or rods that have less than 55 mol% G+C content in their DNA. In general they are aerobic to facultative anaerobic microorganisms. The taxonomy of lactic acid bacteria used in starter cultures has undergone several revisions over the last 40 years or so. Traditionally, their taxonomy was based on morphology and physiology. Today, their taxonomy takes into account molecular characteristics such as mol% G+C content, electrophoretic properties of gene products, DNA:DNA hybridization, sequences of ribosomal RNA, DNA:RNA hybridization, comparative oligonucleotide sequencing, cataloging of the 16S rDNA gene and additional serological work with superoxide dismutase have been important taxonomic tools. Furthermore, phylogenetic relationships among bacteria are accurately determined through sequencing (Stackebrandt et al. 2002, Axelsson 2004). Bifidobacteria were included in the genus *Lactobacillus* but in 1986 they were transferred into the genus *Bifidobacterium* (Scordavi 1986). For a more comprehensive review of taxonomy please see reviews by Stackebrandt et al. (2002), Axelsson (2004), Fox (2011) and Bjorkroth and Koort (2011).

Dairy starter cultures are commonly grouped into mesophilic cultures (optimum temperature of about 30°C) and thermophilic cultures (optimum temperature of about 42–45°C). Their ability to grow at 10 and 45°C can be used to distinguish mesophilic and thermophilic cultures. Microscopic observations, citrate metabolism, and amount of lactic acid and its isomer that is produced can further differentiate most species within these two broad categories. Furthermore, these dairy starter cultures are categorized as defined- or mixed-strain cultures. Defined cultures are selected, defined, pure strains of known cultures. They have been typically isolated from mixed cultures and selected based on their several important characteristics such as resistance to phage, acid production, citrate utilization, good aroma and flavor production. Mixed-strain mesophilic dairy cultures often contain same species of unknown numbers, as well as other species. It is important to note that defined strains of mesophilic cultures in commercial settings are often used as mixture of 2–6 phage unrelated strains; so to a certain extent they are also mixed-strain cultures. In contrast defined strains of thermophilic cultures are used in rotation. Undefined cultures as the name indicates are mixed cultures of undefined strains (Fox et al. 2000, 2004).

Starter cultures are also grouped broadly based on their sugar metabolism as *homofermenters* and *heterofermenters*. In homolactic fermentation, glycolysis or Embden-Meyerhof-Parnas pathway (EMP) is utilized for sugar metabolism. Homolactic fermentation results primarily in the production of lactic acid as the end product under standard conditions (Fig. 2.1). In heterolactic fermentation the 6-phosphogluconate/phosphoketolase pathway is utilized. Heterolactic fermentation results in

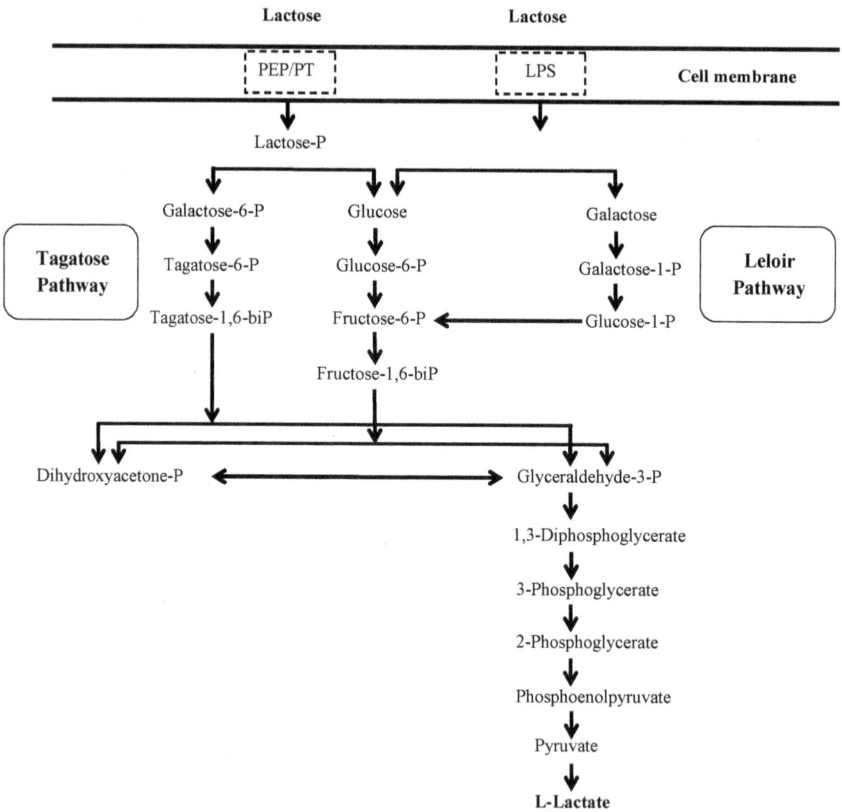

Figure 2.1 Embden-Meyerhof-Parnas (glycolytic) pathway of lactose metabolism by lactic acid bacteria. **PEP/PTS:** Phosphoenolpyruvate-dependent phosphotransferase system, **LPS:** Lactose permease system.

the production of lactic acid, as well as significant amount of other end products, ethanol, acetate and CO_2 (Fig. 2.2). However, it should be noted that end product formation may be significantly altered due to variation in growth conditions (Marth and Steele 2001, Fox et al. 2004).

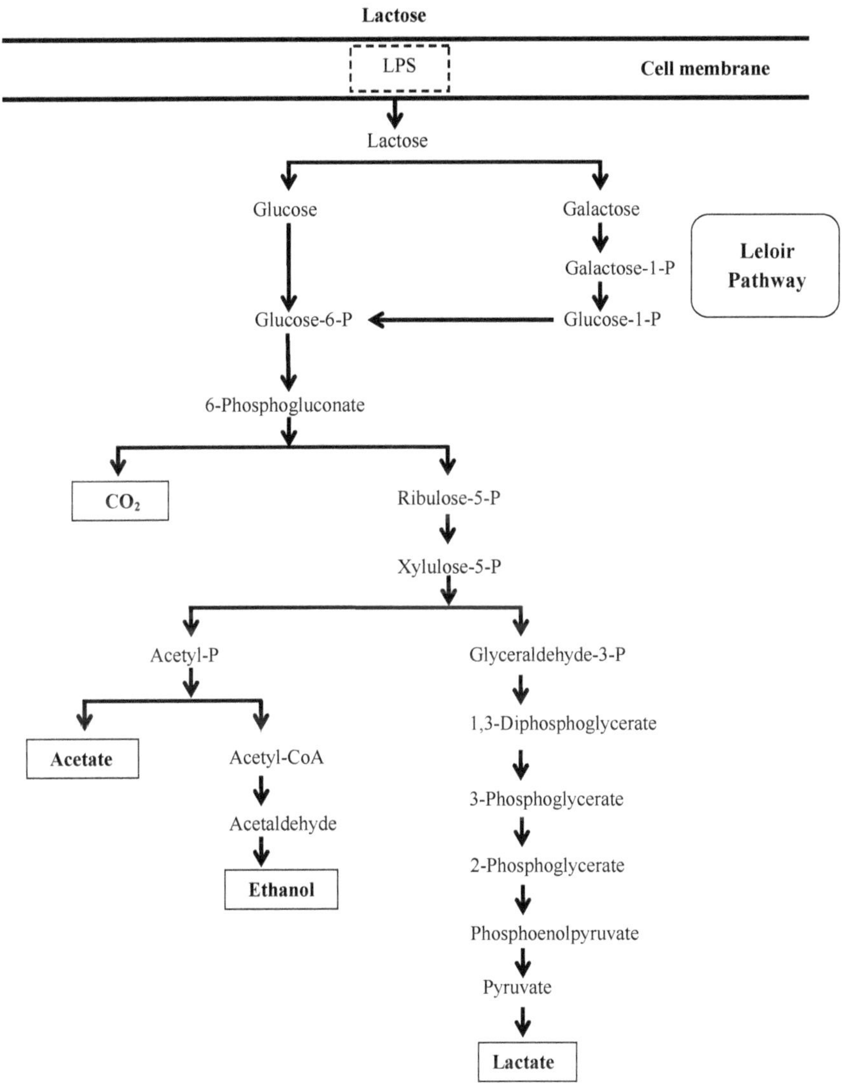

Figure 2.2 Phosphoketolase pathway of lactose metabolism by lactic acid bacteria.

2.3.1 Mesophilic cultures

As mentioned previously, mesophilic cultures have an optimum temperature of about 30°C. Mesophilic dairy starter cultures include primarily *Lactococcus* and *Leuconostoc*. *Pediococcus* are less significant as dairy starter cultures. The main species of *Lactococcus* are *Lactococcus lactis* subsp. *cremoris* and *Lactococus lactis* subsp. *lactis*. These two sub-species can be differentiated from each other by the ability of *L. lactis* subsp. *lactis* to grow at 40°C in 4% NaCl and 0.1% methylene blue milk, at pH 9.2, and by formation of NH_3 from arginine (Stiles and Holzapfel 1997). Furthermore, *L. lactis* subsp. *lactis* contains a glutamate decarboxylase enzyme whereas *L. lactis* subsp. *cremoris* does not. Presence of glutamate decarboxylase provides for the production of γ-aminobutyric acid from glutamate by this organism. *Lactococcus lactis* biovar. *diacetylactis* has the ability to metabolize citrate to produce diacetyl. Of the genus *Leuconostoc* two species are important in dairy fermentation. *Leuconostoc mesenteroides* subsp. *cremoris* and *Leuconostoc lactis*, both heterofermenters. *Leuconost*oc spp. also metabolize citrate to produce CO_2, diacetyl and acetate which are important in eye formation (i.e., Gouda, Edam) and flavor (i.e., Cottage cheese). Organisms that have the ability to metabolize citrate are considered aroma producers. In the past, mesophilic mixed cultures have been classified as L, D, LD or O. This is based on whether they contained *Leuconostoc* spp., or *L. lactis* biovar. *diacetylactis*, or contained both, or no flavor producers, respectively in the final mixture (Cogan and Hill 1993, Fox and McSweeney 2004, Vasiljevic and Shah 2008). Mesophilic mixed cultures are typically 90% acid producers and 10% aroma producers.

2.3.2 Thermophilic cultures

As mentioned previously, thermophilic cultures have an optimum temperature of about 42–45°C. Most important cultures used in production of cultured dairy products which are thermophilic include *Streptococcus thermophilus*, *Lactobacillus delbrueckii* subsp. *bulgaricus*, *Lactobacillus delbrueckii* subsp. *lactis* and *Lactobacillus helveticus*. They may be grown together (i.e., yogurt production) or individually (cheese production). This symbiotic relationship between *S. thermophilus* and *Lb. delbrueckii* subsp. *bulgaricus* is well established. These two organisms are used in the production of yogurt. As a homofermenter *Lb. delbrueckii* subsp. *bulgaricus* will produce lactic acid from lactose, as well as amino acids resulting in proteolysis of casein to stimulate the growth of *S. thermophilus*. Metabolism of *S. thermophilus* produces small amounts of CO_2 as well as formic acid from lactose which then in return will aid in the growth of *Lb. delbrueckii* subsp. *bulgaricus* (Cogan and Hill 1993, Fox and McSweeney 2004, Vasiljevic and

Shah 2008). The symbiotic relationship provides for greater growth rate, and acid production. Other symbiotic relationships of dairy starter cultures are not as well established.

2.3.3 Natural or artisanal starter cultures

With an interest in artisanal cheeses there is also a growing interest in artisanal dairy starter cultures. Traditionally, artisanal cultures were derived from the previous batch of cheese. Previous batch of cheese provided the starter culture thus the inoculum for the next batch of cheese being manufactured. Such cultures particularly were used in countries such as Greece, Italy, France and Switzerland. These cheeses tended to be raw milk cheeses thus the starter culture also relied on the presence of lactic acid bacteria in the raw milk. Composition of these starter cultures were complex, and were variable depending on the temperature of incubation and pH that favored the growth of specific microorganisms. These starter cultures were a mixture of thermophilic as well as mesophilic mixed cultures. They may have been susceptible to phage. As a result they did not provide a product with consistent results (Fox et al. 2000).

Today artisanal cheese is a booming industry; in the U.S. the numbers of artisanal cheese-makers have doubled since 2000. Artisanal cheese implies that the cheeses are made by hand. The fact is that most cheeses have always been made by hand, but the skilled cheese-makers did not identify them as artisans in the past. As the production of artisanal cheeses is growing due to increased interest and demand, today the artisanal cheese-makers are also coming to rely on commercial defined mixed strain cultures for consistency.

2.3.4 Adjunct cultures

Adjunct cultures are typically used in cheeses to provide for additional flavor development. Adjunct cultures can increase the intensity and change the balance of cheese flavor. These adjunct cultures could be thermophilic organisms added to survive the cooking temperatures of the curd and be involved in flavor development later. However, these organisms will not grow and contribute to acid or flavor production at temperatures below 25°C. Mesophilic cultures such as *Lactobacillus casei* and *Lactobacillus paracasei* have also been traditionally used in flavor improvement (i.e., Cheddar cheese). Lack of flavor in cheese today has been a consequence of high quality raw milk (low microbial counts), pasteurization of cheese milk, and improved hygiene in cheese plants. Other non-starter lactic acid bacteria (NSLAB) have been studied as adjunct cultures in cheese flavor development (Thomas 1987, McSweeney et al. 1993, Lynch et al. 1996,

Fitzsimons et al. 1999, Swearingen et al. 2001, de Angelis et al. 2001, Rhea et al. 2004, Cogan et al. 2007, Beresford and Williams 2004). Beresford and Williams (2004) have reported summary of NSLAB found in >50 cheese varieties.

More recently, probiotic microorganisms have been added as adjunct cultures to traditional fermented dairy foods including yogurt, fermented milk cheeses. *Bifidobacterium* spp. and *Lactobacillus acidophilus* have been used as adjunct probiotic cultures due to data accumulating on their health benefits. Exopolysaccharide producing lactic acid bacteria are also being studied as potential adjunct cultures in low fat cheeses to provide for improved texture and better moisture control (Low et al. 1998, Hassan et al. 2005).

2.4 Metabolisms by starter cultures

The main purpose of the starter culture in fermented dairy foods is to bring about the chemical, sensory, and nutritional changes typically associated with that food. Probably the most important feature of lactic acid bacteria is their ability to ferment carbohydrates, and to produce lactic acid as the primary end product or as one of the end products. The next sections will discuss well-known fermentation pathways utilized by lactic acid bacteria and metabolism of other compounds that are significant for fermented dairy foods commercially. It is important to note that lactic acid bacteria may change their metabolism in response to different conditions, therefore, resulting in different end products of fermentation. Most of the time the change is due to altered pyruvate metabolism. Genes for many of these metabolic traits discussed here are encoded in plasmids.

2.4.1 Lactose metabolism

One of the most important characteristics of dairy starter cultures is their metabolism of lactose. Lactose is the main carbohydrate found in milk present at 45–50 g l^{-1}. It is a disaccharide made up of one molecule of glucose and one molecule of galactose that is $\beta 1 \rightarrow 4$ linked. Their ability to ferment lactose is the most important selection criterion for dairy starter cultures.

Dairy starter cultures are categorized into two main categories *homofermentative* or *heterofermentative* organisms based on their ability to ferment sugars. Homofermentative starter cultures include *Lactococcus, Streptococcus* and some of the lactobacilli. These microorganisms metabolize sugars using the glycolytic (or Embden-Meyerhof-Parnas (EMP)) pathway, where one mole of glucose is fermented into two moles of lactic acid and two ATPs are produced. Heterofermentative starter cultures, which include

Leuconostoc and remaining of lactobacilli, ferment one mole of glucose into one mole of lactic acid, one mole of CO_2, and one mole of ethanol, or one mole of acetic acid utilizing the phosphoketolase pathway which yields one ATP. Fermentation of sugars by the phosphoketolase pathway produces only half of the energy that of glycolytic path (Marth and Steele 2001, Vasiljevic and Shah 2008). The differences in metabolic end products have significant impact on the final product.

Disaccharide lactose may enter the cell either as free sugar or as sugar phosphate. The transport mechanism of lactose into the cell is determined largely by the pathway of hydrolysis of the internalized sugar. Thus, lactose is transported into the cell by two distinctly different transport mechanisms. Most efficient of the two mechanisms is the lactose specific phosphoenolpyruvate-dependent phosphotransferase system (PEP-PTS), where lactose is phosphorylated during its transport through the membrane. Once inside the cell, lactose is hydrolyzed by phospho-β-galactosidase. Resulting glucose is metabolized by glycolysis (EMP pathway) into lactic acid. The galactose-6-phosphate that is produced due to the action of phospho-β-galactosidase is converted to tagatose through the tagatose pathway and is converted into trioses which eventually enter the glycolytic pathway to produce lactic acid. The other equally common transport mechanism is the use of a lactose carrier (permease) system where lactose is transported without modification. Once inside the cell, lactose is hydrolyzed by β-galactosidase. Glucose that is produced again is metabolized by glycolysis whereas galactose is converted to glucose through Leloir pathway and further metabolized by glycolysis into lactic acid. In all of these metabolic paths pyruvate is the key intermediate whose conversion results in the production of four moles of lactic acid under normal conditions. It is important to note that some thermophilic starter bacteria such as *S. thermophilus*, *Lb. delbrueckii* subsp. *bulgaricus*, and some *Lb. delbrueckii* subsp. *lactis* ferment only the glucose, and transport the galactose moiety out of the cell in proportion to the amount of lactose transported into the cell through a lactose-galactose antiport system. So the permease system that is involved in the transport of lactose into the cell also exports galactose out of the cell (Marth and Steele 2001, Fox and McSweeney 2004, Vasiljevic and Shah 2008). Lactic acid production by homofermentative lactic starters is discussed in Chapter 8 in detail.

In the heterolactic fermentation lactose also is taken up by the cell through a permease system. Lactose is hydrolyzed into glucose and galactose by β-galactosidase. Galactose is converted to glucose through the Leloir pathway. Glucose is then fermented by the phosphoketolase (PK) pathway to two moles of each lactic acid, ethanol, and CO_2. The isomer of lactic acid produced depends on the organism. *Leuconostoc* and *Lb. delbrueckii* produce D-lactic acid, whereas *Lactococcus* and *S. thermophilus* produce only L-lactic acid. Some organisms such as *Lb. helveticus* produce both isomers (Marth and Steele 2001, Fox and McSweeney 2004, Vasiljevic and Shah 2008).

In the genus *Bifidobacterium* the pathway for metabolism of sugars are different than that for homo and heterofermentative lactic acid bacteria discussed above. Their optimum growth temperature is 36–43°C. Because they also produce lactic acid as their fermentation end product they were (and are) often grouped with lactic acid bacteria even though they are phylogenetically distinct with high G+C content (42–67%) (Scordavi 1986). *Bifidobacterium* degrade hexoses by the fructose-6-phosphate pathway (Fig. 2.3). In this pathway aldolase and glucose-6-phosphate dehydrogenase are absent, instead fructose 6-phosphate phosphoketolase (F6PPK) is present. This enzyme is considered a taxonomic marker for this organism. Fermentation of two moles of glucose produces three moles of acetate and two moles of lactate where pyruvate is converted to L-lactate by lactate dehydrogenase. Pyruvic acid actually can be broken down through two pathways. Either through reduction of pyruvate to form L-lactate by lactate dehydrogenase which activity is controlled by fructose-1,6-diphosphate. Alternatively, the second possible pathway is the degradation of pyruvate to form formic acid and acetyl phosphate a portion of which is further reduced to form ethyl alcohol and regenerate NAD at the expense of

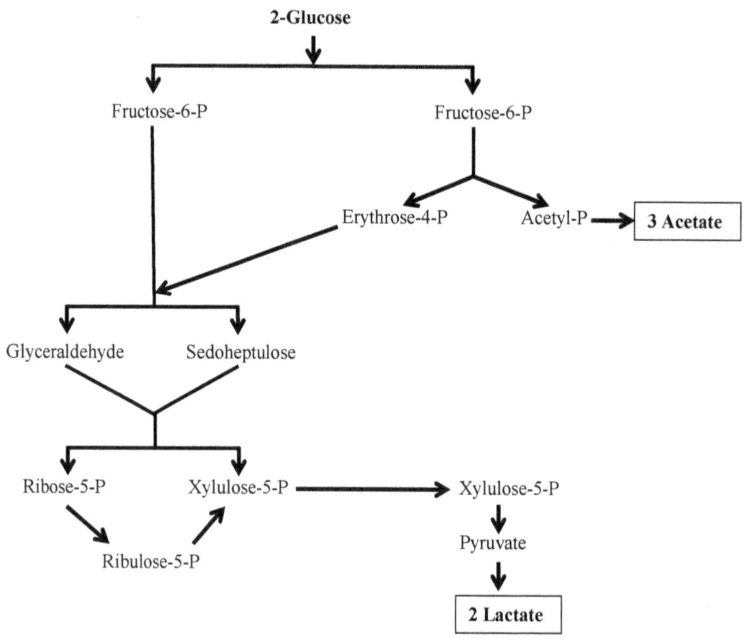

Figure 2.3 Sugar metabolism by bifidobacteria.

lactic acid production. However, different strains of bifidobacteria may have different abilities to metabolize sugars. The proportions of the final fermentation products can vary significantly from one strain to another and even within the same species. Some strains may even produce small quantities of succinic acid and CO_2. The metabolism of inulin type fructans by bifidobacteria changes the acetic acid:lactic acid ratio at the expense of lactic acid. Some bifidobacteria lack mechanisms dedicated to transport monosaccharides, therefore are not able to metabolize simple sugars such as glucose or fructose, although they can metabolize lactose, sucrose or oligofructose (van der Meulen et al. 2004).

2.4.2 Citrate metabolism

Citrate is naturally found in milk at about 8–9 mM. Citrate metabolism by dairy starter cultures is an unstable trait due to the plasmid localization of genes encoding for citrate permease (Hugenholtz 1993, Kempler and McKay 1981). Not all dairy starter cultures are capable of metabolizing citrate (Palles et al. 1998). Citrate is metabolized by certain mesophilic starter cultures such as Citrate$^+$ *Lactococcus* and *Leuconostoc* spp. (*Leu. mesenteroides* subsp. *cremoris* and *Leu. lactis*). Citrate is not metabolized by thermophilic cultures. Studies on citrate permeases have shown that citrate can be taken up by the cells by diverse mechanisms or through a proton motive force generation. Citrate uptake is regulated by pH of the growth media. However, the mechanism of activation of citrate uptake at acidic pH is still not well understood. Citrate metabolism by dairy starter cultures produces acetate, CO_2, diacetyl, acetoin and 2,3 butanediol. Citrate derived pyruvate is the branching point of the formation of these compounds (Fig. 2.4). Although citrate is co-metabolized with lactose, unlike lactose it is not utilized as an energy source. Thus, citrate does not support growth of these microorganisms. CO_2 produced during citrate metabolism provides for the eye formation in certain cheeses such as Dutch type cheeses, whereas diacetyl and acetate are important flavor and aroma compounds in fresh unripened cheeses.

Diacetyl is also an important compound that provides buttery aroma and flavor in many other dairy foods. In cheese such as Comte and Manchego enterococci make up a significant part of the fresh and fully ripened cheese microbiota. It has been reported that the enterococci may play an important role in the aroma and flavor of these cheeses due to citrate catabolism by these organisms (Quintans et al. 2008). However, it is also important to note that different microorganisms have different strategies for citrate conversion (Crow 1990). Even within the same species large variations between strains are observed. For example, within the *L. lactis* species there are large strain differences in acetoin/diacetyl reductase activity (Hugenholtz 1993). Genetic studies of lactic acid bacteria

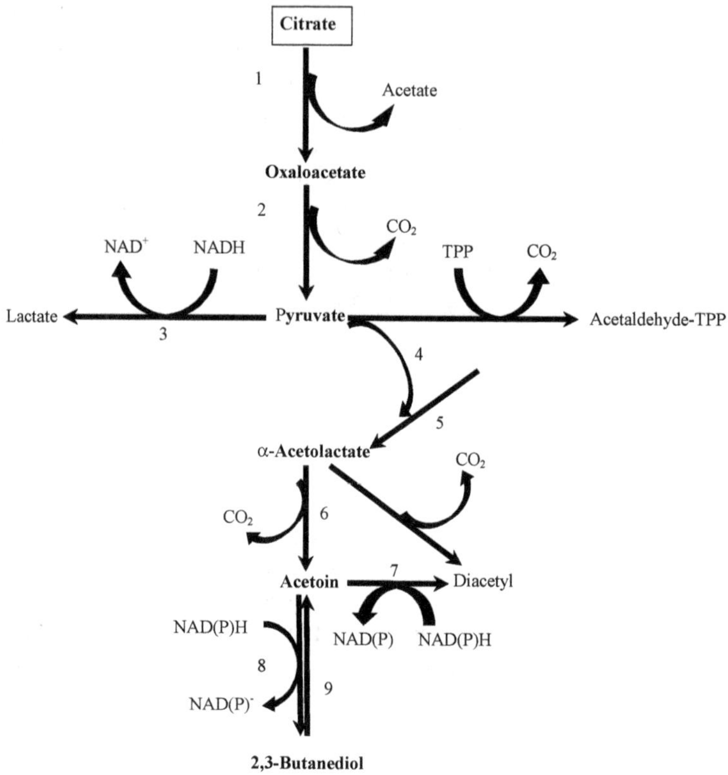

Figure 2.4 Citrate metabolism by dairy starter cultures. **1**: Citrate lyase, **2**: Oxaloacetate decarboxylase, **3**: Lactate dehydrogenase, **4**: Pyruvate decarboxylase, **5**: Acetolactate synthase, **6**: Acetolactate decarboxylase, **7, 8**: Diacetyl acetoin reductase, and **9**: Butanediol dehydrogenase. **TPP**: Thiamine pyrophosphate.

have provided opportunities to genetically engineer *L. lactis* for improved diacetyl production (de Vos 1996). One approach involves inactivation of the gene for α-acetolactate decarboxylase. This enzyme is involved in the conversion of α-acetolactate to acetoin; thus its inactivation results in accumulation of α-acetolactate the immediate precursor for diacetyl. This was achieved by recombinant DNA (rDNA) methods (Swindell et al. 1996) or through selection of naturally occurring mutants through the use of selective media containing leucine but deficient in valine.

Wild-type lactoccocci are unable to grow in this medium deficient in valine. The α-acetolactate decarboxylase mutants in the media continue to grow since they are able to synthesize valine in the presence of leucine. Initially, commercial application of this was limited because commercially used *L. lactis* are auxotrophic for branched chain amino acids (Goupil-Feuillerat et al. 1997). Curic et al. (1999) were successful in transforming the industrially used strains using an rDNA plasmid that provides for the synthesis of enzymes needed for the synthesis of branched chain amino acids. Again after this transformation the α-acetolactate decarboxylase mutants were selected as described above. The variants suitable for commercial production of fermented dairy foods were then obtained through plasmid curing. These variants have been commercially used since they are natural mutants and lack foreign DNA. Recent advances made in determining entire genome of lactic acid bacteria is providing additional important tools to identify genes encoding enzymes that are involved in the pathway of citrate metabolism, such as in the case of *Lb. casei* (Diaz-Muniz et al. 2006). These studies are important in understanding the biosynthetic pathway of citrate metabolism and the genes involved in the regulation of the pathway.

2.4.3 Proteinase activity

Dairy starter lactic acid bacteria are fastidious nutritionally. They require growth promoters such as amino acids as well as vitamins since they have limited abilities to synthesize them themselves (Mierau et al. 1996). Requirements for amino acids are strain specific. These amino acids include glutamate, methionine, valine, leucine, isoleucine and histidine which are required for growth by lactococci. Proteolytic system in lactococci has been extensively studied (Christensen et al. 1999, Broadbent and Steele 2006). Other strains may require phenylalanine, tyrosine, lysine and alanine. Milk is a suitable growth medium for starter cultures; however, it does not have adequate levels of peptides and free amino acids to effectively support growth of lactic acid bacteria (Vasiljevic et al. 2005). Thus, lactic acid bacteria have developed a complex proteolytic system, which includes proteinases which break down caseins to peptides, peptidases which degrade peptides, and transport systems which translocate amino acids, peptides across the cytoplasmic membrane into the bacterial cell (Naes and Nissen-Meyer 1992, Kunji et al. 1996). Proteinases and peptidases produced by starter culture microorganism provide them the ability to utilize casein as an additional source of amino acids and nitrogen for growth and acid production (Smid et al. 1991). Caseins are more susceptible to proteolytic action than whey proteins due to their open and random structure. There are actually four different caseins in milk (α_{s1}-, α_{s2}-, β-, κ-casein; ratio of 4:1:3:1). In addition these microorganisms must have intracellular peptidases to hydrolyze peptides to the constituent amino acids. In lactococci, these peptidases include exopeptidases such as aminopeptidases, tripeptidases and dipeptidases (Broadbent 2001, Sridhar et al. 2005).

The proteolytic activities by the starter lactic acid bacteria also have an important role on the properties of the final dairy product (Khalid and Marth

1990, Broadbent and Steele 2006). Peptides and amino acids produced by these organisms due to degradation of caseins serve as important precursors for the formation of flavor and aroma compounds (Puchades et al. 1989, Ott et al. 2000). Proteinase-negative (Prt^-) strains lack the plasmid which encodes for the proteinase and therefore rely on Prt^+ strains in the mixture for growth. Prt^- cultures although lack proteinase activity they experience less susceptible to phage and antibiotics. They provide better cheese yields and less bitter flavor development. Higher levels of inoculums and longer cheese-making times are required when using Prt^- cultures exclusively (Richardson 1984). Overall, there is no advantage to using Prt^- cultures exclusively, i.e., in cheese-making.

Many dairy starter lactic acid bacteria produce NH_3 from amino acid arginine through the arginine deiminase pathway which can also serve as source of energy for some starter bacteria (Laht et al. 2002) (Fig. 2.5). In this pathway arginine is first converted to citrulline by arginine deiminase with release of NH_3. Citrulline is phosphorylated to carbamoyl phosphate by

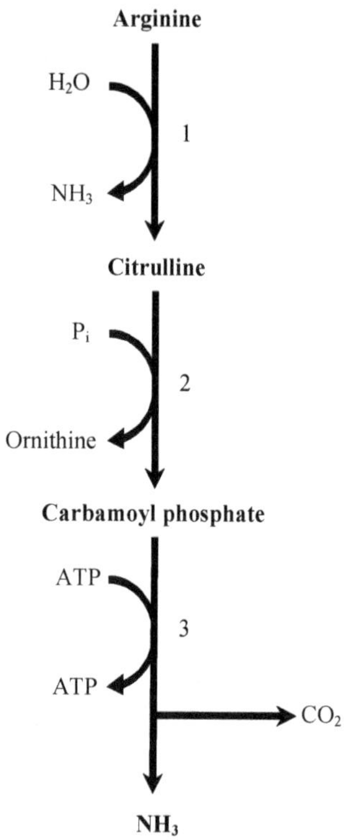

Figure 2.5 Arginine metabolism by arginine deiminase pathway. **1**: Arginine deiminase, **2**: Ornithine carbamoyltransferase, and **3**: Carbamate kinase.

carbamoyltransferase. Carbamoyl phosphate is next hydrolyzed to CO_2 and NH_3 with release of one mole of ATP by the action of carbamate kinase.

2.4.4 Production of exopolysaccharides

Some dairy lactic acid bacteria are capable of secreting extracellular polysaccharides (EPS). Although EPS provides a 'ropy' character, the exact physiological role of these molecules is not clear. EPS contribute to improvement of rheology, texture and mouthfeel, and function as thickening agents in cultured dairy products such as yogurt thus providing a substitute for commercial stabilizers (Welman and Maddox 2003, Ruas-Madiedo et al. 2002, Hassan et al. 2003, 2005). EPS producing cultures have been reported to improve functional properties of low-fat and part-skim Mozzarella cheese (Perry et al. 1997, Low et al. 1998, Petersen et al. 2000), viscoelastic properties of reduced-fat Cheddar cheese (Hassan et al. 2005). EPS also have been of interest due to their potential as prebiotics. Heteropolysaccharides containing gluco- and fructo-oligosaccharides may function as prebiotics and support probiotic organism in fermented dairy foods, and/or have been suggested to manipulate the balance of the gut microflora.

EPS are polymers of repeating units of sugars or sugar derivatives mainly of glucose, galactose and rhamnose of different ratios. They may be branched. They are secreted by the organism into the surroundings media. Unlike capsular polysaccharides they are not attached to the microbial cell. The EPS produced by starter culture bacteria can be classified into two categories, homopolysaccharides or heteropolysaccharides. Homopolysaccharides are repeating units of one type of monosaccharides such as glucose or fructose to produce glucans and fructans. Heteropolysaccharides include gellan and xanthan. They have repeating units that demonstrate little structural similarity and show different linkage patterns. The amount and type of EPS produced is influenced by culture strain, culture conditions and medium composition. Type of carbon source is reported to have a great influence, and may impact EPS composition. *Leu. mesenteroides* and *S. mutans* have been shown to produce homopolysaccharides, whereas heteropolysaccharides are produced by many lactic acid bacteria including *S. thermophilus*, *L. lactis*, and dairy *Lactobacillus* spp. (de Vuyst et al. 2003).

EPS synthesis by dairy starter cultures does not appear to provide an advantage to the growth and metabolism of the microorganisms that produce them. An important intermediate that links the EPS production to glycolysis (EMP pathway) is glucose-6-phosphate. A key enzyme that plays an important role in the conversion of glucose-6-phosphate to glucose-1-phosphate is phosphoglucomutase (PGM). Glucose-6-phosphate is channeled into glycolysis thus lactic acid production. Whereas glucose-1-phosphate serves as the precursor to the formation of sugar nucleotides, sugar nucleotides then are used to form the polysaccharides. The sugars are linked by the action of glycosyltransferases. Once the EPS is formed it is transported across the cell membrane. Further polymerization forms final EPS of many repeating units (Ramos et al. 2001). Genes associated with EPS production may also be plasmid linked. More detailed reviews on EPS production and application have been provided by Broadbent et al. (2003) and Ruas-Madiedo and de los Reyes-Gavilan (2005).

2.5 Inhibitors of starter culture growth, and acid production

There are four main causes of starter culture inhibition and/or slow acid production in the production of fermented dairy foods. These are: (1) presence of antibiotics in the milk, (2) bacteriophage, (3) bacteriocins and (4) natural inhibitors. Of these four, bacteriophage is probably the most important and has the greatest economic consequences in commercial production of foods.

2.5.1 Antibiotics

Antibiotic residues result in milk due to use of antibiotics to treat mastitis in dairy cows. Antibiotics such as penicillin or derivatives are infused into cow's udder during treatment of mastitis. The concentration of antibiotics in milk will decline with each milking. However, milk from antibiotic treated animals must be discarded until the entire antibiotic has been excreted by the animal, which is usually 72 hr. If this is not done, then the milk is contaminated with antibiotics. Different starter cultures have different sensitivities to antibiotics; thermophilic cultures are more sensitive to penicillin and more resistant to streptomycin than mesophilic cultures. Regardless, typically antibiotics in milk will cause slow culture growth and activity during manufacturing of fermented dairy foods. Today a number of tests are available to detect antibiotic residues in milk. In addition better education at the farm level and severe penalties to the farmer who supplies contaminated milk, have nearly eliminated problems associated with antibiotic residues in milk.

Contrary to most starter bacteria, most bifidobacteria are resistant to numerous antibiotics; particularly to nalidixic acid, gentamicin, kanamycin, metronidazole, neomycin, polymixin B and streptomycin. Whereas ampicillin, bacitracin, chloramphenicol, clindamycin, erythromycin, lincomycin, nitrofurantoin, oleandomycin, penicillin G and vancomycin inhibit most species (Scardovi 1986). Resistance of bifidobacteria to antibiotics makes it possible to selectively isolate them from complex flora using antibiotics as selective agents in culture media (Teraguchi et al. 1978,

Wijsman et al. 1989, Sozzi et al. 1990, Munoa and Pares 1998, Shah 2000, Tharmaraj and Shah 2003).

2.5.2 Bacteriocins

Bacteriocins are proteins ribosomally synthesized antimicrobial peptides produced by starter bacteria; although some may be plasmid mediated. Bacteriocins generally have a narrow host range and inhibit only closely related bacteria. Bacteriocins have different mechanism of biosynthesis, structures and mode of action (Barefoot and Nettles 1993, Broadbent 2001). They kill related bacteria by various mechanisms such as inhibiting synthesis of cell-wall, making the membrane of the target cell more permeable, or by inhibiting RNAase or DNAase activity. The bacteriocin producing bacteria also have different mechanisms of self-immunity and gene regulation. The 'self-immunity' of the bacteria that produces the bacteriocin to its own antimicrobial product is what differentiates bacteriocins from antibiotics (Nishie et al. 2012). Genes that code for this immunity are reported to be in close proximity to the other bacteriocin producing and processing genes, or located on the same operon, and next to each other. It is important to note that starter culture bacteria that show bacteriocin resistance do not show cross-resistance with antibiotics.

In the literature, bacteriocins have been grouped into three categories: Class I are called lantibiotics. They are characterized by 19 to more than 50 amino acids, and presence of unusual amino acids, such as lanthionine, methyl-lanthionine, dehydrobutyrine and dehydroalanine. Class I is further sub-divided into Ia and Ib sub-categories. Class Ia consists of cationic and hydrophobic peptides and have flexible structure compared to more rigid class Ib. Class Ib are globular peptides with no net charge. Class II consists of small heat-stable non-lantibiotics, non-modified peptides, and is sub-divided into three sub-categories. Class IIa consists of single peptide bacteriocins. Class IIb are two peptide bacteriocins that have different primary amino acid sequences and both of the peptides need to be fully active. Class IIc consists of bacteriocins secreted by the sec-system. Class III bacteriocins are large heat-labile protein. Less information is available on class III bacteriocins. A fourth class containing complex bacteriocins that contain proteins, lipids and carbohydrates has been proposed, but these complex molecules may be artifacts caused by interaction between the bacteriocin and the cell constituents and/or the growth medium. Class I and II are the best understood and most relevant to food applications (Cleveland et al. 2001). Although bacteriocins are ribosomally synthesized, resulting protein must be post-translationally modified to gain activity. The modification also is needed for secretion and transport of the bacteriocin across the cell membrane. Most class I and II bacteriocins are transported outside of the cell by dedicated ABC transporter system with a few exception that are externalized by the sec-dependent system.

Bacteriocins initially have been isolated from fermented dairy foods that contain lactic acid bacteria, thus they have been consumed for centuries. Bacteriocin production by starter culture bacteria is common. Because lactic acid bacteria are generally regarded as safe (GRAS) organisms over the years, there has been much interest in identifying bacteriocins from lactic acid bacteria that can inhibit spoilage organisms and pathogens in foods. Nisin is probably the best known and studied bacteriocin. It is now approved over 40 countries and has been used as a preservative in foods over 60 years. Nisin is produced by some species of *L. lactis* subsp. *lactis*. It is initially synthesized as 57 amino acid peptide. It is post-translationally modified to 34 amino acids and has a molecular weight of 3353 Da. Nisin is a lantibiotic; it contains lanthionine and beta-methylanthonine. Nisin typically forms dimers and tetramers. It is heat-stable. However, its solubility in the low pH range limits its use in foods. Nisin has a broad spectrum of activity. It has been shown to be inhibitory against *Bacillus, Clostridium, Staphylococcus, Listeria* and *Streptococcus*. Their bactericidal activity is due to the destruction of the proton motive force involved in transport by the cell, which causes release of intracellular components. Nisin is also active against germination of spores of *Clostridium tyrobutyricum* that cause late-gas production in hard natural and processed cheeses. There are different forms of nisin differing in their amino acid substitution. Nisin Z has asparagine instead of histidine at position 27 as is the case with nisin A. Nisin Z is more soluble with greater inhibitory activity.

In mixed starter cultures if a bacteriocin producing strain is present it will reduce the number of strains in these mixed cultures upon sub-culturing over time. Eventually this will result in a mixture of 1 or 2 strains remaining, and these cultures will be more prone to phage attack. Most commercial starter cultures used in commercial preparation today are selected so that they do not produce bacteriocins. Although Ryan et al. (1996) have reported that bacteriocin producing strains can be used in cheese-making.

2.5.3 Other natural inhibitors

In addition to bacteriocins, starter cultures can produce other potential inhibitors such as lactic and acetic acid, and hydrogen peroxide. Milk also contains natural inhibitors, which inhibit the growth of some strains of starter bacteria. These natural inhibitors include immunoglobulins and lactoperoxidase (LP) naturally found in milk. The immunoglobulins can interact with susceptible starter bacteria and cause agglutination of the starter culture. Agglutinated starter bacteria are not evenly distributed and thus cause uneven acid production in the vat (Hicks and Ibrahim

1992). In severe cases the agglutinated starter culture can entrap casein and the mixture will settle to the bottom of the cheese vat. Additional acid produced by the starter organisms will eventually inhibit the culture almost completely. Starter culture agglutination tends to be a problem early spring when the immunoglobulin content of milk is higher, and in products like cultured Cottage cheese (Hicks and Hamzah 1992, Ustunol and Hicks 1994). Cultured Cottage cheese undergoes long incubation times for adequate acid production thus providing the time needed for agglutination reactions to occur. LP requires hydrogen peroxide and thiocyanate for activity. Thiocyanate is naturally present in milk when cows are fed *Brassica* plants. Hydrogen peroxide is produced by some starter bacteria or through xanthine oxidase activity. Hydrogen peroxide will complex with lactoperoxidase, and this complex will oxidize thiocyanate. Resulting intermediates and products are inhibitory to the starter bacteria and some pathogens. Today, inhibition of starter cultures by these inhibitors is unusual because strains are selected accordingly. Other inhibitors such as lysozyme and lactoferrin do not inhibit starter cultures to the same extent to be significant in commercial production of fermented dairy foods (Fox et al. 2000, 2004).

2.5.4 Bacteriophage

Bacteriophage (or phage) is ubiquitous in nature, although sources of phage for the dairy starter cultures still remain unclear. Over the last 70 years, phage has been the major cause of slow acid production or starter failure in the dairy plant. Thus, phage infection can impact manufacturing schedules, and in severe cases results in dead vats. Lactic acid bacteria phage biology is based largely on their impact on *L. lactis* and *S. thermophilus* since these two species have been the most significantly impacted in industrial fermentations (Garvey et al. 1995, Brussow et al. 1998, Capra et al. 2006). Phage is an obligate parasite; it is a virus in most cases consisting of protein and nucleic acid. The majority of phages are members of *Siphoviridae*. Phage morphology consists of a head structure that contains the DNA (genetic information), and a tail structure that is connected to the head. A collar between the head and the tail may also be present. Phage multiplication occurs in one or two ways. Therefore phages are differentiated based on their life cycle as lytic or lysogenic phages and called virulent and temperate, respectively.

In the *lytic cycle*, the phage attaches onto the bacterial cell through phage receptors on the cell surface of the host. This step requires Ca^{2+}. Phage inhibitory media (PIM) which contains phosphate and citrate chelators are used to chelate the Ca^{2+} to prevent the attachment of phage onto the bacterial cell. Once the phage is attached onto the bacterial cell, it injects its DNA into the host cell. The phage replicates itself within the bacterial cell. When phage synthesis is completed the cell lyses releasing new phage particles into the medium (Kashige et al. 2000). The new phage particles that are produced are now available to start the cycle over. Burst size refers to the number of phage particles released and determines the virulence of the phage. With lactococcal phage, the latent period may vary from 10 to 140 min and the burst size may vary from 10 to 300 phage particles. Compared with starter culture bacterial cells, phage multiplication is very rapid. In the *lysogenic cycle,* the adsorption and injection of DNA is similar to that of the lytic cycle, but instead of the phage multiplying within the bacterial cell, the phage inserts its DNA into the bacterial chromosome and multiplies as a part of the chromosome. Most strains of lactic acid bacteria are lysogenic. As a result the host cell is immune to attack by its own phage. Generally, the cells are also immunized against other closely related strains of phage. In certain situations temperate phage can be induced and become lytic phage. Although as to what may cause this induction in a commercial plant setting is not clear, in the laboratory UV light and antibiotic mitomycin C have been shown to induce this transformation. Lysogenic phage is considered the source for lytic phage for most bacteria; however this is not the case for most lactic acid bacteria. Although, many mixed starter cultures contain lysogenic phage, no DNA homology has been reported between this phage and the lytic phage that attack these cultures. In mixed starter cultures normal growth is not affected by its own phage because of the presence of large numbers of acid producing, phage insensitive cells. This is referred to as pseudolysogeny (Garvey et al. 1995, Klaenhammer and Fitzgerald 1994).

To control phage, it is important to identify its source. Raw milk and the starter are considered to be important sources of phage in the dairy plant. Phage capable of infecting lactic acid bacteria has been isolated from raw milk. Its origin is believed to be due to the contamination of the raw milk with wild lactic acid bacteria which allow phage to propagate. Also, lysogenic starter bacteria may be triggered to propagate lytically under certain circumstances. In a dairy processing facility the impact of phage on starter culture can be reduced or eliminated by rotation of phage unrelated strains and use of milk strain cultures (Singh and Klaenhammer 1993, Durmaz and Klaenhammer 1995). Bulk starter preparation should be kept separate from the production area. Use of phage inhibitory media for bulk starter preparation, direct inoculation of the vat with DVS (direct to the vat set) frozen concentrated cultures, proper sanitation, air filtration and positive pressure on the facilities all help minimize phage problems.

Phage is resistant to heat, and can withstand temperatures of 75°C or higher. Therefore medium used for growing starter cultures must be treated at high temperatures (85°C, 30 min recommended) to inactivate any phage that may have contaminated the bulk starter medium. Pasteurization temperatures for milk is not adequate to inactivate phage, therefore raw milk will be a continued source of phage in the dairy facility. Although application of heat and high pressure has been shown to be effective in inhibiting phage this is not practiced commercially (McGrath et al. 2007, Guglielmotti et al. 2012). Efficiency of chlorine and derivatives against phage is reported to be variable (Guglielmotti et al. 2012). Hicks et al. (2004) developed bacteriophage derived peptides for inhibiting phage infection of lactococcal starters. They demonstrated that growth time of starter lactic acid bacteria was significantly prolonged with phage-derived peptides compared to controls grown with no phage-derived peptides. This may provide for an additional barrier to phage infection although not eliminating it completely.

Advances in molecular biology have provided the tools to study the molecular processes involved in infection of starter culture with phage. These studies have led to engineering phage resistance into starter cultures that interferes with the specific steps of phage life cycle. This includes preventing adsorption, which inhibits the adsorption of phage particle to the host bacterial cell, or blocking of the DNA injection into the cytoplasm of the bacterial cell by phage that has successfully attached. Other approaches have included intracellular degradation of the incoming DNA molecules and abortive infection which may encompass a range of mechanisms. In case of lactococci phage resistance may be encoded by plasmid DNA. Through conjugation phage resistance plasmid may be transferred to other *L. lactis* strains. Since conjugation is a 'natural' form of gene transfer this approach to improving lactic acid bacteria starter cultures have not been a concern from regulatory and social perspective. Therefore, phage resistant starter cultures developed by conjugation have been used for years. Today, total loss of a vat and product due to phage is rare. However, there may still be losses due to quality, flavor and texture defects.

There are also desirable aspects to bacteriophage in the production of fermented dairy foods. Beneficial aspects of bacteriophage are reviewed by McGrath et al. (2007). These desirable aspects include elimination of pathogenic bacteria.

2.6 Genomic studies on dairy lactic acid bacteria

Throughout history, lactic acid bacteria were selected for their ability to grow in milk, and metabolize amino acids obtained from caseins through their proteinases. Instability of key industrially important traits in lactococci was observed and reported earlier. Gene transfers from one organism to another such as for lactose metabolism for energy has been significant. However, more focused genetic research on dairy starter cultures began in the 1970s. Early studies in the 1970s showed that many of the unstable traits of lactic acid bacteria were plasmid-linked (McKay 1983). Plasmids are small circular double stranded DNA molecules that exist independent of the bacterial cell chromosome. They may or may not be essential for the bacterial cell growth, and survival, but they have a commercial impact in dairy processing and applications. In *Lactococcus* plasmid genes have been shown to encode for proteinase activity, lactose and citrate metabolism, bacteriocin production and phage resistance (McKay 1983) as discussed earlier in this chapter. With the exception of *Lb. delbrueckii* subsp. *bulgaricus* they are also common in *Lactobacillus* (Marth and Steele 2001, Broadbent and Steele 2006).

The plasmid coded lactococcal proteinases have been extensively studied over the years. These proteinases are closely related to each other and are the most thoroughly characterized genes in this genus. Furthermore, in the 1980s plasmid linkage of phospho-β-galactosidase and proteins involve in lactose transport has been confirmed (Harlander et al. 1984). Later, plasmid linkage of gene coding for tagatose 1,6-biphosphate aldolase was also shown. These are all important in lactose metabolism, thus acid production by dairy starter cultures (de Vos 1996). Today lactose metabolism and its instability is still important due to the widespread use of concentrated cultures where the cultures are grown continuously and concentrated. Citrate metabolism on the other hand is important in flavor and aroma development in dairy products. Uptake of citrate in citrate metabolizing lactococci is also controlled by a plasmid which codes for a permease. Furthermore, several plasmid linked restriction-modification systems and phage resistance mechanism have also been identified. Lactococci also produce a number of bacteriocins that are plasmid-encoded. This was clearly established with bacteriocin diplococcin produced by *L. lactis* subsp. *cremoris* 346. Plasmids are also found in strains of *Lactobacillus*, *Leuconostoc* and *S. thermophilus*.

Ability to re-introduce DNA to the organism of interest is important in improvement of lactic acid bacteria starter cultures. There are number of techniques that are available to researchers to accomplish this. Conjugation is the most commonly used method of transfer of DNA between lactic acid bacteria. In this method plasmid DNA is transferred by cell to cell contact between a donor and a recipient strain. However, not all plasmids may possess the genes needed to carry out this transfer. When the plasmid is transferred through conjugation to plasmid lacking organism, the plasmid associated metabolism can be detected. For example, lactose utilization is transferred through conjugation. Sanders et al. (1986) reported on conjugal strategy for construction of fast-acid producing bacteriophage resistance lactococci for use in dairy fermentation. Transduction is the transfer of DNA by phage. The DNA is packaged in the phage head. During infection of the starter culture cell the DNA is inserted into the cell. This DNA maintains its stability. Temperate phage has been used to transfer chromosomal markers in transduction experiments. It has been reported that both lactose metabolism and proteinase activity genes can be transduced in lactic acid bacteria (McKay and Baldwin 1974, Poolman 1993). Protoplast transformation method of gene transfer is tedious and still somewhat unsatisfactory. Method of choice in most laboratories is electro-transformation or electroporation.

More recent advances in starter cultures are in the area of genomics that involve nucleotide sequence information for complete genomes of these microorganisms (Bolotin et al. 2001, 2004, Makarova et al. 2006, van de Gutche et al. 2006, Smeianov et al. 2007). Knowing the genomic sequence of these microorganisms provides important tools to study the other traits that are commercially important (Sridhar et al. 2005). Currently, numerous lactic acid bacteria have their genome sequence determined or are in the process of being determined. This is highly significant for the dairy industry due to their commercial significance. However, only about half of these genomic sequences are available publicly because of this reason; the other half is not publicly available. Majority of these sequences were contributed to the depository as a part of a joint venture between the US-based Lactic Acid Bacterial Genomics Consortium and the Department of Energy Joint Genome Institute.

It is important to note that although there is much interest and research being conducted on lactic acid bacteria used as dairy starter cultures, cultures containing recombinant DNA (rDNA) are not approved and used commercially in the US. Lactic acid bacteria have GRAS status in the US because of their long history of safe use in foods. So far genetic improvements in starter culture technologies that have been implemented by the dairy industry have been achieved without the introduction and expression of rDNA. Two examples of such improvements include strategies to enhance phage resistance and diacetyl production in lactococci. Advances made in phage resistance and diacetyl production by lactococci have been discussed above in the sections on phage and citrate metabolism, respectively.

Dairy starter cultures are much researched and studied due to their commercial significance. Fermentation and metabolism by dairy starter cultures provide added value to fermented dairy foods. However, it is also important to note that these metabolic activities do not always reach optimum during fermentation. Genetic modification of these cultures will provide the tools to improve these cultures. Recent advances in genetics of lactic acid bacteria are discussed in further detail later in Chapter 3.

2.7 Manufacture of dairy starter cultures

Over the last 100 years there has been a significant increase in the industrial production of fermented dairy foods. Therefore the industrial production of dairy starter cultures has progressed from using former day's whey (or fermented dairy product) for the next process to an exact science of directly inoculating the process vat with highly active starter culture. Traditionally starter systems involved stock culture of liquid or freeze-dried cultures, which then were transferred to mother culture. From the mother culture several intermediate culture transfers made to inoculate the bulk starter tank and eventually the process tank (Hoeier et al. 1999). This process was time-consuming and took several days. It needed skilled personnel, and the potential for contamination was high. Separate culture rooms were needed for culture preparation. Today the concentrated frozen or lyophilized cultures provide direct to the vat inoculation or direct inoculation of the bulk tank. This has provided significant savings in labor and material costs in dairy processing facilities. Minimal sub-culturing and transfers are desirable not to lose plasmids important in dairy fermentation and to be able to maintain the desired balance in mixed cultures. These commercial starter cultures are grown under optimum pH and temperature conditions for the microorganism in an appropriate media (phage inhibitory and with pH control) to optimum numbers. The cells are further concentrated by centrifugation or ultrafiltration, and frozen in liquid nitrogen or lyophilized to contain approximately 5×10^9 cells g^{-1}. Glycerol or sucrose is added as cryoprotectant during freezing or lyophilization. Monosodium glutamate and lactose have also been shown to be effective cryoprotectants. The activity of these cultures is well maintained for minimum six months if the storage is maintained under recommended conditions.

References

Axelsson, L. 2004. Lactic acid bacteria: Classification and physiology. pp. 1–66. *In:* S. Salminen, A. von Wright and A. Ouwehand [eds.]. Lactic Acid Bacteria: Microbiological and Functional Aspects Vol. 1. Marcel Dekker, Inc., New York, NY, USA.

Barefoot, S.F. and C.G. Nettles. 1993. Antibiosis revisited: Bacteriocins produced by dairy starter cultures. J. Dairy Sci. 76: 2366–2375.

Beresford, T. and A. Williams. 2004. The microbiology of cheese ripening. pp. 287–318. *In:* P.F. Fox, P.L.H. McSweeney, T.M. Cogan and T.P. Guinee [eds.]. Cheese: Chemistry, Physics and Microbiology. Elsevier, Amsterdam, The Netherlands.

Bjorkroth, J. and J. Koort. 2011. Lactic acid bacteria—Taxonomy and biodiversity. pp. 45–48. *In:* J.W. Fuquay, P.F. Fox and P.L.H. McSweeney [eds.]. Encyclopedia of Dairy Sciences, 2nd edition. Academic Press, London, England.

Bolotin, A., P. Wincker, S. Mauger, O. Jaillon, K. Malarme, J. Weissenbach, D. Ehrlich and A. Sorokin. 2001. The complete genome sequence of the lactic acid bacterium *Lactococcus lactis* ssp. *lactis* IL1403. Genome Res. 11: 731–753.

Bolotin, A., B. Quinquis, P. Renault, A. Sorokin, S.D. Ehrlich, S. Kulakauskas, A. Lapidus, E. Goltsman, M. Mazur, G.D. Pusch, M. Fonstein, R. Overbeek, N. Kyprides, B. Purnelle, D. Prozzi, K. Ngui, D. Masuy, F. Hancy, S. Burteau, M. Boutry, J. Delcour, A. Goffeau and P. Hols. 2004. Complete sequence and comparative genome analysis of the dairy bacterium *Streptococcus thermophilus*. Nature Biotech. 22: 1554–1558.

Broadbent, J.R. 2001. Genetics of lactic acid bacteria. pp. 243–299. *In:* J.L. Steele and E.H. Marth [eds.]. Applied Dairy Microbiology, 2nd edition. Marcel Dekker Inc., New York, NY, USA.

Broadbent, J.R., D.J. McMahon, D.L. Welker, C.J. Oberg and S. Moineau. 2003. Biochemistry, genetics and application of exopolysaccharide production in *Streptococcus thermophilus*: A review. J. Dairy Sci. 86: 407–423.

Broadbent, J.R. and J.L. Steele. 2006. Proteolytic enzymes of lactic acid bacteria and their influence on bitterness in bacterial-ripened cheeses. pp. 224–243. *In:* K.R. Caldwaller, M.A. Drake and R.J. McGorrin [eds.]. Flavor of Dairy Products. American Chemistry Society, Washington, DC, USA.

Broome, M.C., D.A. Krause and M.W. Hickey. 1990. The use of non-starter lactobacilli in Cheddar cheese manufacture. Aust. J. Dairy Technol. 45: 67–73.

Brussow, H., A. Bruttin, F. Desiere, S. Lucchini and S. Foley. 1998. Molecular ecology and evolution of *Streptococcus thermophilus* bacteriophages—A review. Virus Genes 16: 95–109.

Capra, M.L., A. del L. Quiberoni, H.W. Ackermann, S. Moineau and J.A. Reinheimer. 2006. Characterization of a new virulent phage (MLC-A) of *Lactobacillus paracasei*. J. Dairy Sci. 89: 2414–2423.

Christensen, J.E., E.G. Dudley, J.A. Pederson and J.L. Steele. 1999. Peptidases and amino acid catabolism in lactic acid bacteria. Antonie van Leeuwen. 76: 217–246.

Cleveland, J., T.J. Montville, I.F. Nes and M.L. Chikindas. 2001. Bacteriocins: Safe, natural antimicrobials for food preservation. Int. J. Food Microbiol. 71: 1–20.

Cogan, T.M. and C. Hill. 1993. Cheese starter cultures. pp. 193–255. *In:* P.F. Fox [ed.]. Cheese: Chemistry, Physics and Microbiology. Chapman & Hall, London, UK.

Cogan, T.M., T.P. Beresford, J. Steele, J. Broadbent, N.P. Shah and Z. Ustunol. 2007. Advances in starter cultures and cultured foods. J. Dairy Sci. 90: 4005–4021.

Crow, V. 1990. Properties of 2,3 butanediol dehydrogenase from *Lactococcus lactis* subsp. *lactis* in relation to citrate fermentation. Appl. Environ. Microbiol. 56: 1656–1662.

Crow, V., B. Curry and M. Hayes. 2001. The ecology of non-starter lactic acid bacteria (NSLAB) and their use as adjuncts in New Zealand Cheddar. Int. Dairy J. 11: 275–283.

Curic, M., B. Stuer-Lauridsen, P. Renault and D. Nilsson. 1999. A general method for selection of α acetolactate decarboxylase deficient *Lactococcus lactis* mutants to improve diacetyl formation. Appl. Environ. Microbiol. 65: 1202–1206.

de Angelis, M., A. Corsetti, N. Tosti, M.R. Corbo and M. Gobbetti. 2001. Characterization of non-starter lactic acid bacteria from Italian ewe cheeses based on phenotypic, genotypic and cell wall protein analyses. Appl. Environ. Microbiol. 67: 2011–2020.

de Vos, W.M. 1996. Metabolic engineering of sugar catabolism in lactic acid bacteria. Antonie van Leeuwen. 70: 223–242.

de Vuyst, L., M. Zamfir, F. Mozzi, T. Adriany, V. Marshall, B. Degeest and F. Vaningelgem. 2003. Exopolysaccharide producing *Streptococcus thermophilus* strains as functional starter cultures in production of fermented milks. Int. Dairy J. 13: 707–717.

Diaz-Muniz, I., D.S. Banavara, M.F. Budinich, S.A. Rankin, E.G. Dudley and J.L. Steele. 2006. *Lactobacillus casei* metabolic potential to utilize citrate as an energy source in ripening cheese: A bioinformatics approach. J. Appl. Microbiol. 101: 872–882.

Dudley, E.G. and J.L. Steele. 2005. Succinate production and citrate catabolism by Cheddar cheese non-starter lactobacilli. J. Appl. Microbiol. 98: 14–23.

Durmaz, E. and T.R. Klaenhammer. 1995. A starter culture rotation strategy incorporating paired restriction/modification and abortive infection bacteriophage defences in a single *Lactococcus lactis* strain. Appl. Environ. Microbiol. 61: 1266–1273.

Fitzsimons, N.A., T.M. Cogan, S. Condon and T. Beresford. 1999. Phenotypic and genotypic characterization of non-starter lactic acid bacteria in mature cheddar cheese. Appl. Environ. Microbiol. 65: 3418–3426.

Fox, P.F. 2011. Lactic acid bacteria: An overview. pp. 401–402. *In:* J.W. Fuquay, P.F. Fox and P.L.H. McSweeney [eds.]. Encyclopedia of Dairy Sciences, 2nd edition. Academic Press, London, England.

Fox, P.F., T.P. Guinee, T.M. Cogan and P.L.H. McSweeney. 2000. Fundamentals of Cheese Science. Aspen Publishers, Gaithersburg, MD, USA.

Fox, P.F. and P.L.H. McSweeney. 2004. Cheese: An overview. pp. 1–18. *In:* P.F. Fox, P.L.H. McSweeney, T.M. Cogan and T.P. Guinee [eds.]. Cheese: Chemistry, Physics, and Microbiology, 3rd edition. Elsevier Academic Press, Oxford, UK.

Franz, C.M.A.P., W.H. Holzapfel and M.E. Stiles. 1999. Enterococci at the crossroads of food safety. Int. J. Food Microbiol. 47: 1–14.

Garvey, P., D. van Sinderen, D.P. Twomey, C. Hill and G.F. Fitzgerald. 1995. Molecular genetics of bacteriophage and natural phage defense systems in the genus *Lactococcus*. Int. Dairy J. 5: 905–947.

Goupil-Feuillerat, N., M. Cocaign-Bousquet, J.-J. Godon, S.D. Ehrlich and P. Renault. 1997. Dual role of α-acetolactate decarboxylase in *Lactococcus lactis* subsp. *lactis*. J. Bacteriol. 179: 6285–6293.

Guglielmotti, D., D.J. Mercanti, J.A. Reinheimer and A. del L. Quiberoni. 2012. Efficiency of physical and chemical treatments on the inactivation of dairy bacteriophages. Front. Microbiol. 2: 1–11.

Harlander, S.K., L.L. McKay and C.F. Schachtele. 1984. Molecular cloning of the lactose metabolizing gene from *Streptococcus lactis*. Appl. Environ. Microbiol. 48: 347–351.

Hassan, A.N., S. Awad and K. Muthukumarappan. 2005. Effect of exopolysaccharides-producing cultures on the viscoelastic properties of reduced-fat cheese. J. Dairy Sci. 88: 4221–4227.

Hassan, A., N.R. Ipsen, T. Janzen and K.B. Qvist. 2003. Microstructure and rheology of yogurt made with cultures differing only in the their ability to produce exopolysaccharides. J. Dairy Sci. 86: 1632–1638.

Hicks, C.L., P.A. Clark-Sofko, I. Surjawan and J. O'Leary. 2004. Use of bacteriophage-derived peptides to delay phage infection. Food Res. Int. 37: 115–122.

Hicks, C.L. and B. Hamzah. 1992. Effect of culture agglutination on Cottage cheese yield. Cult. Dairy Prod. J. 27: 4–9.

Hicks, C.L. and S. Ibrahim. 1992. Lactic bulk starter homogenization affects culture agglutination. J. Food Sci. 57: 1086–1091.

Hoeier, E., T. Jansen, C.M. Henriksen, F. Rattray, E. Brockmann and E. Johannsen. 1999. The production, application and action of lactic cheese starter cultures. pp. 99–131. *In:* B.A. Law [ed.]. Technology of Cheesemaking. Sheffield Academic Press, Sheffield, England.

Hugenholtz, J. 1993. Citrate metabolism in lactic acid bacteria. FEMS Microbiol. Rev. 12: 165–178.

IDFA (International Dairy Foods Association). 2011. Cultured dairy products report. 89 p.

Kashige, N., Y. Nakashima, F. Miake and K. Watanabe. 2000. Cloning, sequence analysis and expression of *Lactobacillus casei* phage Pl-1 lysis genes. Arch. Virol. 145: 1521–1534.

Kempler, G.M. and L.L. McKay. 1981. Biochemistry and genetics of citrate utilization in *Streptococcus lactis* spp. *diacetylactis*. J. Dairy Sci. 64: 1527–1536.

Khalid, N.M. and E.H. Marth. 1990. Proteolytic activity by strains of *Lactobacillus plantarum* and *Lactobacillus casei*. J. Dairy Sci. 73: 3068–3076.

Klaenhammer, T.R. and G.F. Fitzgerald. 1994. Bacteriophages and bacteriophage resistance. pp. 106–168. *In:* M.J. Gasson and W.M. de Vos [eds.]. Genetics and Biotechnology of Lactic Acid Bacteria. Blackie Academic and Professional, London, UK.

Kunji, E.R.S., I. Mierau, A. Hagting, B. Poolman and W.N. Konings. 1996. The proteolytic system of lactic acid bacteria. Antonie van Leeuwen. 70: 187–221.

Laht, T.M., S. Kask, P. Elias, K. Adamberg and T. Paalme. 2002. Role of arginine in the development of secondary microflora in Swiss-type cheese. Int. Dairy J. 12: 831–840.

Low, D., J.A. Ahlgren, D. Horne, D.J. McMahon, C.J. Oberg and J.R. Broadbent. 1998. Role of *Streptococcus thermophilus* MR-1C capsular exopolysaccharide on cheese moisture retention. Appl. Environ. Microbiol. 64: 2147–2151.

Lynch, C.M., P.L.H. McSweeney, P.F. Fox, T.M. Cogan and F.D. Drinan. 1996. Manufacture of Cheddar cheeses with and without adjunct lactobacilli under controlled microbiological conditions. Int. Dairy J. 6: 851–867.

Makarova, K., A. Slesarev, Y. Wolf, A. Sorokin, B. Mirkin, E. Koonin, A. Pavlov, N. Pavlova, V. Karamychev, N. Polouchine, V. Shakhova, I. Grigoriev, Y. Lou, D. Rohksar, S. Lucas, K. Huang, D.M. Goodstein, T. Hawkins, V. Plengvidhya, D. Welker, J. Hughes, Y. Goh, A. Benson, K. Baldwin, J.-H. Lee, I. Díaz-Muñiz, B. Dosti, V. Smeianov, W. Wechter, R. Barabote, G. Lorca, E. Altermann, R. Barrangou, B. Ganesan, Y. Xie, H. Rawsthorne, D. Tamir, C. Parker, L. McKay, F. Breidt, J. Broadbent, R. Hutkins, D. O'Sullivan, J. Steele, G. Unlu, M. Saier, T. Klaenhammer, P. Richardson, S. Kozyavkin, B. Weimer and D. Mills. 2006. Comparative genomics of the lactic acid bacteria. Proc. Natl. Acad. Sci. 103: 15611–15616.

Marth, E.H. and J. Steele. 2001. Applied Dairy Microbiology. 2nd edition. CRC Press, Boca Raton, FL, USA.

McGrath, S., G. Fitzgerald and D. van Sinderen. 2007. Bacteriophage in dairy products: Pros and cons. Biotechnol. J. 2: 450–455.

McKay, L.L. 1983. Functional properties of plasmids in lactic streptococci. Antonie van Leeuwen. 49: 259–274.

McKay, L.L. and K.A. Baldwin. 1974. Simultaneous loss of proteinase- and lactose-utilizing enzyme activities in *Streptococcus lactis* and reversal of loss by transduction. Appl. Microbiol. 28: 342–346.

McSweeney, P.L.H., P.F. Fox, J.A. Lucey, K.N. Jordan and T.M. Cogan. 1993. Contribution of the indigenous microflora to the maturation of Cheddar cheese. Int. Dairy J. 3: 613–634.

Mierau, I., E.R.S. Kunji, K.J. Leenhouts, A. Hellendoorn, A.J. Haandrikman, B. Poolman, W.N. Konings, G. Venema and J. Kok. 1996. Multiple-peptidase mutants of *Lactococcus lactis* are severely impaired in their ability to grow in milk. J. Bacteriol. 178: 2794–2802.

Munoa, F.J. and R. Pares. 1988. Selective medium for isolation and enumeration of *Bifidobacterium* spp. Appl. Environ. Microbiol. 54: 1715–1718.

Naes, H. and J. Nissen-Meyer. 1992. Purification and N-terminal amino acid sequence determination of the cell-wall bound proteinase from *Lactobacillus paracasei* subsp. *paracasei*. J. Gen. Microbiol. 138: 313–318.

Nishie, M., J. Nagao and K. Sonomoto. 2012. Antibacterial peptides "bacteriocins": An overview of their diverse characteristics and applications. Biocontrol Sci. 17: 1–16.

Ott, A., J.-E. Germond and A. Chaintreau. 2000. Vicinal diketone formation in yogurt: 13c precursors and effect of branched-chain amino acids. J. Agric. Food Chem. 48: 724–731.

Palles, T., T. Beresford, S. Condon and T.M. Cogan. 1998. Citrate metabolism in *Lactobacillus casei* and *Lactobacillus plantarum*. J. Appl. Microbiol. 85: 147–154.

Perry, D., D.J. McMahon and C.J. Oberg. 1997. Effect of exopolysaccharide-producing cultures on moisture retention in low-fat Mozarella cheese. J. Dairy Sci. 80: 799–805.

Petersen, B.L., R.I. Dave, D.J. McMahon, C.J. Oberg and J.R. Broadbent. 2000. Influence of capsular and ropy exopolysaccharide-producing *Streptococcus thermophilus* on Mozzarella cheese and cheese whey. J. Dairy Sci. 83: 1952–1956.

Poolman, B. 1993. Energy transduction in lactic acid bacteria. FEMS Microbiol. Rev. 12: 125–148.

Porubcan, R.S. and R.L. Sellars. 1979. Lactic starter culture concentrates. pp. 59–92. *In:* H.J. Peppler and D. Perlman [eds.]. Microbial Technology, 2nd edition. Academic Press, New York, NY, USA.

Puchades, R., L. Lemieux and R.E. Simard. 1989. Evolution of free amino acids during the ripening of Cheddar cheese containing added lactobacilli strains. J. Food Sci. 54: 885–888.

Quintans, N.G., V. Blancato and G. Rapizo. 2008. Citrate metabolism and aroma compound production in lactic acid bacteria. pp. 65–88. *In:* B. Mayo, P. Lopez and G. Perez-Martinez [eds.]. Molecular Aspects of Lactic Acid Bacteria for Traditional and New Applications. Research Signpost, Kerala, India.

Ramos, A., I.C. Boels, W.M. de Vos and H. Santos. 2001. Relationship between glycolysis and exopolysaccharide biosynthesis in *Lactococcus lactis*. Appl. Environ. Microbiol. 67: 33–41.

Rhea, M.C., C.M.A.P. Franz, W.H. Holzapfel and T.M. Cogan. 2004. Development of enterococci and production of tyramine during the manufacture and ripening of Cheddar cheese. Irish J. Agric. Food Res. 43: 247–258.

Richardson, G.H. 1984. Proteinase negative cultures for cheesemaking. Cult. Dairy Prod. J. 19: 6–11.

Ruas-Madiedo, P. and C.G. de los Reyes-Gavilan. 2005. Methods for the screening, isolation, and characterization of exopolysaccharides produced by lactic acid bacteria. J. Dairy Sci. 88: 843–866.

Ruas-Mediedo, P., R. Tuinier, M. Kanning and P. Zoon. 2002. Role of exopolysaccharide produced by *Lactococcus lactis* subsp. *cremoris* on the viscosity of fermented milk. Int. Dairy J. 12: 689–695.

Ryan, M., P.M.C. Rea, C. Hill and R.P. Ross. 1996. An application in Cheddar cheese manufacture for a strain of *Lactococcus lactis* producing a novel broad-spectrum bacteriocin, lacticin 3147. Appl. Environ. Microbiol. 62: 612–619.

Sanders, M.E., P.J. Leonhard, W.D. Sing and T.R. Klaenhammer. 1986. Conjugal strategy for construction of fast acid-producing, bacteriophage resistant lactic streptococci for use in dairy fermentations. Appl. Environ. Microbiol. 52: 1001–1007.

Scordavi, V. 1986. The genus *Bifidobacterium*. pp. 1418–1435. *In:* P.H.A. Sneath, N.S. Muis, M.E. Sharpe and J.G. Holt [eds.]. Bergey's Manual of Systematic Bacteriology. Vol. 2. Williams and Wilkins, Baltimore, USA.

Shah, N.P. 2000. Probiotic bacteria: Selective enumeration, and survival in dairy foods. J. Dairy Sci. 83: 1–14.

Singh, W.D. and T.R. Klaenhammer. 1993. A strategy for rotation of different bacteriophage defences in a lactococcal single strain starter culture system. Appl. Environ. Microbiol. 59: 365–369.

Smeianov, V.V., P. Wechter, J.R. Broadbent, J.E. Hughes, B. Rodríguez and J.L. Steele. 2007. Comparative high-density microarray analysis of gene expression during growth of *Lactobacillus helveticus* in milk vs. rich culture medium. Appl. Environ. Microbiol. 71: 3025–3032.

Smid, E.J., B. Poolman and W.N. Konnings. 1991. Casein utilization by lactococci. Appl. Environ. Microbiol. 57: 2447–2452.

Sozzi, T., P. Brigidi, O. Mignot and D. Matteuzzi. 1990. Use of dicloxacillin for the isolation and counting of *Bifidobacterium* from dairy products. Lait 70: 357–361.

Sridhar, V.R., J.E. Hughes, D.L. Welker, J.R. Broadbent and J.L. Steele. 2005. Identification and functional analysis of endopeptidase genes from the genomic sequence of *Lactobacillus helveticus* CNRZ 32. Appl. Environ. Microbiol. 71: 3025–3032.

Stackebrandt, E., W. Frederiksen and G.M. Garrity. 2002. Report of the adhoc committee for the re-evaluation of the species definition in bacteriology. Int. J. Sys. Evol. Microbiol. 52: 1043–1047.

Stiles, M.E. and W.H. Holzapfel. 1997. Lactic acid bacteria of food and their current taxonomy. Int. J. Food Microbiol. 36: 1–29.

Swearingen, P.A., D.J. O'Sullivan and J.J. Warthesen. 2001. Isolation, characterization and influence of native nonstarter lactic acid bacteria on Cheddar cheese quality. J. Dairy Sci. 84: 50–59.

Swindell, S.R., K.H. Benson, H.G. Griffin, P. Renault, S.D. Ehrlich and M.J. Gasson. 1996. Genetic manipulation of the pathway for diacetyl metabolism in *Lactococcus lactis*. Appl. Environ. Microbiol. 62: 2641–2643.

Teraguchi, S., M. Uehara, K. Ogasa and T. Mitsuoka. 1978. Enumeration of bifidobacteria in dairy products. Japan. J. Bacteriol. 33: 753–761.

Tharmaraj, N. and N.P. Shah. 2003. Selective enumeration of *Lactobacillus delbrueckii* ssp. *bulgaricus, Streptococcus thermophilus, Lactobacillus acidophilus,* bifidobacteria, *Lactobacillus casei, Lactobacillus rhamnosus* and Propionibacteria. J. Dairy Sci. 86: 2288–2296.

Thomas, T.D. 1987. Acetate production from lactate and citrate by non-starter bacteria in Cheddar cheese. New Zealand J. Dairy. Sci Technol. 22: 25–38.

Ustunol, Z. and C.C. Hicks. 1994. Use of an enzyme-treated whey based medium to reduce culture agglutination. J. Dairy Sci. 77: 1479–1485.

van de Gutche, M., S. Penaud, C. Grimaldi, V. Barbe, K. Bryson, P. Nicolas, C. Robert, S. Oztas, S. Mangenot, A. Couloux, V. Loux, R. Dervyn, R. Bossy, A. Bolotin, J.-M. Batto, T. Walunas, J.-F. Gibrat, P. Bessieres, J. Weissenbach, S.D. Ehrlich and E. Maguin. 2006. The complete genome sequence of *Lactobacillus bulgaricus* reveals extensive and ongoing reductive evolution. Proc. Nat. Acad. Sci. 103: 9274–9279.

van der Meulen, R., L. Avonts and L. de Vuyst. 2004. Short fractions of oligofructose are preferentially metabolized by *Bifidobacterium animalis* DN-173 010. Appl. Environ. Microbiol. 70: 1923–1930.

Vasiljevic, T., N.P. Shah and P. Jelen. 2005. Growth characteristics of *Lactobacillus delbrueckii* subsp. *bulgaricus* ATCC 11842 as affected by different neutralizers. Aus. J. Dairy Technol. 60: 3–9.

Vasiljevic, T. and N.P. Shah. 2008. Cultured milk and yogurt. pp. 219–251. *In:* R.C. Chandan, A. Kilara and N.P. Shah [eds.]. Dairy Processing and Quality Assurance. Wiley-Blackwell, Ames, IA, USA.

Welman, A.D. and I.S. Maddox. 2003. Exopolysaccharides from lactic acid bacteria: Perspectives and challenges. Trends Biotechnol. 21: 269–274.

Wijsman, M.R., L.P. Johanna, J.M. Hereijgers and F.H. Groote. 1989. Selective enumeration of bifidobacteria in fermented dairy products. Neth. Milk Dairy J. 43: 395–405.

CHAPTER 3

Recent Advances in Genetics of Lactic Acid Bacteria

Nefise Akçelik, Ömer Şimşek and *Mustafa Akçelik**

3.1 Introduction

Lactic acid bacteria (LAB) are a heterogeneous family of microorganisms that can ferment a variety of carbohydrates primarily into lactic acid (Carr et al. 2002). Most of the LAB belongs to the order of *Lactobacillales*, a group of mainly Gram-positive, anaerobic, non-sporulating and acid-tolerant bacteria. Biochemically, LAB include both homofermenters and heterofermenters (Kleerebezem et al. 2003). The former group produces primarily lactic acid through carbohydrate fermentation, while the latter group yields a variety of fermentation by-products including lactic acid, acetic acid, ethanol, carbon dioxide and formic acid (Leroy and de Vuyst 2004). LAB can be sub-classified into seven phylogenetic clades: *Lactococcus, Enterococcus, Oenococcus, Pediococcus, Streptococcus, Leuconostoc* and *Lactobacillus*. The definition of LAB is biological rather than taxonomical, i.e., the LAB do not comprise a monophyletic group of bacteria. Today, it is known that LAB play a crucial role in the world food supply by performing the main bioconversions in fermented dairy products, meats and vegetables. LAB are also used in the production of wine, coffee, silage, cocoa, sourdough and numerous indigenous food fermentations (Leroy and de Vuyst 2004). LAB are indigenous to food-related habitats including plant (fruits, vegetables and cereal grains) and milk environments. In addition, some LAB species are also member of the flora of the mouth, intestine and vagina of the mammalian (Vaughan et al. 2005). Isolates of the same species are often obtained from plant, dairy and animal habitats, implying wide

*Corresponding author

distribution and specialized adaptation of these species to these diverse environments.

Although their primary contribution centers on rapid acid production and acidification of food products, they also contribute to the development of flavor and texture in foods as well as their nutritional values (Smit et al. 2005, 2009). The proteolytic and lipolytic activities, and aroma- and acid-producing abilities of the LAB are regulated by their enzyme systems, and the complex enzymatic reactions determine the final organoleptic, textural and nutritional characteristics of fermented food products (Caplice and Fitzgerald 1999). Some strains of LAB also produce bacteriocins which have enormous potential to inhibit various types of microorganisms in food systems (de Vuyst and Leroy 2007, Cotter et al. 2005). Moreover, LAB are considered to be essential components, playing a large variety of health-promoting functions, such as immunomodulation, intestinal integrity and pathogen resistance (Vaughan et al. 2005). For such reasons strains of some species have traditionally been used as probiotics and added as functional bacteria in various food commodities (Ljungh and Wadström 2006). Recent evidence from *in vitro* systems, animal models and clinical studies suggests that LAB can enhance both specific and non-specific immune responses, possibly by activating macrophages, altering cytokine expression, increasing natural killer cell activity, and/or increasing levels of immunoglobulins (Vitini et al. 2000, Ouwehand et al. 2002, Pena et al. 2005).

The commercial exploitation of LAB as starter and probiotic cultures is economically very significant. Consequently, research on their genetics, physiology and applications has bloomed in the last 25 years (Azcarate-Peril and Klaenhammer 2010, de Vos 2011). Earlier studies focused primarily on the strain selection and the study of individual enzymes or simple metabolic pathways. The genome sequencing of numerous LAB provides an expanded view of their metabolic processes, bioprocessing capabilities and potential roles in health and well-being. Therefore genome sequencing analysis has promoted new area in which functional and comparative genomic studies on LAB are performed.

In this chapter, recent genetic insights in LAB will be discussed within the progress of completed genomic sequencing projects incorporating with comparative and functional genomics studies. Readers may refer to Chapter 2 to obtain more information regarding the taxonomy and biochemical characteristics of LAB.

3.2 Characteristics of LAB genome

At present more than 61 complete genome sequences of LAB strains belonging to 31 different species are available (http://www.genomesonline.org). The published genome sequences of the LAB include *Lactobacillus acidophilus* (2), *Lb. amylovorus* (2), *Lb. brevis* (1), *Lb. buchneri* (1), *Lb. casei* (5), *Lb. crispatus* (1), *Lb. delbrueckii* (4), *Lb. fermentum* (2), *Lb. gasseri* (1), *Lb. helveticus* (2), *Lb. johnsonii* (3), *Lb. kefiranofaciens* (1), *Lb. plantarum* (3), *Lb. reuteri* (2), *Lb. rhamnosus* (3), *Lb. sakei* (1), *Lb. salivarius* (1), *Lb. sanfranciscensis* (1), *Lb. kimchii* (2), *Pediococcus pentosaceus* (1), *Lactococcus lactis* (7), *Streptococcus thermophilus* (5), *Leuconostoc citreum* (1), *Leu. gasicomitatum* (1) and *Leu. mesenteroides* (2). The pathogenic members of *Streptococcus* genus are not included here. The genome features of sequenced LAB strains up to now are presented in Table 3.1.

Like in other bacteria, numerous complete genome sequencing data of LAB indicate that their genomes consist of a core and auxiliary genomes. The essence of a species is core genome which encodes all house-keeping genes necessary for basic cellular functions that are essential for a given species and responsible for maintaining a species identity. LAB are Gram-positive bacteria with low % G+C content with small genomes ranging in size between 1.6 and 3.3 Mb (Table 3.1). The number of genes found in a given LAB/bifidobacterium genome is ranging from 1600 to 3000. Almost all LAB genomes display architectural features of a typical bacterial chromosome such as co-orientation between gene transcription and DNA replication and an asymmetric bias in nucleotide composition of leading and lagging DNA strands (Klaenhammer et al. 2002).

Adaptation to nutritionally rich environments (e.g., milk, plant human and animal gastro-intestinal tract-GIT) results in plasticity and versatility at the genomes of LAB. These changes occur with different genetic events (i.e., mutation, gene duplication, horizontal gene transfer (HGT), gene decay, gene loss and genome rearrangements) to contribute to the present genome shape and structure of LAB species (Altermann et al. 2005, Bolotin et al. 2004). Notably, in the recent genome analysis of two *S. thermophilus* strains, Bolotin et al. (2004) found that 10% of the genes were pseudogenes and non-functional due to frameshifts, nonsense mutation, deletion or truncation. Evidence for genome decay was particularly noted for genes involved in carbohydrate metabolism, uptake and fermentation. In contrast, a specific symporter for lactose was found in *S. thermophilus* that was absent from other pathogenic streptococci. In the case of *Lb. delbrueckii*, the remarkably high number of pseudogenes is indicative of ongoing adaptation and genome specialization (O'Sullivan et al. 2009, Goh et al. 2011).

It is also apparent that HGT has introduced important functions to the genomes of a number of LAB that are expected to promote their competition in these environments. Genes encoding sugar transporters and carbohydrate hydrolyzers can represent a large portion of strain-specific genes that have been acquired by HGT. Although gene decay is obvious in the *S. thermophilus* genome, numerous small genomic islands seem to have been acquired by HGT process. These regions encode a number of important

Table 3.1 General genomic features of typical LAB genomes.

Species-Strain	Genome length, bp	Number ORFs	Plasmids	G+C content (%)	Number of tRNAs	Number of rRNA operons	Reference Genbank No
Lactobacilli							
Lb. acidophilus 30SC	2097766	2134	pRKC30SC1, pRKC30SC2	38	63	12	Gc01676
Lb. acidophilus NCFM	1993564	1970	0	35	61	4	Gc00252
Lb. amylovorus GRL 1112	2126674	2193	0	38	60	4	Gc01545
Lb. amylovorus GRL 1118	1977087	1994	Two without name	38	62	12	Gc01704
Lb. brevis ATCC 367	2340228	2324	pLVIS1, pLVIS2	46	65	5	Gc00436
Lb. buchneri NRRL B-30929	2588309	2541	pCD034-1, pCD034-2, pCD034-3	44	61	5	Gc01732
Lb. casei ATCC 334	2924325	2865	pLSE11	47	59	5	Gc00438
Lb. casei BD-II	3127288	3278	pBD-11	46	59	5	Gc01709
Lb. casei BL23	3079196	3119	0	46	60	5	Gc00822
Lb. casei LC2W	3077434	3237	pLC2W	46	64	5	Gc01710
Lb. casei Zhang	2898335	2923	0	46	59	5	Gc01311
Lb. crispatus ST1	2043161	2100	0	37	64	4	Gc01278
Lb. delbrueckii subsp. *bulgaricus* 2038	1872918	1907	0	50			Gc01634
Lb. delbrueckii subsp. *bulgaricus* ATCC 11842	1864998	2234	0	50	95	9	Gc00394
Lb. delbrueckii subsp. *bulgaricus* ATCC BAA-365	1856951	1865	0	50	98	9	Gc00443

Table 3.1 contd....

Table 3.1 contd.

Species-Strain	Genome length, bp	Number ORFs	Plasmids	G+C content (%)	Number of tRNAs	Number of rRNA operons	Reference Genbank No
Lb. delbrueckii subsp. *bulgaricus* ND02	2131976	2139	LDBND-P	50	94	9	Gc01544
Lb. fermentum CECT 5716	2096777	1125	0	51	54	20	Gc01318
Lb. fermentum IFO 3956	2098684	1915	0	51	59	11	Gc00787
Lb. gasseri ATCC 33323	1894360	1874	0	35	78	6	Gc00447
Lb. helveticus DPC 4571	2080931	1830	0	37	73	4	Gc00690
Lb. helveticus H10	2172383	2052	One without name	37	62	4	Gc01622
Lb. johnsonii DPC 6026	1966342	1840	0	35	52	4	Gc01738
Lb. johnsonii FI9785	1785116	1804	p9785S, p9785L	34	53	4	Gc01114
Lb. johnsonii NCC 533	1992676	1941	0	35	79	6	Gc00171
Lb. kefiranofaciens ZW3	2354088	2222	pWW1, pWW2	37	60	4	Gc01788
Lb. plantarum JDM1	3197759	3026	pLP2000, pLP9000	45	62	16	Gc01008
Lb. plantarum subsp. *plantarum* ST-III	3307936	3117	pST-III	44	64	5	Gc01407
Lb. plantarum WCFS1	3348625	3230	pWCFS101, pWCFS102, pWCFS103	44	62	5	Gc00122
Lb. reuteri F275, JCM 1112	2039414	1901	0	39	68	6	Gc00786
Lb. reuteri SD2112	2316837	2206	0	39	65	6	Gc01917
Lb. rhamnosus GG, ATCC 53103	3005051	2905	0	47	60	5	Gc01116
Lb. rhamnosus K, ATCC 8530	2960339	2978	0	47	60	5	Gc02017

Lb. rhamnosus Lc 705	3033106	3068	One without name	47	60	5	Gc01105
Lb. sakei subsp. *sakei* 23K	1884661	1985	0	41	63	7	Gc00318
Lb. salivarius CECT 5713	2133635	1672	pHN1, pHN2, pHN3	33	87	51	Gc01314
Lb. salivarius subsp. *salivarius* UCC118	2133977	2196	pSF118-20, pSF118-44, pMP118	33	78	7	Gc00362
Lb. sanfranciscensis TMW 1.1304	1375770	1519	pLS1, pLS2	35	61	7	Gc01952
Lactococci							
L. lactis subsp. *lactis* KF147	2635654	2560	pKF147A	35	68	6	Gc01159
L. lactis subsp. *cremoris* NZ9000	2530294	2594	0	36			Gc01316
L. lactis subsp. *lactis* CV56	2518737	2489	pCV56A-pCV56B, pCV56C, pCV56D, pCV56E	35	62	6	Gc01691
L. garvieae ATCC 49156	1950135	2024	0	39	62	5	Gc01924
L. garvieae Lg2	1963964	2045	0	39	62	5	Gc01925
L. lactis subsp. *cremoris* SK11	2598348	2617	pLACR1, pLACR2, pLACR3, pLACR4, pLACR5	36	62	6	Gc00450
L. lactis subsp. *cremoris* A76	2452616	2719	pQA554, pQA549, pQA518, pQA505, pQA549	36	62	6	Gc02039
L. lactis subp. *lactis* Il1403	2365589	2434	plasmid cured	35	62	6	Gc00054
L. lactis subp. *cremoris* MG1363	2529478	2624	plasmid cured	36	62	6	Gc00508

Table 3.1 contd....

Table 3.1 contd.

Species-Strain	Genome length, bp	Number ORFs	Plasmids	G+C content (%)	Number of tRNAs	Number of rRNA operons	Reference Genbank No
Streptococci							
S. thermophilus LMG 18311	1796846	1994	0	39	67	6	Gc00234
S. thermophilus ND03	1831949	1990	0	39	56	5	Gc01558
S. thermophilus JIM 8232	1929905	2230	0	39	67	6	Gc01904
S. thermophilus LMD-9	1864178	1820	0	39	67	6	Gc00451
Pediococci							
P. pentosaceus ATCC 25745	1832387	1843	0	37	55	5	Gc00439
P. claussenii P06, ATCC BAA-344	1966362	1961	pPECL-1 to pPECL-8	37	57	4	Gc02055
Leuconostoc							
Leu. citreum KM20	1896614	1902	pLCK1 to pLCK4	39	69	4	Gc00724
Leu. kimchii IMSNU11154	2101787	2209	Five without name	38	68	4	Gc01279
Leu. gasicomitatum LMG 18811	1954080	1992	0	37	67	4	Gc01488
Leu. kimchii C2	1877273	1935	0	38	68	4	Gc01877
Leu. mesenteroides subsp. *mesenteroides* J18	2016426	2020	pKLE01 to pKLE05	38	71	4	Gc02021
Leu. mesenteroides subsp. *mesenteroides* ATCC 8293	2075763	2119	pLEUM1	38	71	4	Gc00444
Oenococci							
O. oeni ATCC BAA-1163	1753447	1398	0	37	43	3	
O. oeni PSU-1	1780517	1751	0	38	43	3	Gc00440
W. koreensis KACC 15510	1441470	1428	One without name	35	56	5	Gc01876

Data were derived from http://www.genomesonline.org

adaptive traits, which are of industrial relevance such as polysaccharide biosynthesis, bacteriocin production, phage resistance systems, or oxygen tolerance. Interestingly, evidence for HGT was also presented between *S. thermophilus* and other organisms existing at the dairy environment. A 17 kb region was identified that contained multiple copies of IS1191 and a mosaic of fragments with over 90% identity to *Lb. delbrueckii* subsp. *bulgaricus* and *Lb. lactis*. Among them was a unique copy of *met*C that allows methionine biosynthesis, which is a rare amino acid in milk (Bolotin et al. 2004). It has been suggested previously that selected genes involved in sugar transport, catabolic properties and exopolysaccharide synthesis in *Lb. plantarum* have been acquired via HGT (Kleerebezem et al. 2003), as part of the adaptation process of this organism to a diverse number of environments (e.g., plants, cereals, GIT). Evidence supporting HGT for these regions include their grouped position near the origin of replication, lowered % G+C content, and high variability as to the presence or absence of these genes among different strains of *Lb. plantarum* (Kleerebezem et al. 2003, Siezen et al. 2004).

Gene degradation and loss of dispensable functions from ancestral types are the other evidences suggesting evolution of the LAB to nutritionally complex environments. As an example, evolution of *S. thermophilus* to milk environment occurred with the genome degradation of genes that were dispensable (Altermann et al. 2005). Additionally, the most important characteristics of the genome sequence of a Swiss cheese isolate *Lb. helveticus* DPC4571 (Callanan et al. 2008) are a predicted dependency on external supplies of amino acids and co-factors similar to that described for closely related GIT isolates such as *Lb. acidophilus* (Altermann et al. 2005) and *Lb. johnsonii* (Pridmore et al. 2004).

There are numerous examples of gene duplications and multiple copies of related genes predicted to directly important functions in the genomes of the sequenced LAB. Examples include phosphotransferase system (PTS) transporters, β- and phospho-β-galactosidases, lactic dehydrogenases, peptidases, and oligopeptide and amino acid transporters. Also notable are the multiple copies of homologous for mucus-binding (Mub) proteins found in *Lb. gasseri*, *Lb. acidophilus*, *Lb. johnsonii* and *Lb. plantarum*. The predicted Mub proteins ranging in size from 1000 to 4300 amino acids often represent the largest open reading frames (ORFs) in the genomes of many intestinal lactobacilli (Kleerebezem et al. 2003, Pridmore et al. 2004, Altermann et al. 2005, Klaenhammer et al. 2005).

Plasmids are the earliest genetic elements which are commonly found in many members of the lactic acid bacteria (Gasson and Shearman 2003). Today it is well known that their contribution is that plasmid-borne traits are major accessories of phenotypes of industrially important groups such as lactococci (Kim and Mills 2007). Noteworthy lactococcal properties that are plasmid encoded include the production of the PrtP protease (Christensson et al. 2001), abortive infection mechanisms to prevent bacteriophage attack (Boucher et al. 2001), exopolysaccharide biosynthesis (O'Driscoll et al. 2006), and bacteriocin production (Cotter et al. 2005). DNA sequencing of the four plasmids harboured by *L. lactis* strain SK11, a widely used dairy strain, identified a broad repertoire of novel genes that significantly enhance or expand the metabolism, fitness and stress resistance of the bacterium (Siezen et al. 2005). The ability of plasmids to undergo dissemination by conjugation or other processes underlines their potential importance for contributing significant but variable traits to LAB. Plasmids are used for strain development or producing heterologous proteins in the host systems as food-grade (Peterbauer et al. 2011). Here, most attention is paid to the development of different selection markers that meet the requirements. Therefore, with using nisin resistance, lactose utilization or metal resistance phenotypes, many different food-grade plasmids have been constructed (Takala and Saris 2002, Douglas and Klaenhammer 2011). These food-grade plasmids have been used for expressing heterologous proteins or enhance the industrial relevant of LAB strains.

Conjugative transposons are a main type of vehicle regarding antibiotic resistance transport in Gram-positive bacteria. LAB strains also include different sizes of transposons. The most known ones exist in *L. lactis* (Tn5276, Tn5301). In lactococci, they code for nisin (nis) production and sucrose fermentation (sac). These transposons vary in size between 16 and 70 kb and may be inserted into plasmids or chromosome in one or multiple copies. They may mobilize plasmids or chromosomal genes. Mahillon and Chandler (1998) defined insertion sequences (IS) as segments of DNA smaller than 2.5 kb that are capable of inserting at multiple sites in a target molecule. These elements can be as short as 600–700 bp, encoding a transposase. In the genomes of bacteria, the presence of several closely related IS elements facilitate recombination events that may also include sections of unrelated DNA (Thomas and Nielsen 2005). Several mobile elements have been found in lactobacilli, including ISL2 in *Lb. helveticus*, ISL3 in *Lb. delbrueckii*, IS1223 in *Lb. johnsonii*, IS1163 and IS1520 in *Lb. sakei*, and ISLp11 in *Lb. plantarum* (Nicoloff and Bringel 2003). DNA sequence comparisons of various LAB indicated that during cheese manufacturing it was possible for insertion elements to be horizontally transferred, most likely through conjugation (Bourgoin et al. 1998).

3.3 Comparing LAB genome

Accumulation of complete genomes of LAB triggers comparison of the similarities and differences within these groups which is expected to provide an important view of gene content, organization and regulation that contributes to the functionality and taxonomy (Makarova et al. 2006, Zhu et al. 2009, O'Sullivan et al. 2009, Altermann and Klaenhammer 2011). Microbial comparative genomics is an emerging field, providing information for redefining genotypic, and the resulting phenotypic differences between species. Several comparative analyses of LAB genomes included lactobacilli species and have revealed wide phylogenetic and phenotypic diversities of the different species (Canchaya et al. 2006, Makarova et al. 2006), as well as genetic niche-specific adaptation features (Cai et al. 2009, O'Sullivan et al. 2009).

Whole genome comparison of *Lb. acidophilus, Lb. gasseri, Lb. johnsonii* and *Lb. plantarum* shows clearly extensive conservation of gene content and order over the length of the genome among the *Lb. johnsonii* and *Lb. acidophilus* where this substantiates the lack of synteny with *Lb. plantarum. Lb. gasseri* and *Lb. johnsonii* are even more strikingly similar across the length of the genome, except for two apparent chromosomal inversion events in *Lb. gasseri* resulting in a reversal of gene order when compared to the other two closely related species. 83–85% of the proteins were homologs in both genomes between *Lb. gasseri* and *Lb. johnsonii* at comparison of ORFs (Boekhorst et al. 2004, Klaenhammer et al. 2005). Overall, there was a high degree of gene synteny in the three species that have been collectively referred to as members of the *Lb. acidophilus* complex. Differentiation of these three species, particularly *Lb. gasseri* and *Lb. johnsonii*, has been historically difficult using traditional or molecular taxonomic tools. In a separate study, nine LAB genomes covering wide application areas varying from dairy fermentations to wine production were analyzed with comparative genomics approach (Klaenhammer et al. 2005). The homology of multiple protein sequences showed that the streptococci-lactococci branch was basal in the *Lactobacillales* tree and that the *Pediococcus* group was a sister to the *Leuconostoc* group, which supported the paraphyly of the *Lactobacillus* genus. Furthermore, *Lb. casei* was confidently placed at the base of the *Lb. delbrueckii* group which contradicted the earlier classification and suggested a revision of the taxonomy of the *Lactobacillales* (Makarova et al. 2006).

Comparative genomic analysis of LAB also provides opportunity to investigate niche sources of very similar organisms. One of the unique studies in this respect demonstrated that the presence or absence of certain genes involved in sugar metabolism, the proteolytic system and restriction modification enzymes were pivotal in suggesting the niche of a strain. According to the comparisons, nine niche-specific genes were identified of which six were dairy-specific and three were gut-specific with taking 11 fully sequenced LAB strains. These nine genes were suggested as a "barcode" in order to differentiate niche sources by performing wider homology searches to ensure that the gut-specific genes were not present in other dairy organisms and vice versa (O'Sullivan et al. 2009). Su et al. (2012)

employed multilocus sequence analysis and an analysis of host-specific physiological and genetic traits on *Lb. reuteri* LTH2584, a stable member of sourdough microbiota, to assign five sourdough isolates to rodent- or human-specific lineages. Comparative genome hybridization revealed that the model sourdough isolate LTH2584 had genome content very similar to that of the model rodent isolate 100-23. These results demonstrated that sourdough isolates of *Lb. reuteri* were of intestinal origin.

Approximately, 2000–3000 genes exist in most of the LAB genomes. Differences in these genes and/or presence of a core genome, if any, need to be demonstrated. Comparative analysis of 20 completely sequenced *Lactobacillus* genomes showed that the *Lactobacillus* pan genome was found to consist of approximately 14,000 protein-encoding genes while all 20 genomes shared a total of 383 sets of orthologous genes that defined the *Lactobacillus* core genome (Kant et al. 2011). These and other comparative studies confirmed the fact that about one third of the pan genome cannot be accurately annotated and that there exists series of wrongly or poorly annotated genes. Therefore, *in silico* comparative genomics of LAB genomes emphasize that the highest conserved genetic traits are varying biosynthetic and metabolic capabilities. Among the LAB members, glycolysis enzymes are uniformly represented. It seems likely that this is a universal feature of LAB due to their primary energy recovery by glycolysis. A recent transcriptional comparison of global gene expression revealed that genes of the glycolytic pathway were among the most highly expressed within the genome by LAB strains during growth on eight different carbohydrates (Barrangou 2004). Sequence of *Lb. plantarum* genomes revealed that many transporters, particularly PTS transporters to metabolize various carbohydrates from different environments are included (Kleerebezem et al. 2003). In particular, a "lifestyle adaptation island" was identified in the same strains over a 213 kb region that harbored genes involved in sugar transport and metabolism. *Lb. johnsonii*, *Lb. acidophilus* and *Lb. gasseri* genome analysis further showed high numbers of PTS transporters, and only 2 to 3 ABC (ATP-binding cassette) transporters identified for maltose and complex carbohydrates like fructooligosaccharide and raffinose that substantiate these observations (Altermann et al. 2005).

When LAB genomes are compared, it is seen that amino acid uptake systems are more common than sugar and peptide uptake systems. Although *L. lactis* include all amino acid biosynthetic pathways, most of other LAB are deficient for these in varying levels. For instance, *Lb. plantarum* have many synthetic pathways only exception of those for branched chain amino acid synthesis (Kleerebezem et al. 2003), whereas species of the *Lb. gasseri*, *Lb. johnsonii* and *Lb. acidophilus* have limited amino acid biosynthetic capacity (Altermann et al. 2005). The lactobacilli generally encode a large number of peptidases, amino acid permeases, and multiple oligopeptide transporters that could support efficient processing and recovery of amino acids from nutritionally rich environmental sources. However, most of the intestinal lactobacilli (including comparisons with *Lb. gasseri* and *Lb. plantarum*) were not found to encode the cell wall-associated proteinase, PrtP, except for *Lb. acidophilus* and *Lb. johnsonii*. Interestingly, the gene predicted to encode the maturation protein, PrtM, was found in all these lactobacilli genomes (Klaenhammer et al. 2005).

3.4 Sequence to function on LAB genomes

Numerous studies have demonstrated the importance of functional genomics in LAB research. Whole genome sequencing, genome data mining, and comparative genomics provide important clues into possible gene functions, both essential and unique. So far, genomic analyses of LAB have revealed a number of interesting features that are generally considered to be important to the roles of these organisms in bioprocessing or health (Jansen et al. 2002, Klaenhammer et al. 2005, Barrangou et al. 2007, Cogan et al. 2007, de Vos 2011).

In 2002, a novel family of repetitive DNA sequences was identified by Jansen et al. (2002) which was present among both domains of the prokaryotes (Archaea and Bacteria), but absent from eukaryotes or viruses. This repetitive DNA sequences was direct repeats from 21 to 37 bp, interspaced by similarly size characterized as Clustered Regularly Interspaced Short Palindromic Repeats (CRISPR) (Jansen et al. 2002). The genome sequencing projects revealed that CRISPRs existed in an increasing number of bacterial genomes; however, their function in phage resistance was only elucidated in 2007 by altering the CRISPR locus of a dairy strain of *S. thermophilus* DGCC7710 (Barrangou et al. 2007). Functional analysis of *S. thermophilus* showed that these regions along with CRISPRs associated (*cas*) genes had a role in phage resistance. Through adding and deleting spacers, derived isogenic strains were shown to be sensitive or resistant to two virulent bacteriophages isolated from industrial yogurt samples, phage 858 and phage 2972 (Levesque et al. 2005).

One of the efficient tools to investigate the expressed genes of LAB strains in food or intestinal systems is the "*in vivo* expression technology" (IVET). This approach allows the identification promoter elements that are expressed during food fermentation or transit of LAB cultures (Bron et al. 2004). The first IVET studies in LAB were performed in *Lb. plantarum* and 75 inducible genes have thus far been identified including groups encoding nutrient acquisition, intermediate or co-factor biosynthesis and stress responses (de Vos et al. 2004). In another study, the response of *L. lactis* cultures during the cheese production processes was visualized using IVET system (Bachmann et al. 2010). The results revealed a number of genes that were clearly induced in cheese such as *cys*D, *bca*P, *dpp*A, *his*C, *glt*A, *rps*E, *pur*L and *amt*B as well as a number of hypothetical genes, pseudogenes and notably genetic elements located on the non-coding strand of annotated open reading frames. Furthermore genes that are likely to be involved in interactions with bacteria used in the mixed strain starter culture were identified.

Functional genomics contributed to the understanding of the elucidation of cell surface proteins of LAB putatively involved in adhesion to intestinal epithelial cells or the relationship of LAB with the host. For example, *Lb. acidophilus* genome contains 26 genes encoding proteins predicted to anchor at the cell surface, including those that might bind mucus and fibrinogen (Altermann et al. 2005). Functional analysis revealed that three of these proteins mediate adherence to intestinal epithelial cell *in vitro* (Cogan et al. 2007). Analysis of the secretome of *Lb. salivarius* also indicated the role of a cell surface protein (LspA) in adherence. In another interesting approach, adhesion of *Lb. plantarum* WCFS1 to mannose residues, which are commonly found on the surface of eukaryotic cells, was identified. By screening 14 *Lb. plantarum* strains for their mannose adherence capabilities and examining their genotypes using DNA microarrays, two candidate genes involved in adhesion were identified in WCFS strain. Subsequent gene mutations and adhesion analysis confirmed that one protein had a role in mannose adhesion (Pretzer et al. 2005). Comparison of genomes of two *Lb. rhamnosus* strains showed that one of the islands only found in *Lb. rhamnosus* GG contained genes for three secreted LPXTG-like pilins (spaCBA) and a pilin-dedicated sortase. Immunogold electron microscopy showed that the SpaC pilin was located at the pilus tip but also sporadically throughout the structure (Kankainen et al. 2009). The presence of SpaC is essential for the mucus interaction of *Lb. rhamnosus* GG and likely explains its ability to persist in the human intestinal tract longer than *Lb. rhamnosus* LC705 during an intervention trial. In a further study, the phenotypic analysis of a spaCBA pilus knockout mutant in comparison with the wild-type revealed that while SpaCBA pili promote strong adhesive interactions with intestinal epithelial cells, a functional role in balancing IL-8 mRNA expression induced by surface molecules such as lipoteichoic acid was also determined (Lebeer et al. 2012).

LAB are able to survive in the harsh conditions of intestinal system. This feature is related mainly with high competition property and bile salt tolerance. Functional analysis at LAB generated indications for probiotic attributes in these strains. It was reported that *Lb. acidophilus* and *Lb. plantarum* strains originated from intestine were found to be able to metabolize complex carbohydrates not digested by the host enzymes, including fructooligosaccharide (FOS) and raffinose, which may contribute to their ability to compete in the GIT. Functional analyses of FOS utilization in *Lb. paracasei* suggested that FOS might be hydrolyzed extracellularly into fructose and sucrose by a cell wall-bound β-fructosidase with uptake occurring via fructose and sucrose (PTS) transporters (Goh et al. 2006). In contrast, *Lb. acidophilus* FOS utilization was shown to occur via ATP-binding cassette (ABC) transporter and an intracellular β-fructosidase, allowing the organisms to transport FOS into the cell where it is hydrolyzed internally (Barrangou et al. 2003).

Interestingly, Grangette et al. (2005) showed that the presence or absence and the degree of the alanylation of teichoic acids on the cell surface of *Lb. plantarum* could affect the cytokine expression pattern by peripheral blood mononuclear cells (PBMCs) and monocytes. Deletion at *dlt* operon, responsible for D-alanylation of teichoic acids, leads to a substantial reduction in the concentration of polyglycerol phosphate polymers (with D-Ala) in the teichoic acids of the bacterial cell-wall. Notably, this change in the chemical composition is correlated with a reduced secretion of pro-inflammatory cytokines produced by PBMCs, and increased secretion of the anti-inflammatory cytokine IL-10, when exposed to the *dlt*-mutant. Use of the *dlt*-mutant in a murine colitis model was also found to be protective against TNBS-induced colitis. This result provides further evidence that LAB communicate with PBMCs and, for the first time, provides evidence that LAB may induce pro-inflammatory or anti-inflammatory responses based on their cell-wall composition in teichoic acids and perhaps in the display of cell surface bound proteins or polysaccharides, as well. In this regard, a number of reports have already showed that different strains and species of lactobacilli, and other commensal bacteria, can modulate cytokine expression by both human and murine antigen presenting (dendritic) cells (Mohamadzadeh et al. 2005). Overall, the results suggest that variations in bacterial strains and species can direct immunological responses toward pro- or anti-inflammatory responses.

3.5 Conclusion

Today 61 genome sequencing have been completed and more than 100 projects are available as draft. This accumulating information has provided considerable insights into the physiology of these organisms. Thereby it becomes possible to understand their genomic content, explore their capabilities, and handle them for expanded and improved beneficial activities. Thus, comparative analysis has enabled to screen the differences and to explore the unique gene which matched with traits as well as phylogenetic relationship and given clues how the genomes are evolved. Also micro-array analysis of LAB have generated significant contributions to the understanding of global gene expression in response to diverse environmental conditions while functional genomic analysis have facilitated the identification and characterization of genes and gene products critical to cell growth, metabolism, survival, cell communication and probiotic functionality. Application and improvement of "omics" technologies and integrating these with systems and synthetic biology will contribute to a comprehensive mechanistic understanding of physiology of LAB and can only positively impact the ability to utilize these organisms for practical benefit.

References

Altermann, E. and T.R. Klaenhammer. 2011. Group-specific comparison of four lactobacilli isolated from human sources using differential blast analysis. Gen. Nutr. 6: 319–340.

Altermann, E., W.M. Russell, M.A. Azcarate-Peril, R. Barrangou, B.L. Buck, O. McAuliffe, N. Souther, A. Dobsen, T. Doung, M. Callanan, S. Lick, A. Hamrick, R. Cano and T.R. Klaenhammer. 2005. Complete genome sequence of the probiotic lactic acid bacterium *Lactobacillus acidophilus* NCFM. Proc. Nat. Acad. Sci. USA 102: 3906–3912.

Azcarate-Peril, M.A. and T.R. Klaenhammer. 2010. Genomics of lactic acid bacteria: The post-genomics challenge—from sequence to function. pp. 35–56. *In:* F. Mazzi, R.R. Raya and G.M. Vignolo [eds.]. Biotechnology of Lactic Acid Bacteria: Novel Applications. Wiley-Blackwell Publications, Iowa City, Iowa, USA.

Bachmann, H., L. de Wilt, M. Kleerebezem and J.E. van Hylckama-Vlieg. 2010. Time resolved genetic responses of *Lactococcus lactis* to a dairy environment. Environ. Microbiol. 12: 1260–1270.

Barrangou, R., E. Altermann, R. Hutkins, R. Cano and T.R. Klaenhammer. 2003. Functional and comparative genomic analyses of an operon involved in fructooligosaccharide utilization by *Lactobacillus acidophilus*. Proc. Natl. Acad. Sci. USA 100: 8957–8962.

Barrangou, R. 2004. Functional genomic analyses of carbohydrate utilization by *Lactobacillus acidophilus*. Ph.D. Thesis, North Carolina State University, Raleigh, NC, USA.

Barrangou, R., C. Fremaux, H. Deveau, M. Richards, P. Boyaval, S. Moineau, D.A. Romero and P. Horvath. 2007. CRISPR provides acquired resistance against viruses in prokaryotes. Science 315: 1709–1712.

Boekhorst, J., R.J. Siezen, M.C. Zwahlen, D. Vitanovia, R.D. Pridmore, A. Verceneri, M. Kleerebezem, W.M. de Vos, H. Brüssow and F. Desiere. 2004. The complete genomes of *Lactobacillus plantarum* and *Lactobacillus johnsonii* reveal extensive differences in chromosome organization and gene content. Microbiol. 150: 3601–3611.

Bolotin, A., B. Quinquis, P. Renault, A. Sorokin, S.D. Ehrlich, S. Kulakauskas, A. Lapidus, E. Goltsman, M. Mazur, G.D. Pusch, M. Fonstein, R. Overbeek, N. Kyprides, B. Purnelle, D. Prozzi, K. Ngui, D. Masuy, F. Hancy, S. Burteau, M. Boutry, J. Delcour, A. Goffeau and P. Hols. 2004. Complete sequence and comparative genome analysis of the dairy bacterium *Streptococcus thermophilus*. Nat. Biotechnol. 22: 1554–1558.

Boucher, I., E. Emond, M. Parrot and S. Moineau. 2001. DNA sequence analysis of three *Lactococcus lactis* plasmids encoding phage resistance mechanisms. J. Dairy Sci. 84: 1610–1620.

Bourgoin, F., G. Guĕdon, B. Gintz and B. Decaris. 1998. Characterization of a novel insertion sequence IS1194 in *Streptococcus thermophilus*. Plasmid 40: 44–49.

Bron, P.A., C. Grangette, A. Mercenier, W.M. de Vos and M. Kleerebezem. 2004. Identification of *Lactobacillus plantarum* genes that are induced in the gastrointestinal tract of mice. J. Bacteriol. 186: 5721–5729.

Cai, H., R. Thompson, M.F. Budinich, J.R. Broadbent and J.L. Steele. 2009. Genome sequence and comparative genome analysis of *Lactobacillus casei* insights into their niche-associated evolution. Gen. Biol. Evol. 14: 239–257.

Callanan, M., P. Kaleta, J. O'Callaghan, O. O'Sullivan, K. Jordan, O. McAuliffe, A. Sangrador-Vegas, L. Slattery, G.F. Fitzgerald, T. Beresford and R.P. Ross. 2008. Genome sequence of *Lactobacillus helveticus*, an organism distinguished by selective gene loss and insertion sequence element expansion. J. Bacteriol. 190: 727–735.

Canchaya, C., M.J. Claesson, G.F. Fitzgerald, D. van Sinderen and P.W. O'Toole. 2006. Diversity of the genus *Lactobacillus* revealed by comparative genomics of five species. Microbiol. 152: 3185–3196.

Caplice, E. and G.F. Fitzgerald. 1999. Food fermentations: Role of microorganisms in food production and preservation. Int. J. Food Microbiol. 50: 131–149.

Carr, F.J., D. Chill and N. Maida. 2002. The lactic acid bacteria: A literature survey. Crit. Rev. Microbiol. 28: 281–370.

Christensson, C., C.J. Pillidge, L.J.H. Ward and P.W. O'Toole. 2001. Nucleotide sequence and characterization of the cell envelope proteinase plasmid in *Lactococcus lactis* subsp. *cremoris* HP. J. Appl. Microbiol. 91: 334–343.

Cogan, T.M., T.P. Beresford, J.L. Steele, J. Broadbent, N.P. Shah and Z. Ustunol. 2007. Advances in starter cultures and cultured foods. J. Dairy Sci. 90: 4005–4021.

Cotter, P.D., C. Hill and R.P. Ross. 2005. Bacteriocins: Developing innate immunity for food. Nat. Rev. Microbiol. 10: 777–788.

de Vos M., B.E. Hayward, S. Picton, E. Sheridan and D.T. Bonthron. 2004. Novel PMS2 pseudogenes can conceal recessive mutations causing a distinctive childhood cancer syndrome. Am. J. Hum. Gene. 74: 954–964.

de Vos, W.M. 2011. Systems solutions by lactic acid bacteria. Microbiol. Cell Fact. 10(Supp. 1): S2.

de Vuyst, L. and M. Leroy. 2007. Bacteriocins from lactic acid bacteria: Production, purification, and food applications. J. Mol. Microbiol. Biotechnol. 13: 194–199.

Douglas, G.L. and T.R. Klaenhammer. 2011. Directed chromosomal integration and expression of the reporter gene gusA3 in *Lactobacillus acidophilus* NCFM. Appl. Environ. Microbiol. 77: 7365–7371.

Gasson, M.J. and C.A. Shearman. 2003. Plasmid biology, conjugation, and transposition. pp. 25–44. *In:* B.J.B. Wood and P.J. Warner [eds.]. Genetics of Lactic Acid Bacteria. Vol. 3. Kluwer Academic/Plenum Publishers, New York, NY, USA.

Goh, Y.J., C. Zhang, A.K. Benson, V. Schlegel, J.H. Lee and R.W. Hutkins. 2006. Identification of a putative operon involved in fructooligosaccharide utilization by *Lactobacillus paracasei*. Appl. Environ. Microbiol. 72: 7518–7530.

Goh, Y.J., C. Goin, S. O'Flaherty, E. Altermann and R. Hutkins. 2011. Specialized adaptation of a lactic acid bacterium to the milk environment: The comparative genomics of *Streptococcus thermophilus* LMD-9. Microbial. Cell Fact. 10(Supp. 1): 22–24.

Grangette, G., S. Nutten, E. Palumbo, S. Morath, C. Hermann, J. Dewulf, B. Pot, T. Hartung, P. Hols and A. Mercenier. 2005. Enhanced anti-inflammatory capacity of a *Lactobacillus plantarum* mutant synthesizing modified teichoic acids. Proc. Nat. Acad. Sci. USA 102: 10321–10326.

Jansen, R., J.D. Embden, W. Gaastra and L.M. Schouls. 2002. Identification of genes that are associated with DNA repeats in prokaryotes. Mol. Microbiol. 43: 1565–1575.

Kankainen, M., L. Paulin, S. Tynkkynen, I. von Ossowski, J. Reunanen, P. Partanen, R. Satokari, S. Vesterlund, A.P. Hendrickx, S. Lebeer, S.C. de Keersmaecker, J. Vanderleyden, T. Hämäläinen, S. Laukkanen, N. Salovuori, J. Ritari, E. Alatalo, R. Korpela, T. Mattila-Sandholm, A. Lassig, K. Hatakka, K.T. Kinnunen, H. Karjalainen, M. Saxelin, K. Laakso, A. Surakka, A. Palva, T. Salusjärvi, P. Auvinen and W.M. de Vos. 2009. Comparative genomic analysis of *Lactobacillus rhamnosus* GG reveals pili containing a human mucus binding protein. Proc. Natl. Acad. Sci. USA 106: 17193–17198.

Kant, R., J. Blom, A. Palva, R.J. Siezen and W.M. de Vos. 2011. Comparative genomics of *Lactobacillus*. Microbiol. Biotechnol. 4: 323–332.

Kim, J. and D.A. Mills. 2007. Improvement of a nisin-inducible expression vector for use in lactic acid bacteria. Plasmid 58: 275–283.

Klaenhammer, T.R., E. Altermann, F. Arigoni, A. Bolotin, F. Breidt, J. Broadbent, R. Cano, S. Chaillou, J. Deutscher, M. Gasson, M. van de Guchte, J. Guzzo, A. Hartke, T. Hawkins, P. Hols, R. Hutkins, M. Kleerebezem, J. Kok, O. Kuipers, M. Lubbers, E. Maguin, L. McKay, D. Mills, A. Nauta, R. Overbeek, H. Pel, D. Pridmore, M. Saier, D. van Sinderen, A. Sorokin, J. Steele, D. O'Sullivan, W. de Vos, B. Weimer, M. Zagorec and R. Siezen. 2002. Discovering lactic acid bacteria by genomics. Antonie van Leeuwen. 82: 29–58.

Klaenhammer, T.R., R. Barrangou, B.L. Buck, M.A. Azcarate-Peril and E. Altermann. 2005. Genomic features of lactic acid bacteria effecting bioprocessing and health. FEMS Microbiol. Rev. 29: 393–409.

Kleerebezem, M., J. Boekhorst, R. van Kranenburg, D. Molenaar, O.P. Kuipers, R. Leer, R. Tarchini, S.A. Peters, H.M. Sandrink, M.W. Fiers, W. Stiekema, R.M. Lankhorst, P.A. Bron, S.M. Haffer, M.N. Groot, R. Kerhoven, M. de Vries, B. Ursing, W.M. de Vos and R.J. Siezen. 2003. Complete genome sequence of *Lactobacillus plantarum* WCFS1. Proc. Natl. Acad. Sci. USA 100: 1990–1995.

Lebeer, S., I. Claes, H.L. Tytgat, T.L. Verhoeven, E. Marien, I. von Ossowski, J. Reunanen, A. Palva, W.M. de Vos, S.C. Keersmaecker and J. Vanderleyden. 2012. Functional analysis of *Lactobacillus rhamnosus* GG pili in relation to adhesion and immunomodulatory interactions with intestinal epithelial cells. Appl. Environ. Microbiol. 78: 185–193.

Leroy, F. and L. de Vuyst. 2004. Lactic acid bacteria as functional starter cultures for foods fermentation industry. Trends Food Sci. Technol. 15: 67–68.

Levesque, C., M. Duplessis, J. Labonte, S. Labrie, C. Fremaux, D. Tremblay and S. Moineau. 2005. Genomic organization and molecular analysis of virulent bacteriophage 2972 infecting an exopolysaccharide-producing *Streptococcus thermophilus* strain. Appl. Environ. Microbiol. 71: 4057–4068.

Ljungh, A. and T. Wadström. 2006. Lactic acid bacteria as probiotics. Curr. Iss. Intest. Microbiol. 7: 73–89.

Mahillon, J. and M. Chandler. 1998. Insertion sequences. Microbiol. Mol. Biol. Rev. 62: 725–774.

Makarova, K., A. Slesarev, Y. Wolf, A. Sorokin, B. Mirkin, E. Koonin, A. Pavlov, N. Pavlova, V. Karamychev, N. Polouchine, V. Shakhova, I. Grigoriev, Y. Lou, S. Mills, O.E. McAuliffe, A. Coffey, G.F. Fitzgerald and R.P. Ross. 2006. Plasmids of lactococci-genetic accessories or genetic necessities? FEMS Microbiol. Rev. 30: 243–273.

Mohamadzadeh, M., S. Olson, W.V. Kalina, G. Ruthel, G.L. Demmin, K.L. Warfield, S. Barvari and T.R. Klaenhammer. 2005. Lactobacilli activate human dendritic cells that skew T cells toward T helper 1 polarization. Proc. Natl. Acad. Sci. USA 102: 2280–2285.

Nicoloff, H. and F. Bringel. 2003. ISLpl1 is a functional IS30-related insertion element in *Lactobacillus plantarum* that is also found in other lactic acid bacteria. Appl. Environ. Microbiol. 69: 6032–6040.

O'Driscoll, J., F. Glynn, G.F. Fitzgerald and D. van Sinderen. 2006. Sequence analysis of the lactococcal plasmid pNP40: A mobile replicon for coping with environmental hazards. J. Bacteriol. 188: 6629–6639.

O'Sullivan, O., J. O'Callaghan, A. Sangradon-Vegas, O. McAuliffe, L. Slattery, P. Kaleta, M. Callanan, G.F. Fitzgerald, R.P. Ross and T. Beresford. 2009. Comparative genomics of lactic acid bacteria reveals a niche-specific gene set. BMC Microbiol. 9: 50–59.

Ouwehand, A.C., S. Salminen and E. Isolauri. 2002. Probiotics, an overview of beneficial effects. Antonie van Leeuwen. 82: 279–289.

Pena, J.A., A.B. Rogers, Z. Ge, V. Ng, S.Y. Li, J.G. Fox and J. Versalovic. 2005. Probiotic *Lactobacillus* spp. diminish *Helicobacter hepaticus*-induced inflammatory bowel disease in interleukin-10-deficient mice. Infect Immun. 73: 912–920.

Peterbauer, C., T. Maischberger and D. Haltrich. 2011. Food-grade gene expression in lactic acid bacteria. Biotechnol. J. 6: 1147–1161.

Pretzer, G., J. Snel, D. Molenaar, A. Wiersma, P.A. Bron, J. Lambert, W.M. de Vos, R. van der Meer, M.A. Smits and M. Kleerebezem. 2005. Biodiversity-based identification and functional characterization of the mannose-specific adhesion of *Lactobacillus plantarum*. J. Bacteriol. 187: 6128–6136.

Pridmore, R.D., B. Berger, F. Desiere, D. Vilanova, C. Barretto, A.C. Pittet, M.C. Zwahlen, M. Rouvet, E. Altermann, R. Barrangou, B. Mollet, A. Mercenier, T.R. Klaenhammer, F. Arigoni and M.A. Schell. 2004. The genome sequence of the probiotic intestinal bacterium *Lactobacillus johnsonii* NCC 533. Proc. Natl. Acad. Sci. USA 101: 2512–2517.

Siezen, R.J., F.H.J. van Enckevort, M. Kleerebezem and B. Teusink. 2004. Genome data mining of lactic acid bacteria, the impact of bioinformatics. Curr. Opin. Biotechnol. 15: 105–115.

Siezen, R.J., B. Renckens, I. van Swam, S. Peters, R. van Kranenburg, M. Kleerebezem and M.W. de Vos. 2005. Complete sequences of four plasmids of *Lactococcus lactis* subsp. *cremoris* SK11 reveal extensive adaptation to the dairy environment. Appl. Environ. Microbiol. 71: 8371–8382.

Smit, G., B.A. Smit and W.J. Engels. 2005. Flavor formation by lactic acid bacteria and biochemical flavor profiling of cheese products. FEMS Microbiol. Rev. 29: 591–610.

Smit, B.A., W.J. Engels and G. Smit. 2009. Branched chained aldehydes; production and breakdown pathways and relevance for flavor in foods. Appl. Microbiol. Biotechnol. 81: 987–999.

Su, M.S., P.L. Oh, J. Walter and M.G. Ganzle. 2012. Intestinal origin of sourdough *Lactobacillus reuteri* as revealed by phylogenetic, genetic and physiological analysis. Appl. Environ. Microbiol. 78: 6777–6780.

Takala, T.M. and P.E.J. Saris. 2002. A Food-grade cloning vector for lactic acid bacteria based on the nisin immunity gene nisI. Appl. Microbiol. Biotechnol. 59: 467–471.

Thomas, C.M. and K.M. Nielsen. 2005. Mechanisms of and barriers to horizontal gene transfer between bacteria. Nat. Rev. Microbiol. 3: 711–721.

Vaughan, E.E., H.G. Heiling, K. Ben-Amor and W.M. de Vos. 2005. Diversity, vitality and activities of intestinal lactic acid bacteria and bifidobacteria assessed by molecular approaches. FEMS Microbiol. Rev. 29: 477–490.

Vitini, E., S. Alvarez, M. Medina, M. Medici, M.V. de Budeguer and G. Perdigon. 2000. Gut mucosal immunostimulation by lactic acid bacteria. Biocell 24: 223–232.

Zhu, Y., Y. Zhang and Y. Li. 2009. Understanding the industrial application potential of lactic acid bacteria through genomics. Appl. Microbiol. Biotechnol. 83: 597–610.

CHAPTER 4

Biopreservation by Lactic Acid Bacteria

Per E.J. Saris

4.1 Introduction

Apes do not store food, whereas man does. The evolution of food storage resulted in cultural evolution as time was left over for other activities than collecting the daily food. Innovations in food storage made in the early time of mankind are still in use today. Drying, smoking, grilling, boiling, salting and honeying of food are functional food preservation techniques mainly based on heating and/or lowering of the water activity. Biopreservation was also used early in our history, but without the knowledge of the presence of microorganisms, which are pivotal for the process. Nevertheless, the lack of a deeper insight in the major players in biopreservation did not hinder development of many food products, like cheese, beer, wine, vinegar, bread, yogurt and a heap of different fermented vegetables, cereals, meats and fishes. Such food products are palatable, mostly safe, have increased digestibility, vary in structure, smell and taste, and often present increased levels of vitamins in addition to that they preserve longer than the starting raw materials. The pioneering work by Antonie van Leeuwenhoek, Luis Pasteur and many others has shown the critical role of fungi, molds, yeasts, lactic acid and other bacteria in biopreservation. In this review, the focus is on biopreservation by lactic acid bacteria and which molecules and mechanisms are responsible for the preservation effect.

4.2 Biopreservation

Biopreservation can be defined as the extension of shelf-life and food safety by the use of natural or added microbiota and/or their antimicrobial compounds (Stiles 1996). Traditionally, a great number of different foods have been preserved by fermentation. In fermentation, the microorganisms grow in the foods producing organic acids and other compounds with antimicrobial effects resulting in biopreservation. In addition to antimicrobial effects, the microbial activity confer unique flavors, change the texture often in a beneficial way, improve digestibility and may increase the B vitamin content (LeBlanc et al. 2011). No wonder that fermented foods still form even 60% of the diet in industrialized countries (Holzapfel et al. 1995). The majority of fermented foods are fermented by lactic acid bacteria (LAB) with or without yeast, fungi or molds. Fermented foods can be classified into several groups (Table 4.1).

Table 4.1 Classification of fermented foods.

Fermentation type	Example of food product	Microorganisms involved
Fermentation producing a meat substitute from legumes/cereals	Tempe, idli	*Lactobacillus* spp., *Pediococcus* spp., *Rhizopus oryzae*, *Saccharomyces* spp.
High salt meat flavored amino acid/peptide mixtures in liquid or pasta	Soy sauce, miso, patis, nuocmam, bagong, mam	*Aspergillus oryzae*
Lactic acid fermentation	Sauerkraut, fermented cucumber kimchi, fermented sausages, pickles, olives, sour milk, yogurt, cheese, sour dough, bread, idli, enjera, kisra, puto, kishk, trahanas	*Lactobacillus* spp., *Pediococcus* spp., yeasts and moulds
Alcohol fermentations	Wine, beer	*Saccharomyces* sp.
Acetic acid fermentations	Vinegars	Acetic acid bacteria
Leavened and sour-dough bread	Bread	Yeasts, *Lactobacillus* spp.
Alkaline fermentations	Dawa dawa, soumbara, iru, ogiri, kenmina, thua-nao, natto	*Bacillus* spp.

Adapted from Steinkraus (1997).

4.3 Lactic acid bacteria (LAB)

LAB are Gram-positive, catalase-negative, non-motile, non-respiring if not provided with heme (some require in addition menaquinone) and non-sporeforming cocci, and rods, which produce lactic acid as the major end-product of the fermentation of carbohydrate (Lechardeur et al. 2011). Some members of the LAB genera, *Lactobacillus, Leuconostoc, Lactococcus, Pediococcus, Streptococcus, Enterococcus, Carnobacterium, Weisella, Oenococcus, Aerococcus, Tetragenococcus* and *Vagococcus*, carry out a heterofermentation producing carbon dioxide, acetic acid or ethanol (Adams 1999).

LAB growth results in a preservative effect due to reduction of pH, competition for nutrients, and possible production of other antagonistic compounds such as anti-bacterial peptides, anti-fungal peptides, hydroxy-fatty acids, hydrogen peroxide, ethanol, CO_2, diacetyl, acetaldehyde, phenolic compounds, antibiotics, small antibiotic-like molecules, siderophores, enzymes and bacteriophages (Klaenhammer 1988, Stiles 1996, Magnusson et al. 2003) (Table 4.2). In biopreservation the preservative effect cannot usually be contributed to one antimicrobial compound and its concentration. The antimicrobial compounds produced during fermentation may all contribute to the preservative effect; however, the mechanism of antimicrobial action cannot always be satisfactorily explained due to the complex interactive and/or synergistic activity of these different compounds (Corsetti et al. 1998, Niku-Paavola et al. 1999).

Table 4.2 Antimicrobial substances of lactic acid bacteria.

Substances
Organic acids: lactic acid, acetic acid, citric acid, benzoic acid, mevalonic acid, phenyllactic acid
Lactone
Methylhydantoin
Hydrogen peroxide
Ethanol
Carbon dioxide
Diacetyl
Acetaldehyde
Phenolic compounds
Hydroxy-fatty acids
Cyclic peptides: Cyclo(Pro-Pro), Cyclo(Leu-Pro), Cyclo(Tyr-Pro), Cyclo(Met-Pro), Cyclo(His-Pro)
Bacteriocins: nisin, pediocin (>180 known)
Citidine
Deoxycitidine
Reuterin
Reutericyclin
Cell-wall degrading enzymes
Proteases (some inhibit directly and others by releasing antimicrobial peptides)
Siderophores
Bacteriophages

Compiled from Corsetti et al. (1998), Niku-Paavola et al. (1999), Gänzle (2004), Hudson et al. (2005), Schnürer and Magnusson (2005), Beasley et al. (2006), Hammami et al. (2007), Pawlowska et al. (2012) and Cotter et al. (2013).

In general, most LAB are considered as safe mainly due to their long history of safe use in food fermentation, but also as they mostly do not produce toxins and when isolated from patients (blood or inflammation tissue) the patients usually are immunocompromised (Bernardeau et al. 2008). Strains with a history of food usage have a QPS (Qualified Presumption of Safety, in EU) and GRAS (Generally Recognized as Safe, in US) status. LAB may also be added to foods as probiotics for therapeutic advantages (Howarth and Wang 2013). However, it is good to keep in mind that some LAB are pathogenic, play a role in development of caries or spoil foods.

4.3.1 Antimicrobial substances of LAB

The main preservative effect of LAB in biopreservation of foods arises from production of weak organic acids, especially lactic and acetic acid, which may lower the pH of the fermented food to pH 4 to 3. The low pH inhibits many spoilage organisms and pathogens from growing in the foods. Especially, Gram-negative and to a lesser extent Gram-positive bacteria are inhibited, whereas yeasts and molds can grow at a broader pH range. Some yeast can even consume lactic acid resulting in an increase in pH enabling other spoilage organisms to grow. Depending on the substrates, LAB may produce other acids, like benzoic acid and citric acid. A biopreserved food may contain a mixture of these acids presenting synergistic action (Nom and Rombouts 1992). Part of the synergistic action results from the lowering of pH resulting in a larger proportion of the acids in the undissociated form, which has up to several hundred times stronger antimicrobial effect that in the dissociated form (Niku-Paavola et al. 1999). In the undissociated form the acids can penetrate into the cells and dissociate thereby decrease the internal pH. Cells pump out the formed protons in order to maintain a stable intracellular pH consuming ATP energy. Eventually, the cells die due to energy depletion.

Even though the broadest antimicrobial effect arises from the acids produced by LAB in food, other antimicrobial substances can have a strong effect too. Typically, LAB produced antibacterial peptides like bacteriocins have a narrow inhibition spectrum, though exceptions like nisin can be found (Hurst 1981, Takala and Saris 2007). Even with broad spectrum bacteriocins the sensitive bacteria are mostly Gram-positive bacteria. In food fermentation it is more relevant to focus on inhibition of Gram-positive bacteria as the acids produced by LAB will control the Gram-negative bacteria. However, biopreservation includes also addition of antimicrobial substances to foods. Therefore, the bacteriocins active against Gram-negative bacteria are potentially useful in non-fermented foods. Several anti-*Campylobacter* bacteriocins have been purified and characterized. For example, OR-7 and LAB47 bacteriocins by *Lactobacillus salivarius* (Stern et al. 2006, Abbas Hilmi 2010), and E-760 and E 50–52 by *Enterococcus* spp. (Line et al. 2008, Svetoch et al. 2008) were isolated from the chicken intestinal tract and may find usage in biopreservation of foods as well as for lowering colonization of chickens by *Campylobacter*. Antibacterial peptides for inhibition of *Escherichia coli* or *Salmonella* spp. have also been found (Miyamoto et al. 2000, Fayol-Messaoudi et al. 2005, Todorov and Dicks 2005). LAB produce lactic acid and upon isolation of new antibacterial substances against Gram-negative bacteria it is important to exclude the antibacterial effect of lactic acid as it may mask other antibacterial actions (de Keersmaecker et al. 2006). In conclusion, bacteriocins are essential in biopreservation, both produced in the food and as additions, exemplified by nisin, the bacteriocin allowed as a food additive (E234). Many studies demonstrate that bacteriocins can be either produced in the food or added as spent growth medium and be effective in increasing the shelf-life or safety of foods. Characterization, classification, mode of action and developments of bacteriocins for biopreservation are recently reviewed elsewhere (Amalaradjou and Bhunia 2012, Benmechernene et al. 2013, Cotter et al. 2013, O' Shea et al. 2013).

In biopreservation by LAB, the Gram-negative bacteria can be controlled by lactic acid and Gram-positive bacterial food pathogens like *Clostridium botulinum*, *C. perfringens*, *Bacillus cereus*, *Listeria monocytogenes* and *Staphylococcus aureus* by lactic acid and bacteriocins. However, biopreserved food may still be spoiled by the action of fungi, mold and yeast. Therefore, efforts with success have been made to find LAB strains active against these organisms (Schnürer and Magnusson 2005, Corsetti et al. 1998, Pawlowska et al. 2012). The effective substances are diverse, including cyclic peptides, cytidine, deoxycytidine, phenyllactic acid, 3-hydroxylated fatty acids, proteinaceous compounds and unknown substances.

Food transmitted parasites are another hazard in foods. Parasites do not grow in food. Therefore, the efforts to control parasites must be on inactivation or on disturbing the life cycle of parasites once consumed. Fortunately, LAB can affect the viability of certain parasites both in food and in the intestine by many mechanisms which require further study (Porrini et al. 2010, Travers et al. 2011).

Food-borne viruses are similar to parasites in that they do not replicate in food. There are only a few important food viruses like Norovirus and Hepatitis A, and to a lesser extent Rotaviruses, Enteroviruses and Astroviruses, but they cause frequently food outbreaks (Scallan et al. 2011). *Lactobacillus reuteri* producing reuterin has been reported to inhibit viruses (Cleusix et al. 2007). Fermentation of Dongchimi has reduced the infectivity of added murine norovirus by more than 4 log (Lee et al. 2012) showing that biopreservation may be a way to reduce the risk of virus transmittance via vegetables. LAB may also influence the substances of vegetables that are known to inhibit food-borne viruses (Li et al. 2013). A good selection of LAB starters that increase the concentration of virus inactivating substances in foods may be a good choice to further increase the safety of biopreserved foods.

LAB can be used in biopreservation not only for inhibition of unwanted microbes but also for detoxification of harmful substances already present in the raw materials. LAB have been shown to detoxify aflatoxin B_1, ochratoxin A, patulin and deoxynivalenol mycotoxins (Fuchs et al. 2008, El-Nezami et al. 2000, Franco et al. 2011). Clearly, LAB have a potential in increasing food safety by this activity. In addition to detoxification of mycotoxins LAB can bind and partially metabolize heterocyclic aromatic amines and other dietary mutagens (Turbic et al. 2002, Stidl et al. 2008, Nowak and Libudzisz 2009).

4.4 Future development of biopreservation

Biopreservation by LAB is a functional way of preserving food and, in addition to preservation, taste, smell, structure, nutritional adsorption and content may improve. What can be done to further develop biopreservation of foods? Every fermented food has its special features and hazards associated with it. By good strains selection starters can be chosen that are good at inhibiting the pathogens or spoilage organisms found typically in a specific biopreserved food. Inhibition should preferably be with several different types of substances and with different LAB strains and species that may dominate at different stages during the succession of the microbiota during the process of biopreservation and storage. Thereby, many hurdles are present for the pathogens that eventually are present in the processing of the biopreserved food. In addition to selecting strains with inhibitory activities against the pathogens and spoilage organisms attention should be paid on the capacities of the LAB starters to detoxify mycotoxins and other harmful substances present in the chosen biopreserved food. Added value may be included in the biopreserved foods by using starters that increase the B vitamin content of the final product or have a probiotic effect once consumed (Howarth and Wang 2013).

If it is possible to use genetic modification of the starter strains, which is an option in some countries, the possibilities are broadened compared to using selection and screening of natural strains only (Amalaradjou and Bhunia 2013). Recently, Liu et al. (2013) showed that expressing a *Listeria*-binding domain on the surface of a bacteriocin producer rendered the strain more efficient in killing *L. monocytogenes* than the same strain without *Listeria*-binding capacity. Only imagination will set the limit and the regulation of the society on what positive effects could be added to the LAB starters and probiotics.

References

Abbas Hilmi, H.T. 2010. Lactic acid bacteria and their antimicrobial peptides: Induction, detection partial characterization, and their potential applications. PhD Thesis, University of Helsinki, Finland. *In:* Dissertationes Bioscientiarum Molecularium Universitatis Helsingiensis in Viikki, 16/2010.

Adams, M.R. 1999. Safety of industrial lactic acid bacteria. J. Biotechnol. 68: 171–178.

Amalaradjou, M.A. and A.K. Bhunia. 2012. Modern approaches in probiotics research to control foodborne pathogens. Adv. Food. Nutr. Res. 67: 185–239.

Amalaradjou, M.A. and A.K. Bhunia. 2013. Bioengineered probiotics: A strategic approach to control enteric infections. Bioengineered 4: 379–387.

Beasley, S.S., T.J.K. Manninen and P.E.J. Saris. 2006. Lactic acid bacteria isolated from canine faeces. J. Appl. Microbiol. 101: 131–138.

Benmechernene, Z., I. Fernandez-No, M. Kihal, K. Bohme, P. Calo-Mata and J. Barros-Velazquez. 2013. Recent patents on bacteriocins: Food and biomedical applications. Recent Pat. DNA Gene Seq. 7: 66–73.

Bernardeau, M., J.P. Vernoux, S. Henri-Dubernet and M. Guéguen. 2008. Safety assessment of dairy microorganisms: the *Lactobacillus* genus. Int. J. Food Microbiol. 126: 278–285.

Cleusix, V., C. Lacroix, S. Vollenweider, M. Duboux and G. Le Blay. 2007. Inhibitory activity spectrum of reuterin produced by *Lactobacillus reuteri* against intestinal bacteria. BMC Microbiol. 7: 101.

Corsetti, A., M. Gobbetti, J. Rossi and P. Damiani. 1998. Antimould activity of sourdough lactic acid bacteria: Identification of a mixture of organic acids produced by *Lactobacillus sanfransisco* CB1. Appl. Microbiol. Biotechnol. 50: 253–256.

Cotter, P.D., R.P. Ross and C. Hill. 2013. Bacteriocins—A viable alternative to antibiotics? Nat. Rev. Microbiol. 11: 95–105.

de Keersmaecker, S.C.J., T.L.A. Verhoeven, J. Desair, K. Marchal, J. Vanderleyden and I. Nagy. 2006. Strong antimicrobial activity of *Lactobacillus rhamnosus* GG against *Salmonella typhimurium* is due to accumulation of lactic acid. FEMS Microbiol. Lett. 259: 89–96.

El-Nezami, H., H. Mykkänen, P. Kankaanpää, S. Salminen and J. Ahokas. 2000. Ability of *Lactobacillus* and *Propionibacterium* strains to remove aflatoxin B1 from the chicken duodenum. J. Food Protect. 63: 549–552.

Fayol-Messaoudi, D., C.N. Berger, M.-H. Coconnier-Polter, V. Lievin-Le Moal and A.L. Servin. 2005. pH-, lactic acid-, and non-lactic acid-dependent activities of probiotic lactobacilli against *Salmonella enterica* serovar Typhimurium. Appl. Environ. Microbiol. 71: 6008–6013.

Franco, T.S., S. Garcia, E.Y. Hirooka, Y.S. Ono and J.S. dos Santos. 2011. Lactic acid bacteria in the inhibition of *Fusarium graminearum* and deoxynivalenol detoxification. J. Appl. Microbiol. 111: 739–748.

Fuchs, S., G. Sontag, R. Stidl, V. Ehrlich, M. Kundi and S. Knasmüller. 2008. Detoxification of patulin and ochratoxin A, two abundant mycotoxins, by lactic acid bacteria. Food Chem. Toxicol. 46: 1398–1407.

Hammami, R., A. Zouhir, J. Ben Hamida and I. Fliss. 2007. BACTIBASE: A web-accessible database for bacteriocin characterization. BMC Microbiol. 7: 89.

Holzapfel, W.H., R. Geisen and U. Schillinger. 1995. Biological preservation of foods with reference to protective cultures, bacteriocins and food-grade enzymes. Int. J. Food Microbiol. 24: 343–362.

Howarth, G.S. and H. Wang. 2013. Role of endogenous microbiota, probiotics and their biological products in human health. Nutrients 10: 58–81.

Hudson, J.A., C. Billington, G. Carey-Smith and G. Greening. 2005. Bacteriophages as biocontrol agents in food. J. Food Prot. 68: 426–437.

Hurst, A. 1981. Nisin. Adv. Appl. Microbiol. 27: 85–123.

Gänzle, M.G. 2004. Reutericyclin: Biological activity, mode of action, and potential applications. Appl. Microbiol. Biotechnol. 64: 326–332.

Klaenhammer, T.R. 1988. Bacteriocins of lactic acid bacteria. Biochimie 70: 337–349.

LeBlanc, J.G., J.E. Laino, M. Juarez del Valle, V. Vannini, D. van Sinderen, M.P. Taranto, G. Font de Valdez, G. Savoy de Giori and F. Sesma. 2011. B-Group vitamin production by lactic acid bacteria—Current knowledge and potential applications. J. Appl. Microbiol. 111: 1297–1309.

Lechardeur, D., B. Cesselin, A. Fernandez, G. Lamberet, C. Garrigues, M. Pedersen, P. Gaudu and A. Gruss. 2011. Using heme as an energy boost for lactic acid bacteria. Curr. Opin. Biotechnol. 22:143–149.

Lee, M.H., S.-H. Yoo, S.-D. Ha and C. Choi. 2012. Inactivation of feline calicivirus and murine norovirus during Dongchimi fermentation. Food Microbiol. 31: 210–214.

Li, D., L. Baert and M. Uyttendaele. 2013. Inactivation of food-borne viruses using natural biochemical substances. Food Microbiol. doi:10.1016/j.fm.2013.02.009.

Line, J.E., E.A. Svetoch, B.V. Eruslanov, V.V. Perelygin, E.V. Mitsevich, I.P. Mitsevich, V.P. Levchuk, O.E. Svetoch, B.S. Seal, G.R. Siragusa and N.J. Stern. 2008. Isolation and purification of enterocin E-760 with broad antimicrobial activity against gram-positive and gram-negative bacteria. Antimicro. Agents Chemother. 52: 1094–1100.

Liu, S., T.M. Takala, W. Xing, J. Reunanen and P.E.J. Saris. 2013. Cell-mediated killing of *Listeria monocytogenes* by leucocin C producing *Escherichia coli*. Microbiol. Res. doi:pii:S0944–5013(12)00139-5.10.1016/j.micres.2012.11.011.

Magnusson, J., K. Ström, S. Roos, J. Sjögren and J. Schnürer. 2003. Broad and complex antifungal activity among environmental isolates of lactic acid bacteria. FEMS Microbiol. Lett. 219: 129–135.

Miyamoto, T., T. Horie, T. Fujiwara, T. Fukata, K. Sasai and E. Baba. 2000. *Lactobacillus* flora in the cloaca and vagina of hens and its inhibitory activity against *Salmonella enteritidis in vitro*. Poultry Sci. 79: 7–11.

Niku-Paavola, M.L., A. Laitila, T. Mattila-Sandholm and A. Haikara. 1999. New types of antimicrobial compounds produced by *Lactobacillus plantarum*. J. Appl. Microbiol. 86: 29–35.

Nom, M.J.R. and F.M. Rombouts. 1992. Fermentative preservation of plant foods. Appl. Bacteriol. Symp. Suppl. 73: 1365–1478.

Nowak, A. and Z. Libudzisz. 2009. Ability of probiotic *Lactobacillus casei* DN 114001 to bind or/and metabolise heterocyclic aromatic amines *in vitro*. Eur. J. Nutr. 48: 419–427.

O' Shea, E.F., P.D. Cotter, R.P. Ross and C. Hill. 2013. Strategies to improve the bacteriocin protection provided by lactic acid bacteria. Curr. Opin. Biotechnol. 2013 doi:pii: S0958-1669(12)00222-4.

Pawlowska, A.M., E. Zannini, A. Coffey and E.K. Arendt. 2012. Green preservatives: Combating fungi in the food and feed industry by applying antifungal lactic acid bacteria. Adv. Food Nutr. Res. 66: 217–238.

Porrini, M.P., C. Audisio, D.C. Sabaté, C. Ibarguren, S.K. Medici, E.G. Sarlo, P.M. Garrido and M.J. Eguaras. 2010. Effect of bacterial metabolites on microsporidian *Nosema ceranae* and on its host *Apis mellifera*. Parasitol. Res. 107: 381–388.

Scallan, E., R.M. Hoekstra, F.J. Angulo, R.V. Tauxe, M.A. Widdowson, S.L. Roy, J.L. Jones and P.M. Griffin. 2011. Foodborne illness acquired in the United States—major pathogens. Emerg. Infect. Dis. 17: 7–22.

Steinkraus, K.H. 1997. Classification of fermented foods: Worldwide review of household fermentation techniques. Food Control 8: 311–317.

Stern, N.J., E.A. Svetoch, B.V. Eruslanov, V.V. Perelygin, E.V. Mitsevich, I.P. Mitsevich, V.D. Pokhilenko, V.P. Levchuk, O.E. Svetoch and B.S. Seal. 2006. Isolation of a *Lactobacillus salivarius* strain and purification of its bacteriocin, which is inhibitory to *Campylobacter jejuni* in the chicken gastrointestinal system. Antimicro. Agents Chemother. 50: 3111–3116.

Stidl, R., G. Sontag, V. Koller and S. Knasmüller. 2008. Binding of heterocyclic aromatic amines by lactic acid bacteria: Results of a comprehensive screening trial. Mol. Nutr. Food Res. 52: 322–329.

Stiles, M.E. 1996. Biopreservation by lactic acid bacteria. Antonie van Leeuwen. 70: 331–345.

Svetoch, E.A., B.V. Eruslanov, V.V. Perelygin, E.V. Mitsevich, I.P. Mitsevich, V.N. Borzenkov, V.P. Levchuk, O.E. Svetoch, Y.N. Kovalev, Y.G. Stepanshin, G.R. Siragusa, B.S. Seal and N.J. Stern. 2008. Diverse antimicrobial killing by *Enterococcus faecium* E 50–52 bacteriocin. J. Agric. Food Chem. 56: 1942–1948.

Schnürer, J. and J. Magnusson. 2005. Antifungal lactic acid bacteria as biopreservatives. Trends Food Sci. Technol. 16: 70–78.

Takala, T. and P.E.J. Saris. 2007. Nisin: Past, present and future. pp. 181–213. *In:* M.A. Riley and O. Gillor [eds.]. Bacteriocin: Current Research and Application. Cromwell Press, Trowbridge, Wiltshire, UK.

Todorov, S.D. and L.M.T. Dicks. 2005. *Lactobacillus plantarum* isolated from molasses produces bacteriocins active against Gram-negative bacteria. Enz. Micro. Technol. 36: 318–326.

Travers, M.-A., I. Florent, L. Kohl and P. Grellier. 2011. Probiotics for the control of parasites: An overview. J. Parasitol. Res. doi:10.1155/2011/610769.

Turbic, A., J.T. Ahokas and C.A. Haskard. 2002. Selective *in vitro* binding of dietary mutagens, individually or in combination, by lactic acid bacteria. Food Add. Cont. 19: 144–2.

CHAPTER 5

Microbiology of Processed Liquid Milk

*Ebru Şenel** and *Ayşe Gürsoy*

5.1 Introduction

Milk is a good source of nutrients and edible energy for both mammals and wide range of microorganisms. Depending on milking practices (cleaning, hand or machine milking) and the temperature and period of milk storage on the farm, the numbers and types of microorganisms in raw milk differ. Milk drawn aseptically contains fairly low level of microorganisms (i.e., less than 10^3 cfu ml^{-1}), and soon after it leaves the udder, it becomes contaminated by spoilage and/or pathogenic microorganisms (often reaching >10^6 cfu ml^{-1}). This eventually leads to losses in quality of milk and milk products, as well as threatening public health. A total colony count of more than 10^5 cfu ml^{-1} of milk indicates poor sanitation of milk production.

Until end of 19th century, fresh raw milk was allowed to be distributed to the consumers directly in many countries. Today, however, this practice has been largely deserted in most countries where dairy industry is well developed, due to public health concern and has been replaced by processed liquid milk. Processed milk is defined as "liquid milk products processed thermally or non-thermally for direct human consumption". The processed milk industry has developed rapidly in coincidence with fast urbanization where individual ownership of milking animals for personal milk collection is impractical (Boor and Murphy 2002). Since microorganisms and enzymes associated with raw milk threat human health and cause quality problems in finished dairy products, raw milk has to be subjected to an appropriate

*Corresponding author

treatment to ensure food safety and extend the shelf-life of the products (Kelly et al. 2012).

Microorganisms associated with raw milk are discussed in detail in Chapter 1 of this book. This chapter will focus on microbiological aspects of processed liquid milk (i.e., pasteurized, extended shelf-life (ESL), in-container sterilized or UHT milks) and milk-borne diseases or outbreaks associated with processed liquid milk.

5.2 Processing of liquid milk

Heat treatment is the most common way of reducing microbiological load of raw milk and ensuring safety of liquid milk for direct consumption (Kelly et al. 2012). Heat treatment aims to kill all pathogenic microorganisms in raw milk as well as inactivating the large part (i.e., >95%) of contaminating microorganisms (Chandan 2011). Although heat treatment provides food safety in liquid milk, depending on the intensity of heating, application time and type of processing, it causes some changes in milk constituents. It is a well-known fact that caseins are resistant against moderate heat treatments at which whey proteins are largely denatured. Heat treatment at 100°C, for example, causes denaturation of major whey proteins (i.e., β-lactoglobulin and α-lactalbumin) (Chandan 2011). The sensory properties of processed liquid milk are also affected by the severity of heat treatment. Pasteurized milk which is processed at relatively lower temperatures than UHT or ultra-pasteurized milk has a pleasant flavor. However, the latter product is largely characterized with a slightly cooked flavor (Chandan 2011). Overall, the choice of heat treatment applied to liquid milk depends on three factors: (1) the degree of microbial inactivation to ensure food safety, (2) extension of the shelf-life of milk with an acceptable flavor, and (3) the changes in quality of the end product.

Post-heat treatment growth potential of sporeforming bacteria, preference of consumers and characteristics of the market should also be considered when deciding the heat treatment model for processed liquid milk (Kelly et al. 2012). The effectiveness of heat treatment is widely determined by the type and numbers of microorganisms associated with raw milk (Ahmed-Hassan et al. 2009). Table 5.1 shows the common heat treatment applications employed in the dairy industry.

5.2.1 Thermization

Thermization of milk is typically achieved by heat treatment at relatively low temperatures (i.e., at 63–65°C for 15–20 s or at 57–68°C for 15 s). The main purpose of thermization is (a) to kill psychrotrophic bacteria, which may release heat-stable proteases and lipases into milk, (b) to extent storage time of raw milk prior to processing, and (c) to enhance keeping quality

Table 5.1 Types of heat treatment applied to milk and their effects on microbial population.

Process	Process conditions	Heat exchanger	Target	Typical microbial quality
Thermization	57–68°C/15 s	PHE	Psychrotrophs	3–4 log reduction in psychrotrophs
Pasteurization	63°C/30 min	Batch (Vat)	Vegetative pathogens	Total plate count <100,000 cfu ml^{-1}
Pasteurization	72–74°C/15–30 s	PHE	Vegetative pathogens	Total plate count <100,000 cfu ml^{-1}
Heating for yogurt milk	85–95°C/2–30 min	PHE/THE	All pathogens and most spores	-
Preheating for producing milk powder	70°C/5 min or 135°C/1 s	PHE/THE/Other	Depends on treatment	-
Ultra pasteurization (ESL)	120–130°C/<1–5 s	PHE/THE	Vegetative bacteria and most spores	-
UHT (indirect)	135–140°C/3–5 s	PHE/THE/SSHE	All bacteria and spores	Total plate count <100 cfu ml^{-1} after 15 days at 30°C
UHT (direct)	135–140°C/3–5 s	Steam infusion/injection system	All bacteria and spores	Total plate count <100 cfu ml^{-1} after 15 days at 30°C
In-container sterilization	115–120°C/10–20 min	Batch or retorting	All bacteria and spores	Total plate count <100 cfu ml^{-1} after 15 days at 30°C

After: Kelly et al. (2012)

PHE: Plate heat exchanger; **THE**: Tubular heat exchanger; **SSHE**: Scraped surface heat exchanger of the end product (Lewis and Deeth 2009, Rowe and Donaghy 2011, Kelly et al. 2012). Thermized milk can be stored at maximum 8°C for up to three days (IDF 1984). Low temperature inactivation of microorganisms offers advantages for processing UHT milk as well as cheese-making, i.e., preventing blowing in cheese containers. Despite its positive effects on extended storage time of raw milk prior to processing into milk products, thermization is not a fully reliable tool for ensuring food safety as it cannot eliminate milk-borne pathogens completely. *Listeria monocytogenes*, for example, can withstand thermization and may well grow in chilled-stored thermized milk (Fernandes 2009). Similarly, the effect of thermization on the viability of *Mycobacterium bovis* and *Coxiella burnetii* is far limited. Therefore, thermization should not be considered as an alternative treatment to pasteurization (Rowe and Donaghy 2011). In case of applying thermization prior to pasteurization, the germination of *Bacillus cereus* spores is stimulated, simplifying their inactivation by further pasteurization. This process is termed "tyndallization".

5.2.2 Pasteurization

Pasteurization is aimed to make milk and milk products safe by destroying all the vegetative pathogenic microorganisms. The time and temperature conditions of pasteurization are designed to destroy *Mycobacterium tuberculosis* and *C. burnetii* which are known as the most heat-resistant pathogens in milk. Pasteurization process is performed in two ways: batch and continuous operations. In batch operations, milk is heated at lower temperatures (i.e., 63–65°C) for longer times (i.e., 30 min). This process is also called low-temperature long-time (LTLT) process. In continuous pasteurization systems, milk is subjected to heat treatment at higher temperatures (i.e., 72°C) for shorter time (i.e., 15 s). This process is called high-temperature short-time (HTST) pasteurization. In continuous system, the milk is passed through a plate heat exchanger and kept in holding tubes. Pasteurization systems are designed to provide a 5 log reduction of the microbial load of raw milk (de Jong 2008). Although pasteurization aims to destroy the most thermo-tolerant pathogen, *C. burnetii*, due to the differences in activation kinetics of different strains in raw milk, the changes in pasteurization conditions may be considered. For example, a slight increase in pasteurization temperature and/or a slight extension of holding time (i.e., 20–30 s) is recommended to inactivate heat-resistant strains of *L. monocytogenes*, *E. coli* and *Campylobacter* spp. (Kelly and O'Shea 2003). In Australia, for example, dairy factories apply pasteurization norms varying from 72°C for 15 s to 78–80°C for 25 s. In the case of suspicion of presence of *Mycobacterium avium* subsp. *paratuberculosis* (MAP) a more intensive heating is applied (Lewis and Deeth 2009).

5.2.3 Extended shelf-life (ESL) milk

Ultra-pasteurization (i.e., 120–135°C for 1–4 s) is one of the approaches to extend the shelf-life of processed liquid milk (Mehta 1980). It is a continuous heat treatment between HTST pasteurization and UHT sterilization. The differences between HTST pasteurization, ultra-high pasteurization and UHT sterilization are given in Table 5.2. The extended shelf-life (ESL) milk is defined as "milk showing a negative reaction to the peroxidase test" and must be labeled as "high temperature pasteurized milk" (EU 1992). A broader definition made by Rysstad and Kolstad (2006) was as follows: "ESL products are products that have been treated in a manner to reduce the microbial count beyond normal pasteurization, packaged under extreme hygienic conditions, and which have a defined prolonged shelf-life under refrigeration conditions".

Ultra-pasteurization is applied to milk before or after packaging. All microorganisms including vegetative cells and endo-spores are destroyed after this treatment (Public Health Services, PMO 2009). The aroma and flavor of ESL milk is slightly less preferable by the consumers comparing to the HTST pasteurized milk (Chapman and Boor 2001). Commercial ESL milk has a shelf-life varying from 30 days to 90 days at 4°C.

The combination of microfiltration and pasteurization has proved to be a very efficient way of producing ESL milk (Fernandes 2009, Fernández-Garcia et al. 2013). Microfiltration is a common membrane technology employed in removing raw milk microflora and improves organoleptic properties of dairy products. Cross-flow microfiltration is operated at lower temperatures and makes efficient control of microbial growth possible, leading to extending shelf-life of liquid milk (Fernández-Garcia et al. 2013). In microfiltration applications, inorganic membranes (i.e., ceramic membranes) are commonly used (Fernandes 2009). The major advantages of ceramic membranes with 1.4 μm pore size are the improvement of microbiological quality and the taste of the processed liquid milk. Reduction degree (RD) of microfiltration is 4 for total bacterial count and 2.3–3.7 for spores. On the other hand, microfiltration itself cannot guarantee to produce pathogen-free milk and, therefore, it must be combined with an appropriate heat treatment.

Micro-sieving is a novel technology which has a potential to use to reduce the number of bacteria in milk at low transmembrane pressures (Fernández-Garcia et al. 2013). Another approach to produce ESL milk is to use bacteriocins in addition to heat treatment. Addition of nisin at low concentrations (i.e., 40 IU ml^{-1}) followed by heat treatment at 90°C for 15 s was found to be more effective on microbial reduction than heat treatment at 72°C for 5 s. Such ESL milk was reported to be stored for >15 days with fairly low deterioration (Lewis and Deeth 2009).

Table 5.2 The comparison of different heat treatments to milk.

	HTST pasteurization	Ultra-pasteurization (ESL milk)	UHT
Reference enzymes	Phosphatase (–)	Phosphatase (–)	Phosphatase (–)
	Lactoperoxidase (+)	Lactoperoxidase (–)	Lactoperoxidase (–)
Storage conditions	Refrigerated	Refrigerated	Room temperature
Packaging	Clean	Ultra-clean or aseptic	Aseptic
Shelf-life	10–14 days	30–60 days	>6 months
Flavor	Little cooked flavor	Mild cooked flavor	Definite cooked flavor
Lactulose (mg l^{-1})	Trace	20–≤40	80–500
Furosine (mg g^{-1} protein)	Trace	200	400–1200
α-lactalbumin denaturation (%)[a]	~5	~5	~30–80
β-lactoglobulin denaturation (%)[b]	~13	~22	~60–100
Immunoglobulin denaturation (%)	~67	~100	~100

[a]Assuming concentration in raw milk: 1200 mg l^{-1}
[b]Assuming concentration in raw milk: 3000 mg l^{-1}
Adapted from Lewis and Deeth (2009)

5.2.4 Sterilization (in-container and UHT)

High heat treatment applied in the manufacture of long shelf-life liquid milk is called "sterilization". In this process, all pathogenic and non-pathogenic microorganisms and spores are destroyed, as well as most milk enzymes. Therefore, sterilized milk can be stored for several months at ambient temperatures (Lewis and Deeth 2009, Chavan et al. 2011).

In-container sterilization of milk, the temperatures vary from 105°C to 120°C for 20–40 min (Boor and Murphy 2002). In this process, milk is filled into cans or bottles, closed with crimped metal cap and then sterilized in a batch or continuous retort. The aim of this process, also called conventional sterilization, is to ensure a 9 log reduction in the spores of thermophilic bacteria or 12 log reduction of *Clostridium botulinum* (Boor and Murphy 2002). The major drawback of in-container sterilization is the development of cooked flavor triggered by Maillard reaction in milk. The shelf-life of in-container sterilized milk can be extended up to one year.

More common way of producing sterile and shelf-stable liquid milk is ultra high temperature (UHT) treatment of milk. UHT process uses continuous flow of milk and is carried out at temperatures varying from 130°C to 150°C (in general 140–145°C) with holding times of 1 to 8 s (usually 1–2 s) (Boor and Murphy 2002). UHT sterilization enables about 9-log reduction in endospores of thermophilic bacteria with fairly less chemical and sensory changes compared to in-container sterilization process. UHT process can be applied directly or indirectly. In direct system, high pressure steam is injected directly into the milk to reach desired temperatures. Afterwards, the sterilized milk is cooled under vacuum to remove excess water (Fernandes 2009). Indirect system includes a heat conducting barriers such as a plate heat exchanger or a tubular heat exchanger, separating the heating medium from milk (Boor and Murphy 2002). The heat regeneration in indirect sterilization system is much higher than that in direct system (>90% vs. ~50%) (Lewis and Herpell 2000). After heating, the milk is rapidly cooled down by a heat exchanger and filled in carton or glass bottles aseptically. In order to prevent re-contamination during aseptic filling, extra measures should be taken (Chavan et al. 2011).

5.3 Microflora of processed liquid milk

Factors determining shelf-life and quality of pasteurized milk include microbiological and enzymatic profile of raw milk, types of microorganisms and spores withstanding heat treatment and types and incidences of microorganisms contaminated to pasteurized milk after processing. Critical level of microbial counts for detectable sensory deterioration in pasteurized milk is 10^6–10^7 cfu ml^{-1} (Muir 1996). Pasteurized milk may contain two groups of microorganisms: spoilage or pathogenic. Major source of spoilage microorganisms in pasteurized milk is the post-pasteurization contamination. Heat-resistant bacteria surviving pasteurization are another source of microorganisms in processed liquid milk. Gram-negative psychrotrophic bacteria are the most common post-production contaminants in pasteurized milk. These bacteria widely contaminate milk during filling. Gram-negative psychrotrophs show certain degrees of resistance against sanitation solutions and may well-colonize on stainless steel surfaces and rubber gaskets that contact with milk. Shelf-life of contaminated pasteurized milk varies depending on contamination level, storage temperature and time, and generation time of contaminant microorganism. *Serratia, Enterobacter, Citrobacter, Hafnia, Pseudomonas, Alcaligenes* and *Flavobacterium* are among the most common spoilage microorganisms in pasteurized milk (Fernandes 2009). *Pseudomonas* spp. can spoil pasteurized milk stored at 4 or 7°C when they are present in milk at level of 10^3 cfu ml^{-1} (Stepaniak and Abrahamsen 1995). Common species of *Pseudomonas* isolated from pasteurized milk include *P. fluorescens, P. putida, P. maltophilia, P. aureofaciens, P. cepacia* and *P. pseudomallei* (Eneroth et al. 2000, Deeth et al. 2002). Psychrotrophs can grow very rapidly under refrigeration conditions; thus, very low level of initial contamination of these bacteria may shorten the shelf-life of pasteurized milk. Cousin (1982) reported that the number of bacteria doubled at every six hours and exceeded 10^6 cfu ml^{-1} after seven days of cold storage. Common sensory problems caused by *Pseudomonas* spp. in pasteurized milk are development of rancidity, bitterness, fruity and unclean off-flavors, as well as coagulation. Wessel et al. (1989) isolated the strains of *Enterobacter cloacae, Serratia rubidaea, Serratia marcescens, Klebsiella oxytoca* and *Klebsiella pneumoniae* from pasteurized milk.

Some thermophilic bacteria can survive on the surfaces of heat-exchangers during pasteurization. While the count of bacteria is rather low at the beginning of process, a rapid increase in the number of bacteria towards the end of process is observed. Thus, the product is contaminated before being removed from pasteurizer. The load of bacteria can reach 10–100 times of initial count in raw milk (Rademacher et al. 1996).

Among the thermoduric Gram-positive bacteria withstanding pasteurization conditions are *Microbacterium* spp., *Micrococcus* spp., *Enterococcus* spp., *Streptococcus* spp., *Lactobacillus* spp. and *Corynebacterium* spp. *Coryneform, Micrococci* and *Streptococci* cannot grow in milk stored below 6°C (Muir 1996, Varnam and Sutherland 2001). On contrary, *Bacillus* spp. are dominant group of bacteria in pasteurized milk stored at 5°C or below. Especially, heat-resistant *Bacillus* spores germinate and grow under cold conditions causing aroma defects and, in extreme cases, coagulation through extracellular enzymes such as proteinases, lipases and lecithinases which they secrete.

Zhou et al. (2008) screened the microbiological profiles of 54 whole pasteurized milk samples collected from Chinese markets in spring and autumn. Out of 102 isolates 92 isolates were identified as *Bacillus cereus*, nine isolates were identified as *Bacillus thuringiensis* and one isolate was found to belong to *Bacillus mycoides*. Incidence probability of *B. cereus* in the samples was 71.4% in spring and 33.3% in autumn. The authors also found that six pasteurized milk samples had enterotoxin genes including *hblA*, *hblC*, *hblD*, *nheA*, *nheB* and *nheC*.

Totally 458 pasteurized whole milk, pasteurized low fat milk and pasteurized double cream samples collected from three different Danish dairy plants were analyzed by Larsen and Jorgensen (1997). The prevalence rate of *B. cereus* in the samples stored at 7°C for eight days was 72% in summer samples and 56% in winter samples. Although *Bacillus* spp. is primarily responsible for quality losses in pasteurized milk, psychro-tolerant, endo-sporeforming *Paenibacillus* spp. cause microbiological problems in HTST-pasteurized milk and, thus, reduce shelf-life of the product (Fromm and Boor 2004, Ranieri et al. 2009). In a comprehensive study, 490 psychro-tolerant, endo-sporeforming bacteria were identified in pasteurized milk by DNA sequence-based sub-typing (Raineri and Boor 2009, Raineri et al. 2009). The researchers demonstrated that the microbial flora of the samples showed a time-dependent characteristic. In the first seven days of storage, more than 85% of the isolates were found to belong to *Bacillus* spp. However, in the second and third weeks of cold-storage, *Paenibacillus* spp. became dominant in the samples reaching 92% of total isolates. Similar findings were reported by Huck et al. (2008) who investigated 336 milk samples from four different dairy plants (i.e., 76.2% of the isolates were *Paenibacillus* spp. and 23.8% were *Bacillus* spp.).

5.3.1 Sources of contamination in pasteurized milk

Process equipment

The primary source of contamination of pasteurized milk is filling equipment. The psychrotrophic bacteria usually penetrate to the filler from vacuum or bulk tanks. In the case that the tanks are used for storing pasteurized milk prior to filling, the tanks may well be a source of contamination as well. Microscopic gaps on tank surfaces decrease the efficiency of cleaning and sanitation. Especially *Pseudomonas* spp. are prone to adsorb on tank and plate surfaces. Some strains of *Pseudomonas* spp. form polysaccharide fibrils and colonize on tank surfaces. An efficient sanitation is needed to remove these colonies from the surfaces.

Highly hydrophobic spores of some *Bacillus* spp. (i.e., *B. thermoleovorans*, *B. coagulans*, *B. pumilis*) and *Geobacillus stearothermophilus* adhere on the stainless steel surfaces and when the conditions are suitable, the spores germinate and then colonize quickly (Faille et al. 2001, Parkar et al. 2001).

Air

Bacteria occupy the largest stake of air microflora (i.e., 85% of total air flora) which are mainly Gram-positive types, followed by molds (~10% of total flora) and yeast (~5% of total flora). Air enters into plants through ventilation system, floor drains, personnel and other openness. Products are exposed to contamination by air mostly during packaging. Besides, residue of product in areas which are not cleaned properly is also an important source of air contamination. Although the numbers of psychrotrophic bacteria in air is fairly low, it may still pose a problem in the processed milk (Fredsted et al. 1996). The generation time of Gram-negative psychrotrophic bacteria is 4 to 5 hours and the number of this group of bacteria can reach at 10^7 cfu ml^{-1} within 7–10 days at 7°C. Milk containing psychrotrophic bacteria at level of <10 cfu ml^{-1}, is spoiled within 7–11 days under refrigerated storage (Eneroth et al. 1998).

Packaging materials

Production conditions of packaging materials have been improved greatly during the last decades. The number of bacteria associated with unit carton surfaces is <5000 cfu cm^{-2}. Gram-positive bacteria are the most frequently present group of bacteria on ESL milk packages (Fredsted et al. 1996, Mayr et al. 2004a). On the other hand, re-contamination of milk packages by Gram-negative psychrotrophic bacteria during filling has also been reported by Eneroth et al. (1998). Generally, flavored pasteurized milk is spoiled faster than unflavored pasteurized milk. It was demonstrated that the chocolate powder used in the production of chocolate-flavored pasteurized milk stimulated the growth of bacteria in milk but it did not introduce additional microbes into the milk. It was reported that the growth of *L. monocytogenes* in chocolate milk was more pronounced than skim and whole milk and whipping cream.

5.4 Microflora of ESL liquid milk

As stated in the previous sections, classical HTST pasteurization is ineffective to eradicate bacterial spores or all thermoduric non-sporeforming bacteria (Lewis and Deeth 2009, Tomasula et al. 2011). Therefore, heat treatment at higher temperatures (i.e., >100°C) for shorter times (i.e., 1–4 s) is applied to extend the shelf-life of processed liquid milk.

Although the rate of bacterial inactivation is greatly enhanced in the ESL milk compared to HTST pasteurized milk, spores of *Bacillus* spp. may still survive in the end product. It was reported that the presence of psychrotrophic *Bacillus* spp. at a level of 13–130 spores l^{-1} caused spoilage of ESL milk stored under cold conditions for 3 to 4 weeks (Mayr et al. 2004b). The most common sporeforming bacteria in ESL milk heat-treated at 127°C for 2 s by direct method was *Bacillus licheniformis* (73%) followed by *Bacillus subtilis* (6%), *Bacillus cereus* (6%) and *Brevibacillus brevis* (3%). Blake et al. (1995) found that the dominant sporeforming *Bacillus* spp. in ESL milks heat-treated at 132°C for 4 s by direct injection method consisted of *B. coagulans*, *B. licheniformis* and *B. cereus*. *Rhodococcus* spp., *Anquinibacter* spp., *Arthrobacter* spp., *Microbacterium* spp., *Enterococcus* spp., *Staphylococcus* spp., *Micrococcus* spp. and coryneform bacteria are re-contaminant non-sporeforming Gram-positive bacteria frequently isolated from ESL milk heat-treated at 127°C for 5 s. Anaerobic sporeformers and Gram-negative bacteria are rarely present in ESL milk.

As with HTST pasteurized milk, the most common sources of re-contamination to ESL milk are air and packaging materials. Schmidt et al. (2012) monitored the changes in microflora of micro-filtered and pasteurized ESL milks collected from three different regions stored at 4°C or 10°C for 29 days. Application of microfiltration decreased the microbial load of the milk remarkably (from 5–6 log_{10} cfu ml^{-1} to 1 cfu ml^{-1}). *B. cereus* and *Paenibacillus* spp. were the major spoilage groups present in ESL milks. Also, post-process contaminants included *Acinetobacter*, *Chryseobacterium* spp., *Psychrobacter* spp. and *Sphingomonas* spp.

Packaging materials used for ESL milk are disinfected but they are not sterilized. Fungi including *Penicillium* spp. and aerobic sporeforming bacteria (e.g., *Bacillus megaterium*, *B. cereus*, *B. licheniformis*, *B. flexus*, *B. pumilus*, *Paenibacillus macerans*, *P. polymyxa* and *P. pabuli*) were isolated from paperboards. Pirttijärvi et al. (1996) determined fungi in ESL milk stored up to 45 days under cold conditions.

5.5 Microflora of sterilized and UHT milk

UHT processing of milk provides a shelf-stable product with minimal chemical damages compared to in-container sterilization. Some bacteriological and chemical indices are used to evaluate the effects of heat treatment on the microflora and chemical compounds of milk. F_0, for example, is defined as the time (i.e., min) at 121°C which is required to achieve desired microbiological effect in sterilized milk. B* is a parameter used to measure the bacteriological effect of heat treatment at reference temperature of 135°C. C* is a chemical index used in the assessment of heating efficiency. To ensure the safety of sterilized liquid milk, it must be achieved a minimum of F_0 of 3 min in the sterilization process (Fernandes 2009). Desirable properties in UHT milk are maintained by minimum conditions with B* of 1 and C* of 1 (Lewis and Deeth 2009, Chavan et al. 2011). When B*, for example, equals to 1 in a process which requires a holding time of 10.1 s at 135°C, it provides 9 decimal reduction in counts of thermophilic spores.

Raw milk allocated for production of UHT milk should not be stored at 4°C for more than 48 hr, as the concentration of heat-stable enzymes released from psychrotrophic bacteria is likely to increase. These enzymes may cause perceivable changes in UHT milk such as development of bitter or rancid flavor and age-gelation (Lewis and Deeth 2009). There is a correlation between the level of bacterial proteases and time of gelation during storage of UHT milk. In case of use of low quality raw milk in the manufacture of UHT milk (i.e., high somatic cell count, bacterial count or plasmin activity), the temperature of sterilization should be increased to *ca.* 150°C to minimize the risk of gelation or bitterness in the end product (Topçu et al. 2006). The flat-sour defect in UHT milk is often associated with heat-stable spores of *G. stearothermophilus* and *B. licheniformis* (Boor and Fromm 2006). *Bacillus sporothermodurans* was first detected in UHT milk in 1985 in Europe (Petterson et al. 1996). This bacterium can reach maximum cell count of 10^5 cfu ml^{-1} when stored at 30°C for five days and no spoilage associated with this bacterium in UHT milk has been reported so far (Chavan et al. 2011).

Recently, a filamentous fungus, *Fusarium oxysporum*, has been found to cause an off-flavor in UHT milk within a few weeks. Typical indication of this fungus in UHT milk is the swollen packages. The major source of contamination of *F. oxysporum* is air in filling chamber (Lewis and Deeth 2009).

5.6 Public health concern related with processed liquid milk

The consumption and preferences of consumers for processed liquid milk change from one country to another. For example, 92.9% of the liquid milk produced in the UK in 2003 was pasteurized milk, followed by UHT milk (5.7%) and in-container sterilized milk (1.4%) (Lewis and Deeth 2009). In Australia, the consumption rates of the same segments were 91.9%, 0% and 8.1%, in the same order. On the contrary, in European countries, except for the UK, the processed liquid milk market is dominated by the UHT milk. Milk is a suitable source of food-borne illnesses. Therefore, strict measures should be taken to ensure food-safety (Griffiths 2009). In the late 1940s, before pasteurization became a common industrial application, two bacteria, *Salmonella typhi* and *Streptococcus pyogenes* causing typhoid and scarlet fever, respectively, were responsible for 50–80% and 15–27%

of milk-borne diseases, in the same order. Dairy-related diphtheria and tuberculosis caused by *Corynebacterium diphtheria* and *Mycobacterium tuberculosis*, respectively, were also common. Although the main source of milk-borne diseases is raw milk, some outbreaks have been found to be related to pasteurized milk. For example, Grant et al. (2002) screened 567 commercial pasteurized milk samples for the presence of *Mycobacterium avium* subsp. *paratuberculosis* and found that 1.8% of the samples evaluated were contaminated with this sub-species. *M. avium* subsp. *paratuberculosis* is known to be responsible for Chrone's disease in humans (named as Johne's disease in ruminants) (Griffiths 2009). This organism is capable of surviving HTST pasteurization (at 72°C for 15 s) and can be present in processed milk as post-pasteurization contaminant (Ryser 2012). For a full inactivation of *M. avium* subsp. *paratuberculosis*, the holding time of milk during HTST treatment was recommended to extend from 15 s to 25 s (Grant et al. 1999).

Campylobacter spp. survive unlikely pasteurization. However, in 1979, more than 2500 school children in England were affected by pasteurized milk contaminated by *Campylobacter jejuni* (Jones 1981). The possible reason of this outbreak was thought to be insufficient pasteurization of raw milk. *Staphylococcus aureus* which produces heat-stable enterotoxin cannot withstand classical pasteurization. Staphylococcal enterotoxin A (SEA) can remain active after heat treatment at 121°C for 28 min and causes intoxication in humans (Fernandes 2009). SEA outbreaks originated from chocolate-flavored milk or low fat milk were reported in the USA and Japan. Rall et al. (2008) detected this pathogen in 20.4% of the thermally-treated milk samples at level of 8.7×10^3 cfu ml^{-1}, indicating the insufficiency of thermal processing. *E. coli* O157:H7 is a heat-labile pathogen and its presence in pasteurized milk indicates defective pasteurization or post-pasteurization contamination.

Clostridium perfringens, a sporeforming pathogenic organism, produces heat-stable spores that survive pasteurization. However, spores of this bacterium are not able to grow under refrigeration conditions. *Salmonella* spp. cannot survive pasteurization. Therefore, presence of this organism in pasteurized milk indicates either insufficient pasteurization or post-pasteurization contaminations. In the past, a number of outbreaks related with consumption of pasteurized (Adams et al. 1984, Ryan et al. 1987) or certified raw milk or milk products (Mazurek et al. 2004, Lind et al. 2007) contaminated with *Salmonella* spp. were reported.

Unless insufficient pasteurization is applied to milk or contamination occurs after pasteurization, pasteurized milk is unlikely to contain *L. monocytogenes* (Ryser 1998). However, another *Listeria* species (*L. innocua*) was detected in 0.9% (Maura et al. 1993), 10.7% (Garayzabál et al. 1986)

Table 5.3 Major outbreaks associated with pasteurized milk.

Etiology	Location	Year	Number of cases	Cause(s)
Staphylococcus aureus	USA	1914–1942	29	PPC*
Campylobacter jejuni	UK	1978–1984	27	Improper pasteurization PPC
	USA	1986	33	Inadequate pasteurization
	USA	2000	3	Unknown
	USA	2005	200	Unknown
	USA	2006	1644	Unknown
Salmonella spp.	USA	1978	23	PPC
	USA	1984	16	Inadequate pasteurization
	USA	1985	1600	PPC
	USA	2000	93	PPC
Salmonella Typhimurium	USA	1975	49	Unknown
	USA	1993	6	Unknown
	USA	2002	116	Unknown
Salmonella Newport	USA	2004	100	Unknown
Shigella flexneri type 2	USA	1966	97	PPC
Eschericia coli	USA	1994	18	PPC
Listeria monocytogenes	USA	1983	49	Unknown
	USA	1994	45	PPC
	USA	2007	5 (3 deaths)	Unknown
Yersinia enterocolitica	USA	1975	217	Poor raw material handling
	USA	1976	38	PPC
	USA	1982	172	PPC
	Europe	1985	36	PPC
	USA	1995	10	PPC
	USA	2011	16	Unknown
Bacillus cereus	Europe	1980	280	Unknown

Data extracted from Olsen et al. (2004), Kelly et al. (2012)
*Post pasteurization contamination

and 2.0% (Ahrabi et al. 1998) of pasteurized milks collected from Brazilian, Spanish and Turkish dairy processing plants, respectively.

Similar to other pathogens, *Yersinia enterocolitica* cannot withstand classical pasteurization conditions, and presence of this organism in pasteurized milk is a sign of post-production contamination (Lovett et al. 1982). The most recent *Y. enterocolitica* outbreak was recorded in 1995, in New Hampshire and Vermont, in the USA with ten patients aged between six months and 44 years-old were hospitalized. The major sources of contamination of this organism to pasteurized milk are dairy pigs and rinse of bottles with untreated water (Acker et al. 2000). According to Center for Disease Control and Prevention (CDC-USA), between 1998 and 2010, totally 29 dairy-related outbreaks were recorded in the USA and nine of these outbreaks were linked with commercial pasteurized milks (www.cdc.gov). Number of people affected from these products was 2200 with three deaths. Major outbreaks associated with pasteurized milk are summarized in Table 5.3.

References

Ackers, M.L., S. Schoenfeld, J. Markman, M.G. Smith, M.A. Nicholson, W. de Witt, D.N. Cameron, P.M. Griffin and L. Slutsker. 2000. An outbreak of *Yersinia enterocolitica* 0:8 infections associated with pasteurized milk. J. Infect. Dis. 181: 1834–1837.

Adams, D., S. Well, R.F. Brown, S. Gregorio, L. Townsend, J.W. Scags and M.W. Hinds. 1984. Salmonellosis from inadequately pasteurised milk: Kentucky: Morbidity and mortality. Weekly Report 33: 504–505.

Ahmed-Hassan, N.B., M.O. Mohamed-Abdalla and A.A.A. Mohamed-Nour. 2009. Microbiological quality of heat-treated milk during storage. Pakistan J. Nutr. 8: 1845–1848.

Ahrabi, S.S., S. Ergüven and A. Günalp. 1998. Detection listeria in raw and pasteurized milk. Cent. Eur. J. Public Health 6: 254–255.

Blake, M.R., B.C. Weimer, D.J. McMahon and P.A. Savello. 1995. Sensory and microbial quality of milk processed for extended shelf life by direct steam injection. J. Food Protect. 58:1007–1013.

Boor, K. and H. Fromm. 2006. Managing microbial spoilage in dairy industry. pp. 171–193. *In:* C. De and W. Blackburn [eds.]. Food Spoilage Microorganisms. Woodhead Publishing Ltd., Cambridge, UK.

Boor, K.J. and S.C. Murphy. 2002. Microbiology of market milks. pp. 91–122. *In:* R.K. Robinson [ed.]. Dairy Microbiology Handbook: The Microbiology of Milk and Milk Products. Wiley-Interscience, New York , USA.

Chandan, R.C. 2011. Dairy ingredients for food processing: An overview. pp. 3–33. *In:* R.C. Chandan and A. Kilara [eds.]. Dairy Ingredients for Food Processing. Wiley-Blackwell, Ames, Iowa, USA.

Chapman, K.W. and K.J. Boor. 2001. Acceptance of 2% ultra-pasteurized milk by consumers 6 to 11 years old. J. Dairy Sci. 84: 951–954.

Chavan, R.S., S.R. Chavan, C.D. Khedkar and A.H. Jana. 2011. UHT milk processing and effect of plasmin activity on shelf life: A review. Comp. Rev. Food Sci. Food Safety 10: 251–268.

Cousin, M.A. 1982. Presence and activity of psychrotrophic microorganisms in milk and dairy products: A review. J. Food Protect. 45: 172–207.

de Jong, P. 2008. Thermal processing of milk. pp. 2–34. *In:* T.J. Britz and R.K. Robinson [eds.]. Advanced Dairy Science and Technology. Blackwell Publishing, Oxford, UK.

Deeth, H.C., T. Khusniati, N. Datta and R.B. Wallace. 2002. Spoilage patterns of skim and whole milks. J. Dairy Res. 68: 228–242.

Eneroth, Å., A. Christiansson, J. Brendehaug and G. Molin. 1998. Critical contamination sites in the production line of pasteurised milk, with reference to the psychrotrophic spoilage flora. Int. Dairy J. 8: 325–331.

Eneroth, Å., S. Ahrné and G. Molin. 2000. Contamination routes of Gram-negative spoilage bacteria in the production of pasteurised milk, evaluated by randomly amplified polymorphic DNA (RAPD). Int. Dairy J. 10: 325–331.

EU. 1992. Council Directive 92/46/EEC of 16 June 1992 laying down the health rules for the production and placing on the market of raw milk, heat-treated milk and milk based products. Official Journal of the European Communities 35: 1–32.

Faille, C., F. Fontaine and T. Benezech. 2001. Potential occurrence of adhering living *Bacillus* spores in milk product processing lines. J. Appl. Microbiol. 90: 892–900.

Fernandes, R. 2009. Microbiology Handbook Dairy Products. Leatherhead Publishing and Royal Society Chemists, Surrey, Cambridge, UK.

Fernández-Garcia, L., S. Alvarez-Blanco and F.A. Riera-Rodriguez. 2013. Microfiltration applied to dairy streams: removal of bacteria. A review. J. Sci. Food Agric. 93: 187–196.

Fredsted, L.B., G. Rysstad and T. Eie. 1996. Pure-Lac™: the new milk with protected freshness and extended shelf-life. pp. 104–125. *In:* The Proceedings of the IDF Symposium on Heat Treatments and Alternative Methods, Special Issue No. 9602. International Dairy Federation, Brussels, Belgium.

Fromm, H.I. and K.J. Boor. 2004. Characterization of pasteurized fluid milk shelf-life attributes. J. Food Sci. 69: M200–M214.

Garayzabàl, J.F.F., L.D. Rodriguez, J.A.V. Boland, J.L.B. Cancelo and G.S. Fernandez. 1986. *Listeria monocytogenes* in pasteurized milk. Can. J. Microbiol. 32: 149–150.

Grant, I.R., H.J. Ball and R.T. Rowe. 1999. Effect of higher pasteurization temperatures, any longer holding times at 72°C, on the inactivation of *Mycobacterium paratuberculosis* in milk. Let. Appl. Microbiol 28: 461–468.

Grant, I.R., H.J. Ball and R.T. Rowe. 2002. Incidence of *Mycobacterium paratuberculosis* in bulk raw and commercially pasteurized cows' milk from approved dairy processing establishments in the United Kingdom. Appl. Environ. Microbiol. 28: 461–468.

Griffiths, M.W. 2009. Food safety issues and the microbiology of milk and dairy products. pp. 147–167. *In:* N. Heredia, I. Wesley and S. Garcia [eds.]. Microbiologically Safe Foods. John Wiley & Sons, Hoboken, New Jersey, USA.

Huck, J.R., M. Sonnen and K.J. Boor. 2008. Tracking heat-resistant, cold-thriving fluid milk from farm to packaged product. J. Dairy Sci. 91: 1218–1228.

IDF. 1984. The thermization of milk. Document No. 182. International Dairy Federation, Brussels, Belgium.

Jones, P.H., A.T. Willis, D.A. Robinson, M.B. Skirrow and D.S. Josephs. 1981. *Campylobacter enteritis* associated with the consumption of free school milk. J. Hygien. Camb. 87: 155–162.

Kelly, A.L. and N. O'Shea. 2003. Pasteurizers, design and operation. pp. 2237–2244. *In:* H. Roginski, J.W. Fuquay and P.F. Fox [eds.]. Encyclopedia of Dairy Sciences. Academic Press, London, UK.

Kelly, A.L., N. Data and H.C. Deeth. 2012. Thermal processing of dairy products. pp. 265–298. *In:* Da-Wen, Sun [ed.]. Thermal Food Processing New Technologies and Quality Issues. Taylor & Francis, Boca Raton, FL, USA.

Larsen, H.D. and K. Jorgensen. 1997. The occurrence of *Bacillus cereus* in Danish pasteurized milk. Int. J. Food Microbiol. 34: 179–186.

Lewis, M.J. and H.C. Deeth. 2009. Heat treatment of milk. pp. 168–204. *In:* A.Y. Tamime [ed.]. Milk Processing and Quality Management. Wiley-Blackwell, Oxford, UK.

Lewis, M.J. and N. Herpell. 2000. Continuous Thermal Processing of Foods: Pasteurization and UHT Sterilization. Aspen Publishers, Gaithersburg, MD, USA.

Lind, L., J. Reeser, K. Stayman, M. Deasy, W. Moll, A. Weltman, W. Urdaneta, S. Ostroff, W. Chirdon, E. Campagnolo and T. Chen. 2007. *Salmonella typhimurium* infection associated with raw milk and cheese consumption Pennsylvania: Morbidity and mortality. Weekly Report 44: 1161–1164.

Lovett, J., J.G. Bradshaw and J.T. Peeler. 1982. Thermal inactivation of *Yersinia enterocolitica* in milk. Appl. Environ. Microbiol. 44: 517–519.

Maura, S.M., M.T. Destro and B.D.G.M. Franco. 1993. Incidence of *Listeria* species in raw and pasteurised milk produced in Sao Paulo, Brazil. Int. J. Food Microbiol. 19: 229–237.

Mayr, R., K. Gutser, M. Busse and H. Seiler. 2004a. Gram positive non-sporeforming recontaminants are frequent spoilage organisms of German retail ESL (Extended Shelf Life) milk. Milchwiss. 59: 262–266.

Mayr, R., K. Gutser, M. Busse and H. Seiler. 2004b. Indigenous aerobic sporeformers in high heat treated (127°C, 5 s) German ESL (Extended Shelf Life) milk. Milchwiss. 59: 143–146.

Mazurek, J., E. Salehi, D. Propes, J. Holt, T. Bannerman, L.M. Nicholson, M. Bundesen, R. Duffy and R.L. Moolenaar. 2004. A multistate outbreak of *Salmonella enterica* serotype Typhimurium infection linked to raw milk consumption-Ohio, 2003. J. Food Protect. 67: 2165–2170.

Mehta, R.S. 1980. Milk processed at ultra-high temperature—a review. J. Food Protect. 43: 212–225.

Muir, D.D. 1996. The shelf life of dairy products: I. Factors influencing raw milk and fresh products. J. Soc. Dairy Technol. 49: 24–32.

Olsen, S.J., M. Ying, M.F. Davis, M. Deasy, B. Holland, L. Iampietro, C.M. Baysinger, F. Sassano, L.D. Polk, B. Gormley, M.J. Hung, K. Pilot, M. Orsini, S. Van Duyne, S. Rankin, C. Genese, E.A. Bresnitz, J. Smucker, M. Moll and J. Sobel. 2004. Multidrug-resistant *Salmonella* Typhimurium infection from milk contaminated after pasteurization. Emerg. Infect. Dis. 10: 932–935.

Parkar, S.G., S.H. Flint, J.S. Palmer and J.D. Brooks. 2001. Factors influencing attachment of thermophilic bacilli to stainless steel. J. Appl. Microbiol. 90: 901–908.

Petterson, B., F. Lembke, P. Hammer, E. Stackebrandt and F.G. Priest. 1996. *Bacillus sporothermodurans*, a new species producing highly heat-resistant endospores. Int. J. Syst. Bacteriol. 46: 759–764.

Pirttijärvi, T.S.M., T.H. Graeffe and M.S. Salkinoja-Salonen. 1996. Bacterial contaminants in liquid packaging boards: Assessment of potential for food spoilage. J. Appl. Microbiol. 81: 445–458.

Public Health Service-PMO. 2009. Grade "A" Pasteurized Milk Ordinance. Department of Health and Human Services, Food and Drug Administration, USA.

Rademacher, B., W. Walenta and H.G. Kessler. 1996. Contamination during pasteurisation by biofilms of thermophilic streptococci. pp. 26–33. *In:* The Proceedings of the IDF Symposium on Heat Treatment and Alternative Methods. Special Issue No. 9602. International Dairy Federation, Brussels, Belgium.

Rall, W.L.M., F.P. Vieira, R. Rall, R.L. Vieitis, A. Fernandes, J.M.G. Candeias, K.F.G. Cardoso and J.P. Araujo. 2008. PCR detection of staphylococcal enterotoksin genes in *Staphylococcus aureus* strains isolated from raw and pasteurized milk. Vet. Microbiol. 132: 408–413.

Ranieri, M.L. and K.J. Boor. 2009. Bacterial ecology of high-temperature, short-time pasteurized milk processed in the United States. J. Dairy Sci. 92: 4833–4840.

Ranieri, M.L., J.R. Huck, M. Sonnen, D.M. Barbano and K.J. Boor. 2009. High temperature short time pasteurization temperatures inversely affect bacterial numbers during refrigerated storage of pasteurized fluid milk. J. Dairy Sci. 92: 4823–4832.

Rowe, M. and J. Donaghy. 2011. Microbiological aspects of dairy ingredients. pp. 59–101. *In:* R.C. Chandan and A. Kilara [eds.]. Dairy Ingredients for Food Processing. Wiley-Blackwell, Ames, Iowa, USA.

Ryan, C.A., M.K. Nickels, N.T. Hargrett-Bean, M.E. Potter, T. Endo, L. Mayer, C.W. Langkop, C. Gibson, R.C. McDonald, R.T. Kenney, N.D. Puhr, P.J. McDonald, R.J. Martin, M.L. Cohen and P.A. Blake. 1987. Massive outbreak of antimicrobial-resistant salmonellosis traced to pasteurized milk. JAMA 258: 3269–3274.

Ryser, E.T. 1998. Public health concerns. pp. 263–404. *In:* E.H. Marth and J.L. Steele [eds.]. Applied Dairy Microbiology. Marcel Dekker, New York, NY, USA.

Ryser, E.T. 2012. Safety of dairy products. pp. 127–145. *In:* O.A. Oyarzabal and S. Backert [eds.]. Microbial Food Safety: An Introduction. Springer Science+Business Media, New York, NY, USA.

Rysstad, R. and J. Kolstad. 2006. Extended shelf life milk—advances technology. Int. Dairy J. 59: 85–96.

Schmidt, V.S.J., V. Kaufmann, U. Kulozik, S. Scherer and M. Wenning. 2012. Microbial biodiversity, quality and shelf life of microfiltered and pasteurized extended shelf life (ESL) milk from Germany, Austria and Switzerland. Int. J. Food Microbiol. 154: 1–9.

Stepaniak, L. and R.K. Abrahamsen. 1995. Effect of sampling and storage temperature on microbiological quality of pasteurized milk. Milchwis. 50: 22–26.

Tomasula, P.M., S. Mukhopadhyay, N. Data, A. Porto-Fett, J.E. Call, J.B. Luchansky, J. Renye and M. Tunick. 2011. Pilot-scale crossflow-microfiltration and pasteurization to remove spores of *Bacillus anthracis* (Sterne) from milk. J. Dairy Sci. 94: 4277–4291.

Topçu, A., E. Numanoğlu and I. Saldamlı. 2006. Proteolysis changes in lactose-hydrolysed UHT milks during storage. Milchwiss. 62: 410–415.

Varnam, A.H. and J.P. Sutherland. 2001. Milk and Milk Products—Technology, Chemistry and Microbiology. Aspen Publishers, Gaithersburg, MD, USA.

Wessel, D., P.J. Jooste and J.F. Mosters. 1989. Psychrotrophic, proteolytic and lipolytic properties of *Enterobacteriaceae* isolated from milk and dairy products. Int. J. Food Microbiol. 9: 79–83.

Zhou, G., H. Liu, J. He, Y. Yuan and Z. Yuan. 2008. The occurrence of *Bacillus cereus, B. thuringiensis* and *B. mycoides* in Chinese pasteurized full fat milk. Int. J. Food Microbiol. 121: 195–200.

CHAPTER 6

Cheese Microbiology

*Manuela Pintado, Adriano Gomes da Cruz** and *Patricia B. Zacarchenco Rodrigues de Sá*

6.1 Introduction

Microorganisms play significant role in the development of well-balanced aroma/flavor and texture in cheese and therefore give cheese a dynamic nature. Cheese contains a complex microflora including bacteria, yeasts and molds. Cheese microflora is divided into two basic groups namely starter cultures and secondary flora (Cogan 2002). Both groups basically contribute to the formation of desired aroma/flavor and textural characteristics in cheese. Cheese starters ferment milk sugar (lactose) to yield lactic acid and other organic acids which reduce the milk pH (Beresford 2003). *Lactococcus lactis, Streptococcus thermophilus, Lactobacillus helveticus* and *Lactobacillus delbrueckii* are the most common lactic acid bacteria used in the manufacture of cheese. At the early stages of ripening, the counts of starter lactococci are around 10^6–10^{10} cfu g^{-1}. However, these figures decline rapidly within the first weeks of ripening at 2–16°C. Among the factors affecting the counts of starter lactococci are salt level of cheese, low pH, insufficient level of fermentable carbohydrate and low ripening temperature. Autolytic properties of starter bacteria, salt tolerance and resistance against phage attacks determine the level of decrease in the counts of starter bacteria.

The microbial examination of ripening curd has shown that in addition to deliberately added lactic acid bacteria (LAB), bacterially-ripened cheeses also contain large populations of adventitious ('contaminant') bacteria that gain access to cheese through milk or the milk-processing environment. This secondary microflora has a decisive role during cheese ripening and normally are referred to as non-starter lactic acid bacteria (NSLAB), other

*Corresponding author

bacterial groups, yeasts and molds which can grow internally or externally in most cheese varieties (Cogan and Beresford 2002).

The NSLAB composition of cheese is influenced by several factors, including milk, heat treatment and equipment sanitation in the manufacturing plant (Broadbent et al. 2011). The NSLAB group is particularly heterogeneous with lactobacilli being mostly represented by *Lactobacillus farciminis* among obligately homofermentative species, *Lactobacillus casei*, *Lactobacillus paracasei*, *Lactobacillus plantarum*, *Lactobacillus pentosus*, *Lactobacillus curvatus* and *Lactobacillus rhamnosus* among facultatively heterofermentative species and *Lactobacillus fermentum*, *Lactobacillus buchneri*, *Lactobacillus parabuchneri* and *Lactobacillus brevis* among obligately heterofermentative species. The non-*Lactobacillus* species of NSLAB commonly isolated from cheese during ripening are *Pediococcus acidilactici*, *Pediococcus pentosaceus*, *Enterococcus durans*, *Enterococcus faecalis*, *Enterococcus faecium* and also leuconostocs with the same species that act as starter cultures (Settanni and Moschetti 2010). NSLAB are of crucial importance in determining the final texture and flavor of the cheese. Cheese contains lactose at very low concentrations. Therefore, NSLAB utilize carbohydrates deriving from glycomacropeptides of caseins and glycoproteins deriving from fat globule membranes; heterofermentative NSLAB may ferment pentoses liberated from lysed lactic acid bacteria. Modern sanitation practices help keeping NSLAB to very low levels in young cheese, but these bacteria inevitably begin to grow and will reach high numbers within a few months of ripening (Broadbent et al. 2011). NSLAB also have probiotic potential and induced beneficial health effects at the gut mucosa *in vivo* (Burns et al. 2012).

There are many factors including water activity, salt, pH and organic acids, ripening temperature and redox potential affecting the growth of microorganisms in cheese. Water activity of cheese (a_w 0.917–0.988) is usually lower than the optimum water activity of starter bacteria. However, at these a_w levels, most of the non-starter lactic acid bacteria and contaminant microorganisms are able to grow in cheese (Beresford 2007). The bacteria, molds and yeasts are ordered based on their optimum a_w requirements as follows: $a_{w\ bacteria} > a_{w\ yeasts} > a_{w\ molds}$, with minimum a_w values of 0.92, 0.83 and 0.75 in the same order. Osmophilic yeasts can withstand a_w values lower than 0.60 (Cogan 2002). In general, lactic acid bacteria can grow at lower a_w values than other bacterial groups. The minimum a_w values for *L. lactis*, *S. thermophilus*, *Lb. helveticus* and *Propionibacterium freudenreichii* subsp. *shermanii* are as follows 0.93, >0.98, >0.96 and 0.96, respectively (Fox et al. 2000, Beresford et al. 2001).

Salt reduces the a_w value of the cheese and hence generate an inhibitory effect on starter cultures and spoilage/pathogenic microorganisms. The level of salt to provide complete microbial inactivation in cheese depends on the structure and water content of the cheese. In brined-type cheeses, the salt concentration of the cheese surface is higher than the interior part of the cheese block, especially at the early stages of ripening. Therefore, microorganisms with higher salt resistance can grow on the surface of this type of cheeses. Many *Corynebacterium* spp. and *Micrococcus* spp. can grow in medium supplemented with 15% NaCl, *Staphylococcus* spp. can grow at lower salt levels (i.e., 10% NaCl) (Cogan 2002). *Brevibacterium linens* and *Debaryomyces hansenii* are associated with surface-ripened cheeses and are able to grow at 15% NaCl level (a_w = ~0.916).

In general, majority of bacteria are able to grow well at neutral pHs. Very few bacterial groups (e.g., *Lactobacillus* spp.) can grow at pH values lower than 4 (Jay 2000, Ray 2004). *Corynebacterium* spp. and *Micrococcus* spp. cannot grow at pH values lower than 5.5–6.0. On contrary, molds and yeasts may well grow at lower pHs (i.e., <4.5). Comparing with bacteria, yeasts and molds can grow within a much wider pH range (between 1.7 and 10.0) and variation in pH does not affect the growth of yeasts and molds remarkably (Heperkan 2010). Some organic acids also affect the viability of bacteria in cheese. Some weak organic acids (i.e., sorbic and propionic acids) can pass through bacterial cell membrane and ionized inside the cells. The protons released inside the cells acidify the cells and hence cause bacterial lysis. Major organic acids in cheese are lactic, acetic and propionic acids. Propionic acid is an effective inhibitor against molds in cheese.

Ripening temperature is an important tool for controlling microbial growth in cheese. Mesophilic and thermophilic lactic acid bacteria which are commonly used in cheese-making, have optimum growth temperatures of 35°C and 55°C, respectively. In the selection of starter strains for pasta-filata type cheeses and Cheddar-type cheeses, apart from the optimum growth temperatures, the resistance to stretching/scalding or cheddaring temperatures of the bacteria should be taken into consideration. For example, *L. lactis* subsp. *lactis* may well grow during cheddaring at 38°C, but many strains of *L. lactis* subsp. *cremoris* are inhibited at this temperature (Fox et al. 2000). Ripening temperature should ideally control the growth of secondary flora in cheese to prevent quality losses and should prevent the growth of spoilage/pathogenic microorganisms (Beresford et al. 2001). Except for the Swiss-type cheeses in which growth of propionic acid bacteria is promoted at 22–25°C, most of the cheese varieties are ripened at 6–15°C (Beresford 2007).

Redox potential of cheese is around –250 mV and during ripening it reduces gradually. The exact mechanism(s) of decrease in redox potential of cheese is still unknown. There is a remarkable difference between the redox potentials of the inner and outer parts of the cheese. For example, while the redox potential of the surface of Camembert cheese was +330 mV, this figure was –330 mV in the center of the cheese (Abraham et al.

2007). The differences between the redox potentials of different parts of the cheese determine the dominant flora of the cheese. Overall, anaerobic and/or facultatively aerobic microorganisms can easily grow inside the cheese and aerobic microorganisms such as *Brevibacterium* spp., *Bacillus* spp. and *Micrococcus* spp. grow on the surface of the cheese.

6.2 Microbiology of Brined cheeses

The term "Brined cheeses" is related to group of cheese varieties ripened and preserved in brine for a considerable amount of time, until the moment of consumption. They are traditionally produced under various names in the Balkan and Eastern Mediterranean countries (Alichanidis and Polychroniadou 2008).

Brined cheeses are generally rindless and are produced in the form of blocks of various shapes and sizes (cubes, bricks, or segments) immersed in brine or salted whey. Preservation in brine (pickle) is the main characteristic of this group of cheeses. The brine serves to preserve the cheese and prevents it drying out. Consequently, the composition and properties of these cheeses and those of the brine are interrelated (El Soda and Abd-El Salam 2002).

Several sub-categories of cheeses are covered by brined cheeses; indeed this category includes soft and semi-hard varieties, and nearly all of them are made by rennet coagulation. In general, brined cheeses can be made from different kinds of milk, but sheep's milk is preferred for most of these varieties; independent of the milk used in the manufacture, the final cheese retain the white color of these milks (El Soda et al. 2011). White-brined cheeses is a sub-category of brined cheeses that includes cheeses made from curds that are not subjected to any heat treatment, like Feta and similar cheese varieties produced in Greece and Turkey. They have unique sensory and textural characteristics. Their flavor is slightly acidic and salty that sometimes turns to rancid and piquant (Moatsou and Govaris 2011).

Feta is the major white-brined cheese produced worldwide, followed by Domiati cheese. Other brined cheeses such as Haloumi, Medaffara/Magdola, Chenakh, Turkish white cheese, Teleme and Brinza are produced in much smaller amounts (Moatsou and Govaris 2011).

6.2.1 Microbial groups in brined cheeses

White-brined cheeses are matured for long periods in brine, and thus the dominant microfloras make a significant contribution to the maturation process and, to a degree, regulate the quality of the final product. In addition, the safety and shelf-life of the final product is highly dependent on the microflora present in cheese (Bintsis and Papademas 2002).

The microbiota present in brined cheeses is diversified and covers all kind of microbial groups with preferably high salt resistance. The brined-cheese varieties made from unpasteurized, thermized, or in some cases pasteurized milk may contain non-starter lactic acid bacteria (NSLAB), originating from raw milk or post-heat treatment contamination of milk. The majority of NSLAB in white-brined cheeses are mesophilic lactobacilli. NSLAB also contain *Pediococcus* spp., *Enterococcus* spp. and *Leuconostoc* spp. Most of the NSLAB are salt- and acid-tolerant facultative anaerobic bacteria and can grow easily in cheese. The number of NSLAB increases rapidly after pressing and salting of cheese, reaching up to 10^9 cfu g^{-1} during ripening. *Lactobacillus plantarum*, *Lactobacillus paracasei* subsp. *paracasei*, *Lactobacillus hilgardii*, *Lactobacillus brevis*, *Lactobacillus paraplantarum*, and *Lactobacillus pentosus* are the most commonly isolated lactobacilli from white-brined cheeses made from goat's or sheep's milk. Lactic acid bacteria are dominant in Domiati cheese. Streptococci are predominant during the early stage of pickling, and lactobacilli prevail at the later stages of ripening. *Enterococcus faecalis*, *L. lactis*, *Lactobacillus casei*, *Lb. plantarum*, *Lb. brevis* and *Lactobacillus fermenti* are the most common microbial groups present in Domiati cheese. Several non-lactic acid bacteria and yeasts are also found in this cheese variety. The principal psychrotrophs found in Domiati cheese are from the genera *Pseudomonas*, *Aeromonas* and *Rhodotorula*. Salt-resistant yeasts growing on the surface of Feta cheese during ripening may contribute to ripening (Fernandes 2009). Non-starter lactic acid bacteria are also present and have distinct contribution: a positive aspect is the contribution to the flavor, texture and pH of the product while a negative aspect is to cause spoilage (Broadbent et al. 2011). The counts of psychrotrophic bacteria tend to increase in white-brined cheeses during the first weeks of maturation, and then their numbers fluctuate depending on the initial microbial load in milk or degree of contamination during the production stages. *Pseudomonas* spp., *Aeromonas* spp. and *Acinetobacter* spp. are among the genera of psychrotrophs most frequently found in white-brined cheeses.

In traditional practices, brined cheeses are manufactured without starter culture and in this sense, we have a heterogeneous quality of the final products; however, this picture has been changing and the use of starter has been tested to the effect in the final parameters of the product. Since pasteurized milk is used widely in the large-scale manufacture of white-brined cheese, the use of defined starter cultures is essential. The selection, maintenance and use of starter cultures are, perhaps, the most important aspects of cheese-making, particularly in the context of modern mechanized processes for which predictability and consistency are essential. Lactic acid bacteria (LAB) are predominant in white-brined cheeses and the main isolates are *L. lactis* subsp. *lactis*, *L. lactis* subsp. *cremoris*, *Lb. casei*, *Enterococcus faecalis* var. *liquefaciens* and *Leu. paramesenteroides*. However, it is prudent to comment that the strains used must be tolerant to high levels of salt present in the product besides having a proteolytic activity. In this context, the salt tolerant strains of *L. lactis* subsp. *lactis* and *cremoris* and thermophilic cultures such as *S. thermophilus* and *Lb. delbrueckii* subsp. *bulgaricus* are widely used as starters in the manufacture of brined cheeses (Özer 1999).

The choice of the starter cultures influences the final characteristics of the product: thermophilic cultures are inhibited at refrigeration temperatures, which results in constant pH values along with the ripening besides negative impact on the flavor (i.e., bitter taste) and texture (i.e., fragile structure) of the product while mesophilic cultures show a gradual increase along the storage periods (Özer 1999). In the same extent, it is essential to perform preliminary tests for an adequate selection of lactobacilli for use as 'adjunct cultures' as the development of desirable flavors or defective products seems to be strain-dependent rather than species-dependent (Bintsis and Papademas 2002).

6.2.2 Pathogens in brined cheeses

The microbiological quality of cheese is closely related to the manufacture and, as unpasteurized milk is still in use in the manufacture of white-brined cheeses, the initial microbiological load of the milk determines the quality of the final product. Raw-milk cheese generally has higher bacterial counts than the pasteurized-milk cheese during early stages of ripening, but both cheeses have similar bacterial counts thereafter (Fernandes 2009).

Coliform bacteria are frequently found in brined cheeses in large numbers mainly due to the use of raw milk and poor sanitary conditions during small-scale cheese-making. At the later stages of ripening, the counts of coliforms are reduced to negligible levels under usual conditions of ripening of white-brined cheeses. The presence of coliforms is largely responsible for the blowing defect in white-brined cheeses (El Soda and Abd-El Salam 2002). The most common defect of Feta and related cheeses is 'early blowing', a defect that is characterized by the presence of large gas holes in the cheese, which, in addition, has a spongy texture; this defect is due to coliforms and/or yeasts growing in excessive numbers (Özer 1999).

The pathogens including *Yersinia enterocolitica, Staphylococcus aureus* and *Listeria monocytogenes* also may be present in white-brined cheeses (Bintsis and Papademas 2002). The survival of *Y. enterocolitica* in brined cheeses depends on the final pH of the product. At pH values of >4.5, *Y. enterocolitica* can survive up to 30 days under cold conditions. However, this period may be as short as four days if the acidification develops fastly. *Listeria monocytogenes* may survive in Feta and related cheeses up to the point of retail sale. *Staphylococcus aureus* can survive in white-brined cheeses, especially in the presence of yeasts. Even under unsuitable conditions for microbial growth (i.e., low pH and high salt levels) a mutual stimulation between yeasts and *Staphylococcus aureus* is evident. *Staphylococcus aureus* can tolerate a wide range of salt concentration varying from 2.5% to 15% and thus is the major pathogenic organism likely present in Domiati cheese. *Salmonella typhi* is another pathogenic bacteria frequently isolated from Domiati cheese with 10% NaCl.

Yeasts are not among the predominant microflora of white-brined cheeses and present at low levels in brined cheeses. Yeasts may have an important role in the formation of flavor, through enhancing proteolysis and, therefore, they are recommended for inclusion in the starter culture for the manufacture of Teleme cheese. On the other hand, excessive yeast growth will cause softening and discoloration of brined cheeses, a condition that is usually associated with an unpleasant yeasty or ester-like odor or gas formation; in the case of white-brined cheeses, swelling of the cans can be caused by yeasts that ferment lactose, e.g., *Kluyveromyces* spp. In addition, yeasts can increase the pH of the cheese surface, thus spurring the growth of *Staphylococcus aureus* and possibly other pathogenic and/or spoilage bacteria (Fernandes 2009, Bintsis and Papademas 2002). Molds are frequently found in brined cheeses, but except for the mycotoxigenic potential, there is no matter for the public health, but changes in the aroma and flavor, in the end product have been reported (Özer 1999). The genera *Penicillium, Mucor, Aspergillus, Cladosporium* and *Fusarium* have been isolated from Teleme, Feta, Turkish white and Domiati cheeses, and there is a concern that some species, including *Penicillium cyclopium, Penicillium viridicatum, Aspergillus flavus* and *Aspergillus ochraceus* are able to produce mycotoxins.

6.3 Microbiology of Hard cheese

The terms hard cheese or very hard cheese refer to cheeses that are firm or very firm, respectively, and require some form of pressure to break apart. Hard and very hard cheeses have upper limits for moisture content and lower limits for fat content—usually expressed as fat-in-dry matter (FDM) (Farkye 2004). Hard cheeses are appreciated around the world, being present in several types of culinary preparations such as Parmesan (pasta) and Cheddar (sandwiches). The most commonly known members of this group of cheese are Permiggiano-Reggiano, Grana Padano, Asiago, Romano, Cheddar, Chesire, Derby, Gloucester, Liecester, Cantal, Leiden, Graviera, Manchego, Ras and Idiazabal cheeses.

Hard cheeses are produced by enzymatic coagulation and have a ripening periods varying from about three weeks to more than two years. Generally, the duration of ripening is inversely related to the moisture content of the cheese. Many varieties are consumed at any stage of maturation, depending on the flavor preferences of consumers and economic factors. The unique characteristics of each variety develop during ripening as a result of a complex set of biochemical reactions. The flavor, aroma and texture of the mature cheese are predetermined by the manufacturing process, especially by the levels of moisture and NaCl and pH, residual coagulant activity, the type of starter, and, in many cases, by the secondary microflora (added or adventitious).

Ripening is a very complex series of biochemical reactions, which may be divided into three principal groups: glycolysis, which represents the catabolism of lactose to acid lactic; lipolysis, which is related to the catabolism of fatty acids; and proteolysis, representing the catabolism of proteins, peptides and amino acids (McSweeney 2011). Combination of these biochemical reactions determines the flavor, texture and functionality of the final product. Metabolism of lactose, lactate and citrate and related events are caused by living microorganisms (starter and/or non-starter), while lipolysis and proteolysis are catalyzed mainly by enzymes from the coagulant, milk, starter bacteria, adventitious non-starter bacteria, and, usually, secondary (adjunct) cultures (Sousa et al. 2001). Readers are recommended to refer to Chapter 7 for more information about cheese ripening.

6.3.1 Starter cultures for hard cheeses

In the manufacture of hard-type cheeses, mesophilic (e.g., *Lactococcus* spp. and *Enterococcus* spp.) and thermophilic (e.g., *S. thermophilus* and various *Lactobacillus* species) lactic starters are used. Defined-strain starter cultures for making hard-type cheeses usually contain *S. thermophilus* and *Lb. helveticus* (Powell et al. 2011). The starter lactic acid bacteria contribute to the production of acid during manufacturing that provides an adequate environment exerting a positive control to redox potential, pH, moisture and salt content allowing the enzyme activity from the rennet.

Excessive acid production negatively affects the cooking performance of the hard cheeses. Therefore, it is essential to control the growth of bacteria for preventing excessive acid production. It is important to note that an extensive survival of the starter can cause accumulation of bitter peptides which have negative contribution to the cheese flavor and, consequently, depreciate the commercial value. It is also believed that intracellular peptidases released from the starter bacteria with the advance of cheese ripening are able to hydrolyze bitter peptides to smaller non-bitter peptides and amino acids (Cogan and Beresford 2002).

In Cheddar- and Dutch-type cheeses, L(+) lactate is isomerized to a racemic mixture (D/L-lactate). Racemization has no effect on cheese flavor but if the concentration of D(–) lactate is too high, it may favor the formation of undesirable crystals of calcium lactate on the cheese surface. Oxidation of lactic acid to acetic acid is dependent on the concentration of oxygen in the curd and permeability of packaging materials and occurs at fairly low levels (McSweeney 2011).

Semi-hard and hard cheeses are potential sources of biogenic amine (BA) production. Tyramine and histamine produced by the decarboxylation of the amino acids—tyrosine and histidine—by lactic acid bacteria during cheese ripening are the most common biogenic amines present in hard-type cheeses. The presence of these amines at high levels in cheese has been associated with the increase in blood pressure, flusing and headaches (Fernandes 2009).

Storage temperature is the most important factor contributing to BA formation and the effects of temperature abuse on BA formation have been studied extensively (Linares et al. 2012). Synthesis of BA is possible only when three conditions converge: (i) availability of the substrate amino acids; (ii) presence of microorganisms with the appropriate catabolic pathway activated; and (iii) environmental conditions favorable to the decarboxylation activity. These conditions depend on several factors such as milk treatment (i.e., pasteurization), use of starter cultures, NaCl concentration, time, and temperature of ripening and preservation, pH, temperature, or post-ripening technological processes (Linares et al. 2011).

6.4 Microbiology of mold-ripened cheeses

Mold-ripened cheeses are divided into two distinct groups: surface-ripened cheeses (i.e., Camembert, Brie, Carrè de l Est ripened by *Penicillium camemberti*) and interior-ripened cheeses (i.e., blue cheeses, Roquefort, Stilton, Danablu, Gorgonzola, Edelpikäse and Mycella cheeses where *Penicillium roqueforti* grows in the curd fissures). *P. roqueforti* is unique in that it is relatively resistant to salt, low oxygen levels, and high carbon dioxide tension. Roquefort cheese is made with sheep's milk in the region of Roquefort, France, and its version made with cow's milk is called "Bleu" (blue) cheese in other areas of France. According to Fox et al. (2000) and Nath (1992), blue cheeses contain no more than 46% moisture, 29.5 to 30.5% fat, 20 to 21% protein, and 4.5 to 5% salt.

Blue cheese can be produced with homogenized milk (Iowa method) or non-homogenized milk (Minnesota method). Raw or pasteurized milk is homogenized at 2000 psi at 32 to 43.3°C. Roughly 0.5% of mesophilic lactic acid starter culture containing *L. lactis* subsp. *lactis* var. *diacetylactis* is added to the milk at 32°C. The mold spores can be added to the milk in the cheese tank before adding rennet at a ratio of 0.00025:1 of milk (4 oz 1000 lbs^{-1} or 113.4 g 453 kg^{-1} of milk). The curd is cut, allowed to rest for five minutes, and carefully stirred for another five minutes. The mixture is then stirred for another sixty minutes at a temperature of 31 to 32°C. The acidity of the whey rises from 0.11 to 0.14% lactic acid. Before the whey is drained, the temperature is increased to 33°C for two minutes. All the whey is drained and the curd is molded. If the mold spores have not yet been added to the milk, they can be added at this stage. About 900 g of salt and 28.35 g of spores can be added at this stage to each 45 kg of curd. The cheese is then pierced and rolled over with 15 min intervals for two hours, allowed to drain for 24 hr at 22°C, removed from the mold, dry salted, and allowed to ripen inside a chamber at 15.6°C and 85% moisture. Salt is added once daily for four days. The cheese is rolled for 20 days, and packaged and ripened for three to four months at 2–4°C (Fox et al. 2000, Nath 1992, Kosikowski and Mistry 1997). The addition of mold spores to the milk restricts the use of its whey in other dairy products.

6.4.1 Starter cultures for mold-ripened cheeses

Starter cultures employed for the manufacture of blue-veined and mold-ripened cheeses contain white and blue molds (*Penicillium* spp.) (Bockelmann 2010). *P. roqueforti* generally produces greenish-blue colonies throughout the cheese that become dark green as the cheese ages. A white mutant of this mold was developed for the Nuworld cheese. Dried spores or spores in brine are added to the milk or spread on the surface of the cheese. Air circulation must be carried out on the cheese to allow aeration and fungal growth. In blue and similar cheeses, acidity develops slowly. The curd is not pressed, so the resulting texture allows carbon dioxide to escape and oxygen to enter. *P. roqueforti* lineages can grow, however, slowly even in atmospheres containing 5% oxygen and 8% salt. The optimum temperature for the growth of *P. roqueforti* is 20–25°C but it can grow in temperatures ranging from 5 to 35°C. Mycelium production is high in pH range of 4.5 to 7.5, but the molds tolerate pH values as low as 3.0 and as high as 10.5. Different lineages isolated from cheeses and starter cultures have different salt tolerances. Mold growth is evident 8 to 10 days after inoculation and complete after 30 to 90 days. Intense proteolysis raises the pH of the cheese from 4.5–4.7 to 6.0–7.0, 16 to 18 weeks after the development of the mold. Molds have both lipolytic and proteolytic activity. Citrulline, ornithine, aminobutyric acid, histamine, tyramine and tryptamine are formed in the blue-veined cheeses, in addition to fatty acids, aldehydes, ammonia, alcohols, amines and other acids (Ardö 2007, Nath 1992, Bockelmann 2010).

P. camemberti is employed in the manufacture of Camembert cheese. This mold species can grow at 5–35°C, with optimum growth at 20°C. The optimum moisture requirement of *P. camemberti* is 85–95%. This mold species show rather low salt tolerance (satisfactory growth at 2% NaCl and complete inactivation at 5% NaCl). Under optimum conditions, *P. camemberti* actively control the growth of other molds present in cheese (Gueguen 1992). *Geotrichum candidum*—a white mold—is also employed in the manufacture of Camembert cheese and is largely responsible for the development of characteristic aroma/flavor of this cheese variety. This mold is also involved in the microflora of Romadur, Tilsit, Bel Paese, Quark and Quargelin cheeses. It forms a white layer on the surface of Saint Nectaire cheese. *G. candidum* shows optimum growth at pH 5.6–7.0 and is able to synthesize lipases and caseinolytic enzymes (Gueguen and Schmidt 1992). During maturation of cheese, *G. candidum* neutralizes cheese acidity and contributes to proteolysis and lipolysis. *Rhizomucor rasmussen* and *Mucor mucedo* are the mold species that are used in the manufacture of some local cheeses native to Norway and southern Europe. Salting influences the growth of both *P. camemberti* and *P. roqueforti* negatively, and *G. candidum* is more sensitive to salt than *P. camemberti*.

Gamonéu cheese is a type of blue cheese manufactured by traditional methods in the Asturias, northeast Spain. It is made with a mixture of cow's, goat's and sheep's milk. Coagulation occurs at 22–24°C in three to four hours after the addition of animal rennet. Later, lactic acid or mold starter cultures are added. The cheeses are smoked for three to four weeks and ripened at 9–10°C in caves for three to four months. Gonzalez de Llano et al. (1992) assessed many characteristics of this cheese variety including microbial populations. Lactic acid bacteria prevailed in the interior of the cheese. Enterococcus and micrococcus counts were greater towards the end of the ripening process, indicating their importance in determining cheese characteristics during this phase of ripening. The most common lactic acid bacteria found were *Lb. plantarum* followed by *Lb. casei*. Streptococcus and leuconostocs counts were low. Other lactic acid bacteria present were *L. lactis* subsp. *lactis*, *Leuconostoc mesenteroides* subsp. *mesenteroides* and *Leuconostoc paramesenteroides*. Lysis of the lactic acid bacteria is among the phenomena that occur during ripening, which releases proteases and lipases that contribute to the ripening process, in addition to the lactic acid that was produced before autolysis. Additionally, lactic and non-lactic acid bacteria may be inhibited by *P. roqueforti*, which produces inhibitory substances. Engel et al. (1982) studied the synthesis of penicillin by this mold. In this sense, probiotics in blue cheeses may have limited viability because of their sensitivity to the substances produced by the said mold. However, this limitation can be overcome by selecting appropriate species.

Cabrales is a famous Spanish blue cheese of which characteristics are determined by the action of *P. roqueforti*. Cabrales cheese is separated from other blue-veined cheeses as the mold spores are not added during the manufacturing process. Instead, they come from the milk or from the caves where the cheese ripens after drying (Florez et al. 2007). Florez et al. (2007) found that the counts of *G. candidum* lineages were similar to those of *P. roqueforti*, suggesting that the former has an important role in the ripening of this blue cheese variety.

6.4.2 Spoilage microorganisms associated with blue-veined cheeses

As shown above, there are many examples of blue-veined cheeses made with raw milk without a starter culture. However, many cheeses ripened by fungi are produced with pasteurized milk and commercial lactic acid cultures, and use of *Penicillium* species as the secondary culture. In general, yeasts are not a part of starter or adjunct culture in cheese-making, but they grow as contaminants in cheeses and reach counts in excess of 10^6 cfu g^{-1}. Depending on the yeast species or lineage, yeasts can have a negative impact on cheese because their metabolites change the flavor and texture of the final product (Viljoen et al. 2004). Yet, some yeasts contribute positively to the final characteristics of the cheese, as mentioned above regarding the Cabrales and Gamonéu cheeses.

Brown stains, usually caused by microbes, are a common defect of blue cheeses. They may be caused by yeasts, molds, thermophilic bacteria, or even some *P. roqueforti* lineages in cheeses with long ripening stages. Some yeasts responsible for this defect are salt-tolerant, so they are present in high counts (10^4 to 10^6 cfu g^{-1}) in the brine used in the manufacturing process. The molds that contaminate blue cheeses include *Penicillium commune* or *Penicillium nalgiovense* and *Penicillium caseifulvum*, which frequently contaminate the curd, brine and facility surfaces. Good manufacturing processes reduce these problems (Ardö 2007).

6.5 Microbiology of Pasta-Filata Cheeses

Stretched-curd cheeses, such as Mozzarella, are characterized by being stretched in hot water, which gives the curd its characteristic texture before the addition of salt. These cheeses are classified as semi-soft because during the manufacturing, the curd is heated to 55°C or more and submitted to mechanical treatment. Stretching makes the curd fibrous and malleable. Most of the stretched-curd cheeses originated from the Mediterranean region. The most important member of this group is Mozzarella, which originated in southeast Italy and was originally made from buffalo milk.

Buffalo Mozzarella is manually shaped into balls weighing 100 to 300 g. This cheese variety is still manufactured traditionally in Italy, but the majority of the Mozzarella manufacturers in the world use pasteurized, semi-skimmed cow's milk in the production of Mozzarella cheese which is also called pizza cheese, cheese for pizza, or in the North American market, low-moisture part-skim Mozzarella. This type of Mozzarella has a salt content of 1.5 to 1.7%, which is higher than that of buffalo Mozzarella (Fox et al. 2000, Kindstedt et al. 2010, Nath 1992).

The manufacturing of Mozzarella for pizza requires standardizing pasteurized cow's milk to 1.8% fat. Higher fat contents (3.6% or more) are used for manufacturing fresh Mozzarella, also known as table Mozzarella. Pizza cheese is manufactured with 1% to 2% thermophilic starter cultures. These starter cultures contain *Lactobacillus* spp. and *S. thermophilus.* Lactobacilli are not used in the production of fresh Mozzarella because the acidification cannot be as high as that for Mozzarella for pizza. Proteolytic enzymes of lactobacilli may contribute to the functionality of the final product by mildly hydrolyzing casein. In general, the addition of rennet causes the milk to coagulate after it becomes slightly acidic. The curd is cut and treated thermally at 41°C. The whey is drained and the desired texture of the curd is achieved by acidifying it to a pH of 5.1 to 5.3. Mozzarella is stretched in water at about 70°C. The curd is mechanically treated until the desired texture and shape are achieved. Pizza Mozzarella curd receives a more intense mechanical treatment than fresh Mozzarella curd. The curd can be salted during stretching or shaping or by submersion in brine. The hot, plastic curd is usually shaped into rectangular blocks and quickly cooled in cold water or brine. The resulting Mozzarella is generally consumed within a few weeks after manufacture. Prolonged ripening is not desired since it changes the functional properties of the cheese. String cheese is manufactured from Mozzarella, Cheddar, or similar cheeses by heating and stretching the plastic curd and forming long strings one to two centimeters in diameter. These strings are salted in brine, cut in convenient sizes, and packaged. These cheeses are usually marketed for children (Fox et al. 2000, Kindstedt et al. 2010, Nath 1992).

Provolone, a cheese made from cow's milk characteristically shaped like a pear, originated in southeastern Italy. Rennet paste may be used in its manufacture, resulting in a more piquant cheese (piquant Provolone) than normal (sweet Provolone), which is made with rennet extract. Provolone is ripened for two to six months. Caciocavallo is a hard Italian cheese made with cow's milk. The manufacturing process is similar to that of Provolone. The curd is stretched in hot water and salted in brine. Caciocavallo is ripened for three to four months or longer (i.e., more than twelve months, when grated). In Europe, there are yet other varieties of stretched cheeses made from sheep's milk and smoked, such as Ostiepok cheese, native to the Czech

Republic and Slovakia (Fox et al. 2000, Kindstedt et al. 2010, Nath 1992). Stretched-curd and high-salt cheeses are considered specific cheese families because of their unique stretching and ripening in brine characteristics, respectively. However, these two families are interior ripened by the same microbial agents (Kosikowski and Mistry 1997, Fox et al. 2000, Kindstedt et al. 2010, Nath 1992).

6.5.1 Starter cultures for pasta-filata cheeses

The transformation of milk into curd and later into cheese is a complex process that can be divided into two parts: manufacturing (from 5 to 24 hours) and ripening (from two weeks to two years, depending on variety). Microorganisms, whether natural contaminants, starter culture components, or adjunct culture components, play an important role in both stages of pasta-filata cheese manufacturing. These microorganisms directly or indirectly affect the biochemical, chemical and physical processes that occur in the milk, curd and cheese (Kosikowski and Mistry 1997, Piraino et al. 2008, Fox et al. 2000, Kindstedt et al. 2010).

Many studies on the technological properties of lactic acid bacteria isolated from traditional pasta-filata cheeses such as *S. thermophilus, Streptococcus macedonicus, Lb. helveticus, Lb. delbrueckii* subsp. *lactis, Enterococcus* spp. and non-starter lactic acid bacteria (NSLAB) have been published (Piraino et al. 2008, Kindstedt et al. 2010). Piraino et al. (2008) studied the microbiota present in the natural starter culture used for manufacturing cheeses like Caciocavallo, Provolone and Scamorza to determine their technological properties. Sicilian Caciocavallo is an Italian protected-designation-of-origin (PDO) cheese made with raw milk without commercial starter cultures or the use of whey culture. Licitra et al. (2007) studied the microbiota of another Italian, stretched-curd, PDO cheese salted in brine called Ragusano and identified forty different microbial groups, including *L. lactis, Enterococcus* spp., *S. thermophilus, S. macedonicus* and various mesophilic lactobacilli. This cheese is produced in farms in northeast Sicily with raw milk and ripened for six to twelve months. Instead of a commercial starter culture, it uses the lactic acid bacteria coming from raw milk and/or contaminated from the surfaces of the equipment and utensils used in the cheese-making facilities. The microbiological diversity of raw milk for the production of Fiordilatte di Agerola—a stretched-curd cheese produced from raw milk—is high. Coppola et al. (2006) isolated 272 strains from this variety and identified them with molecular techniques. Most of the isolates belonged to the species *L. lactis* and *Lb. helveticus*. Likewise, Parente et al. (1997) analyzed microorganisms isolated from stretched-curd cheeses from Basilicata, southeast Italy, and identified dominant lactobacilli as *Lb. helveticus* and dominant cocci as *Lactococcus* or *Enterococcus*. Gobbetti et al. (2002) studied the microorganisms present in the stretched-curd cheese Caciocavallo Pugliese and its whey culture, and found that completely ripened cheeses had mesophilic bacterial counts of eight log cycles and presumptive staphylococcus counts of six log cycles, which may pose a risk to their safety. These authors also detailed the activities of different proteases in cheese ripening and the production of free fatty acids by esterase from the bacteria present in the product, emphasizing the importance of lipolysis to the characteristic flavor and aroma of ripened cheeses.

6.5.2 Spoilage and pathogenic microorganisms associated with pasta-filata cheeses

Spoilage and pathogenic microorganisms can contaminate the stretched-curd cheeses from multiple sources. According to Fusco et al. (2012), the risk of pathogenic or spoilage bacteria contamination is greater for high-moisture (greater than 60%), low-salt stretched-curd cheeses with high water activity; even a pH of 5.2 is not capable of inhibiting many spoilage and pathogenic microorganisms. These authors point out that although stretching can reduce the count of unwanted microorganisms, its target is to give the curd the desirable texture, not to ensure its safety. If heating is inadequate or sub-lethal, it may result in incomplete inactivation of pathogens. The injured or stressed bacteria are a potential risk and may recover their virulence under the suitable conditions. The use of pasteurized milk or clean raw milk (i.e., microfiltered milk), the use of appropriate starter or adjunct cultures with antagonistic activity to pathogenic microorganisms, and performing good manufacturing practices may contribute to cheese safety and limit the risk of contamination by, for example, *Eschericia coli* O157:H7.

The brine used for manufacturing stretched-curd and other types of cheeses needs to be periodically submitted to decontamination treatments such as boiling, microfiltration, or addition of preservatives, or even be replaced. Yeasts are the main contaminants of brine. The most common yeasts found in Mozzarella and brine in which Mozzarella was kept from the region of Basilicata were *Kluyveromyces marxianus*, *K. lactis*, *Debaryomyces hansenii*, *Candida kefir*, *C. famata*, *C. colliculosa* and *C. catenulate* (Romano et al. 2001). *Saccharomyces cerevisiae* was the most common yeast in the cheeses Caciocavallo Podolico and buffalo Mozzarella.

6.6 Microbiology of Soft (Fresh) Cheeses

The term soft cheese indicates that the consistency of cheese is soft to touch or to pressure applied between fingers, being directly related to the moisture content of the cheese; in this case the cheeses which have higher moisture levels are softer than that with low moisture levels (Farkye and Vedamuthu 2002). Several examples of soft cheeses are available around the world, being part of ordinary diet of the people: Cottage, cream cheese, Minas cheese (Brazil), whey cheese (Portugal) among others.

From a technological point of view, majority of soft cheeses are not submitted for ripening, being ready to consumption just after the processing and stored by refrigeration temperatures. Whole milk, skim milk, cream, whey and combinations thereof are used in the manufacturing of these products. The use of rennet coagulation (employed in most soft cheeses) or direct acidification is a common practice at the dairy industry. Direct acidification is more controllable than biological acidification and, unlike starters, is not susceptible to bacteriophage infection. However, enzymes from starter bacteria are essential in cheese ripening and chemical acidification is used mainly for cheese varieties for which texture is more important than flavor.

6.6.1 Starter cultures used for soft (fresh) cheese-making

Fresh cheeses, with limited shelf-life, have the primary proteolysis, which is performed by the coagulating agents and, to a lesser extent, plasmin, residual coagulants, and enzymes from the starter organisms (Sousa et al. 2001). The starter cultures added during the production of fresh cheeses are mesophilic group including *L. lactis* subsp. *lactis* and *L. lactis* subsp. *cremoris*, with different capacity of producing citrate (Lucey et al. 2011). Diacetyl is a major product of citrate metabolism by lactococci and is desired in many fresh cheese varieties such as Cottage cheese.

6.6.2 Pathogenic and spoilage microorganisms associated with soft (fresh) cheeses

High moisture levels of soft cheeses make them more susceptible to pathogenic and spoilage microorganisms. Several reports have presented the prevalence of *E. coli*, *L. monocytogenes*, *Salmonella* spp. and *S. aureus* in soft cheeses ready for consumption (Kousta et al. 2010). In this sense, the use of pasteurized milk in the manufacture of fresh soft cheeses is essential. As a general rule, good sanitation of the cheese plants, heat treatment of milk, addition of salt, quality of starters noted by the acid production as well as strict control of the operation parameters as storage and processing temperature are important factors to be taken in account along with the cheese processing (Farkye and Vedamuthu 2002). Gram-negative, psychrotrophic microorganisms such as *Pseudomonas* spp., coliform bacteria and, molds and yeasts have been frequently associated with spoilage of fresh cheeses (Fernandes 2009).

6.7 Probiotic cheeses

It is clear that the probiotics are dominated by yogurts and fermented milks, but scientific studies suggest that cheese can be a good but underused alternative food vehicle for the delivery of viable probiotic bacteria conferring health benefits in the host, with specific advantages compared with fermented milks and yogurts such as high cell viability (Ong and Shah 2009, Özer et al. 2009, Grattepanche et al. 2008). Cheese provides a valuable alternative to fermented milks and yogurts as a food vehicle for probiotic delivery, due to certain potential advantages. It creates a buffer against the high acidic environment in the gastrointestinal tract, and thus creates a more favorable environment for probiotic survival throughout the gastric transit, due to higher pH and dense matrix which offer additional protection (Cruz et al. 2009).

Probiotic bacteria used in cheese-making include several species: *Lactobacillus acidophilus, Lb. casei, Lb. johnsonii, Lb. rhamnosus, Lb. reuteri, Lb. delbrueckii* subsp. *bulgaricus, Bifidobacterium bifidum, B. longum, B. brevis, B. infantis* and *B. animalis*. These bacteria are sensitive to low pH and high dissolved oxygen values besides presenting limited proteolytic activity which results in a limited role as starter culture in cheese processing (Granato et al. 2010).

Fresh cheeses appear to be ideally suited to serve as a carrier for probiotic bacteria since they are unripened and their shelf-life is rather limited (Heller et al. 2003). Examples of probiotic fresh cheeses were published elsewhere (Gomes et al. 2011, Özer and Kirmaci 2011, Abadia-Garcia et al. 2013). However, in contrast to the short shelf-life of probiotic fermented milks, yogurts and fresh cheeses, hard cheeses such as Cheddar have long ripening period of up to 1–2 years; hence the development of probiotic cheese requires stringent selection of probiotic strains to maintain viability in the cheese throughout processing, maturation and storage period till consumption (Phillips et al. 2006). In a long-term study carried out by Jatila and Matilainen (2008) from Valio Ltd. (Finland), *Lactobacillus* GG (LGG) counts in hard cheese at the beginning of sales period were followed for five years. The LGG counts remained at the level of 2×10^7 cfu g^{-1} throughout the years with no adverse effect on the overall quality of the product. Cheddar cheese was shown to be a good medium for the growth of *Lb. paracasei* inoculated into cheese milk at rather lower concentrations (i.e., 0.5%) (Ross et al. 2005). Similarly, Daigle et al. (1999) found that *B. infantis* survived very well in Cheddar cheese packed in vacuum-sealed bags kept at 4°C for 84 days and remained above 3×10^6 cfu g^{-1} cheese. It is difficult to generalize the growth trend of probiotic bacteria in cheese as it is affected by a number of parameters including cheese type, production practices of cheese, strain or species of probiotics and symbiosis of probiotics with other cheese microflora. Zehntner (2008), for example, tested two strains of *Lb. gasseri* for their suitability as probiotic additive to semi-hard cheese (Tilsit-type cheese and Swiss-type cheese) together with the starter cultures. It was shown that one strain (strain K7, a bacteriocin-gassericin K7-producing strain) had the ability to remain in concentrations above 10^6 cfu g^{-1} during entire 90 d storage, but the other strain lost its viability within a shorter period.

The major challenge associated with the application of probiotic cultures in the manufacture of probiotic cheese is that the survival of probiotic bacteria during ripening period cannot be predicted with accuracy. Biochemical changes occurring inside the cheese environment such as decreasing water activity, sometimes together with a decrease in pH, create a hostile and stressful environment for the probiotic bacteria (Cruz et al. 2009). Despite this, several ripened probiotic cheeses have been developed, with or without changes in the proteolytic and lipolytic profiles, exerting a positive effect on the overall quality of the cheese (Karimi et al. 2011). An extensive review of technological and health aspects of probiotic cheeses has been published by Özer and Kirmaci (2011). More information about the strains used in the manufacture of probiotic cheeses and characteristics of the final products are available in Chapter 9.

References

Abadia-Garcia, L., A. Cardador, S.T.M. del Campo, S.M. Arvízu, E. Castaño-Tostado, C. Regalado-González, B. García-Almendarez and S.L. Amaya-Llano. 2013. Influence of probiotic strains added to Cottage cheese on generation of potentially antioxidant peptides, anti-listerial activity, and survival of probiotic microorganisms in simulated gastrointestinal conditions. Int. Dairy J. 33: 191–197.

Abraham, S., R. Cachon, B. Colas, G. Feron and J. de Coninck. 2007. E_h and pH gradients in Camembert cheese during ripening: Measurements using microelectrodes and correlations with texture. Int. Dairy J. 17: 954–960.

Alichanidis, E. and A. Polychroniadou. 2008. Characteristics of major traditional regional cheese varieties of East-Mediterranean countries: A review. Le Lait 88: 410–495.

Ardö, Y. 2007. Introduction: What are Blue cheese varieties? pp. 284–286. *In:* P.L.H. McSweeney [ed.]. Cheese Problems Solved. CRC Press/Woodhead Publishing Limited, Cambridge, UK.

Beresford, T.P., N.A. Fitzsimons, N.L. Brennan and T.M. Cogan. 2001. Recent advances in cheese microbiology. Int. Dairy J. 11: 259–274.

Beresford, T.P. 2003. Non-starter lactic acid bacteria (NSLAB) and cheese quality. pp. 448–469. *In:* G. Smit [ed.]. Dairy Processing: Improving Quality. CRC Press/Woodhead Publishing Ltd., Cambridge, UK.

Beresford, T. 2007. What factors affect microbial growth in cheese? pp. 119–123. *In:* P.L.H. McSweeney [ed.]. Cheese Problems Solved. Woodhead Publishing Ltd., Cambridge, UK.

Bintsis, T. and P. Papademas. 2002. Microbiological quality of white-brined cheeses: A review. Int. J. Dairy Technol. 55: 113–120.

Bockelmann, W. 2010. Secondary cheese starter cultures. 2010. pp. 193–230. *In:* B.A. Law and A.Y. Tamime [eds.]. Technology of Cheesemaking. Blackwell Publishing, Oxford, UK.

Broadbent, J.R., M.F. Budinich and J.L. Steele. 2011. Non starter acid lactic bacteria. pp. 639–644. *In:* J.W. Faquay [ed.]. Encyclopedia of Dairy Sciences. Elsevier Science, London, UK.

Burns, P., F. Cuffiaa, M. Milesia, G. Vinderola, C. Meiinardia, N. Sabbagh and E. Hynesa. 2012. Technological and probiotic role of adjunct cultures of non-starter lactobacilli in soft cheeses. Food Microbiol. 30: 45–50.

Cogan, T.M. 2002. Microbiology of cheese. pp. 306–314. *In:* H. Roginski, J.W. Fuquay and P.F. Fox [eds.]. Encyclopedia of Dairy Sciences. Academic Press, London, UK.

Cogan, T.M. and P. Beresford. 2002. Microbiology of hard cheese. pp. 515–560. *In:* R.K. Robinson [ed.]. Dairy Microbiology Handbook. John Wiley and Sons, New York, NY, USA.

Coppola, S., V. Fusco, R. Andolfi, M. Aponte, G. Blaiotta, D. Ercolini and G. Moschetti. 2006. Evaluation of microbial diversity during the manufacture of Fior di Latte di Agerola, a traditional raw milk pasta-filata cheese of the Naples area. J. Dairy Res. 73: 264–272.

Cruz, A.G., F.C.A. Buriti, C.H.B. Souza, J.A.F. Faria and S.M.I. Saad. 2009. Probiotic cheese: Health benefits, technological and stability aspects. Trends Food Sci. Technol. 20: 344–354.

Daigle, A., D. Roy, G. Belanger and J.C. Vuillemard. 1999. Production of probiotic cheese (Cheddar-like cheese) using enriched cream fermented by *Bifidobacterium infantis*. J. Dairy Sci. 82: 1081–1091.

El Soda, M. and M.H. Abd El-Salam. 2002. Cheeses matured in brine. 406–411. *In:* H. Roginski, J.W. Fuquay and P.F. Fox [eds.]. Encyclopedia of Dairy Sciences. Elsevier, London, UK.

El Soda, M., S. Awad and M.H.A. El-Salam. 2011. Cheeses matured in brine. pp. 790–794. *In:* J.W. Fuquay [ed.]. Encyclopedia of Dairy Sciences. Elsevier, Sciences, London, UK.

Engel, G., K.E. von Milczewskli, D. Prokopek and M. Teuber. 1982. Strain-specific synthesis of mycophenolic acid by *Penicillium roqueforti* in blue-veined cheese. Appl. Environ. Microbiol. 43: 1034–1040.

Farkye, N.Y. and R. Vedamuthu. 2002. Microbiology of soft cheeses. pp. 479–515. *In:* R.K. Robinson [ed.]. Dairy Microbiology Handbook. John Wiley and Sons, New York, NY, USA.

Farkye, N.Y. 2004. Cheese technology. Int. J. Dairy Technol. 57: 91–98.

Fernandes, R. 2009. Microbiology Handbook. Dairy Products. Leatherhead Food International Ltd. Surrey, UK.

Florez, A.B., P. Alvarez-Martin, T.M. Lopez-Diaz and B. Mayo. 2007. Morphotypic and molecular identification of filamentous fungi from Spanish blue-veined Cabrales cheese, and typing of *Penicillium roqueforti* and *Geotrichum candidum* isolates. Int. Dairy J. 17: 350–357.

Fox, P.F., T.P. Guinee, T.M. Cogan and P.L.H. McSweeney. 2000. Fundamentals of Cheese Science. Aspen Publishers, Maryland, USA.

Fusco, V., M. Riccardi and G.M. Quero. 2012. Thin agar layer versus most probable number PCR to enumerate viable and stressed *Escherichia coli* O157:H7 and application in a traditional raw milk pasta filata cheese. Int. J. Food Microbiol. 159: 1–8.

Gobbetti, M., M. Morea, F. Baruzzi, M.R. Corbo, A. Matarante, T. Considine, R. Di Cagno, T. Guinee and P.F. Fox. 2002. Microbiological, compositional, biochemical and textural characterisation of Caciocavallo Pugliese cheese during ripening. Int. Dairy J. 12: 511–523.

Gomes, A.A., S.P. Braga, A.G. Cruz, R.S. Cadena, P.C. Lollo, C. Carvalho, J. Amaya-Farfán, J.A.F. Faria and H.M.A. Bolini. 2011. Effect of inoculation level of *Lactobacillus acidophilus* in probiotic cheese on the physicochemical features and sensory performance compared with commercial cheeses. J. Dairy Sci. 10: 4777–4786.

Gonzalez de Llano, D., M. Ramos, A. Rodriguez, A. Montilla and M. Juarez. 1992. Microbiological and physicochemical characteristics of Gamonedo bluec during ripening. Int. Dairy J. 2: 121–135.

Granato, D., G.F. Branco, A.G. Cruz, J.A.F. Faria and N.P. Shah. 2010. Probiotic dairy products as functional foods. Comp. Rev. Food Sci. Safety 9: 455–470.

Grattepanche, F., S. Miescher-Schwenninger, L. Meile and C. Lacroix. 2008. Recent developments in cheese cultures with protective and probiotic functionalities. Dairy Sci. Technol. 88: 421–444.

Gueguen, M. 1992. Les moisissures. pp. 325–370. *In:* R.J. Hermie, J. Lenoir and F. Weber [eds.]. Les Groupes Microbiens d'Intérêt Laitier. Cepil, Paris, France.

Gueguen, M. and J.L. Schmidt. 1992. Levures and *Geotrichum*. pp. 165–220. *In:* H.R.J. Hermie, J. Lenoir and F. Weber [eds.]. Les Groupes Microbiens d'Intérêt Laitier. Cepil, Paris, France.

Heller, K.J., W. Bockelmann, J. Schrezenmeir and M. deVrese. 2003. Cheese and its potential as a probiotic food. pp. 203–225. *In:* E.R. Farnworth [ed.]. Handbook of Fermented Functional Foods. CRC Press, Boca Raton, FL, USA.

Heperkan, D. 2010. Gıdalarda mikroorganizmaların çoğalması ve çoğalmayı etkileyen faktörler. pp. 50–67. *In:* O. Erkmen [ed.]. Gıda Mikrobiyolojisi, Efil Yayinevi, Ankara, Turkey.

Jatila, H. and K. Matilainen. 2008. Probiotic cheese quality. Proceedings of 5th IDF Symposium on Cheese Ripening Symposium, IDF, 9–13 March, Bern, Switzerland. 86 p.

Jay, J.M. 2000. Modern Food Microbiology. 6th ed. Aspen Publishers Inc., Gaithersburg, Maryland, USA.

Karimi, R., A. Mortazavian and A.G. Cruz. 2011. Viability of probiotic microorganisms in cheese during production and storage: A review. Dairy Sci. Technol. 91: 283–308.

Kindstedt, P.S., A.J. Hillier and J.J. Mayes. 2010. Technology, biochemistry andf of pasta filata/pizza cheese. pp. 330–359. *In:* B.A. Law and A.Y. Tamime [eds.]. Technology of Cheesemaking. Wiley-Blackwell Publishing, West Sussex, UK.

Kosikowski, F.V. and V.V. Mistry. 1997. Cheese and Fermented Milk Foods. Vol. I. Origins and Principles, 3rd ed., F.V. Kosikowski L.L.C., Wesport, CT, USA.

Kousta, M., M. Mataragas, P. Skandamis and E.H. Drosinos. 2010. Prevalence and sources of cheese contamination with pathogens at farm and processing levels. Food Control 21: 805–815.

Licitra, G., J.C. Ogier, S. Parayre, C. Pediliggieri, T.M. Carnemolla, H. Falentin, M.N. Madec, S. Carpino and S. Lortal. 2007. Variability of bacterial biofilms of the "Tina" wood vats used in the Ragusano cheese-making process. Appl. Environ. Microbiol. 13: 6980–6987.

Linares, D.M., M. Martín, V. Laderoa, M.A. Alvareza and M. Fernándeza. 2011. Biogenic amines in dairy products. Crit. Rev. Food Sci. Nut. 51: 691–703.

Linares, D.M., B. Del Río, V. Ladero, N. Martínez , M. Fernández, M.C. Matin and M.A. Alvarez. 2012. Factors influencing biogenic amines accumulation in dairy products. Front Microbiol. 28: 180.

Lucey, J.A. 2011. Rennet-induced coagulation of milk. pp. 579–584. *In:* J.W. Fuquay [ed.]. Encyclopedia of Dairy Sciences. Elsevier Sciences, London, UK.

McSweeney, P.L.H. 2011. Biochemistry of cheese ripening. pp. 667–674. *In:* J.W. Fuquay [ed.]. Encyclopedia of Dairy Sciences. Elsevier Sciences, London, UK.

Moatsou, G. and A. Govaris. 2011. White-brined cheeses: A diachronic exploitation of small ruminants milk in Greece. Small Rumin. Res. 101: 113–121.

Nath, K.R. 1992. Cheese. pp. 161–255. *In:* Y.H. Hui [ed.]. Dairy Science and Technology Handbook. Vol. 2. Product Manufacturing. Wiley-VCH, New York, NY, USA.

Ong, L. and N.P. Shah. 2009. Probiotic Cheddar cheese: Influence of ripening temperatures on survival of probiotic microorganisms, cheese composition and organic acid profiles. LWT- Food Sci. Technol. 42: 1260–1268.

Özer, B.H. 1999. Microflora of white-brined cheeses. pp. 397–403. *In:* R.K. Robinson, C.A. Batt, P.D. Patel [eds.]. Encyclopaedia of Food Microbiology. Academic Press, London, UK.

Özer, B., H.A. Kırmacı, E. Şenel, M. Atamer and A.A. Hayaloglu. 2009. Improving the viability of *Bifidobacterium bifidum* BB-12 and *Lactobacillus acidophilus* LA-5 in white-brined cheese by microencapsulation. Int. Dairy J. 19: 22–29.

Özer, B.H. and H.A. Kirmaci. 2011. Technological and health aspects of probiotic cheese. pp. 1–42. *In:* R.D. Foster [ed.]. Cheese: Types, Nutrition and Consumption. Nova Science Publishers, Inc., Hauppauge, NY, USA.

Parente, E., M.A. Rota, A. Ricciardi and F. Clementi. 1997. Characterization of natural starter cultures used in the manufacture of Pasta Filata cheese in Basilicata (Southern Italy). Int. Dairy J. 27: 775–783.

Phillips, M., K. Kailasapathy and L. Tran. 2006. Viability of commercial probiotic cultures (*L.acidophilus, Bifidobacterium* sp., *L. casei, L. paracasei* and *L. rhamnosus*) in cheddar cheese. Int. J. Food Microbiol. 108: 276–280.

Piraino, P., T. Zotta, A. Ricciardi, P.L.H. McSweeney and E. Parente. 2008. Acid production, proteolysis, autolytic and inhibitory properties of lactic acid bacteria isolated from pasta filata cheeses: A multivariate screening study. Int. Dairy J. 18: 81–92.

Powell, I.B., M.C. Broome and G.K.Y. Limsowtin. 2011. Starter cultures: General aspects. pp. 552–558. *In:* J.W. Fuquay [ed.]. Encyclopedia of Dairy Sciences. Elsevier Science, London, UK.

Ray, B. 2004. Fundamental Food Microbiology, 3rd ed., CRC Press, New York, NY, USA.

Romano, P., A. Ricciardi, G. Salzano and G. Suzzi. 2001. Yeasts from water buffalo Mozzarella, a traditional cheese of the Mediterranean area. Int J. Food Microbiol. 69: 45–51.

Ross, R.P., C. Desmond, G.F. Fitzgerald and C. Stanton. 2005. Overcoming the technological hurdles in the development of probiotic foods. J. Appl. Microbiol. 98: 1410–1417.

Settanni, L. and G. Moschetti. 2010. Non-starter lactic acid bacteria used to improve cheese quality and provide health benefits. Food Microbiol. 27: 691–697.

Sousa, M.J., Y. Ardö and P.L.H. McSweeney. 2001. Advances in the study of proteolysis during cheese ripening. Int. Dairy J. 11: 327–345.

Viljoen, B.C., A. Knox, L.R. Beuchat, T. Deak, M. Malfeito-Ferreira, T.K. Hansen, A. Hugo, M. Jakobsen, V. Loureiro, A. Lourens-Hattingh and R. Vasdinnyei. 2004. An inter-laboratory evaluation of selective media for the detection and enumeration of yeasts from blue-veined cheese. Int. J. Food Microbiol. 13: 9–14.

Zehntner, U. 2008. Behaviour of the probiotic strain *Lactobacillus gasseri* K7 in ripened semi-hard cheese. Agrarforschung 15: 194–197.

CHAPTER 7

Primary Biochemical Events During Cheese Ripening

*A.A. Hayaloglu** and *P.L.H. McSweeney*

7.1 Introduction

Cheese ripening is a complex set of biochemical events that involves at least three flavor generating pathways: (i) lipolysis, (ii) proteolysis and (iii) metabolism of residual lactose and of lactate and citrate. It is affected by many factors including type and amount of coagulant, presence of starter, adjunct starter and non-starter microorganisms, addition of enzymes used to accelerate ripening and environmental conditions during manufacturing and ripening. Many different cheese varieties are essentially similar at the end of manufacturing stage in terms of chemical composition and texture; however, a number of changes occur during ripening based on ripening conditions and these influence the flavor and aroma of cheese. Changes to the texture and flavor of the cheeses which take place during manufacturing and ripening are primary reactions and largely affect biochemistry of cheese. Also, residual clotting enzyme, starter or non-starter lactic acid bacteria (NSLAB), fungi and their enzymes influence cheese quality. As it is a wide research field, people from many different disciplines study on cheese. So, collaborative studies have been done in many laboratories to understand the biochemical and physical changes during ripening.

Any change during manufacture can cause changes in the ripening characteristics of cheese. So, to evaluate a cheese on a plate, one should take into account many factors from farm to plate. In this chapter, ripening agents and biochemical reactions including proteolysis, lipolysis and metabolism

*Corresponding author

of residual lactose and of lactate and citrate during ripening of cheeses are reviewed with examples given of different varieties of cheeses.

7.2 Ripening agents in cheese

7.2.1 Proteolytic agents

Ripening agents are coagulant (rennet or rennet substitute), indigenous enzymes from milk, enzymes from starter or non-starter microorganisms and from secondary starter microorganisms. The ripening agents and their roles are summarized in Fig. 7.1. Chymosin (EC 3.4.23.4) is a common coagulant for cheese manufacture and is extracted from the abomasums of milk-fed calves, kids or lambs. The role of the coagulant is to destabilize the colloidal structure of casein by cutting the peptide bond between Phe_{105}-Met_{106} in κ-casein (Fig. 7.2) (Fox and McSweeney 1996, Fox et al. 1996). This peptide bond is more sensitive to chymosin action than the other bonds in milk proteins (Sousa et al. 2001). Most of coagulant used in cheese manufacture is lost in whey due to its high solubility in water; however, a low level of the coagulant activity (about 4–6%) is retained in the curd based on factors including type of coagulant, cooking temperature of the curd, draining pH and the level of moisture in the curd (Fox et al. 1993, Upadhyay et al. 2004a). Residual rennet plays a crucial role in the initial proteolysis of casein. Residual rennet contributes to the hydrolysis of other casein fractions such as α_{s1}-, α_{s2}- or β-caseins during ripening. NaCl concentration of more than 5% strongly inhibits the action of chymosin on β-casein, but it is completely inhibited by 10% NaCl (Fox and McSweeney 1996). α_{s1}-Casein is cleaved

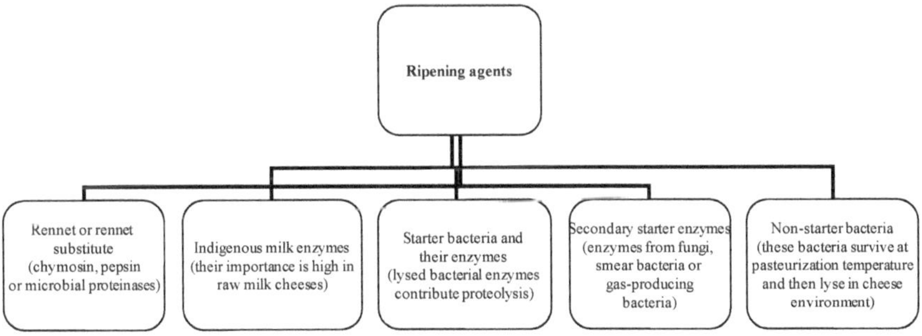

Figure 7.1 Ripening agents in cheese.

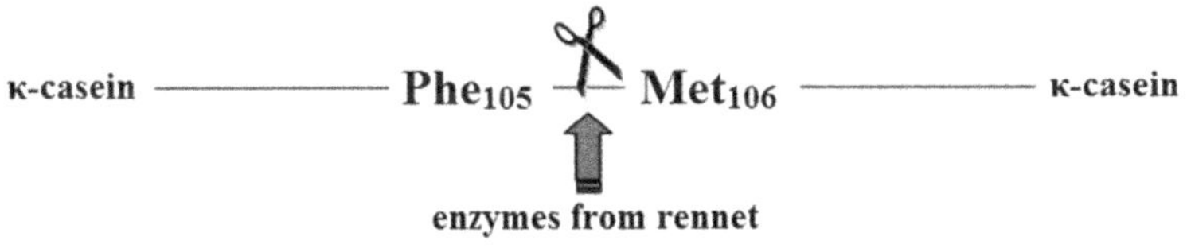

Figure 7.2 Hydrolysis of κ-casein by rennet. From Hayaloglu and Özer (2011).

by chymosin at Phe_{23}-Phe_{24} and the fragment (f1-23) is then hydrolyzed by starter bacteria (Crow et al. 1993, Law et al. 1993). The hydrolysis of α_{s1}-casein is not as negatively influenced by salt concentration as is the hydrolysis of β-casein. Indeed, NaCl at 5% in cheese will stimulate the hydrolysis of α_{s1}-casein and it was reported that α_{s1}-casein was hydrolyzed even at 20% NaCl (Fox et al. 1993). α_{s2}-casein is more resistant to hydrolysis by chymosin and relatively few bonds are susceptible to the action of this enzyme (Fox et al. 1993, Fox and McSweeney 1996). Although *para*-κ-casein has several potential chymosin cleavage sites, it does not appear to be hydrolyzed either in solution or in the cheese environment (Upadhyay et al. 2004a). Other clotting agents sometimes used in cheese manufacture include pepsin (EC 3.4.23.1) and it also contributes to cheese ripening; however, its proteolytic activity is higher than that of chymosin. The ratio of chymosin:pepsin is high (up to 9:1) in milk-fed calves or young ruminant, the ratio can reduce 1:1 if the ruminant is fed with foods other than milk.

Due to a limited supply of traditional calf rennet, several proteases from animal, microbial or plant origin were investigated as substitute for calf rennet. For this purpose, proteases from *Rhizomucor miehei*, *Rhizomucor pusillus* and *Cryphonectria parasitica* have been used in commercial practice (Fox and McSweeney 1996). The coagulant from *R. miehei* is the most used and has close specificity to chymosin. *C. parasitica* proteinase has a higher degree of proteolysis on ovine casein than does *R. miehei* protease (Trujillo et al. 2000). Also, genetically modified microorganisms have been used for the production of calf chymosin by commercial companies such as DSM (the Netherlands) or Chr. Hansen (Denmark). *Kluyveromyces marxianus* var. *lactis*, *Aspergillus niger* var. *awamori* or *Escherichia coli* have been used to express the gene cloning for calf chymosin (Fox et al. 2000). Dried flowers of *Cynara cardunculus* have been used for centuries to coagulate milk for certain varieties in Spain and Portugal (i.e., Iberian Peninsula) including Serra da Estrela, Guia and Los Pedroches (Sousa et al. 2001). The activity of extracts from these flowers has been studied extensively in solutions of bovine, ovine and caprine caseins (Sousa et al. 2001, Sousa and Malcata 2002). The extract contains two proteinases, cardosin A and cardosin B, and their proteolytic specificity resembles that of chymosin (Sousa and Malcata 2002).

Enzymes from lactic acid bacteria (LAB) including mesophilic lactococci and leuconostocs, thermophilic lactobacilli and *Streptococcus thermophilus* contribute to secondary proteolysis in cheese. These microorganisms are used for the manufacture of cheese as starter cultures for acid production or as secondary starters for development of texture or flavor. The principal role of the starter culture is to produce acid and decrease in pH; however, secondary starter cultures do some other tasks including development of flavor, modifying ripening, or changing color and/or appearance of cheese. The LAB are weakly proteolytic, but they have a broad spectrum of proteinase/peptidase system that catalyzes the formation of various peptides and amino acids (Visser 1993, Bockelmann 1995, Fox et al. 1996, Fox and McSweeney 1996).

Biochemical, molecular and genetic studies have focused on starter lactococci as this genus is commonly used in many varieties of cheese (Meijer 1997, Rijnen et al. 1999, Ayad et al. 2000, Yvon et al. 2000). The genus *Lactococcus* has a cell-envelope associated proteinase [CEP, lactocepin (EC 3.4.21.96)] and this enzyme hydrolyzes caseins and casein-derived peptides, but the specificities of the enzymes vary depending on gene. The LAB have different peptidase, aminopeptidase, iminopeptidase, dipeptidase, tripeptidase and endopeptidase activities (for details see Sousa et al. 2001, Yvon and Rijnen 2001, Parente and Cogan 2004). Some thermophilic lactobacilli have been used as starter or adjunct cultures in the manufacture of various cheeses and their enzyme systems generally resemble those of *Lactococcus* spp. The most common thermophilic lactobacilli are *Lb. helveticus, Lb. acidophilus, Lb. delbrueckii* subsp. *bulgaricus* and *Lb. delbrueckii* subsp. *lactis*. *S. thermophilus* is used for the manufacture of a number of cheeses including Swiss (Parente and Cogan 2004) and brine-ripened cheeses like Feta-type or Turkish white (Beyaz peynir) cheese (Bintis and Papademas 2002, Hayaloglu et al. 2002).

The population of starter bacteria declines during the first stage of ripening, and NSLAB increase and gradually dominate the viable microflora of the most hard cheeses. NSLAB are usually present in cheese at low numbers (<50 cfu g^{-1}) at the first stage of ripening, grow rapidly to reach about 10^7 cfu g^{-1} within four weeks and then their number remains unchanged thereafter (Folkertsma et al. 1996). Extracellular enzymes from lysed NSLAB in cheese have an importance to proteolysis, especially secondary proteolysis. Volatile profiles and levels of proteolysis are generally higher in experimental cheeses made using NSLAB adjuncts or from raw milk (which develop higher numbers of NSLAB) than in control cheeses (Lau et al. 1991, McSweeney et al. 1993, Lynch et al. 1997, Gomez et al. 1999, Hayaloglu and Brechany 2007, Hayaloglu et al. 2010).

Secondary microorganisms play an important role in proteolysis and flavor development in cheese and, in varieties where they are present, they can dominate ripening. The most common microorganisms used as secondary starters in cheese are *Lactobacillus* spp. (e.g., *Lb. helveticus, Lb. casei* when used as adjuncts), yeasts (*Kluyveromyces lactis, Saccharomyces* spp., *Debaryomyces hansenii*), molds (*Penicillium roqueforti, P. camemberti, Geotrichum candidum*), *Propionibacterium* spp., *Micrococcus* spp., *Enterococcus* spp., and *Brevibacterium linens*. These microorganisms have a broad range of enzyme systems which catabolize both milk fat and casein-derived peptides and contribute to ripening indices and flavor development. The ripening process is very fast in surface-ripened and/or mold-ripened cheeses due to the complex bacterial or fungal flora on their surface. The complex bacterial flora including *Arthrobacter* spp., *Brevibacterium* spp., *Corynebacterium* spp., *Micrococcus* spp. and *Staphylococcus* spp. produces the pungent aroma and red-orange color of smear-ripened cheeses.

A broad spectrum of yeasts was isolated from Brie and Camembert cheeses as reported by Viljoen et al. (2003). Two yeast species (*Debaryomyces hansenii* and *Yarrowia lipolytica*) were most commonly isolated, but other species were also found including *Torulaspora delbrueckii, Rhodotorula mucilaginosa, R. minuta* and various species of *Candida* (Viljoen et al. 2003). However, yeasts contribute to spoilage or flavor in some cheeses especially surface or brine-ripened cheeses (Fleet 1990, Cantor et al. 2004) due to their tolerance to low pH and high NaCl concentration and their high proteolytic ability (Wyder and Puhan 1999a,b). *Penicillium roqueforti* and *P. camemberti* secrete aspartyl and metalloproteinases and these enzymes act on α_{s1}- and β-casein. These molds also produce amino peptidases, carboxypeptidases and intracellular acid proteinases (Fox et al. 1996, Sousa et al. 2001). *Brevibacterium linens* has a strong (both extracellular and intracellular) proteolytic enzyme system and lysis of *B. linens* cells liberates intracellular enzymes (Rattray and Fox 1999). *Brevibacterium* is commonly used for the ripening of smear cheese due to its high proteolytic activities, contribution of flavor of smear-ripened cheeses and characteristic appearance like bright orange pigments (Bockelmann and Hoppe-Seyler 2001) although added strains may become out-competed by wild microorganisms.

7.2.2 Lipolytic agents

Oxidative reactions are quite limited in cheese due to its low redox potential and low levels of polyunsaturated fatty acids (PUFA) in milk fat. Due to low level of oxidative changes in cheese, many studies focused on hydrolytic changes (lipolysis) in cheese. Levels of lipolysis in cheese are not extensive except for Blue-type and some varieties made using rennet paste which also contains lipases. Lipolytic agents in cheese originate from milk, coagulant, starter, secondary starter and non-starter bacteria and, in some cases, exogenous lipases.

Raw milk contains an indigenous lipoprotein lipase (LPL) which is associated with the casein micelles. Any damage to the milk fat globule membrane by agitation, pumping, homogenization, foaming or improper milking techniques can activate LPL by allowing it access to its substrate (Fox et al. 2000, Collins et al. 2003). Bovine milk contains 10–20 nmol l^{-1} lipase; however, it is activated to some extent by damaging processes mentioned above. High lipase activity is observed in raw milk cheeses (McSweeney et al. 1993, Shakeel-Ur-Rehman et al. 2000, Buffa et al. 2001);

however, its activity is limited in pasteurized milk cheeses due to extensive inactivation of lipase on heating milk at 72°C for 15 s (Deeth and FitzGerald 1995). For the complete inactivation of lipase, higher temperatures (78°C for 10 s) are required (Collins et al. 2004). LPL is not specific for a type of fatty acid, but is specific for residues esterified at the *sn*-1 and *sn*-3 positions of mono-, di- or triglicerydes. Thus, short-chain fatty acids are liberated preferentially by the action of LPL, especially in raw milk cheeses, since butyric acid is found predominantly at the *sn*-3 position.

None of the coagulants used in the manufacture of cheese contains lipase except for rennet paste which contains pregastric esterase (PGE) and is used for the manufacture of some cheeses (e.g., Provolene, Pecorino Romano). PGE is responsible for a piquant flavor that develops in these cheeses (Fox and Stepaniak 1993, Collins et al. 2004). PGE is highly specific for short chain fatty acids esterified at the *sn*-3 position and high concentration of highly flavored short- and medium-chain acids are produced after PGE action (Fox and Stepaniak 1993, Collins et al. 2004). There are some differences amongst PGEs from calf, kid or lamb and they produce cheeses with slightly different flavor characteristics (Fox et al. 2000). Rennet paste is prepared using abomasums of some ruminants including calves, kids or lambs during or just after suckling. The rennet paste is not widely used due to often uncontrolled hygiene during its manufacture. In Turkey, home-made rennet is used in the manufacture of Tulum cheeses (as reviewed by Hayaloglu et al. 2007). For this purpose, cleaned young calves' stomachs are air-dried while shaded from the sun, cut into slices, and then placed in whey containing ca. 10% (w/v) NaCl. After 1–2 weeks, the rennet extract (prepared by blending macerated stomach slices in the NaCl solution) is clarified (filtered) using a cotton cloth and the filtrate is used as a coagulant.

Lipases and esterases from LAB contribute to lipolysis in cheese and this was shown by making starter-free cheese using gluconic acid-δ-lactone as acidifying agent by Reiter et al. (1967). The lipolytic activity of LAB (e.g., *Lactococcus* spp., *Lactobacillus* spp.) is weaker than those of many other bacteria (e.g., *Pseudomonas, Acinetobacter, Flavobacterium*). Small differences are present between cheese starters; *Lactococcus lactis* subsp. *cremoris* has a higher lipolytic activity than *L. lactis* subsp. *lactis* (Kamaly et al. 1990). Also, there are inter-strain differences between enzymes released via lysis of *L. lactis* subsp. *cremoris* AM2 or *L. lactis* subsp. *cremoris* HP (Collins et al. 2004). These authors pointed out that highly autolytic strain (AM2) developed significantly higher levels of some free fatty acids in cheese than less autolytic strain (HP). It was reported that the lipolysis in Cheddar or Dutch-type cheeses originates from the lipase or esterase activities of LAB. Opportunities can be provided by increasing the numbers of starter bacteria or extending ripening time of cheese (Upadhyay et al. 2004a). Also, a number of studies have been performed on the lipolytic and esterolytic activities of *Lb. helveticus, Lb. acidophilus, Lb. delbrueckii* subsp. *bulgaricus, Lb. delbrueckii* subsp. *lactis, Lb. plantarum, Lb. casei* subsp. *casei, Lb. fermentum* and *S. thermophilus* (Upadhyay et al. 2004a).

Extensive lipolysis has been recorded for surface or mold-ripened cheeses due to highly lipolytic activity of some bacteria (e.g., *Brevibacterium linens, Corynebacterium* spp., *Microbacterium* spp.), molds (e.g., *Penicillium roqueforti, P. camemberti, Geotrichum candidum*) or yeasts (*Debaryomyces hansenii, Yarrowia lipolytica*). In Blue-type of cheeses, levels of free fatty acid may be increased to 25% by action of fungi. However, acids formed can be neutralized by increasing pH (due to lactate metabolism and formation of alkali) and the new pH value may be optimum for lipase activity (Fox et al. 2000, Collins et al. 2003). For example, *P. roqueforti* produces two extracellular lipases and these have pH optima of 7.5–9.5, while a lipase produced from *P. camemberti* is optimally active at pH 9.0 (McSweeney 2004).

7.3 Proteolysis in cheese

Proteolysis in cheese is a series of biochemical reactions catalyzed by the action of coagulant enzymes, indigenous milk proteinases, and enzymes from starters, non-starter and secondary starter organisms (Fox 1989, Visser 1993). Primary proteolysis of the caseins in cheese is usually catalyzed by the coagulant retained in the curd, the level of which is dependent on pH at drainage, cook temperature and moisture content of the curd. Apart from the activity of coagulant, plasmin and enzymes from starter or non-starter microorganisms break down the caseins to lower molecular weight peptides and amino acids (Fig. 7.3). Proteolysis has been expressed by determining the amount of nitrogen soluble in water, pH 4.6 buffers, 12% trichloroacetic acid or 5% phosphotungstic acid. Also, urea-PAGE patterns of caseins or RP-HPLC profiles of peptides during ripening are routinely determined. A number of factors including manufacturing process, salt concentration, coagulant type or ratio, starter type, presence of non-starter microorganism, ripening time or temperature influence the proteolysis levels of cheese. So, different varieties of cheese have different ripening indices expressed by nitrogenous fractions and concentrations of amino acids (Table 7.1).

Primary proteolysis is mainly catalyzed by residual coagulant and, to a lesser extent, plasmin, while secondary proteolysis results from the action of proteinases and peptidases of starter or non-starter bacteria that convert casein to lower molecular weight peptides and free amino acids (Kok 1993, Visser 1993, Broadbent et al. 2002). Visser and de Groot-Mostert (1977)

Table 7.1 Proteolysis levels in some cheese varieties.

Type of cheese	Characteristics	WSN or pH 4.6-SN (% of TN)	TCA-SN (% of TN)	PTA-SN (% of TN)	Total FAA (mg Leu g^{-1} cheese)	References
Civil	Brined	5.6–8.4	2.5–3.5	-	-	Cambaztepe et al. (2009)
Beyaz	Brined	20.9	13.5	3.2	10.1	Hayaloglu et al. (2005)
Otlu (Herby)	Brined	26.5	10.2	5.0	-	Tarakci et al. (2005)
Kasar	Pasta-filata	33.3	4.9	1.4	0.112[a]	Sahan et al. (2008)
Kasar	Pasta-filata	16.6	3.2	1.6	-	Yasar and Guzeler (2011)
Kasar	Pasta-filata	18.7	9.0	-	18.4	Hayaloglu (2009)
Kuflu	Blue-type	21.9[b]	16.9	-	19.3	Hayaloglu et al. (2008)
Ezine	Brined	20.0	8.9	2.6	0.5	Hayaloglu and Karabulut (2013)
Feta	Brined	19.5	16.1	3.8	-	Michaelidou et al. (2003)
Feta	Brined	19.5	16.0	5.0	-	Pappa and Anifantakis (2001)
Tilsit	Smear ripened	22.6	-	-	-	Hannon et al. (2004)
Kefalogreviera	Hard	19.9	11.1	6.0	-	Michaelidou et al. (2007)
White-brined	Brined	25.9	16.5	2.8	-	Tzanetakis et al. (1995)
Xinotyri goat	Hard, acid-curd	6.8	5.1	2.0	0.6	Bontinis et al. (2012)
Cacioricotta	Goat	11.3	10.8		5.3	Faccia et al. (2012)
Feta	Brined	18.5	15.8	4.2	-	Moatsou et al. (2002)
Gorgonzola-type	Blue	40.3	18.8	8.9	-	Seratlic et al. (2011)
Gorgonzola	Blue	53.7	44.7		15.4	Gobbetti et al. (1997)
Roquefort	Blue	72.8	51.3	29.1	0.36[c]	Fernandez-Salguero (2004)
Stilton	Blue	56.0	41.6	16.5	0.27[c]	Fernandez-Salguero (2004)
Stilton	Blue	62.4	23.0	4.8	-	Madkor et al. (1987)

Picon Bejes-Tresviso	Blue	70.8	62.0	21.8	-	Prieto et al. (2000)
Camembert	Surface-ripened	18.2	-	-	21.0	O'Sullivan et al. (2005)
UF White cheese	UF brine-ripened	24.9	8.9	-	0.67	Al-Otaibi and Wilbey (2005)
Manchego	Brined	15.6	10.0	2.8	-	Gomez et al. (1999)
Manchego	Brined	19.8	14.8	4.7	6.4	Pavia et al. (2000)
Cheddar	Hard	27.5	-	3.6	-	Law et al. (1992)
Cheddar	Hard	20.4	-	1.1	7.7	Hannon et al. (2003)
Malatya	Boiled and brined curd	7.7	1.5	0.9	0.13	Hayaloglu and Karabulut (2013)
Halloumi	Boiled and brined curd	2.7	1.2	0.9	0.04	Hayaloglu and Karabulut (2013)

[a] Expressed as absorbance values at 507 nm, [b] Expressed as pH 4.6-SN (% of TN), [c] Expressed as g Tyr 100 g^{-1} of cheese

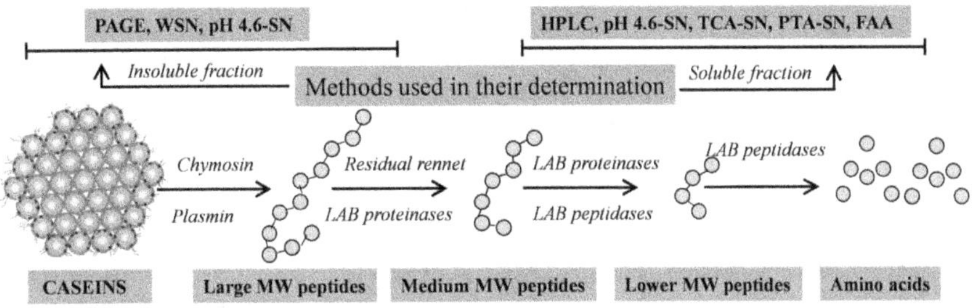

Figure 7.3 Degradation of casein and determination of proteolysis in cheese. From Hayaloglu and Özer (2011). With permission of Sidas Publishing.

reported that the level of β- and α_{s1}-casein hydrolysis (on urea-PAGE) in six mo-old aseptic rennet-free cheeses were 30% and 50%, respectively. Lactic acid bacteria have limited hydrolysis of casein in comparison to coagulant and initial hydrolysis of casein is achieved by coagulant; however, casein-derived peptides are hydrolyzed by lactic acid bacteria (Hayaloglu et al. 2004). The hydrolysis of κ-casein leads to coagulation of milk due to loss of colloidal stability of the milk proteins. Cleavage of κ-casein at Phe_{105}-Met_{106} yields *para*-κ-casein (f1–105) which remains in the curd with α_{s1}-, α_{s2}- and β-caseins and glycomacropeptides (f106–169) which are lost in the whey.

The casein fractions, including α_{s1}-, α_{s2}- and β-caseins, are not hydrolyzed during the coagulation of milk, but are hydrolyzed during cheese ripening (Fox et al. 2000). Cleavage of α_{s1}-casein at Phe_{23}-Phe_{24} by chymosin during ripening may cause some softening in cheese texture early in ripening (Lawrence et al. 1987, Fenelon et al. 1999) although changes in calcium equilibrium are probably of more significance (O'Mahony et al. 2005). The small peptide formed (f1–23) is converted to lower molecular weight (MW) peptides or amino acids by starter or non-starter proteinases and peptidases. Relatively little hydrolysis of β-casein occurs in cheese but chymosin action has been detected at the site Leu_{192}-Tyr_{193} under conditions of low ionic strength. β-casein is also hydrolyzed by plasmin at the sites Lys_{28}-Lys_{29}, Lys_{105}-His_{106} and Lys_{107}-Glu_{108}, yielding β-CN(f29–209) (γ_1-CN), α-CN(f106–209) (γ_2-CN) and β-CN(f108–209) (γ_3-CN). α_{s2}-casein and *para*-κ-casein are very resistant to hydrolysis by chymosin (Fox et al. 2000) although α_{s2}-casein is degraded by plasmin during ripening.

Due to increased demand for calf chymosin because of increased cheese production, a search commenced for rennet substitutes in the mid-20th century. For this purpose, alternative animal (lamb or kid rennets and porcine or chicken pepsins), plant (examples enzymes from fig or *Cynara cardunculus* L.) and microbial rennets or enzymes produced by genetically engineered microorganisms have been developed. These enzymes have been used for the manufacture of different varieties of cheese alone or in mixtures by some researchers (e.g., Alichanidis et al. 1984, Guinee and Wilkinson

1992, Fox and McSweeney 1997, Broome and Limsowtin 1998, Fox et al. 2000, Sousa et al. 2001, Sousa and Malcata 2002). Hydrolytic behavior of microbial enzymes is similar to chymosin; however, some differences are detected in proteolysis in cheese by determining soluble nitrogen fractions and gel electrophoretic patterns. Some microorganisms including *Bacillus subtilis, B. cereus, B. polymyxa, Penicillium expansum, Rhizomucor miehei, R. pusillus, Cryphonectria parasitica, Fusarium moliniforme, Aspergillus versicolor* or *Irpex lactis* have been studied for the extraction of possible milk clotting enzymes; however, studies have concentrated on *R. miehei, R. pusillus* and *Cryphonectria parasitica*. Enzymes from these three microorganisms have been successfully used for many years in the manufacture of various cheeses (Yasar and Guzeler 2011). It has been emphasized that enzymes of microbial origin have higher levels of proteolytic action than chymosin and have non-specific actions on caseins (Ustunol and Hicks 1990).

Studies showed that the cheeses made using microbial rennets had different levels of proteolysis and sensory changes during ripening in comparison to the cheeses made using calf chymosin (Alichanidis et al. 1984, Yetismeyen et al. 1998, Yasar and Guzeler 2011). Much more β-casein hydrolysis has been detected in Feta (Alichanidis et al. 1984), Cheddar (Bogenrief and Olson 1995, Broome et al. 2006) and Kasar (Yasar and Guzeler 2011) cheeses made using enzymes from *C. parasitica*. Proteinases from *C. parasitica* produce a series of degradation products on β-casein with lower electrophoretic mobilities than β-casein. Yasar and Guzeler (2011) determined that the protease from *C. parasitica* hydrolyzed β-casein more than recombinant chymosin and *R. miehei* protease (Fig. 7.4). This protease

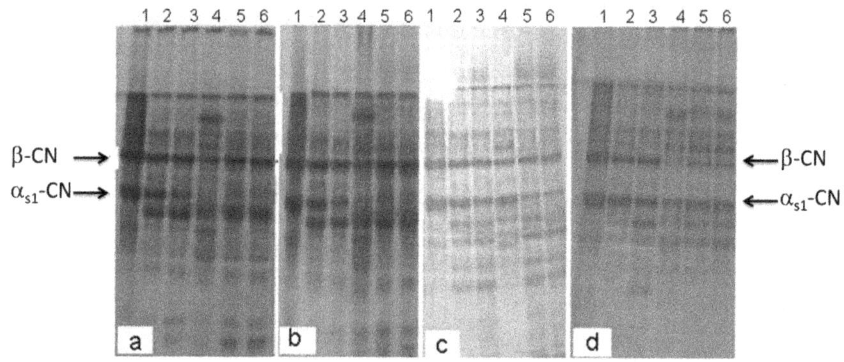

Figure 7.4 Urea-polyacrylamide gel electrophoretograms of Kasar cheeses calf rennet (a), chymosin (b), *Rhizomucor miehei* protease (c) and *Cryphonectria parasitica* protease (d). Lane 1: Na-caseinate; lanes 2–6: Kasar cheeses after 1, 15, 30, 60 or 90 days of ripening period, respectively. From Yasar and Guzeler (2011). With permission of Wiley & Sons Inc.

Color image of this figure appears in the color plate section at the end of the book.

also cleaves κ-casein at Ser_{104}-Phe_{105} rather than Phe_{105}-Met_{106}, which is cleaved by chymosin and *R. miehei* proteinase (Sousa et al. 2001).

In addition to the specificity on casein and their clotting strength, the thermal stability of these enzymes is different (Fig. 7.5). Genetically engineered microorganisms have been used increasingly since 1990s for coagulant production and the gene for chymosin has been cloned and inserted into microorganisms such as *Kluyveromyces marxianus* var. *lactis*, *Aspergillus niger* var. *awamori*, *A. nidulans*, *Escherichia coli*, *Saccharomyces cerevisiae* and *Trichoderma reesei* (Fox et al. 2000, Sousa et al. 2001). This led to the development of fermentation-produced chymosins which are commercially available under the names Maxiren (DSM Food Specialties, The Netherlands) and Chymax (Chr. Hansen, Denmark). Similar cheese-making performances and technological properties including cheese yield, chemical composition of the resultant cheeses, levels of total solids in the whey, ripening indices and electrophoretic patterns of the cheeses made using calf rennet or recombinant chymosin have been determined by performing comparative studies (Yasar and Guzeler 2011). Recently, fermentation-produced camel chymosin has been studied and compared with calf chymosin by Bansal et al. (2009). These authors found no significant differences between cheeses made using these coagulants in terms of gross chemical composition, pH and individual free amino acids with the exception of Ile, His and Lys; however, primary proteolysis was lower in the cheese made by camel chymosin and large quantitative differences were also determined between the cheeses (Bansal et al. 2009).

A more recent study on camel chymosin was performed by Møller et al. (2012) to compare its proteolytic and kinetic specificity with bovine chymosin by using capillary electrophoresis and reversed-phase HPLC-MS techniques. It was found that these enzymes had identical specificities, cleaving the Phe_{105}-Met_{106} bond of κ-casein to produce *para* κ-casein and caseinomacropeptide. The authors also studied the kinetic parameters of these enzymes by Michaelis-Menten model validation and reported that

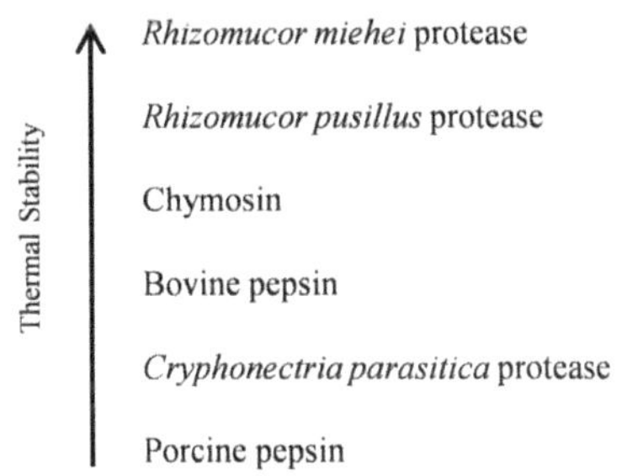

Figure 7.5 Thermal stability of coagulant used in cheese manufacture. From Hayaloglu and Ozer (2011). With permission of Sidas Publishing.

camel chymosin bound κ-casein with ~30% lower affinity (K_M) and exhibited a 60% higher turnover rate (k_{cat}), resulting in ~15% higher catalytic efficiency (k_{cat}/K_M) as compared to bovine chymosin (Møller et al. 2012). In another study, Kappeler et al. (2006) pointed out that camel chymosin exhibited a 70% higher clotting activity for bovine milk compared to bovine chymosin and has only 20% of the unspecific protease activity of bovine chymosin. The authors also reported that camel enzyme is more thermostable than bovine chymosin.

Milk contains a plasmin system consisting of plasmin, plasminogen, plasmin inhibitors, plasminogen activators and inhibitors of plasminogen activators. Most of plasmin and/or plasminogen in milk are retained in cheese and contribute to cheese ripening under suitable conditions for plasmin action (Fox et al. 1993). Plasmin activity may differ based on cheese variety, depending on the heat treatment applied to the milk and pH of the curd or cheese. Higher plasmin activities were determined in Mozzarella which scalded at 55–60°C or Swiss cheeses (cooked at ~54°C) than in Cheddar which is cooked at 39–40°C. This may be explained by high resistance of plasmin against heat treatment and thermal inactivation of heat-labile inhibitors (Upadhyay et al. 2004a). The heat stability of plasmin may be of some advantage for cheese biochemistry, e.g., degradation of β-casein under controlled conditions. Due to their high pH, plasmin is more active in surface mold-ripened and Blue-type cheeses and contributes to proteolysis more in these varieties than internal bacterially ripened varieties such as Cheddar (Schlesser et al. 1992, Sousa and McSweeney 2001, Gripon 2002).

In mold-ripened cheeses, lactate metabolism and production of alkaline compounds from proteolysis are more intense than in bacterially-ripened cheeses and the pH value increases during ripening to close to 7.5 which favors for plasmin action (Sousa and McSweeney 2001). Extensive degradation in β-casein and accordingly formation of γ-casein in Kuflu cheese with an average pH value of 6.3, were reported (Hayaloglu et al. 2008). Due to the effects of plasmin on cheese ripening, some attempts have been made by addition of plasmin into cheese milk. Addition of plasmin resulted in 20% increase in the level of water-soluble nitrogen (WSN) in Cheddar cheese (Farkye and Fox 1992), and more hydrolysis in β-casein in Mozzarella cheese (Somers et al. 2002), and primary proteolysis (pH 4.6-soluble nitrogen fractions and urea-PAGE patterns) was more intense than in controls (O'Farrell et al. 2002). Use of urokinase, which is a plasminogen activator, increased the breakdown of β-casein, formation of peptides as determined by RP-HPLC and soluble nitrogen fractions in several studies (Barrett et al. 1999, Upadhyay et al. 2004a,b, 2007).

The principal role of starter culture is the production of lactic acid during cheese manufacture, which reduces the pH of the cheese milk and curd.

The common bacteria used in the manufacture of cheese are mesophilic *Lactococcus* and *Leuconostoc* species and thermophilic *Lactobacillus* species and *S. thermophilus*. Although these bacteria need some amino acids and complex nutritional requirements, milk does not contain these amino acids at sufficient levels for the growth of starter bacteria (Fox et al. 1993). These bacteria have intracellular enzymes which are released into the cheese by lysis and contribute to the development of cheese flavor. The principal proteinase of the most lactic acid bacteria is lactocepin. This enzyme is associated with the cell-wall and degrades caseins when the organism is growing in milk; however, major role of this enzyme in cheese ripening is the degradation of intermediate-sized peptides produced from the caseins by plasmin or chymosin (Upadhyay et al. 2004a).

Lactic acid bacteria have a wide range of proteolytic enzymes (see Fig. 7.6) including oligoendopeptidases, di- and tri-peptidases, aminopeptidases and a number of proline-specific peptidases. Proline-specific peptidases have an importance as the caseins are rich in this amino acid residue. Other peptidases cannot efficiently hydrolyze proline-containing peptides due to proline's ring structure (Upadhyay et al. 2004b,

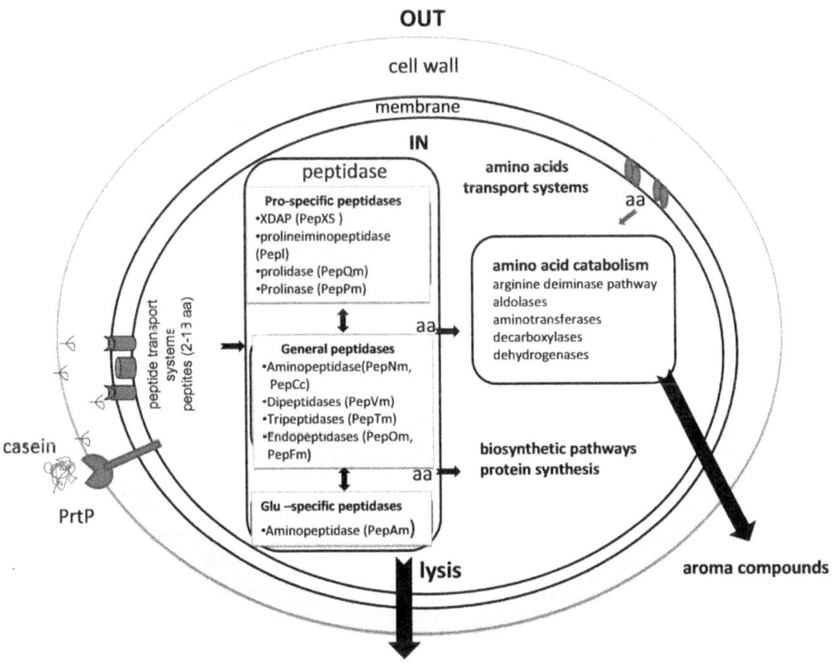

Figure 7.6 Catabolism of nitrogenous compounds in cheese during ripening and enzyme system of lactic acid bacteria. From Parente and Cogan (2004). With permission of Elsevier B.V.

Color image of this figure appears in the color plate section at the end of the book.

McSweeney 2007). Comparison of the relative contribution of enzymes from starter and coagulant to cheese ripening was done by Visser and de Groot-Mostert (1977) and these researchers pointed out that α_{s1}-casein was degraded in the cheeses manufactured using starter bacteria; however, β-casein was resistant to hydrolysis in the same cheeses. These workers also showed that the caseins were very slowly degraded in the cheeses manufactured without coagulant. So, it can be concluded from these results that the initial hydrolysis of caseins in cheeses is catalyzed by enzymes from coagulant and plasmin from the milk.

Effects of starter bacteria on the casein fractions were studied by Hayaloglu et al. (2004, 2005) and the urea-PAGE electrophoretograms of 90 day-old Turkish white-brined cheeses are given in Fig. 7.7. These workers showed that the defined starter cultures contributed to the hydrolysis of α_{s1}-casein in Turkish white-brined cheese at different manner (Hayaloglu et al. 2004, 2005). In the cheese samples, limited hydrolysis of β-casein was observed; however, degradation products from α_{s1}-casein were different in each sample. Also, significant differences were determined in the peptide profiles, concentrations of individual free amino acid and soluble nitrogen fractions in the cheeses during ripening (Hayaloglu et al. 2004, Hayaloglu 2007). Similarly, it was reported that the starter cultures significantly influenced the RP-HPLC peptide profiles of Croatian Krk cheese (Pavlinic et al. 2010). Accordingly, adjunct cultures including *Lb. delbrueckii* subsp. *lactis, Lb. paracasei* subsp. *paracasei, Lb. casei, Lb. plantarum* and *Enterococcus faecium* increased the levels of individual free amino acids and contributed to the flavor development in Domiati cheese (Awad et al. 2010).

Proteolytic systems of non-starter lactic acid bacteria are generally similar to those of starter bacteria and these appear to supplement of starter in producing similar molecular weights peptides and similar profiles of amino acids (Sousa et al. 2001). *Lb. casei* accelerated proteolysis in Cheddar cheese and gave different proteolysis pattern when compared with cheese samples made only with lactococcal starters (El-Soda et al. 1981, 1982). Hannon et al. (2003) made Cheddar cheese with an autolytic *Lb. helveticus* strain as adjunct and the authors emphasized that the autolytic bacteria increased the levels of free amino acids, pH 4.6-soluble and 5% phosphotungstic acid-soluble nitrogen fractions and RP-HPLC peptide profiles during ripening. Dramatic differences among peptide profiles of autolytic or non-autolytic starters are shown in Fig. 7.8. In general, starter or non-starter lactic acid bacteria significantly influenced proteolysis indices especially levels of individual free amino acids (Fig. 7.9) and RP-HPLC peptide profiles, but do not cause any significant effects on gross chemical composition (Law and Wigmore 1983, Law et al. 1992, 1993, Lane and Fox 1996, 1997, Fenelon et al. 1999, Shakeel-Ur-Rehman et al. 1999, Hayaloglu 2007).

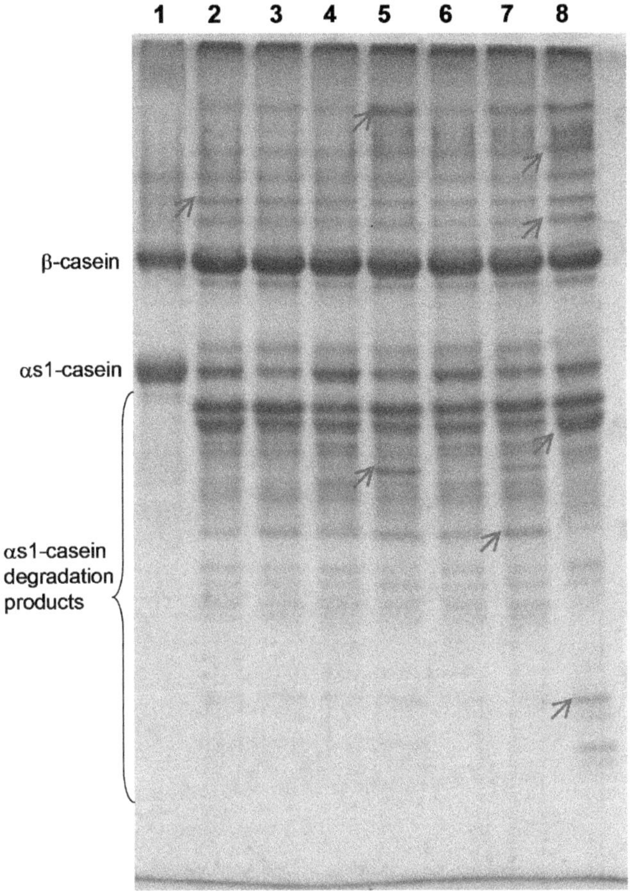

Figure 7.7 Urea-polyacrylamide gel electrophoresis electrophoretograms of Turkish White-brined cheese manufactured using defined lactococcal starters after 90 days of ripening. 1: Sodium caseinate, 2: *L. lactis* subsp. *lactis* UC317, 3: *L. lactis* subsp. *lactis* NCDO763, 4: *L. lactis* subsp. *cremoris* HP, 5: *L. lactis* subsp. *cremoris* SK11, 6: *L. lactis* subsp. *lactis* UC317 plus *L. lactis* subsp. *cremoris* HP, 7: *L. lactis* subsp. *lactis* NCDO763 plus *L. lactis* subsp. *cremoris* SK11, and 8: Starter-free. From Hayaloglu and Özer (2011). With permission of Sidas Publishing.

Color image of this figure appears in the color plate section at the end of the book.

The proteolytic systems of secondary microorganisms in cheese often play significant roles in cheese ripening. *Penicillium roqueforti* (Blue cheese), *P. camemberti* (surface mold-ripened cheeses like Camembert, Brie), *Brevibacterium linens, Arthrobacter* and other coryneform bacteria (smear-ripened cheeses) and *Propionibacterium freudenreichii* (Swiss cheeses) have been used traditionally for the manufacture and ripening of certain varieties. These microorganisms produce extracellular proteinases or peptidases

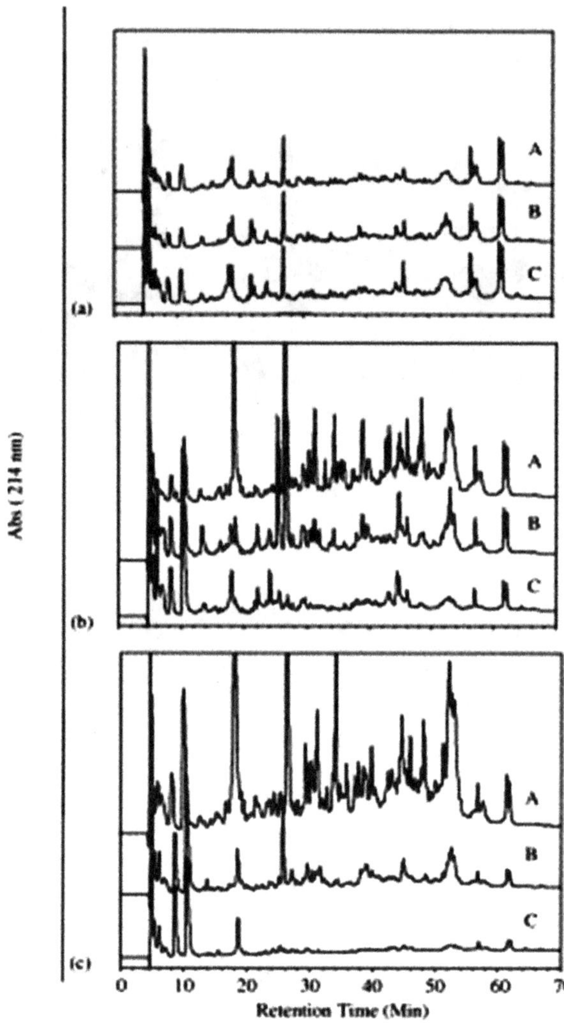

Figure 7.8 RP-HPLC chromatograms of the pH 4.6 soluble nitrogen extracts of cheeses made with starter system A, B or C at (a) 1 d, (b) 2 months and (c) 8 months of ripening. Starter system A contained a blend of two *Lactococcus lactis* strains (223 and 227) which had low levels of autolysis. System B was identical to A but included an adjunct of a highly autolytic strain of *Lactobacillus helveticus* (DPC4571). System C consisted only of strain DPC4571 as starter. From Hannon et al. (2003). With permission of Elsevier B.V.

which contribute to the ripening of these varieties (Gagnaire et al. 1999, Rattray and Fox 1999, Fox et al. 2000, Cantor et al. 2004). In smear-ripened cheeses, yeasts initially dominate the surface of cheese and catabolize lactate resulting in an increase in the pH of the cheese surface and a pH gradient

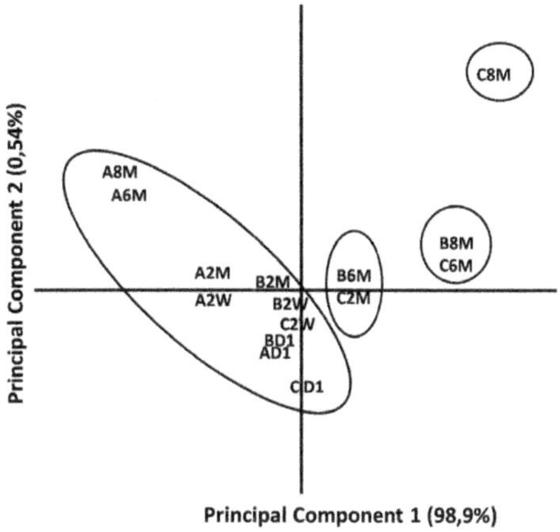

Figure 7.9 Principal Component Analysis (PCA) showing the first two principal components of individual free amino acid analysis of Cheddar cheeses made with starter system A, B or C at 1 d, 2 weeks, and 2, 6 and 8 months of ripening. Samples are clustered according to HCA. For details of starter systems refer to the legend of Fig. 7.8. From Hannon et al. (2003). With permission of Elsevier B.V.

form from surface to center of the cheese. As a result of the contribution of yeasts and production of some vitamins including niacin, riboflavin, pantothenic acid, the surface of the cheese forms a useful environment for bacterial growth, especially coryneform bacteria (Fox et al. 2000, Beresford and Williams 2004). Yeasts are commonly isolated from different varieties of cheese including brined-cheeses. *Debaryomyces hansenii* was the dominant species during the ripening of Danbo cheese, while *Trichosporon, Rhodotorula* and *Candida* spp. were isolated at the beginning of maturation (Peterson et al. 2002). These microorganisms grow at high salt concentrations, e.g., 24% (w/v). *Debaryomyces, Kluyveromyces* and *Saccharomyces* are the genera most commonly isolated from the surface of mold-ripened or smear-ripened cheeses. Of these genera, the species most commonly isolated include *Debaryomyces hansenii, Kluyveromyces marxianus, Yarrowia lipolytica* and *Torulopsis* spp.

Due to their contribution to cheese ripening, some species of yeasts have been used as adjunct starters in the manufacture of some varieties of cheese. Ferreira and Viljoen (2003) reported that added yeasts including *D. hansenii* and *Yarrowia lipolytica* contributed to Cheddar cheese aroma and accelerated the ripening of Cheddar. The authors also pointed out that these yeasts did not inhibit the starter bacteria used in the production. These yeasts had similar impact on the ripening characteristics of Danablu cheese during ripening (van den Tempel and Jakobsen 2000). *Saccharomyces cerevisiae* was used in the manufacture of Mycella, a Danish Blue-type cheese, and *S. cerevisiae* contributed to proteolysis but did not result in competition with the mold (Hansen et al. 2001). Bintis et al. (2003) used extracts from *D. hansenii* and *S. cerevisiae* in Feta cheese manufacture and the authors observed that the extracts increased the proteolytic, lipolytic and esterolytic activities, shortened the ripening time and contributed to the aroma of the cheeses. *D. hansenii* and *S. cerevisiae* also contributed to the accumulation of amino acids in Blue-type cheeses (Klein et al. 2002).

7.4 Lipolysis in cheese

Fats in foods undergo many biochemical changes including hydrolytic and oxidative degradation. Oxidative reactions in cheese is not important due to low redox potential of cheese; however, lipolysis and fatty acids catabolism have relatively higher importance in Cheddar, Dutch, Swiss and some Italian-type cheeses which contain moderate to high levels of free fatty acids. Lipolysis refers to the hydrolysis of triglycerides in cheese during ripening and occurs in most natural cheese varieties to varying extents. Generally, Italian, Blue and certain mold- and smear-ripened type cheeses undergo extensive lipolysis. Swiss-type and Cheddar cheeses undergo intermediate levels of lipolysis, while Dutch-types undergo relatively low levels of lipolysis. Despite the occurrence of lipolysis in many cheese varieties, it is only in recent years that its importance to cheese flavor has become well recognized. Due to some difficulties in extraction of fatty acids from cheese and lack of clear analytical methodology in the past, scientists did not attempt to quantify lipolysis in cheeses in detail. The drawback has been decreased by development of more reliable methods principally based on gas chromatography and mass spectroscopy for advanced studies.

Lipolysis occurs through the hydrolysis of the ester bond between a fatty acid and glycerol in a triacylglycerol (TAG) by lipases (Deeth and FitzGerald 1995, Collins et al. 2003). The action of these enzymes on a representative TAG is shown in Fig. 7.10. Both esterases and lipases are classified as lipolytic enzymes. Esterases hydrolyze acyl ester chains between 2 and 8 carbon atoms in length, while lipases hydrolyze those acyl ester chains of 10 or more carbon atoms. Esterases hydrolyze soluble substrates in aqueous solutions while lipases hydrolyze emulsified substrates. There are also some differences between the two enzymes in terms of type of substrate and enzyme kinetics. Esterases have classical Michaelis–Menten type kinetics while lipases, since they are activated only in the presence of a hydrophobic/hydrophilic interface, display interfacial Michaelis–Menten type kinetics (Collins et al. 2003).

Changes in lipolysis in mold ripened cheeses are pronounced during ripening and work has been concentrated on these types of cheese in the literature due to their complex myco- and microflora (Fox et al. 1996, Collins et al. 2004). Molds and yeasts can produce a large diversity of lipases which

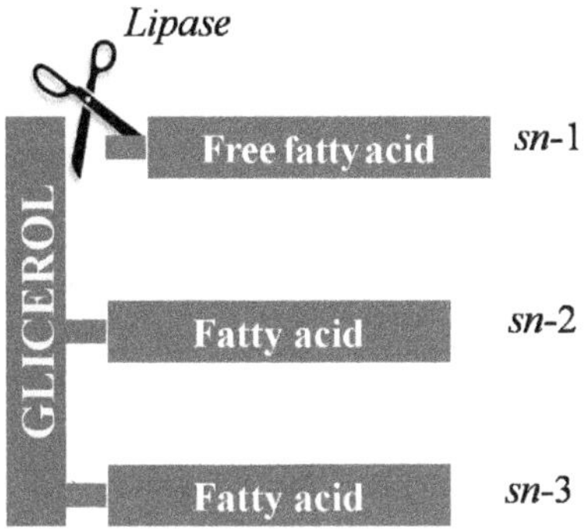

Figure 7.10 Lipase action on a triacylglycerol. From Hayaloglu and Özer (2011). With permission of Sidas Publishing.

Color image of this figure appears in the color plate section at the end of the book.

are active at interface between fat globules and continuous phase. Lipases produced by these microorganisms are not very specific, but can hydrolyze the TAGs more or less rapidly according to their molecular weight (Spinnler and Gripon 2004), releasing free fatty acids to the cheese matrix easily. The pH (approx. 6.0 or above) at the surface or center of these cheeses may be suitable for lipase activity. Two lipases secreted by *P. roqueforti* have optimal activity at alkaline pH values (pH 9.0–9.5) and at neutral pH (pH 7.0–8.0). The increase in pH in some cheeses during ripening partly originates from the catabolism of lactate and proteolysis (Godinho and Fox 1982, Zarmpoutis et al. 1996, 1997, Gobbetti et al. 1997, Hansen et al. 2001, Hayaloglu et al. 2008). Many factors including water activity (a_w), fat contents, salt concentration and salting method, salt gradient in cheese matrix, pH and pH gradient in cheese, oxidation-reduction potential of the cheese, ripening temperature, presence and/or concentration of antimicrobials produced by microorganisms, cheese variety, manufacturing and ripening conditions of cheese influence lipolysis. So, the level of lipolysis and the concentrations of liberated free fatty acids differ considerably between cheese varieties as shown in Table 7.2. The levels of fatty acids in internal bacterially ripened cheeses vary between 100–5000 mg kg^{-1}, while levels are greater than 30,000 mg kg^{-1} in mold-ripened cheeses. Hard type cheeses including Cheddar, Roncal, Kasar and Tulum contain around 10,000 mg kg^{-1} free fatty acids.

Short-chain fatty acids which are formed by lipolysis are present in milk and these contribute directly to cheese flavor. In addition to their direct contribution to cheese flavor, fatty acids serve as important precursors for many volatile flavor compounds produced by a series of biochemical pathways given in Fig. 7.11. Esters, which are responsible for fruity flavors in cheese, are formed through two enzymatic reactions: direct esterification and, perhaps more importantly, alcoholysis. Esterification

Table 7.2 Total free fatty acids (FFA) in some varieties of cheeses.

Cheese variety	Total FFA (mg kg^{-1} cheese)	References
Beyaz peynir	395	Akin et al. (2003)
	6878	Guler (2003)
	221	Özer et al. (2011)
Blue	35230	Woo et al. (1984)
Brick	2350	Woo et al. (1984)
Cheddar	9492	Kilcawley et al. (2001)
Civil	2318	Hayaloglu and Karabulut (2013)
Edam	356	Woo et al. (1984)
Ezine	2113	Hayaloglu and Karabulut (2013)
Gruyere	961	Zerfiridis et al. (1984)
Hellim	779	Hayaloglu and Karabulut (2013)
Kasar	13442	Guler (2003)
Kefalogreviera	235	Katsiąri et al. (2001)
Limburger	4187	Woo et al. (1984)
Mahon	8743	de la Funte et al. (1993)
Manchego	7285	Fernandez-Garcia et al. (1994)
Mihalic	2857	Hayaloglu and Karabulut (2013)
Mozzarella	363	Woo and Lindsay (1984)
Picante	57	Freitas et al. (1999)
Port Salut	700	Woo et al. (1984)
Roquefort	32453	Woo et al. (1984)
Romano	6754	Woo and Lindsay (1984)
Roncal	8178	de la Funte et al. (1993)
Serra da Estrela	880	Partidario et al. (1998)
Teleme	1384	Mallatou et al. (2003)
Traditional Feta	8712	Georgala et al. (2005)
Tulum	17358	Guler (2003)
	1797	Yilmaz et al. (2005)
Urfa	1436	Hayaloglu and Karabulut (2013)
	836	Atasoy and Turkoglu (2009)
Van Otlu	2602	Hayaloglu and Karabulut (2013)

is the formation of esters from alcohols and carboxylic acids, whereas alcoholysis is the production of esters from alcohols and acylglycerols or from alcohols or acyl-CoA (Liu et al. 2004). Methyl ketones are formed by enzymatic oxidation of free fatty acids to β-ketoacids and their consequent decarboxylation to methyl ketones (McSweeney and Sousa 2000); these compounds contribute to the pungent aroma of Blue cheese (Ortigosa et al. 2001). Aldehydes are produced by the catabolism of fatty acids or amino acids via decarboxylation or deamination. Aldehydes are transitory compounds and do not accumulate in cheese because they are transformed rapidly to alcohols or to the corresponding acids (Dunn and Lindsay 1985).

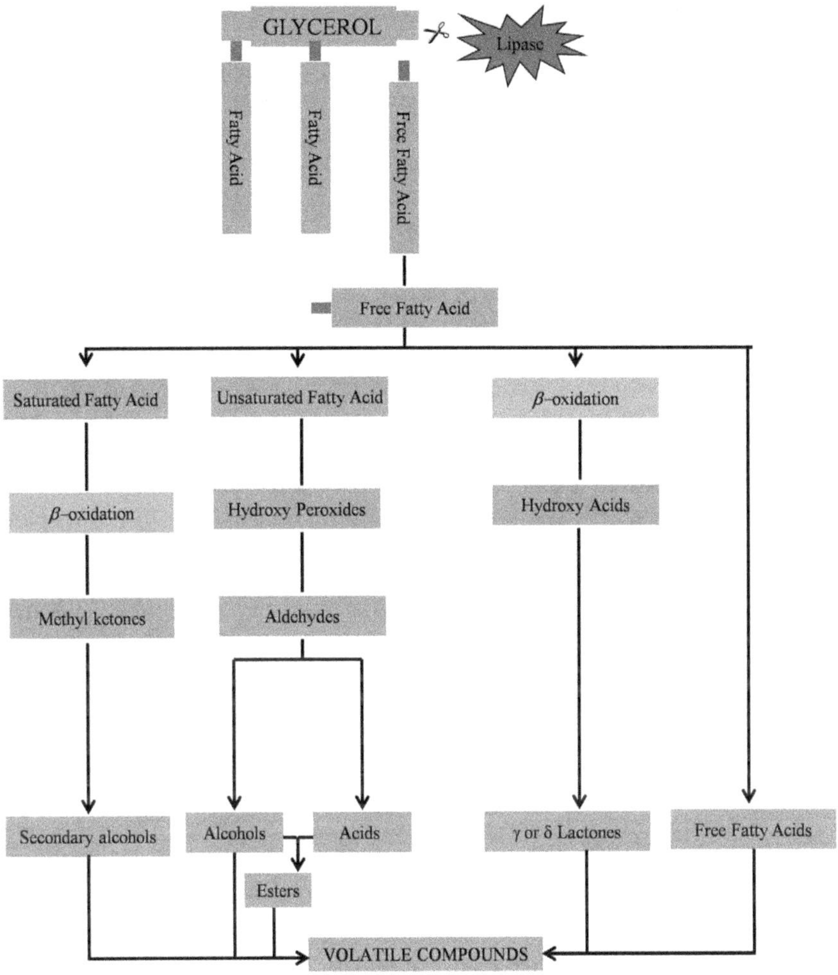

Figure 7.11 Fatty acid metabolism in cheese and volatiles originated from the fatty acids. From Molimard and Spinnler (1996), Hayaloglu and Özer (2011). With permission of Sidas Publishing.

Color image of this figure appears in the color plate section at the end of the book.

Alcohols are synthesized via many metabolic pathways, i.e., metabolism of lactose and amino acids, degradation of methyl ketones and linoleic and linolenic acids (Molimard and Spinnler 1996). Fatty acid lactones are cyclic compounds formed through the intramolecular esterification of a hydroxyacid; both γ- and δ-lactones have been found in cheese (McSweeney 2007).

7.5 Metabolism of lactose, lactate and citrate

Glycolysis of lactose to lactate is the fastest biochemical event which is almost completed during manufacturing; however, it has been studied less than proteolysis or lipolysis by the scientists working in the field. Glycolysis involves only fermentation of lactose but metabolism of lactate and citrate are commonly included under this heading (Fox et al. 1993, 1996). Metabolism of lactose to lactate by the starter microorganisms is essential and most lactate is lost (about 98%) in the whey during manufacture and cheese curd contains low level of lactose at the beginning of manufacture (Fox et al. 1993). The residual lactose (~2%) is transformed into lactate by lactic acid bacteria with a concomitant reduction in pH, which varies with cheese type (Fox 2002). Full metabolism of lactose is preferred to avoid the development of undesirable microorganisms in cheese. The action of these microorganisms can be restricted by increasing the salt-in-moisture level and starter action may be stopped quickly. The uptake of salt by cheese curd is faster in dry-salted varieties (e.g., Cheddar, Tulum) than in brine-salted varieties (e.g., Turkish white-brined—Beyaz peynir—Feta). So, the pH is a slightly higher in dry-salted cheeses than in the cheeses ripened under brine, due to the quick stopping of lactose fermentation in these varieties (McSweeney 2004, 2007, McSweeney and Fox 2004). The pH reduces about 2.0–2.5 unit after manufacture and dropped to 4.4–4.8 in brine-ripened varieties (see for detail Hayaloglu et al. 2002, 2005, Abd El-Salam and Alichanidis 2004). Another factor affecting lactose fermentation in the curd during cheese manufacture is type of starter and their salt tolerance. *L. lactis* subsp. *cremoris* is more salt-sensitive than *L. lactis* subsp. *lactis* which in turn is more sensitive than non-starter lactic acid bacteria (Fox et al. 1993). However, the growth of starter lactococci decreases above 5% salt-in-moisture level and the ripening rate of the cheese is slowed down.

The reduction of pH during manufacture of cheeses ripened under brine is vital process; so, to protect cheese against undesirable microorganisms, the pH of the cheese should be below 5.0–4.9 before packaging in tin-plated cans (Hayaloglu and Özer 2011). Low pH can also increase the level of residual coagulant in curd; higher level of coagulant is retained at lower pH. Fermentation of lactose is almost stopped after manufacture in cheeses ripened under brine (due to the high salt concentration of the brine). Also, metabolism of lactate or citrate is not continued during ripening in these cheeses which have close texture and no gas-blowing is desirable at any stage of ripening (Hayaloglu and Özer 2011). Lactate is metabolized to acetate, ethanol, formate, etc. by lactic or propionic acid bacteria or fungi (Fox et al. 2000). Work on determination of lipolysis or volatile composition in Feta (Vafopoulou et al. 1989, Kandarikis et al. 2001, Kondyli et al. 2002) and Beyaz peynir (Özer et al. 2011) showed that acetic acid is one of the principal carboxylic acids in these varieties and it is thought that acetic acid is produced by the action of halotolerant lactic acid bacteria and yeasts.

Lactate metabolism is a crucial event in Swiss-type and Camembert-type cheeses. In Swiss-type cheese, lactate is metabolized by *Propionibacterium freudenreichii* to propionic and acetic acid, water and CO_2 (Fig. 7.12). The CO_2 formed migrates through the cheese mass and accumulates as cheese eyes which is a characteristic feature for this type of cheese (McSweeney and Fox 2004). Lactate metabolism is also observed in Emmental and some other Swiss cheeses (e.g., Gruyere, Appenzeller, Comte), resulting in a characteristic taste, aroma and texture in these cheeses (Brennan et al. 2004, Fröhlich-Wyder and Bachmann 2004). A relatively low proportion of the CO_2 produced by propionic acid bacteria in Swiss cheese is visible in the eyes (~15%); half remains dissolved in cheese and about 35% diffuses out from the cheese surface (Fröhlich-Wyder and Bachmann 2004). Extensive catabolism of lactate also occurs in surface-mold ripened cheeses (e.g., Camembert, Brie) and the catabolism of lactate causes an increase in the pH of the surface of the cheeses which forms a pH gradient from the surface to the core (McSweeney and Fox 2004, McSweeney 2007). The high pH at the surface causes migration of calcium phosphate from center to cheese surface where it precipitates; loss of calcium from the cheese body results in a characteristic softening of surface mold-ripened cheeses which develop an almost liquid consistency when mature (Fig. 7.13). When the catabolism of lactate is complete, *P. camemberti* metabolizes proteins and produces ammonia (McSweeney 2007). Oxidative metabolism of lactate also occurs

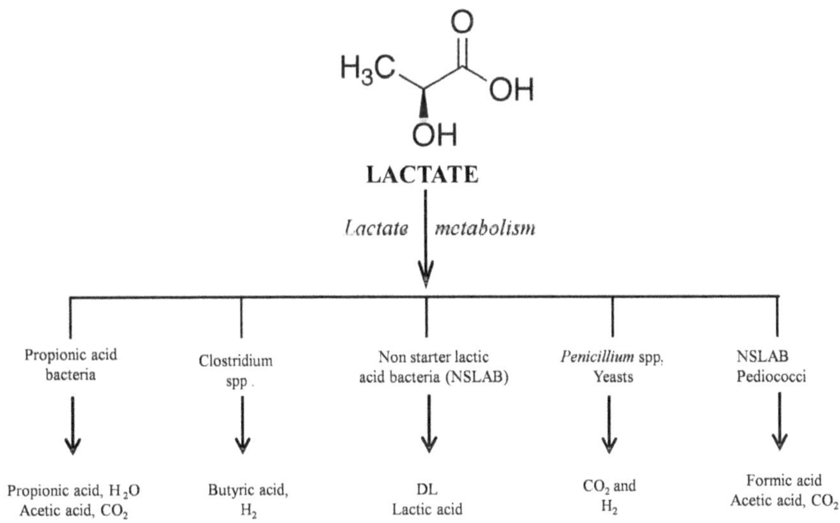

Figure 7.12 Lactate metabolism during cheese ripening. From Hayaloglu and Özer (2011). With permission of Sidas Publishing.

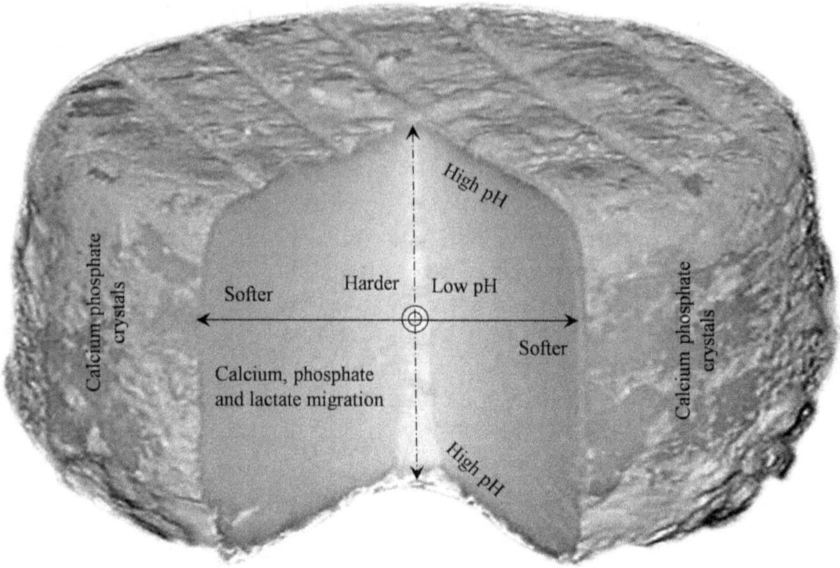

Figure 7.13 A representative demonstration of pH gradient, lactate metabolism and texture changes in a Camembert-type cheese.

Color image of this figure appears in the color plate section at the end of the book.

in surface smear-ripened cheeses (e.g., Tilsit, Limburger), in which lactate is decomposed by yeasts and the resulting environment is made more favorable for Gram-positive microorganisms of the smear. In Blue cheeses, lactate metabolism also occurs throughout the cheese mass due to growth of *P. roqueforti* in fissures inside the cheese.

Metabolism of citrate is caused by citrate-positive (Cit^+) microorganisms and is important in Dutch-type cheeses which are characterized with eye formation. Although milk contains a relatively low level of citrate (ca. 1.8 g l^{-1}) and most of it is soluble rather than being associated with colloidal casein phase of milk. Although the majority of citrate in milk is lost in whey, its concentration increases proportionally due to concentration of milk solids during cheese-making. Citrate is an important precursor for flavor compounds in Dutch-type cheeses (Parente and Cogan 2004, McSweeney and Fox 2004, McSweeney 2007). In Dutch-type cheeses, citrate is metabolized (pathway is shown in Fig. 7.14) with the production of diacetyl, acetoin, 2,3-butanediol and CO_2 by Cit^+-strains of *Lactococcus* or by *Leuconostoc mesenteroides* subsp. *cremoris* and *Leuconostoc lactis*, although citrate may also be metabolized by some facultatively heterofermentative lactobacilli that are common components of the NSLAB flora, but not thermophilic lactobacilli and *S. thermophilus* (Fox et al. 2000). Although citrate metabolism is a desirable process in Dutch-type cheese to complete

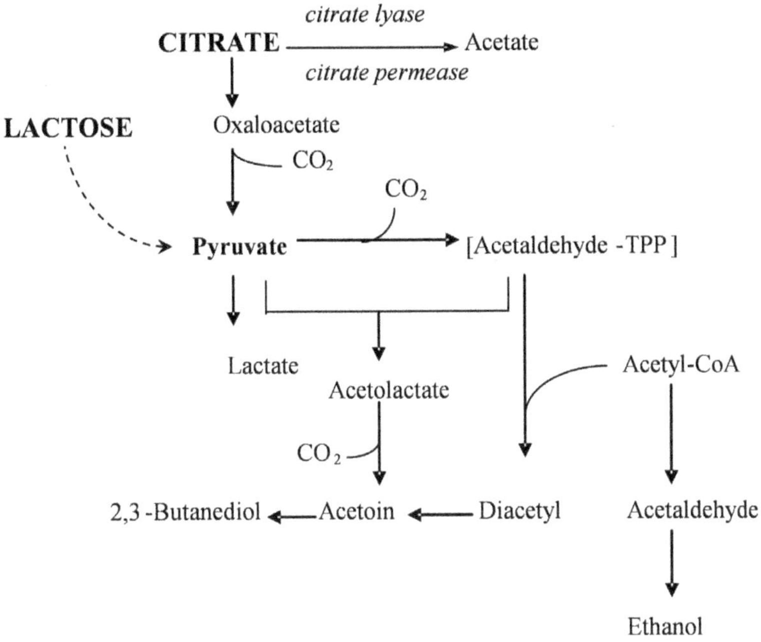

Figure 7.14 Citrate metabolism in Dutch-type cheeses. From Hayaloglu and Özer (2011). With permission of Sidas Publishing.

eye formation, these openings are considered defects in Cheddar, Cottage or varieties ripened under brine (Fox et al. 1993, Kandarikis et al. 2001, Hayaloglu et al. 2002, McSweeney and Fox 2004, Özer et al. 2011).

References

Abd El-Salam, M.H. and E. Alichanidis. 2004. Cheeses varieties ripened in brine. pp. 227–249. *In:* P.F. Fox, P.L.H. McSweeney, T.M. Cogan and T.P. Guinee [eds.]. Cheese: Chemistry, Physics and Microbiology. Vol. 1. Elsevier Academic Press, London, UK.

Akin, N., S. Aydemir, C. Kocak and M.A. Yildiz. 2003. Changes free fatty acid contents and sensory properties of white pickled cheese during ripening. Food Chem. 80: 77–83.

Alichanidis, E., E.M. Anifantakis, A. Polychroniadou and M. Nanou. 1984. Suitability of some microbial coagulants for Feta cheese manufacture. J. Dairy Res. 51: 141–147.

Al-Otaibi, M.M. and R.A. Wilbey. 2005. Effect of chymosin and salt reduction on the quality of ultrafiltrated white-salted cheese. J. Dairy Res. 72: 234–242.

Atasoy, A.F. and H. Turkoglu. 2009. Lipolysis in Urfa cheese produced from raw and pasteurized goats' and cows' milk with mesophilic or thermophilic cultures during ripening. Food Chem. 115: 71–78.

Awad, S., N. Ahmed and M. El-Soda. 2010. Influence of microfiltration and adjunct culture on quality of Domiati cheese. J. Dairy Sci. 93: 1807–1814.

Ayad, E.H.E., A. Verheul, J.T.M. Wouters and G. Smit. 2000. Application of wild starter cultures for flavour development in pilot plant cheesemaking. Int. Dairy J. 10: 169–179.

Bansal, N., M.A. Drake, P. Piraino, M.L. Broe, M. Harboe, P.F. Fox and P.L.H. McSweeney. 2009. Suitability of recombinant camel (*Camelus dromedarius*) chymosin as a coagulant for Cheddar cheese. Int. Dairy J. 19: 510–517.

Barrett, F.M., A.L. Kelly, P.L.H. McSweeney and P.F. Fox. 1999. Use of exogenous urokinase to accelerate proteolysis in Cheddar cheese during ripening. Int. Dairy J. 9: 421–427.

Beresford, T. and A. Williams. 2004. The microbiology of cheese ripening. pp. 287–317. *In:* P.F. Fox, P.L.H. McSweeney, T.M. Cogan and T.P. Guinee [eds.]. Cheese: Chemistry, Physics and Microbiology. Vol. 1. Elsevier Academic Press, London, UK.

Bintis, T., A. Vafopoulou-Mastrojiannaki, E. Litopoulou-Tzanetaki and R.K. Robinson. 2003. Protease, peptidase and esterase activities by lactobacilli and yeast isolates from Feta cheese brine. J. Appl. Microbiol. 95: 68–77.

Bintis, T. and P. Papademas. 2002. Microbiological quality of white-brined cheeses: A review. Int. J. Dairy Technol. 55: 113–120.

Bockelmann, W. 1995. The proteolytic system of starter and non-starter bacteria: Components and their importance for cheese ripening. Int. Dairy J. 5: 977–994.

Bockelmann, W. and T. Hoppe-Seyler. 2001. The surface flora of bacterial smear-ripened cheeses from cow's and goat's milk. Int. Dairy J. 11: 307–314.

Bogenrief, D.D. and N.F. Olson. 1995. Hydrolysis of β-casein increases Cheddar cheese meltability. Milchwiss. 50: 678–682.

Bontinis, T.G., H. Mallatou, E. Pappa, T. Massouras and E. Alichanidis. 2012. Study of proteolysis, lipolysis and volatile profiles of a traditional Greek goat cheese (Xinotyri) during ripening. Small Rum. Res. 105: 193–201.

Boutrou, R., A. Sepulchre, G. Pitel, C. Durier, L. Vassal, J.-C. Gripon and V. Monnet. 1998. Lactococcal lysis and curd proteolysis: Two predictable events important for development of cheese flavour. Int. Dairy J. 8: 609–616.

Brennan, N.M., T.M. Cogan, M. Loessner and S. Scherer. 2004. Bacterial surface ripened cheeses. pp. 199–225. *In:* P.F. Fox, P.L.H. McSweeney, T.M. Cogan and T.P. Guinee [eds.]. Cheese: Chemistry, Physics and Microbiology. Vol. 1. Elsevier Academic Press, London, UK.

Broadbent, J.R., M. Barnes, C. Brennand, M. Strickland, K. Houck, M.E. Johnson and J.L. Steele. 2002. Contribution of *Lactococcus lactis* cell envelope proteinase specifity to peptide accumulation and bitterness in reduced–fat Cheddar cheese. Appl. Environ. Microbiol. 68: 1778–1785.

Broome, M.C. and G.K.Y. Limsowtin. 1998. Milk coagulants. Aus. J. Dairy Technol. 53: 188–190.

Broome, M.C., X. Xu and J.J. Mayes. 2006. Proteolysis in Cheddar cheese made with alternative coagulants. Aus. J. Dairy Technol. 61: 85–87.

Buffa, M., B. Guamis, M. Pavia and A.J. Trujillo. 2001. Lipolysis in cheese made from raw, pasteurized or high pressure-treated goats' milk. Int. Dairy J. 11: 175–179.

Cambaztepe, F., S. Cakmakci and E. Dagdemir. 2009. Effect of some technological parameters on microbiological, chemical and sensory qualities of Civil cheese during ripening. Int. J. Dairy Technol. 62: 541–548.

Cantor, M.D., T. van den Tempel, T.K. Hansen and Y. Ardö. 2004. Blue cheese. pp. 175–198. *In:* P.F. Fox, P.L.H. McSweeney, T.M. Cogan and T.P. Guinee [eds.]. Cheese: Chemistry, Physics and Microbiology. Vol. 1. Elsevier Academic Press, London, UK.

Collins, Y.F., P.L.H. McSweeney and M.G. Wilkinson. 2003. Evidence of a relationship between autolysis of starter bacteria and lipolysis in Cheddar cheese during ripening. J. Dairy Res. 70: 105–113.

Collins, Y.F., P.L.H. McSweeney and M.G. Wilkinson. 2004. Lipolysis and catabolism of fatty acids in cheese. pp. 374–389. *In:* P.F. Fox, P.L.H. McSweeney, T.M. Cogan and T.P. Guinee [eds.]. Cheese: Chemistry, Physics and Microbiology. Vol. 1. Elsevier Academic Press, London, UK.

Crow, V.L., T. Coolbear, R. Holland, G.G. Pritchard and F.G. Martley. 1993. Starters as finishers: Starter properties relevant to cheese ripening. Int. Dairy J. 3: 423–460.

de la Feunte, M.A., J. Fontecha and M. Juarez. 1993. Fatty acid composition of the triglyceride and free fatty acid fractions in different cows-, ewes- and goats-milk cheeses. Z. Lebensm. Unters. Forsch. 196: 155–158.

Deeth, H.C. and C.H. FitzGerald. 1995. Lipolytic enzymes and hydrolytic rancidity in milk and milk products. pp. 247–308. *In:* P.F. Fox [ed.]. Advanced Dairy Chemistry—2: Lipids, 2nd ed. Chapman and Hall, London, UK.

Dunn, H.C. and R.C. Lindsay. 1985. Evaluation of the role of microbial Strecker–derivated aroma compounds in unclean–type flavours of Cheddar cheese. J. Dairy Sci. 68: 2859–2874.

El-Soda, M., M.J. Desmazeaud, S. Aboudonia and A. Badran. 1982. Acceleration of cheese ripening by the addition of extracts from *Lactobacillus helveticus, Lactobacillus bulgaricus* and *Lactobacillus lactis* to the cheese curd. Milchwiss. 37: 325–327.

El-Soda, M., M.J. Desmazeaud, S. Aboudonia and N. Kamal. 1981. Acceleration of cheese ripening by the addition of whole cells or cell free extracts from *Lactobacillus casei* to the cheese curd. Milchwiss. 36: 140–142.

Faccia, M., G. Picariello, A. Trani, P. Loizzo, G. Gambacorta, C. Lamacchia and A. di Luccia. 2012. Proteolysis of Cacioricotta cheese made from goat milk coagulated with caprifig (*Ficus carica sylvestris*) or calf rennet. Eur. J. Food Res. Technol. 234: 527–533.

Farkye, N.Y. and P.F. Fox. 1992. Contribution to plazmin to Cheddar cheese ripening: Effect of added plasmin. J. Dairy Res. 59: 209–216.

Fenelon, M.A., M.P. Ryan, M.C. Rea, T.P. Guinee, R.P. Ross, C. Hill and D. Harrington. 1999. Elevated temperature ripening of reduced fat cheddar cheese made with or without lacticin 3147-producing starter culture. J. Dairy Sci. 82: 10–22.

Fernandez-Garcia, E., R. Lopez-Fandino, L. Alonso and M. Ramos. 1994. The use of lipolytic and proteolytic enzymes in the manufacture of Manchego-type cheese from ovine and bovine milk. J. Dairy Sci. 77: 2139–2149.

Fernandez-Salguero, J. 2004. Internal mold-ripened cheese: Characteristics, composition and proteolysis of the main European Blue vein varieties. Italian J. Food Sci. 16: 437–445.

Ferreira, A.D. and B.C. Viljoen. 2003. Yeasts as adjunct starters in matured Cheddar cheese. Int. J. Food Microbiol. 86: 131–140.

Fleet, G.H. 1990. Yeasts in dairy-products. J. Appl. Bacteriol. 68: 199–211.

Folkertsma, B., P.F. Fox and P.L.H. McSweeney. 1996. Acceleration of Cheddar cheese ripening at elevated temperatures. Int. Dairy J. 6: 1117–1134.

Fox, P.F. 1989. Proteolysis during cheese manufacture and ripening. J. Dairy Sci. 72: 1379–1400.

Fox, P.F. and L. Stepaniak. 1993. Enzymes in cheese technology. Int. Dairy J. 3: 509–533.

Fox, P.F., J. Law, P.L.H. McSweeney and J. Wallace. 1993. Biochemistry of cheese ripening. pp. 389–438. *In:* P.F. Fox [ed.]. Cheese: Chemistry, Physics and Microbiology. Vol. 1. Chapman and Hall, London, UK.

Fox, P.F. and P.L.H. McSweeney. 1996. Proteolysis in cheese during ripening. Food Rev. Int. 12: 457–509.

Fox, P.F., T.P. O'Connor, P.L.H. McSweeney, T.P. Guinee and N.M. O'Brien. 1996. Cheese: Physical, chemical, biochemical and nutritional aspects. Adv. Food Nutr. Res. 39: 163–328.

Fox, P.F. and P.L.H. McSweeney. 1997. Rennets: Their role in milk coagulation and cheese ripening. pp. 1–49. *In:* B.A. Law [ed.]. Microbiology and Biochemistry of Cheese and Fermented Milk. Blackie Academic & Professional, London, UK.

Fox, P.F., T.P. Guinee, T.M. Cogan and P.L.H. McSweeney. 2000. Fundamentals of Cheese Science. An Aspen Publication, Gaithershurg, Maryland, USA.

Fox, P.F. 2002. Biochemistry of cheese ripening. pp. 320–326. *In:* H. Roginski, J.W. Fuquay, P.F. Fox [eds.]. Encyclopedia of Dairy Sciences. Vol. 1. Academic Press, London, UK.

Freitas, A.C., A.E. Pintado, M.E. Pintado and F.X. Malcata. 1999. Role of dominant microflora of Picante cheese on proteolysis and lipolysis. Int. Dairy J. 9: 593–603.

Fröhlich-Wyder, M.T. and H.P. Bachmann. 2004. Cheeses with propionic acid fermentation. pp. 141–156. *In:* P.F. Fox, P.L.H. McSweeney, T.M. Cogan and T.P. Guinee [eds.]. Cheese: Chemistry, Physics and Microbiology. Vol. 1. Elsevier Academic Press, London, UK.

Gagnaire, V., D. Mollé, T. Sørhaug and J. Léonil. 1999. Peptidases of dairy propionic acid bacteria. Lait 79: 43–57.

Georgala, A., E. Moschopoulou, A. Aktypis, T. Massouras, E. Zoidou, I. Kandarikis and E. Anifantakis. 2005. Evaluation of lipolysis during the ripening of traditional Feta cheese. Food Chem. 93: 73–80.

Gobbetti, M., R. Burzigotti, E. Smacchi, A. Corsetti and M. de Angelis. 1997. Microbiology and biochemistry of Gorgonzola cheese during ripening. Int. Dairy J. 7: 519–529.

Godinho, M. and P.F. Fox. 1982. Ripening of Blue cheese: Influence of salting rate on proteolysis. Milchwiss. 36: 476–478.

Gomez, M.J., E. Rodrigues, P. Gaya, M. Nunez and M. Medina. 1999. Characteristics of Manchego cheese manufactured from raw and pasteurized ovine milk and with defined-strain or commercial mixed-strain starter cultures. J. Dairy Sci. 82: 2300–2307.

Gripon, J.-C. 2002. Mold-ripened cheeses. pp. 401–406. *In:* H. Roginski, J.W. Fuquay and P.F. Fox [eds.]. Encyclopedia of Dairy Science. Vol. 1. Academic Press, London, UK.

Guinee, T.P. and M.G. Wilkinson. 1992. Rennet coagulation and coagulants in cheese manufacture. J. Soc. Dairy Technol. 45: 94–104.

Guler, Z. 2003. The determination of free fatty acids in pickled White cheese, Kasar and Tulum cheeses by using alumina adsorption method. pp. 85–90. *In:* Proceedings of GAP III. Agricultural Congress, 2–3 October, Sanliurfa, Turkey.

Hannon, J.A., M.G. Wilkinson, C.M. Delahunty, J.M. Wallace, P.A. Morrissey and T.P. Beresford. 2003. Use of autolytic starter systems to accelerate the ripening of Cheddar cheese. Int. Dairy J. 13: 313–323.

Hannon, J.A., M.J. Sousa, S. Lillevang, A. Sepulchre, W. Bockelmann and P.L.H. McSweeney. 2004. Effect of defined-strain surface starters on the ripening of Tilsit cheese. Int. Dairy J. 14: 871–880.

Hansen, T.K., T.V.D. Tempel and C.M. Jakobsen. 2001. *Saccharomyces cerevisiae* as a starter culture in Mycella. Int. J. Food Microbiol. 69: 101–111.

Hayaloglu, A.A., M. Guven and P.F. Fox. 2002. Microbiological, biochemical and technological properties of Turkish White cheese "Beyaz Peynir". Int. Dairy J. 12: 635–648.

Hayaloglu, A.A., M. Guven, P.F. Fox, J.A. Hannon and P.L.H. McSweeney. 2004. Proteolysis in Turkish White-brined cheese made with defined strains of *Lactococcus*. Int. Dairy J. 14: 599–610.

Hayaloglu, A.A., M. Guven, P.F. Fox and P.L.H. McSweeney. 2005. Influence of starters on chemical, biochemical, and sensory changes in Turkish White-brined cheese during ripening. J. Dairy Sci. 88: 3460–3474.

Hayaloglu, A.A. 2007. Comparisons of different single strain starter cultures for their effects on ripening and grading of Beyaz cheese. Int. J. Food Sci. Technol. 42: 930–938.

Hayaloglu, A.A. and E.Y. Brechany. 2007. Influence of milk pasteurization and scalding temperature on the volatile compounds of Malatya, a farmhouse Halloumi-type cheese. Lait 87: 39–57.

Hayaloglu, A.A., P.F. Fox, M. Guven and S. Cakmakci. 2007. Cheeses of Turkey: 1. Varieties ripened in goat-skin bags. Lait 87: 79–95.

Hayaloglu, A.A., E.Y. Brechany, K.C. Deegan and P.L.H. McSweeney. 2008. Characterization of chemistry, biochemistry and volatile profiles of Kuflu cheese, a mold-ripened variety. LWT- Food Sci. Technol. 41: 1323–1334.

Hayaloglu, A.A. 2009. Volatile composition and proteolysis in traditionally produced mature Kashar cheese. Int. J. Food Sci. Technol. 44: 1388–1394.

Hayaloglu, A.A., K.C. Deegan and P.L.H. McSweeney. 2010. Effect of milk pasteurization and curd scalding temperature on proteolysis in Malatya, a Halloumi-type cheese. Dairy Sci. Technol. 90: 99–110.

Hayaloglu, A.A. and B.H. Özer. 2011. Penirde olgunlasma [Ripening in Cheese]. pp. 173–209. *In:* A.A. Hayaloglu and B.H. Özer [eds.]. Peynir Biliminin Temelleri [Principles of Cheese Sciences]. Sidas Medya, Izmir, Turkey.

Hayaloglu, A.A. and I. Karabulut. 2013. Characterization and comparison of free fatty acid profiles of eleven varieties of Turkish cheeses. Int. J. Food Prop. 16: 1407–1416.

Kamaly, K.M., K. Takayama and E.H. Marth. 1990. Acylglycerol acylhydrolase (lipase) activities of *Streptococcus lactis, Streptococcus cremoris,* and their mutants. J. Dairy Sci. 73: 280–290.

Kandarikis, I., G. Moatsou, A.I.K. Georgala, S. Kaminarides and E. Anifantakis. 2001. Effect of draining temperature on the biochemical characteristics of Feta cheese. Food Chem. 72: 369–378.

Kappeler, S.R., H.J.M. van den Brink, H. Rahbek-Nielsen, Z. Farah, Z. Puhan, E.B. Hansen and E. Johansen. 2006. Characterization of recombinant camel chymosin reveals superior properties for the coagulation of bovine and camel milk. Biochem. Biophysic. Res. Comm. 342: 647–654.

Katsiari, M.C., L.P. Voutsinas, E. Alichanidis and I.G. Roussis. 2001. Lipolysis in reduced sodium Kefalogreviera cheese made by partial replacement of NaCl with KCl. Int. Dairy J. 11: 193–197.

Kilcawley, K.N., M.G. Wilkinson and P.F. Fox. 2001. A survey of lipolytic and glycolytic end-products in commercial Cheddar enzyme-modified cheese. J. Dairy Sci. 84: 66–73.

Klein, N., A. Zourari and S. Lortal. 2002. Peptidase activity of four yeast species frequently encountered in diary product—Comparison with several dairy bacteria. Int. Dairy J. 12: 853–861.

Kok, J. 1993. Genetics of proteolytic enzymes of lactococci and their role in cheese flavour development. J. Dairy Sci. 76: 2056–2064.

Kondyli, E., M.C. Katsiari, T. Masouras and L.P. Voutsinas. 2002. Free fatty acids and volatile compounds of low-fat Feta-type cheese made with commercial adjunct culture. Food Chem. 79: 199–205.

Lane, C.N. and P.F. Fox. 1996. Contribution of starter and adjunct lactobacilli to proteolysis in Cheddar cheese during ripening. Int. Dairy J. 6: 715–728.

Lane, C.N. and P.F. Fox. 1997. Role of starter enzymes during ripening of Cheddar cheese made from pasteurised milk under controlled microbiological conditions. Int. Dairy J. 7: 55–63.

Lau, K.Y., D.M. Barbano and R.R. Rasmussen. 1991. Influence of pasteurization of milk on protein breakdown in Cheddar cheese during aging. J. Dairy Sci. 74: 727–740.

Law, B.A. and A.S. Wigmore. 1983. Accelerated ripening of Cheddar cheese with a commercial proteinase and intracellular enzymes from starter streptococci. J. Dairy Res. 50: 519–525.

Law, J., G.F. Fitzgerald, C. Daly, P.F. Fox and N.Y. Farkye. 1992. Proteolysis and flavour development in Cheddar cheese made with the single starter strains *Lactococcus lactis* ssp. *lactis* UC317 or *Lactococcus lactis* ssp. *cremoris* HP. J. Dairy Sci. 75: 1173–1185.

Law, J., G.F. Fitzgerald, T. Uniacke-Lowe, C. Daly and P.F. Fox. 1993. The contribution of Lactococcal starter proteinases to proteolysis in Cheddar cheese. J. Dairy Sci. 76: 2455–2467.

Lawrence, R.C., L.K. Creamer and J. Gilles. 1987. Texture development during cheese ripening. J. Dairy Sci. 70: 1748–1760.

Liu, S.-Q., R. Holland and V.L. Crow. 2004. Esters and their biosynthesis in fermented dairy products: A review. Int. Dairy J. 14: 923–945.

Lynch, C.M., P.L.H. McSweeney, P.F. Fox, T.M. Cogan and F.D. Drinan. 1997. Contribution of starter lactococci and non-starter lactobacilli to proteolysis in Cheddar cheese with a controlled microflora. Lait 77: 441–459.

Madkor, S., P.F. Fox, S.I. Shalabi and N.H. Metwalli. 1987. Studies on the ripening of Stilton Cheese: Proteolysis. Food Chem. 25: 13–29.

Mallatou, H., E. Pappa and T. Massooras. 2003. Changes in free fatty acids during ripening of Teleme cheese made with ewes', goats', or a mixture of ewes' and goats' milk. Int. Dairy J. 13: 211–219.
McSweeney, P.L.H., P.F. Fox, J.A. Lucey, K.N. Jordan and T.M. Cogan. 1993. Contribution of the indigenous microflora to the maturation of Cheddar cheese. Int. Dairy J. 3: 613–634.
McSweeney, P.L.H. and M.J. Sousa. 2000. Biochemical pathways for the production of flavour compounds in cheese during ripening: A review. Lait 80: 293–324.
McSweeney, P.L.H. 2004. Biochemistry of cheese ripening: Introduction and overview. pp. 347–360. *In:* P.F. Fox, P.L.H. McSweeney, T.M. Cogan and T.P. Guinee [eds.]. Cheese: Chemistry, Physics and Microbiology. Vol. 1. Elsevier Academic Press, London, UK.
McSweeney, P.L.H. and P.F. Fox. 2004. Metabolism of residual lactose and of lactate and citrate. pp. 361–371. *In:* P.F. Fox, P.L.H. McSweeney, T.M. Cogan and T.P. Guinee [eds.]. Cheese: Chemistry, Physics and Microbiology. Vol. 1. Elsevier Academic Press, London, UK.
McSweeney, P.L.H. 2007. Cheese manufacture and ripening and their influence of cheese flavour. pp. 1–25. *In:* B.C. Weimer [ed.]. Improving the Flavour of Cheese. CRC Press, Cambridge, England.
Meijer, W.C. 1997. Expression and release of proteolytic enzymes of *Lactococcus lactis*—Ripening of UF cheese. NIZO, PhD Thesis, Ede, The Netherlands.
Michaelidou, A., M.C. Katsiari, E. Kondyli, L.P. Voutsinas and E. Alichanidis. 2003. Effect of a commercial adjunct culture on proteolysis in low-fat Feta-type cheese. Int. Dairy J. 13: 179–189.
Michaelidou, A., M.C. Katsiari, L.P. Voutsinas, A. Polychroniadou and E. Alichanidis. 2007. Effect of multiple-species starters on peptide profile and free amino acids in low-fat Kefalograviera-type cheese. Food Chem. 104: 800–807.
Moatsou, G., T. Massouras, I. Kandarakis and E. Anifantakis. 2002. Evaluation of proteolysis during the ripening of traditional Feta cheese. Lait 82: 601–611.
Molimard, P. and H.E. Spinnler. 1996. Compounds involved in the flavour of surface mold–ripened cheeses: Origins and properties. J. Dairy Sci. 79: 169–184.
Møller, K.K., F.P. Rettary, J.C. Sørensen and Y. Ardö. 2012. Comparison of the hydrolysis of bovine κ-casein by camel and bovine chymosin: A kinetic and specificity study. J. Agric. Food Chem. 60: 5454–5460.
O'Farrell, I.P., J.J. Sheehan, M.G. Wilkinson, D. Harrington and A.L. Kelly. 2002. Influence of addition of plasmin or mastitic milk to cheese milk on quality of smear-ripened cheese. Lait 82: 305–316.
Ortigosa, M., P. Torre and J.M. Izco. 2001. Effect of pasteurization of ewe's milk and use of an indigenous starter culture on the volatile compounds and sensory characteristics of Roncal cheese. J. Dairy Sci. 84: 1320–1330.
O'Mahony, J.A., J.A. Lucey and P.L.H. McSweeney. 2005. Chymosin-mediated proteolysis, calcium solubilization, and texture development during the ripening of Cheddar cheese. J. Dairy Sci. 88: 3101–3114.
O'Sullivan, N.A., M.J. Sousa, D. O'Grady-Walsh, T. Uniacke, A.L. Kelly and P.L.H. McSweeney. 2005. Ripening of Camembert-type cheese made from caprine milk using calf rennet or kid rennet as coagulant. Int. J. Dairy Technol. 58: 13–18.
Özer, B., H.A. Kirmaci, A.A. Hayaloglu, M. Akcelik and N. Akkoc. 2011. The effects of incorporating wild-type strains of *Lactococcus lactis* into Turkish white brined cheese (*Beyaz peynir*) on the fatty acid and volatiles content. Int. J. Dairy Technol. 64: 494–501.
Pappa, H.C. and E.M. Anifantakis. 2001. Effect of different concentrated cultures on the proteolysis and organoleptic characteristics of Feta cheese. Milchwiss. 56: 384–387.
Parente, E. and T.M. Cogan. 2004. Starter cultures: general aspects. pp. 123–147. *In:* P.F. Fox, P.L.H. McSweeney, T.M. Cogan and T.P. Guinee [eds.]. Cheese: Chemistry, Physics and Microbiology. Vol. 1. General Aspects, 3rd edition. Elsevier, Amsterdam, The Netherlands.

Partidario, A.M., M. Barbosa and L.V. Boas. 1998. Free fatty acids, triglycerydes and volatile compounds in Serra da Estrela cheese—Changes throughout ripening. Int. Dairy J. 8: 873–881.

Pavia, M., A.J. Trujillo, B. Guamis and V. Ferragut. 2000. Proteolysis in Manchego-type cheese salted by brine vacuum impregnation. J. Dairy Sci. 83: 1441–1447.

Pavlinic, I., N. Mikulec, S. Kalit, I. Jakasa, J. Havranek, M. Tudor and N. Antunac. 2010. Influence of starter culture on peptide profiles of Krk cheese. Milchwiss. 65: 262–266.

Peterson, K.M., S. Westall and L. Jespersen. 2002. Microbial succession of *Debaryomyces hansenii* during the production of Danish surface-ripened cheese. J. Dairy Sci. 85: 478–486.

Prieto, B., I. Franco, J.M. Fresno, A. Bernardo and J. Carballo. 2000. Picon Bejes-Tresviso blue cheese: An overall biochemical survey throughout the ripening process. Int. Dairy J. 10: 159–167.

Rattray, F.P. and P.F. Fox. 1999. Aspects of enzymology and biochemical properties of *Brevibacterium linens* relevant to cheese ripening: A review. J. Dairy Sci. 82: 891–909.

Reiter, B., T.F. Fryer, A. Pickering, H.R. Chapman, R.C. Lawrence and M.E. Sharpe. 1967. The effect of microbial flora on the flavour and free fatty acid composition of Cheddar cheese. J. Dairy Res. 34: 257–271.

Rijnen, L., S. Bonneau and M. Yvon. 1999. Genetic characterization of the major lactococcal aromatic aminotransferase and its involvement in conversion of amino acids to aroma compounds. Appl. Environ. Microbiol. 65: 4873–4880.

Sahan, N., K. Yasar, A.A. Hayaloglu, O.B. Karaca and A. Kaya. 2008. Influence of fat replacers on chemical composition, proteolysis, texture profiles, meltability and sensory properties of low-fat Kashar cheese. J. Dairy Res. 75: 1–7.

Schlesser, J.E., S.J. Schmidt and R. Speckman. 1992. Characterization of chemical and physical changes in Camembert cheese during ripening. J. Dairy Sci. 75: 1753–1760.

Seratlic, S., Z.N. Miloradovic, Z.T. Radulovic and O.D. Macej. 2011. The effect of two types of mold inoculants on the microbiological composition, physicochemical properties and protein hydrolysis in two Gorgonzola-type cheese varieties during ripening. Int. J. Dairy Technol. 64: 408–416.

Shakeel-Ur-Rehman, A.H. Pripp, P.L.H. McSweeney and P.F. Fox. 1999. Assessing the proteolytic and cheese ripening properties of single strains of *Lactococcus* in miniature cheeses. Lait 79: 361–383.

Shakeel-Ur-Rehman, P.L.H. McSweeney, J.M. Banks, E.Y. Brechany, D.D. Muir and P.F. Fox. 2000. Ripening of Cheddar cheese made from blends of raw and pasteurized milk. Int. Dairy J. 10: 33–34.

Somers, J.M., T.P. Guinee and A.L. Kelly. 2002. The effect of plasmin activity and cold storage of cheesemilk on the composition, ripening and functionality of Mozzarella-type cheese. Int. J. Dairy Technol. 55: 5–11.

Sousa, M.J. and P.L.H. McSweeney. 2001. Studies on the ripening of Cooleeney, an Irish farmhouse Camembert-type cheese. Ir. J. Agric. Food Res. 40: 83–95.

Sousa, M.J., Y. Ardo and P.L.H. McSweeney. 2001. Advances in the study of proteolysis during cheese ripening. Int. Dairy J. 11: 327–345.

Sousa, M.J. and F.X. Malcata. 2002. Advances in the role of a plant coagulant (*Cynara cardunculus*) *in vitro* and during ripening of cheese from several milk species. Lait 82: 151–170.

Spinnler, H.-E. and J.-C. Gripon. 2004. Surface mold-ripened cheese. pp. 157–174. *In:* P.F. Fox, P.L.H. McSweeney, T.M. Cogan and T.P. Guinee [eds.]. Cheese: Chemistry, Physics and Microbiology. Vol. 1. Elsevier Academic Press, London, UK.

Tarakci, Z., H. Durmaz, E. Sagun and H. Sancak. 2005. Influence of brine concentration on chemical, microbiological and sensory characteristics of herby cheese. Ind. Vet. J. 82: 279–282.

Trujillo, A.J., B. Guamis, J. Laencina and M.B. Lopez. 2000. Proteolytic activities of some milk clotting enzymes on ovine casein. Food Chem. 71: 449–457.

Tzanetakis, N., A. Vafopoulou-Mastrojiannaki and E. Litopoulou-Tzanetaki. 1995. The quality of white-brined cheese from goat's milk made with different starters. Food Microbiol. 12: 55–63.

Upadhyay, V.H., P.L.H. McSweeney, A.A.A. Magboul and P.F. Fox. 2004a. Proteolysis in cheese during ripening. pp. 391–433. *In:* P.F. Fox, P.L.H. McSweeney, T.M. Cogan and T.P. Guinee [eds.]. Cheese: Chemistry, Physics and Microbiology. Vol. 1. Elsevier Academic Press, London, UK.

Upadhyay, V.K., M.J. Sousa, P. Ravn, H. Israelsen, A.L. Kelly and P.L.H. McSweeney. 2004b. Use of exogenous streptokinase to accelerate proteolysis in Cheddar cheese during ripening. Lait 84: 527–538.

Upadhyay, V.K., P. Ravn, H. Israelsen, M.J. Sousa, A.L. Kelly and P.L.H. McSweeney. 2007. Acceleration of proteolysis during ripening of Cheddar-type cheese using a streptokinase-producing strain of Lactococcus. J. Dairy Res. 73: 70–73.

Ustunol, Z. and C.L. Hicks. 1990. Effect of milk-clotting enzymes on cheese yield. J. Dairy Sci. 73: 8–16.

Vafopoulou, A., E. Alichanidis and G. Zerfiridis. 1989. Accelerated ripening of Feta cheese, with heat–shocked cultures or microbial proteinases. J. Dairy Res. 56: 285–296.

van den Tempel, T. and M. Jakobsen. 2000. The technological characteristics of *Debaryomyces hansenii* and *Yarrowia lipolitica* and their potential as starter cultures for production of Danablu. Int. Dairy J. 10: 263–270.

Viljoen, B.C., A.R. Khoury and A. Hatingh. 2003. Seasonal diversity of yeasts associated with white-surface mold-ripened cheeses. Food Res. Int. 36: 275–283.

Visser, F.M.W. and A.E.A. de Groot-Mostert. 1977. Contribution of enzymes from rennet, starter bacteria and milk to proteolysis and flavour development in Gouda cheese. 4. Protein breakdown: A gel electrophoretical study. Neth. Milk Dairy J. 31: 247–264.

Visser, S. 1993. Proteolytic enzymes and their relation to cheese ripening and flavor. An overview. J. Dairy Sci. 76: 329–350.

Woo, A.H., S. Kollodge and R.C. Lindsay. 1984. Quantification of major free fatty acids in several cheese varieties. J. Dairy Sci. 67: 874–878.

Woo, A.H. and R.C. Lindsay. 1984. Concentrations of major free fatty acids and flavour development in Italian cheese varieties. J. Dairy Sci. 67: 960–968.

Wyder, M.T. and Z. Puhan. 1999a. Investigation of yeast flora in smear ripened cheeses. Milchwiss. 54: 330–333.

Wyder, M.T. and Z. Puhan. 1999b. Role of selected yeasts in cheese ripening: An evaluation in aseptic curd slurries. Int. Dairy J. 9: 117–124.

Yasar, K. and N. Guzeler. 2011. Effects of coagulant type on the physicochemical and organoleptic properties of Kashar cheese. Int. J. Dairy Technol. 64: 372–379.

Yetişmeyen, A., A. Çimer, M. Özer, S. Odabaşı and O. Deveci. 1998. Ultrafiltrasyon tekniği ile salamura Beyaz peynir üretiminde kalite üzerinde değişik maya enzimlerinin etkileri. Gıda 23: 3–9.

Yilmaz, G., A. Ayar and N. Akin. 2005. The effect of microbial lipase on the lipolysis during the ripening of Tulum cheese. J. Food Eng. 69: 269–274.

Yvon, M., E. Chambellon, A. Bolotin and F. Roudot–Algaron. 2000. Characterisation and role of the branched-chain aminotransferase (BcaT) isolated from *Lactococcus lactis* subsp. *cremoris* NCDO763. Appl. Environ. Microbiol. 66: 571–577.

Yvon, M. and L. Rijnen. 2001. Cheese flavour formation by amino acid catabolism. Int. Dairy J. 11: 185–201.

Zarmpoutis, I.V., P.L.H. McSweeney, J. Beechinor and P.F. Fox. 1996. Proteolysis in the Irish farmhouse Blue cheese, Chetwynd. Irish J. Agric. Food Res. 35: 25–36.

Zarmpoutis, I.V., P.L.H. McSweeney and P.F. Fox. 1997. Proteolysis in blue-veined cheeses: An intervarietal study. Irish J. Agric. Food Res. 36: 219–229.

Zerfiridis, G.K., A. Vafopolou-Mastrogiannaki and E. Litopoulou-Tzanetaki. 1984. Changes during ripening of commercial Gruyere cheese. J. Dairy Sci. 67: 1397–1405.

CHAPTER 8

Microbiology and Biochemistry of Yogurt and Other Fermented Milk Products

Barbaros Özer

8.1 Introduction

The history of fermented milks dates back to early civilizations around 10,000 BC where milking animals were first domesticated. Today, a great number of fermented milks are produced at local or industrial scale. Tamime and Robinson (2007) classified the fermented milks into three groups: (1) lactic fermentation products (mesophilic, thermophilic and probiotic), (2) yeast/lactic fermentation products, and (3) mold/lactic fermentation products. Yogurt and kefir which have economical importance are the most popular members of groups 1 and 2, respectively. Koumiss—a fermented milk product native to Central Asia—has also gained popularity during the last two decades owing to its high nutritive value and therapeutic properties. Until now, numerous scientific researches have been conducted to fully understand the microbiological and technological properties, and health-promoting effects of fermented milks. The microbiology, biochemistry and health-promoting properties of yogurt, kefir and koumiss will be discussed hereafter. Surely, there are many more fermented milk products which have different microbiological, chemical, physical and sensory characteristics around the globe. The following are recommended for further reading regarding processing technologies and characteristics of other fermented milk products including Nordic/Scandinavian fermented milk products and cultured/sour/fermented cream (Fonden et al. 2006, Lyck et al. 2006, Tamime 2002).

8.2 Yogurt

Production of lactic acid, aroma compounds and exopolysaccharides are the major metabolic activities of yogurt starter bacteria (*Streptococcus thermophilus* and *Lactobacillus delbrueckii* subsp. *bulgaricus*). All these metabolic activities are directly related to lactose metabolism of yogurt bacteria. During the last two decades, great efforts have been made to develop metabolically engineered starter strains so that consumer-oriented modification in yogurt is possible. Resulting from these efforts, galactose utilization in galactose-negative (Gal^-) strains of *S. thermophilus* and *Lb. delbrueckii* subsp. *bulgaricus*, for example, has been achieved. Similarly, strains producing more L(+) lactic acid, which is more easily digested by human body, have been developed by replacing *IdhL* gene with *IdhD* gene. Another remarkable success of the genetic engineering in yogurt bacteria is the development of exopolysaccharide-producing strains by genetically modifying the central sugar metabolism of these organisms. The recent developments in lactose metabolism of yogurt starter bacteria will be discussed below.

8.2.1 Taxonomy and genetic properties of Streptococcus thermophilus *and* Lactobacillus delbrueckii *subsp.* bulgaricus

S. thermophilus belongs to the thermophilic group of the lactic acid bacteria (Pearce and Flint 2002). This bacterium was first characterized as lactic streptococci by Orla-Jensen in 1919 (Farrow and Collins 1984). The basic characteristics of *S. thermophilus* are as follows: Gram-positive, facultatively anaerobic, cytochrome-, oxidase- and catalase-negative, non-motile, non-sporeforming and homofermentative. *S. thermophilus* has a spherical to ovoid morphology with 0.7–0.9 µm in diameter and occurs in pairs and chains. This bacterium grows optimally at 40–45°C with a minimum of 20–25°C and a maximum of 47–50°C (Pearce and Flint 2002). *S. thermophilus* bears relatively high heat treatment up to 60°C for 30 min (Tamime and Robinson 2007). *S. thermophilus* is an alpha-hemolytic species of the viridans group. It is able to ferment lactose, fructose, sucrose and glucose, but not capable of utilizing arginine. The majority of *S. thermophilus* strains rarely utilize galactose. Lactic acid, acetaldehyde and diacetyl are the major outputs of lactose fermentation by *S. thermophilus* and the predominant lactic acid isomer produced by this organism is L(+) lactic acid.

Despite its phenotypic and genotypic similarities to the other lactic acid bacteria, *S. thermophilus* does not fit any systematic grouping. Previously, *S. thermophilus* was described as a sub-species of *Streptococcus salivarius* (Hutkins and Morris 1987). Recently, based on the DNA-DNA homology studies, *S. thermophilus* has been re-classified as a distinct streptococcal species. *S. thermophilus* does not have an N group antigen; therefore, it is unique among the other streptococci.

Guanidine+cytosine (mol G+C) content of DNA ranges between 37.2 and 40.3% (Zourari et al. 1992). Some strains are able to produce exopolysaccharides (EPS^+) and require B group vitamins and some amino acids including glutamic acid, histidine, methionine, cysteine, valine, leucine, isoleucine, tryptophan, arginine and tyrosine for enhanced growth rates. The cell-wall peptidoglycan type is Lys-Ala_{2-3} (Tamime and Robinson 2007). Due to its resistance against relatively high heat treatments, it is widely used in the manufacture of yogurt and some Italian and Swiss-type cheeses in association with one or more lactobacillus species.

S. thermophilus is abundantly present in dairy environment. Isolation of *S. thermophilus* from dairy environment is achieved conventionally on the basis of incubation temperature. At >40°C, the bacteria isolated from raw milk contain *S. thermophilus*. Recent developments of species-specific DNA probes have made positive identification of suspect colonies possible. The most common niche where *S. thermophilus* grows as biofilm is the regeneration section of plate heat exchanger pasteurizers (Pearce and Flint 2002).

The *S. thermophilus* genome is fairly smaller than its nearest relative in the lactic acid bacteria (an average of 1.75 and 1.82 Mb in *S. thermophilus* ST1 and A054, respectively, vs. 2.35 Mb of *Lactococcus lactis*). Up until now, more than 100 DNA sequence entries have been reported to be listed in GenBank. *S. thermophilus* may show a genetic instability after a serial sub-culturing on solid media. Pébay et al. (1993) found several loci which were unstable or exhibiting sequence polymorphism in *S. thermophilus*. The genetic instability of *S. thermophilus* affects the colony morphology as well. Four variants of *S. thermophilus* CNRZ368, for example, differing in size, opacity and shape in colonies were isolated by Pébay et al. (1993). Up to now, five restriction and modification enzymes encoded chromosomally have been characterized in *S. thermophilus*. Genetic characterization of phosphoenolpyruvate (PEP)-dependent phosphotransferase (PTS) system of *S. thermophilus* (Vaughan et al. 2001), protein and peptide utilization (Fernandez-Espla et al. 2000), polysaccharide production (Almiron-Roig et al. 2000), the stress response system, and phage resistance mechanisms (Burrus et al. 2001) have been studied extensively. Fundamental understanding of EPS genetics and biosynthesis in *S. thermophilus* has increased remarkably during the last decades.

Comparing to other lactic acid bacteria, *S. thermophilus* have much fewer plasmids and *S. thermophilus* plasmids are far less important on the metabolic functions compared to mesophilic lactococci plasmids. Majority of *S. thermophilus* strains are plasmid-free and the largest *S. thermophilus* plasmid described is 25.5 kb (Pearce and Flint 2002). No correlation exists between plasmid size and metabolic functions of the host organism. For example, curing the organism of 25.5 kb plasmid did not affect the host's ability to utilize carbohydrate or modify its antibiotic resistance. On the contrary, the absence of 6.9 kb plasmid affects the host's ability to resist antibiotic and phage as well as altering cell morphology and milk coagulation rate. Comparative genomics and multilocus sequencing analysis reveal that *S. thermophilus* is still undergoing a process of regressive evolution towards a specialized bacterium for growth in milk (Hols et al. 2005). Although *S. thermophilus* has a well-developed nitrogen metabolism, its carbohydrate metabolism has been subjected to a high level of degeneracy due to a paucity of carbon sources in milk (Hols et al. 2005). An obvious gene decay in *S. thermophilus* genome evolution resulted in nearly loss of its cell adhesion ability which is a common feature in pathogenic streptococci. On the other hand, the traits of this bacterium including polysaccharide biosynthesis, bacteriocin production, restriction-modification systems or oxygen tolerance have been improved due probably to horizontal gene transfer (Hols et al. 2005).

As with *S. thermophilus*, *Lb. delbrueckii* subsp. *bulgaricus* was also first described by Orla-Jensen (Farrow and Collins 1984) and initially named *Thermobacterium bulgaricum*. Currently, this bacterium is classified as a sub-species of *Lactobacillus delbrueckii*. *Lb. delbrueckii* subsp. *bulgaricus* is one of the three sub-species of *Lb. delbrueckii* and is relatively resistant to acidity. *Lb. delbrueckii* subsp. *bulgaricus* is an obligatory homofermentative (represented in Group I or A) and capable of fermenting carbohydrate including lactose, glucose and fructose and rarely galactose and mannose (Axelsson 1998). *Lb. delbrueckii* subsp. *bulgaricus* is Gram-positive, facultatively anaerobic, non-sporeforming and non-motile. The cells are rod and rounded ends, of 0.5–0.8×2–9 μm, and occur singly or in short chains. In old cultures, it may form extremely long chains. It has a high growth temperature (up to 48–50°C) and slight growth occurs at <10°C. The cell-wall peptidoglycan type is Lys-$_D$Asp (Tamime and Robinson 2007). Among other lactobacilli, *Lb. delbrueckii* subsp. *bulgaricus* is considered unique because of its atypical G+C content (ranging from 49 to 51%) (van de Gutche et al. 2006). This difference is mainly due to very important differences at codon position 3 (GC3) (65.0% in *Lb. delbrueckii* subsp. *bulgaricus* as compared with 25.0% and 24.4% in *Lb. acidophilus* and *Lb. johnsonii*, respectively) (van de Gutche et al. 2006). Since the evolution at codon position 3 is in general much faster than at codon 1 and 2 positions, the high GC3 values in *Lb. delbrueckii* subsp. *bulgaricus* may indicate that this sub-species is still in an active phase of evolution. Comparative genomic analyses also revealed that the *Lb. delbrueckii* subsp. *bulgaricus* genome has undergone a rapid reductive evolution as gene loss and metabolic simplification, known to be the central trend of evolving lactic acid bacteria (LAB) (Hao et al. 2011). On the other hand, one should not underestimate that the correlation between GC3 and overall GC content may be lost or changed. The major products of lactose metabolism in this organism are D(–) lactate and acetaldehyde. Some strains are able to produce exopolysaccharides (EPS^+). The proteolytic capacity of *Lb. delbrueckii* subsp. *bulgaricus* is relatively higher than that of *S. thermophilus* and is able to produce amino acids that are essential for the growth of *S. thermophilus*.

The genome of *Lb. delbrueckii* subsp. *bulgaricus* is estimated to be 2.3 Mb. Phenotypic and several genetic traits of *Lb. delbrueckii* subsp. *bulgaricus* were investigated in detail by Germond et al. (2003) and van de Gutche et al. (2006). These included 16S rDNA sequence mutations, expression of β-galactosidase and of the cell-wall-anchored protease, the characterization of the lactose operon locus and of the sequence of lacR gene, galactose metabolism, and the distribution of insertion sequence. Glucose metabolism of *Lb. delbrueckii* subsp. *bulgaricus* have been laid open (Cochu et al. 2003). The genetic profiles of both *Lb. delbrueckii* subsp. *bulgaricus* and its bacteriophages have also been described (Rantsiou et al. 1999, Serror et al. 2002).

The growth characteristics of *S. thermophilus* and *Lb. delbrueckii* subsp. *bulgaricus* are presented in Table 8.1.

8.2.2 Symbiosis between yogurt starter bacteria

S. thermophilus and *Lb. delbrueckii* subsp. *bulgaricus* exhibit in the milk an interaction that is mutually favorable and not obligatorily characterized by the fact that each bacterium produces one or more substances that stimulates the growth of the other (de Souza Oliveira et al. 2012). This relationship is termed "symbiosis". *S. thermophilus*, for example, does not possess substantial extracellular proteolytic activity and the amino acid and free peptide content of milk is not high enough to promote its full growth (Zourari et al. 1992). *Lb. delbrueckii* subsp. *bulgaricus* produces free amino acids from casein fractions and these amino acids are used by *S. thermophilus* for its growth (Fira et al. 2001). The type and number of amino acids stimulating the growth of *S. thermophilus* is strain-dependent (Letort and Juillard 2001). Leusine, lysine, aspartic acid, histidine and valine are the principal amino acids that are synthesized by *Lb. delbrueckii* subsp. *bulgaricus* to stimulate the growth of *S. thermophilus*. During the winter/autumn months, glycine, isoleucine, tyrosine, glutamic acid, methionine, as well as five amino acids mentioned above, are essential (Tamime and Robinson 2007). While some strains of *S. thermophilus* require valine, glycine and histidine for their growth, some strains show weak growth in the presence of these amino acids.

Table 8.1 Growth characteristics of *S. thermophilus* and *Lb. delbrueckii* subsp. *bulgaricus*.

Characteristics	*S. thermophilus*	*Lb. delbrueckii* subsp. *bulgaricus*
G+C (%)	37.2–40.3	49–51
Lactic acid isomer(s)	L(+)	D(–)
Growth		
10°C	–	–
45°C	+	+
Requirement for		
Thiamine	nd	–
Riboflavine	nd	+
Prydoxal	nd	–
Folic acid	nd	+
Thymidine	nd	+
Vit B_{12}	nd	+
Carbohydrate utilization		
Aesculin	+	–
Amygdalin	+	–
Cellobiose	+	–
Fructose	+	+
Galactose	–	–
Lactose	+	–
Maltose	+	–
Mannose	nd	–
Melezitose	nd	–
Melibiose	–	–
Raffinose	–	–
Ribose	–	–
Salicin	+	–
Sucrose	nd	–
Trehalose	p/w	–

Data compiled from Tamime and Robinson (2007), Hammes and Vogel (1995), Heller (2001), Hardie and Whilley (1995).

(+) positive reaction by 90% or more strains, (–) negative reaction by 90% or more strains, (p/w) positive or weak reaction by 11–89%. nd: no data available.

Apart from amino acids released by *Lb. delbrueckii* subsp. *bulgaricus*, some di- and oligopeptides naturally present in milk can also promote the growth of *S. thermophilus*. Especially, lysyl and hisdityl peptides play an important role in this mechanism (Nakamura et al. 1991). Additionally, some water soluble vitamins, pyridine and pyrimidine have positive effects on the growth of *S. thermophilus*.

In order to obtain high *S. thermophilus* counts in yogurt, pre-treatment of milk with proteolytic enzymes prior to fermentation to increase the concentration of free peptides/amino acids is recommended (Marshall et al. 1982).

The growth of *Lb. delbrueckii* subsp. *bulgaricus* is stimulated by formic acid liberated by *S. thermophilus* (Zourari et al. 1992). Formic acid is also largely produced in milk during heat treatment at >100ºC. It was reported that when milk was heated at 100ºC for 10 min, the formic acid concentration increased from 1.7 mg kg^{-1} to 39 mg kg^{-1} (Kikuchi et al. 1984). Kikuchi et al. (1984) also found that addition of sodium formate into milk at a level of 20 mg kg^{-1} led to 2.7 times increase in the acid production by *Lb. delbrueckii* subsp. *bulgaricus* and the number of this bacteria was ten times higher than that of the control yogurt. There is an adverse relationship between the oxygen content of the medium and formic acid concentration. At high oxygen concentrations, the growth of *Lb. delbrueckii* subsp. *bulgaricus* is decreased (Driessen 1981). Formic acid production in milk is only possible under the condition of <4 mg O_2 l^{-1}. When formic acid is lacking, ribonucleic acid (RNA) synthesis is depressed. Horiuchi and Sasaki (2012) demonstrated that acid production by mixture of *Lb. delbrueckii* subsp. *bulgaricus* 2038 and *S. thermophilus* 1131 was remarkably accelerated by reducing dissolved oxygen to almost 0 mg kg^{-1} in the yogurt mix.

In addition to formic acid, some other compounds produced by *S. thermophilus* during fermentation stimulate the growth of *Lb. delbrueckii* subsp. *bulgaricus*. CO_2 produced during the urea metabolism of *S. thermophilus* was shown to be an effective agent stimulating the growth of *Lb. delbrueckii* subsp. *bulgaricus* (Ascon-Reyes et al. 1995). Other compounds that are promoting the growth of *Lb. delbrueckii* subsp. *bulgaricus* are purine, pyrimidine, adenine, guanine, urasyl, fumaric acid, oxaloacetic acid and cystein. The concentration of CO_2 required by *Lb. delbrueckii* subsp. *bulgaricus* for the production of lactic acid at optimum level is >30 mg kg^{-1} and *S. thermophilus* is capable of producing CO_2 at concentrations higher than this level (Tinson et al. 1982). The CO_2 produced by *S. thermophilus* is also used as a precursor of aspartic acid production. Level of CO_2 produced by *S. thermophilus* varies depending on the types of milk used. Production of CO_2 in goat's milk is higher than in the other milk types. Although, it is not a common practice, in some cases sodium formate or sodium carbonate may be added into milk to stimulate the formic acid and/or CO_2 formation. Symbiotic relationship between yogurt bacteria yields considerably higher lactic acid and β-galactosidase activity than single culture (Ustok et al. 2007). Until now, very few studies have been conducted to investigate the symbiosis between yogurt bacteria on molecular and regulatory levels. Herve-Jimenez et al. (2009) studied the kinetics of the transcriptomic and proteomic modifications of *S. thermophilus* LMG 18311 in response to the presence of *Lb. delbrueckii* subsp. *bulgaricus* ATCC 11842 during growth in milk. Authors showed that seventy seven different genes or proteins, implicated mainly in the metabolism of nitrogen, nucleotide base and iron, varied specifically in co-culture. The expression of genes potentially encoding iron-chelating *dpr* as well as that of the *fur* (*per* R) regulatory genes increased, indicating a reduction in the intracellular iron concentration, possibly in response to H_2O_2 production by *Lb. delbrueckii* subsp. *bulgaricus* (Herve-Jimenez et al. 2009). In the co-culture of *Lb. delbrueckii* subsp. *bulgaricus* and *S. thermophilus*, high ATP is required for the growth and maintenance of biomass at the beginning of fermentation with high energy demand of enzyme induction during lag phase (de Souza Oliveira et al. 2012). Addition of inulin as a prebiotic to the growth medium resulted in reduction in these requirements, making biomass synthesis and maintenance less energy-consuming.

As discussed above, the positive effect of the associative growth of yogurt bacteria on the metabolic activities of both yogurt bacteria is out of question. However, a competition between these two organisms also occurs during fermentation of milk. The balance between competition and symbiosis determines the fate of metabolic products and of balance between rods and cocci. Depending upon the variation in the balance between yogurt bacteria, the textural and sensoric properties of yogurt may change, without affecting the lactic acid production.

Yogurt fermentation includes four successive stages (Beal and Corrieu 1991). These are:

(i) *Phase 1*. Stationary phase (lag phase) where both yogurt bacteria do not show remarkable multiplication.
(ii) *Phase 2*. Rapid growth of *S. thermophilus* during the first 80–100 min of fermentation. At this stage, *S. thermophilus* covers 90–95% of total bacterial population (log phase).
(iii) *Phase 3*. Growth of *Lb. delbrueckii* subsp. *bulgaricus* is stimulated and the count of *S. thermophilus* declines gradually. *S. thermophilus* represents 70–75% of total bacterial population (late log phase).
(iv) *Phase 4*. Stationary phase where the growth of both bacteria is decelerated.

The balance between *Lb. delbrueckii* subsp. *bulgaricus* and *S. thermophilus* is widely influenced by the pH, incubation temperature and total solids level of milk. Overall, low pH and high incubation temperatures stimulates the growth of *Lb. delbrueckii* subsp. *bulgaricus* (Beal and Corrieu 1991, Lankes et al. 1998). Radke-Mitchel and Sandine (1986) investigated the effect of incubation temperature on the ratio of rods and cocci. They reported that incubating milk at 37°C, 42°C or 45°C yielded rods to cocci ratios of 1:2.2, 1:8 and 1:2.4, respectively. Acidity is another factor determining the ratio between yogurt bacteria. While the growth of *S. thermophilus* is optimal at pH 6.5, *Lb. delbrueckii* subsp. *bulgaricus* grows optimally at pH 5.8. Method of total solids elevation also affects the fermentation kinetics. Özer and Robinson (1999) who studied the behavior of yogurt starter bacteria in strained yogurt concentrated either by ultrafiltration (UF) or by reverse osmosis (RO), found that concentration of milk by membrane techniques led to increase in inhibitory substances present in milk and to a decrease in water activity, both resulted in deceleration of bacterial multiplication.

8.2.3 Metabolism of yogurt starter bacteria

Carbohydrate metabolism

The lactic acid bacteria meet their energy requirements from fermentation of carbohydrates (i.e., milk lactose). Both yogurt bacteria ferment lactose homofermentatively. The principal step in lactose fermentation by thermophilic yogurt starters is the transport of lactose into the bacterial cell. The mechanism of lactose transport from outer source into the cell in *S. thermophilus* is somehow different from that of other dairy starter bacteria, e.g., lactococci. Lactococci possess a specific lactose transport system called phosphoenolpyruvate dependent (PEP)-phosphotransferase system (PTS) (Marshall and Tamime 1997). This transport system involves an enzyme, phospho-β-galactosidase (β-P-gal), which catalyses further hydrolysis of lactose-6-phosphate formed during lactose transport into glucose and galactose (Zourari et al. 1992). Four proteins (in sequential order: enzyme II, III, I and HPr) are also involved in translocating the lactose from outside to inside of the cytoplasmic membrane (Tamime and Robinson 2007, Monnet et al. 1996). Vast majority of *S. thermophilus* strains do not possess PEP-PTS system or β-P-gal, and lactose transfer in *S. thermophilus* is achieved by a lactose permease (LacS), which operates as a lactose-galactose antiporter or as a galactoside-proton symport system (de Vin et al. 2005, Foucaud and Poolman 1992, Poolman 1993). During the lactose uptake in *S. thermophilus*, no metabolic energy is required in lactose-galactose exchange mechanism (Poolman 1993). Rate of lactose uptake via antiport system in *S. thermophilus* is adversely affected by expression of galactokinase gene (*galK*), since galactose would no longer be available for the antiport reaction.

Hydrolyzation of lactose into glucose and galactose inside the cell is carried out by β-galactosidase (β-gal) (Poolman et al. 1992). Only glucose moiety is metabolized further via the Embden-Meyerhof-Parnas pathway to L(+) lactate by *S. thermophilus*. In the presence of excess lactose, galactose is excreted into the medium in equimolar amounts with the lactose uptake, resulting in a galactose-negative (*gal*$^-$) phenotype (Svensson et al. 2007, Hutkins et al. 1985). The excretion of the *gal*$^-$ phenotype of *S. thermophilus* has been attributed either to a defect in the induction mechanism for galactokinase (GalK), which seems to be the rate-limiting enzyme of the Leloir pathway or an energetically favorable reaction of the lactose transport system (Hutkins et al. 1985, de Vos 1996). On the other hand, characterization of galactose-fermenting (*gal*$^+$) mutants of *S. thermophilus* have been reported by various authors under appropriate selective conditions such as limited lactose and excess galactose concentrations (Hutkins et al. 1985).

The *gal*$^+$ strains of *S. thermophilus* can metabolize galactose excreted into the medium via Leloir pathway, involving the enzymes galactokinase (GalK), galactose-1-phosphate-uridyl transferase (GalT), uridyldiphosphate-4-epimerase (GalE), and mutarotase (GalM) (Zourari et al. 1992, de Vin et al. 2005). In fact, non-galactose fermenting (*gal*$^-$) strains of *S. thermophilus* (i.e., strain CNRZ 302) contain structurally intact genes for the Leloir pathway but these genes are weakly transcribed, if any (van den Bogaard 2002). Independently isolated *gal*$^+$ mutants contain mutations in the *gal* promoter region. The expression of these activated *gal* genes is under control of the apoinducer GalR, probably with a galactose or a derivative thereof as inducer (de Vos 1996). Today, it may well be possible to obtain stable mutants of *S. thermophilus* that can ferment galactose. This provides many advantages to the dairy applications where accumulation of galactose seems to be problematic, i.e., growth of heterofermentative lactic acid bacteria, cheese browning during baking. Accumulation of toxic galactitol in human tissue cells which is a result of excess galactose consumption can also be reduced by efficiently fermenting galactose in dairy foods (Hiratsuka and Li 1992).

Lb. delbrueckii subsp. *bulgaricus* is known to be responsible for the post-fermentation acidification in yogurt. Post-fermentation acidification by yogurt lactobacilli is somehow an uncontrollable process which leads to excess lactic acid accumulation in the end product, and, therefore, impairs the sensory quality. In order to overcome this handicap, the *lacZ* gene from *Lb. delbrueckii* subsp. *bulgaricus* has been targeted for engineering studies (Schmidt et al. 1989). Adams et al. (1994) detected a series of cold-sensitive mutations in the *lacZ* gene using an *Eschericia coli* expression system and random mutagenesis. Deletion affecting *LacZ* gene of *Lb. delbrueckii* subsp. *bulgaricus* offers an option to reduce over-acidification by lactobacilli. Germond et al. (1995) found that these deletions involve the presence of new insertion sequence (IS) element (ISL3). Strains of *Lb. delbrueckii* subsp. *bulgaricus* carrying these deletions cannot utilize lactose as a carbon source for generating energy (Delley et al. 1990).

Lactic acid isomers

Lactic acid is the major end product of lactose catabolism by yogurt bacteria. Lactic acid is a chiral molecule which exists as L(+) lactic acid and D(–) lactic acid (Benthin and Villadsen 1995). *S. thermophilus* exclusively produces L(+) lactic acid from pyruvate; on the contrary, in *Lb. delbrueckii* subsp. *bulgaricus* more than 90% of pyruvate is converted into D(–) lactate (Tamime and Robinson 2007). Atypical production of L(+) lactic acid by

Lb. delbrueckii subsp. *bulgaricus* has also been reported (Rogout et al. 1989). Less extensively, some strains of yogurt bacteria are also capable of producing the third lactic acid isomer called DL(±) lactic acid (Herrero et al. 2004). These strains contain both types of *ldh* genes coding for lactate dehydrogenase, designated *ldhL* and *ldhD* (de Vos 1996). Biosynthesis of lactic acid is stereospecific and the racemate may result from the combined action of D- and L-lactate dehydrogenase (LDH) or from a single dehydrogenase combined with the racemase. LDH is located in the cytoplasm of the bacterial cell and its activity is dependent upon nicotinamide adenine dinucleotide (NAD)/reduced nicotinamide adenine dinucleotide (NADH). NAD is regenerated from NADH during conversion of pyruvic acid to lactic acid. D-LDH from *Lb. delbrueckii* subsp. *bulgaricus*, a homodimer with 332 amino acid residues and a molecular mass of 36 kDa per subunit, acts at the last step of the glycolytic pathway under anaerobic conditions, allowing re-oxidation of NAD, which is necessary for glycolysis (Le Bras and Garel 1991). Vinals et al. (1995) demonstrated that the structure of LDH of *Lb. delbrueckii* subsp. *bulgaricus* constitutes of subunits of α/β structure with a catalytic domain (i.e., consisting of a histidine residue along with arginine and phenylalanine) and a co-enzyme binding domain.

During the early stages of yogurt fermentation, accumulation of L(+) lactic acid is more pronounced, resulting from glycolytic activity of the fast growing *S. thermophilus* at this stage. This is followed by rapid increase in D(–) lactic acid at the later stages of fermentation concomitantly with the rapid multiplication of *Lb. delbrueckii* subsp. *bulgaricus*. The balance between D(–) and L(+) lactic acid isomers in yogurt is strain-dependent and yogurt usually contains 45–60% L(+) lactic acid and 40–55% D(–) lactic acid (Tamime and Robinson 2007). The ratio of L(+) to D(–) lactic acid can be used as a quality parameter of yogurt. As the D(–) lactic acid concentration increases, the sharpness of yogurt becomes more pronounced. Therefore, changing the ratio of L(+) lactic acid to D(–) lactic acid may offer an advantage to modify end product in accordance with the consumers' demand. Sharp and acidic yogurt, for example, should contain a low ratio of L(+) to D(–) lactic acid and vice versa. Both isomers are absorbed from human intestinal tract, although the rate of metabolism of the L(+) lactic acid is higher than D(–) lactic acid (Alm 1982). Excess D(–) lactic acid may cause some metabolic disorders and, therefore, World Health Organization (WHO) recommends limited uptake of D(–) lactic acid through foods, i.e., <100 mg per body weight. Especially, it is highly recommended to avoid consuming infant formula containing D(–) and/or DL(±) lactic acid isomers.

To achieve reduction in D(–) lactic acid in yogurt, *ldhD* gene from *Lb. delbrueckii* subsp. *bulgaricus* has been replaced by gene conversion with the *ldhL* gene from *S. thermophilus* resulting in a yogurt starter that exclusively produce L(+) lactic acid. Germond et al. (1995) proposed a reduction in LDH activity in *Lb. delbrueckii* subsp. *bulgaricus* to prevent excess accumulation of D(–) lactic acid. Environmental factors such as pH, incubation temperature and storage temperature can affect ratio of lactic acid isomers (Alm 1982). In the probiotic yogurt stored at <10°C, the L(+) lactic acid isomer becomes dominant and increasing storage temperature causes changes in the ratio of L(+) lactic acid to D(–) lactic acid in favor of D(–) lactic acid (Lee et al. 1998). Lactic acid isomers have been reported to have inhibitory effects on *Lb. delbrueckii* subsp. *bulgaricus* (Benthin and Villadsen 1995). It is well-established that L(+) lactic acid is more inhibitory on *Lb. delbrueckii* subsp. *bulgaricus* than D(–) lactic acid.

Exopolysaccharide (EPS) production

Both *S. thermophilus* and *Lb. delbrueckii* subsp. *bulgaricus* have exopolysaccharide-producing (EPS^+) strains. EPS produced by yogurt bacteria may have different monosaccharide composition (Table 8.2). The structure of the repeating unit of heteropolysaccharides (HEPS) produced by *S. thermophilus* was first determined by Doco et al. (1990). Until now, several repeating unit structures of HEPS produced by both yogurt starter bacteria have been revealed through 1D and 2D ^{1}H-NMR spectroscopy (de Vuyst et al. 2001, Leeflang et al. 2000).

The quantity of EPS produced by yogurt bacteria in pure strain cultures can vary considerably (Bouzar et al. 1997). It was reported that the EPS concentrations in milk cultures may range from 50 to 3000 mg l^{-1} for *S. thermophilus* and from 60 up to 2100 mg l^{-1} for *Lb. delbrueckii* subsp. *bulgaricus* (Bouzar et al. 1997, Grobben et al. 1996). The chemical composition, chain length, and structure of the repeating units of the HEPS produced by yogurt bacteria together with the molar mass and radius of gyration of the EPS molecule determine the texture-promoting properties (Ruas-Mediedo and de los Reyes-Gavilian 2005, Laws and Marshall 2001). No definite correlation can be present between the EPS concentrations and apparent viscosities of yogurt since the composition, charge, spatial distribution and rigidity of EPS from various strains may well differ as well as their ability to interact with proteins. On the other hand, in most cases, the EPS with high molecular mass and relatively stiff chain structure give higher viscosities in the product. Tuinier et al. (1999) demonstrated that chain stiffness is directly related to the type of linkages between subunits and β-(1→4) linkages resulted in stiffer chains than α(1→4) or β-(1→3). The EPS have high relative molecular masses for *Lb. delbrueckii* subsp. *bulgaricus* (0.5×10^6 Da) and for *S. thermophilus* (1×10^6 Da) (Doco et al. 1990, Cerning et al. 1986). The structure of a HEPS produced by yogurt bacteria is strain specific (Table 8.3).

Table 8.2 Repeating units and monosaccharide composition of heteropolysaccharides produced by *S. thermophilus* and *Lb. delbrueckii* subsp. *bulgaricus*.

Structure of the repeating unit	Strain	Glc	NAc Glc	Gal	NAc Gal	Rha	Fuc	Reference
Tetrasaccharide	*S. thermophilus* SFi39	2		2				[1]
	S. thermophilus SY89 & SY102	2		2				[2]
	S. thermophilus CNCMI 733	1		2	1			[3]
	S. thermophilus SFi6	1		2	1			[4]
	S. thermophilus IMDO 01, IMDO 02, IMDO 03, NCFB 859 & 21	1		2	1			[5]
	S. thermophilus SFi20	1		2	1			[6]
Pentasaccharide	*Lb. delbrueckii* subsp. *bulgaricus* 291	3		2				[7]
Hexasaccharide	*S. thermophilus* SFi 12	1		3		2		[1]
	S. thermophilus S3			2		1		[7]
Heptasaccharide	*S. thermophilus* Rs & Sts			5		2		[8]
	S. thermophilus EU 20	2		3		2		[5]
	S. thermophilus NCFB 2393	1		3	1	2		[9]
	S. thermophilus OR 901			5		2		[10]
	Lb. delbrueckii subsp. *bulgaricus* rr	1		5		1		[11]
	Lb. delbrueckii subsp. *bulgaricus* LY03, 24 & 25	1		5		1		[5]
Octasaccharide	*S. thermophilus* MR-1C			5		2	1	[12]

Sources: [1]van Kranenburg et al. (2000), [2]Marshall et al. (2001a), [3]Doco et al. (1990), [4]Stingele et al. (1999), [5]Marshall et al. (2001b), [6]Navarini et al. (2001), [7]Faber et al. (2001a), [8]Faber et al. (1998), [9]Almiron-Roig et al. (2000), [10]Bubb et al. (1997), [11]Gruter et al. (1993), [12]Low et al. (1998)

Table 8.3 Primary structure of heteropolysaccharide produced by various strains of *S. thermophilus* and *Lb. delbrueckii* subsp. *bulgaricus*.

Strains	Primary structure
Lb. delbrueckii subsp. *bulgaricus*[1,2] rr LY03 24 25	β-D-Gal*p* β-D-Gal*p* α-L-Rha*p* 1 1 1 ↓ ↓ ↓ 3 4 3 →2)-α-D-Gal*p*-(1→3)- β-D-Glc*p*-(1→3)- β-D-Gal*p*-(1→4)- α-D-Gal*p*-(1→)
Lb. delbrueckii subsp. *bulgaricus*[3] 291	β-D-Gal*p*-(1→4)- β-D-Glc*p* 1 ↓ 6 →4)- β-D-Gal*p*-(1→4)-β-D-Glc*p*-(1→4)- α-D-Glc*p*-(1→4)-
S. thermophilus[4–9] CNCMI 733 Sfi6 IMDO 01 IMDO 02 IMDO 03 NCFB 859 & 21 Sfi 20	→3)-β-D-Gal*p*-(1→3)- β-D-Glc*p*-(1→3)-α-D-Gal*p*NAc-(1→ 6 ↑ 1 α-D-Gal*p*
S. thermophilus[10] Sfi12	β-D-Gal*p* 1 ↓ 4 →2)-α-L-Rha*p*-(1→2)-α-D-Gal*p*-(1→3)-α-D-Glc*p*-(1→3)-α-D-Gal*p*-(1→3)-α-L-Rha*p*-(1→
S. thermophilus[1,10] Sfi39 SY 89 SY 102	β-D-Gal*p* 1 ↓ 4 →3)-α-D-Glc*p*-(1→3)-β-D-Glc*p*-(1→3)-β-D-Gal*f*-(1→
S. thermophilus[11,12] OR 901 Rs Sts	β-D-Gal*p*-(1→6)-β-D-Gal*p* 1 ↓ 4 →3)-α-D-Gal*p*-(1→3)-α-L-Rha*p*-(1→2)-α-L-Rha*p*-(1→2)-α-D-Gal*p*-(1→3)-α-D-Gal*p*-(1→
S. thermophilus[13] MR-1C	β-D-Gal*p*-(1→6)-β-D-Gal*p* L-Fuc 1 1 ↓ ↓ 4 3 →3)-α-D-Gal*p*-(1→3)-α-L-Rha*p*-(1→2)-α-L-Rha*p*-(1→2)-α-D-Gal*p*-(1→3)-α-D-Gal*p*-(1→
S. thermophilus[14] S3	β-D-Gal*f*2Ac$_{0.4}$ 1 ↓ 6 →3)-β-D-Gal*p*-(1→3)-α-D-Gal*p*-(1→3)-α-L-Rha*p*-(1→2)-α-L-Rha*p*-(1→2)-α-D-Gal*p*-(1→
S. thermophilus[6] EU20	6)-β-D-Gal*p*-(1→6)-α-D-Gal*p*-(1→3)-β-L-Rha*p*-(1→4)β-D-Glc*p*-(1→6)-α-D-Gal*f*-(1→6)-β-D-Glc*p*-(1→ 2 ↑ 1 α-L-Rha*p*

Sources: [1]Marshall et al. (2001b), [2]Gruter et al. (1993), [3]Faber et al. (2001a), [4]Doco et al. (1990), [5]Stingele et al. (1996), [6]Marshall et al. (2001a), [7]Navarini et al. (2001), [8]Degeest et al. (2001), [9]Tiwari and Misra (2007), [10]Lemoine et al. (1997), [11]Faber et al. (1998), [12]Bubb et al. (1997), [13]Low et al. (1998), [14]Faber et al. (2001b)

HEPS production is an energy-intensive process, and yogurt bacteria have a limited number of catabolic pathways from which it can derive energy (Broadbent et al. 2003). This is more pronounced in *S. thermophilus*.

HEPS is synthesized from activated nucleotide sugars and low levels of sugar precursors may be a limiting factor on EPS production. Therefore, recent genetic approaches to enhance the EPS production by *S. thermophilus* and *Lb. delbrueckii* subsp. *bulgaricus* have been concentrated on central sugar metabolism of these bacteria. According to Levander et al. (2002) and Svensson et al. (2005), EPS production by lactic acid bacteria could be improved by altering the levels of enzymes in the central metabolism that influence the production of precursor nucleotide sugars. In *S. thermophilus*, which possesses the Leloir pathway, UDP glucose (UDPglu) and UDP galactose (UDPgal) can be formed from either the glucose or the galactose moiety of lactose (Levander et al. 2002). Phosphoglucomutase (PGM) is the key enzyme of Leloir pathway and catalyses interconversion of glucose-6-phosphate and glucose-1-phosphate. It was demonstrated that *S. thermophilus* mutants lacking PGM activity were able to produce the same amount of EPS when they were grown on lactose (Levander and Rådström 2001). This indicates the importance of galactose metabolism through Leloir pathway for supplying EPS precursors.

HEPS biosynthesis and secretion involve a complex genetic organization, and specific *eps/cps* genes as well as a number of housekeeping genes for synthesis of sugar nucleotides are required in this mechanism. *S. thermophilus* Sfi6 was the first strain in which the genes involved in HEPS biosynthesis were described. Stingele et al. (1996, 1999) revealed that the 14.5-kb *eps* gene cluster *epsABCDEFGHIJKLM* was responsible for EPS production in *S. thermophilus* Sfi6. Later, the complete *cps* gene cluster *cps ABCDEFGHIJKL* (11.2-kb) was also identified in *S. thermophilus* NCFB 2393 (Almiron-Roig et al. 2000, Griffin et al. 1996). Instability of HEPS production by both mesophilic and thermophilic lactic acid bacteria have been reported (Cerning et al. 1990, Faber et al. 1998). It was demonstrated that the spontaneous loss of slime-producing trait from mesophilic lactic acid bacteria has been related to the plasmid encoded genes (van Kranenburg et al. 2000). This is not the case for *S. thermophilus* and *Lb. delbrueckii* subsp. *bulgaricus* as both bacteria do not contain such plasmids (Vescovo et al. 1989). The genetic instability of HEPS in thermophilic yogurt bacteria could be due to existing mobile genetic elements like insertion sequences or to a generalized genetic instability, including DNA deletions and re-arrangements (de Vuyst et al. 2001, Stingele et al. 1996, Bourgoin et al. 1999, Germond et al. 2001). Additionally, *S. thermophilus* may produce exocellular glycohydrolyses which are able to degrade the polysaccharide (Cerning et al. 1990). On contrary, some strains of *S. thermophilus* (e.g., strain ST 111) were reported to produce high molecular mass EPS which is more stable against hydrolyzation (Vaningelgem et al. 2004).

The production of EPS is growth-related and bacterial growth conditions including medium composition (carbon and nitrogen sources), pH and temperature strongly influence the production of HEPS by *S. thermophilus* and *Lb. delbrueckii* subsp. *bulgaricus* (de Vuyst and Degeest 1999, Petry et al. 2000). In general, an excess of carbohydrates in combination with a nutrient limitation, such as nitrogen or phosphorus, stimulates EPS production (Cerning et al. 1990, Bianchi-Salvadori 1997). Bacterial growth phase also affects the biosynthesis of HEPS. It was reported that optimal pH and temperature for HEPS production by *Lb. delbrueckii* subsp. *bulgaricus* is close to optimal growth pH and temperature of this bacteria (Moreira et al. 2000). Lactose concentration of regular milk is sufficient enough for EPS production and no lactose supplementation is required for the further increase in EPS production by yogurt bacteria (Fajardo-Lira et al. 1997). On contrary, lactose deficiency may cause reduction in EPS production by thermophilic lactic acid bacteria. Modification of growth medium by adding salts, sucrose or β-casitone yields higher amounts of EPS production by *Lb. delbrueckii* subsp. *bulgaricus* (Mende et al. 2012, Abdi et al. 2012). Biosynthesis of EPS requires a series of enzymes and likely occurs via lipidic intermediary. Figure 8.1 demonstrates the EPS production pathway in *Lb. delbrueckii* subsp. *bulgaricus* NCFB 2772. Grobben et al. (2000) reported that supplementation of milk with Mn^{+2} and phosphate had positive effect on EPS production by yogurt bacteria. Similarly, addition of neutralizers (e.g., NH_4OH) into growth medium stimulated the EPS production by some strains of *Lb. delbrueckii* subsp. *bulgaricus* without affecting the primary structure of the EPS (Vasilijevic et al. 2005).

As mentioned above, EPS^+ strains of *S. thermophilus* and *Lb. delbrueckii* subsp. *bulgaricus* are used in stirred-type yogurt production to obtain desirable body and texture characteristics (Hess et al. 1997, Hassan et al. 1996). Increasing total solids content of milk, addition of stabilizers (when permitted), and heat treatment are among the well-known technological approaches to improve the texture of fermented milk products. However, recent consumer demands on products with reduced fat, low sugar and low cost, and legal barriers against the use of additives in food formulas such as stabilizers in many countries, provide an advantage for the EPS naturally produced by starter culture.

The texture of yogurt type fermented products made with EPS^+ strains is determined by the interaction between proteins and polysaccharides (Hess et al. 1997). The EPS secretion leads to physical cell adhesion with the formation of a very complex three-dimensional reticulum, in the meshes of which the casein micelles and fat globules remain bound (Bianchi-Salvadori 1997). The binding of hydration water by EPS filaments causes a reduction in the amount of free water molecules and, therefore, increases the concentration of EPS in the serum (Duboc and Mollet 2001). Viscosity of yogurt type products made with EPS^+ strains of yogurt bacteria is affected by the surface charge of the EPS. The viscosity of a yogurt with neutral EPS

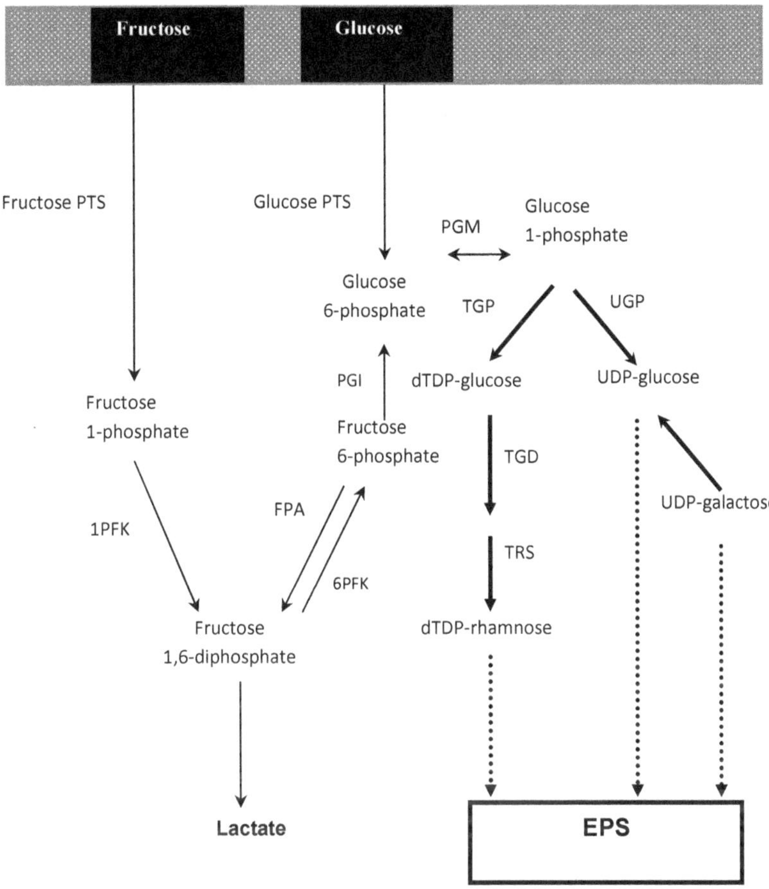

Figure 8.1 Proposed EPS production pathway in *Lb. delbrueckii* subsp. *bulgaricus* NCFB 2772.
After: Bianchi-Salvadori (1997). Reproduced with permission from International Dairy Federation (IDF).

increased with time and about 10 times higher than the yogurt made with non-EPS producing strains (EPS⁻), as reported by Duboc and Mollet (2001). In general, the neutral EPS contribute to the viscosity but not the elasticity and, on the contrary the negatively charged EPS increase the elasticity but do not have any effect on the viscosity development. This is due to the fact that negatively charged EPS interact with positively charged casein particles by electrostatic binding, increasing the network moduli (especially storage modulus, G′). The interacted EPS with caseins are weakly dispersed in the serum phase and, therefore, have little effect on the viscosity development. EPS production ability of yogurt bacteria may decrease as a function of time. Especially, with the increase in sub-culturing, the EPS biosynthesis capacity of strains may be weakened (Cerning 1988).

It has also been suggested that EPS from food-grade bacteria may confer some health benefits such as cholesterol-lowering effect (Pigeon et al. 2002), immunomodulating and anti-tumoral activities (Kitazawa et al. 1998) and prebiotic effects (Dal Bello et al. 2001). Makino et al. (2006) demonstrated that EPS produced by *Lb. delbrueckii* subsp. *bulgaricus* OLL 1073R-1 had an immunomodulatory effect on the human body. However, when this strain was used in combination with *S. thermophilus* OLS 3059, this effect was not observed. β-galactosidase (β-gal) from yogurt bacteria may serve as a suitable tool for the chemoenzymatic synthesis of therapeutic oligosaccharides (Black et al. 2012). The functions of murine macrophages were reported to be enhanced by phosphopolysaccharide produced by *Lb. delbrueckii* subsp. *bulgaricus* OLL 1073R-1 (Nishimura-Uemura et al. 2003). EPS produced by yogurt bacteria play a significant role in the protection of microbial cells against phagocytosis, phage attacks, antibiotics, toxic compounds (e.g., toxic metal ions, sulphur dioxide, ethanol), osmotic stress, and bacteriocins such as nisin (de Vuyst and Degeest 1999, Tiwari and Misra 2007).

Acetaldehyde production

The control of aroma and flavor of yogurt is important for quality control and market success of the end product. The characteristic flavor of the final product (over and beyond that provided by the sharpness of the lactic acid) is mainly attributed to acetaldehyde (Cheng 2010, Martin et al. 2011). The ideal concentration of acetaldehyde in regular yogurt is ca. 10–25 mg kg^{-1}. However, its ratio to other carbonyl compounds such as acetoin, diacetyl and acetone has not been firmly established (Tamime and Robinson 2007). Acetaldehyde production by homofermentative lactic acid bacteria other than yogurt bacteria has been well documented by de Vos and Hugenholtz (2004). On contrary, although tremendous researches have been dedicated to fully understand the flavor chemistry of yogurt, a standard flavor profile for yogurt is still a matter of conjecture. The production of acetaldehyde by yogurt bacteria seems to be strain dependent. No consensus has been reached on which yogurt bacterium is principally responsible for the acetaldehyde production. While some researchers have claimed that *S. thermophilus* produce more acetaldehyde than *Lb. delbrueckii* subsp. *bulgaricus*, others have reported the opposite (Ott et al. 2000).

Yogurt bacteria have diverse shunts for the production of acetaldehyde, which is involved in the metabolism of carbohydrates, proteins and nucleic acids (Fig. 8.2) (Bongers et al. 2005). In all of these cases, the acetaldehyde is a by-product or intermediate, which upon accumulation is toxic and, therefore, must be excreted. Despite a number of metabolic pathways possible for acetaldehdye production, it seems that carbohydrate utilization,

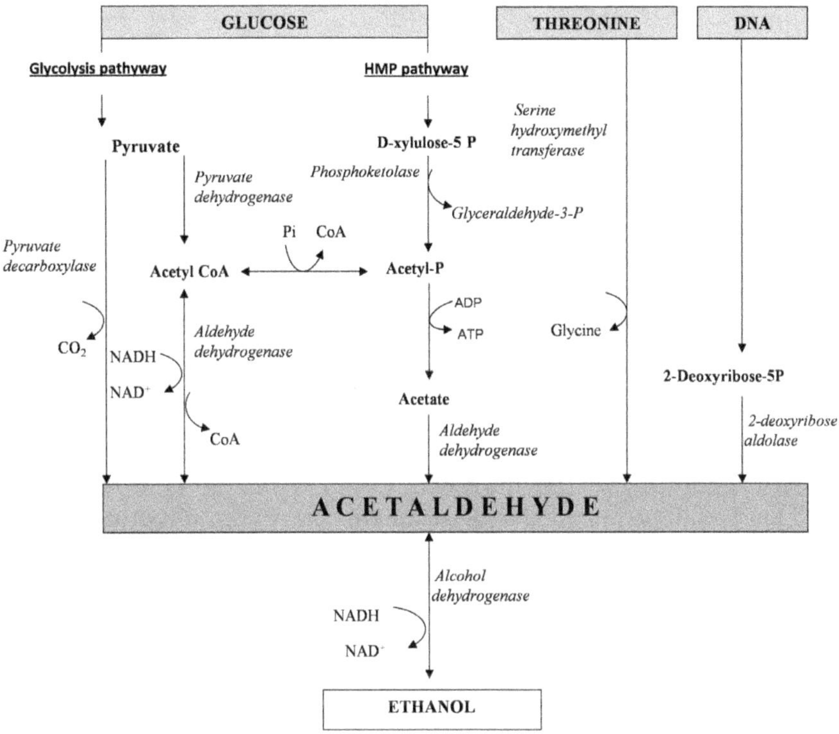

Figure 8.2 Possible metabolic routes for acetaldehyde production in lactic acid bacteria. Based on Chaves et al. (2002) and Özer (2006).

and, in particular, the sources and fates of pyruvate, the key intermediate for several pathways, are considered to be of primary importance (Hugenholtz 1993). During yogurt fermentation, acetaldehyde can be directly produced from lactose metabolism as a result of pyruvate decarboxylation. It can be produced either directly via pyruvate decarboxylase or pyruvate oxidase, or indirectly through the formation of the intermediate acetyl CoA by pyruvate dehydrogenase or pyruvate formate lyase (Chaves et al. 2002). Control of flavor compounds production via these pathways is of special relevance during the later stages of the fermentation and/or cold storage. Additionally, thymidine may be degraded to acetaldehyde. Another possible way of acetaldehyde production is based on conversion of amino acids into acetaldehyde via pyruvate. Threonine aldolase also acts on threnoine yielding acetaldehyde and glycine (Chaves et al. 2002). The possible metabolic pathways of acetaldehyde production by yogurt bacteria are discussed below.

Formation of acetaldehyde from glucose

Glucose oxidation together with the utilization of other carbon sources appears to be the most intriguing area of investigation in acetaldehyde chemistry. This is because of the potentially varied pathways of pyruvate utilization. The most likely route to acetaldehyde formation in yogurt bacteria is conversion of pyruvate which is generated through Embden-Meyerhof-Parnas pathway to acetaldehyde by α-carboxylase (Tamime and Robinson 2007, Zourari et al. 1992). An alternative route involves the action of pyruvate dehydrogenase on pyruvate resulting in the formation of acetyl-CoA which can further be either catalyzed or reduced by an aldehyde dehydrogenase to yield acetaldehyde (Lees and Jago 1976a,b, 1977). Regarding aldehyde dehydrogenase and α-carboxylase activities in yogurt bacteria, various scientific data are available. Lees and Jago (1976a) showed the presence of aldehyde dehydrogenase activity in four strains of each yogurt bacteria. On the contrary, both Manca de Nadra et al. (1987) and Raya et al. (1986) failed to detect neither aldehyde dehydrogenase nor α-carboxylase activities in two strains of *S. thermophilus* and two strains of *Lb. delbrueckii* subsp. *bulgaricus*. Therefore, it is difficult to suggest that formation of acetaldehyde from pyruvate metabolism is likely, as the enzymes involving in this metabolism are rarely present in *S. thermophilus* and *Lb. delbrueckii* subsp. *bulgaricus*.

Acetaldehyde production via hexose monophosphate (HMP) pathway is unlikely since only few strains of *S. thermophilus* possess aldehyde dehydrogenase which catalyses the biosynthesis of acetaldehyde from acetyl-CoA or acetate that are generated by phosphotransacetylase and acetate kinase, respectively, via HMP shunt.

Formation of acetaldehyde from threonine

Lees and Jago (1976b) presented the initial importance for the production of acetaldehyde from threonine by threonine aldolase (EC 2.1.2.1). Threonine aldolase is known to catalyze the conversion of threonine to glycine, with acetaldehyde as a by-product in yogurt bacteria (Sagui et al. 2008). Methionine may serve as a precursor of threonine and, therefore, leads to acetaldehyde accumulation by the same mechanism. The proposed conversion pathway of methionine to acetaldehyde and glycine includes S-adenosyl-homocysteine, L-homocysteine, cysthione, L-homoserine, homoserine phosphate and threonine intermediates (Özer and Atasoy 2002). In single cultures, threonine aldolase activity in lactobacilli is more pronounced than in streptococci (Özer and Atasoy 2002). However, in the presence of amino acid supplementation, or in mixed yogurt culture, *S. thermophilus* greatly increases its threonine aldolase activity (Varga 1998).

Wilkins et al. (1986) labelled threonine to verify that this amino acid was a precursor of acetaldehyde in yogurt. Some 2% of the acetaldehyde was accounted for via L-[^{14}C] threonine. Since methionine was not taken into account in this study, one may expect a greater proportion of the total acetaldehyde to be generated via this mechanism.

Ott et al. (2000) investigated acetaldehyde formation by *S. thermophilus* and *Lb. delbrueckii* subsp. *bulgaricus* during fermentation of cow's milk using ^{13}C-labelled threonine and ^{13}C-labelled glucose. They found that over 90% and almost 100% acetaldehyde originated from glucose during fermentation by *Lb. delbrueckii* subsp. *bulgaricus* and *S. thermophilus*, respectively, taking into account both singly and doubly labelled acetaldehyde. Both microorganisms showed threonine aldolase activity and formed labelled acetaldehyde from ^{13}C-labelled threonine during the fermentation of milk. It was shown that the activity of the streptococcal threonine aldolase decreased with an increase in temperature to normal incubational levels (Marranzini et al. 1989). On the contrary, activity of threonine aldolase in *Lb. delbrueckii* subsp. *bulgaricus* remained almost intact when the incubation temperature was changed (Zourari et al. 1992). Since yogurt is manufactured at 43°C, acetaldehyde is probably produced mainly by *Lb. delbrueckii* subsp. *bulgaricus* (Bianchi-Salvadori 1997).

Supplementation of milk with methionine and/or threonine may, in principle, lead to increase in acetaldehyde level in yogurt. Özer and Atasoy (2002) supplemented milk with methionine (10 and 30 mg 100 ml^{-1} of milk) and threonine (5 and 10 mg 100 ml^{-1} of milk) and monitored the acetaldehyde accumulation in the resulting yogurt made with ropy culture. They concluded that increasing the levels of methionine and threonine added into milk greatly stimulated the acetaldehyde production by mix ropy culture. Similarly, addition of threonine into yogurt milk at much lower concentrations (1 or 3 mg l^{-1} of milk) was reported to cause an increase in acetaldehyde synthesis by yogurt bacteria (Baranowska 2006). Kurultay et al. (2005) reported that methionine supplementation at a level of 100 mg kg^{-1} increased the acetaldehyde production by some strains of *S. thermophilus*.

Chaves et al. (2002) investigated the process of acetaldehyde formation by *S. thermophilus* by focusing on one specific reaction for this mechanism catalyzed by serine hydroxymethyltransferase (SHMT), encoded by the *glyA* gene. In *S. thermophilus*, SHMT also possesses threonine aldolase activity. The authors screened several wild-type *S. thermophilus* strains for acetaldehyde production in the presence and absence of L-threonine. Supplementation of the growth medium with L-threonine led to an increase in acetaldehyde production. Similar observations were made by Tong et al. (2012) who investigated the acetaldehyde production in *S. thermophilus* MGD4-7 in the presence of L-threonine in growth medium. To investigate the physiological role of SHMT, a *glyA* mutant was constructed by gene disruption. Inactivation of *glyA* resulted in a severe reduction in threonine aldolase activity and complete loss of acetaldehyde formation during fermentation. In contrast, the threonine aldolase activity in and acetaldehyde production by a *S. thermophilus* strain in which the *glyA* gene was cloned under the control of a strong promoter (P_{LacA}), increased during fermentation. This indicated that in *S. thermophilus*, SHMT constitutes the main pathway for acetaldehyde formation (Chaves et al. 2002).

High glycine concentration in the growth medium is a limiting factor for the threonine aldolase activity in both *S. thermophilus* and *Lb. delbrueckii* subsp. *bulgaricus*. Schmidt et al. (1989) studied the inhibitory effect of glycine on different strains of yogurt bacteria. The authors demonstrated that a glycine level of 25 μmole added into growth medium containing 125 μmole of threonine resulted in a range of inhibition from 39.1 to 98.1% for *S. thermophilus* and from 2.1 to 18.9% for *Lb. delbrueckii* subsp. *bulgaricus*. Goat's milk contains high level of glycine and, therefore, acetaldehyde accumulation in yogurt made from goat's milk is relatively lower than yogurts from other major milk species due to high level of threonine aldolase inhibition (Rysstad et al. 1990). Rodriguez-Serrano et al. (2002) demonstrated that acetaldehyde production by mixed yogurt starters including *S. thermophilus* NCFB 2075 and *Lb. delbrueckii* subsp. *bulgaricus* NCFB 2074 in ultrafiltered whey was higher than the unconcentrated milk. This was attributed to the different protein types leading to different glycine contents, which, in turn, could result in a lower activity of threonine aldolase in milk than in concentrated whey.

The inhibitory effect of glycine on threonine aldolase is strain dependent (Wilkins et al. 1986). High salt concentration and presence of divalent cations such as Cu^{+2}, Zn^{+2}, Fe^{+2} and Co^{+2} also cause inhibition in threonine aldolase activity (Schmidt et al. 1989, Wilkins et al. 1986). Maximum threonine aldolase activity is observed at 40°C and pH 6.5 for *Lb. delbrueckii* subsp. *bulgaricus* and at 30°C (less at 37–42°C) for *S. thermophilus* (Tamime and Robinson 2007, Zourari et al. 1992).

Acetaldehyde production from nucleic acids

The formation of acetaldehyde has also been attributed to 2-deoxyriboaldolase which is possessed by at least four strains of *S. thermophilus* and one strain of *Lb. delbrueckii* subsp. *bulgaricus* (Raya et al. 1986). This enzyme, along with thymidine phosphorylase and deoxyribomutase, is involved in catabolic pathways of DNA and produce acetaldehyde as a by-product (Raya et al. 1986, Varga 1998). The substrate of this pathway, 2-deoxyribose-5-phosphate, is probably a catabolite of thymidine. The enzyme, however, also serves in the oxidation of deoxyribose in *Salmonella* Typhimurium and other bacteria which uses this substance as an energy source. This particular pathway is not expected to be of primary importance in the control of acetaldehyde production during the short period of fermentation.

Further degradation of acetaldehyde to ethanol by alcohol dehydrogenase is rarely possible in yogurt as this enzyme has not been detected in *Lb. delbrueckii* subsp. *bulgaricus*, but has been found in at least two strains of *S. thermophilus* (Lees and Jago 1976a, Raya et al. 1986).

Proteolytic activity

Although the yogurt bacteria are moderately proteolytic, they may cause significant level of proteolysis during fermentation of yogurt. Milk proteins (particularly caseins) are the major sources of nitrogen for yogurt starter bacteria. According to Loones (1989), free amino acid content doubles in yogurt within 24 hours of production, and the proteolysis continues during storage of yogurt, doubling free amino acids again in 21 days storage at 7°C. Most strains of *S. thermophilus* and *Lb. delbrueckii* subsp. *bulgaricus* possess proteinase activity (Kalantzopoulous et al. 1990). Proteinase from *Lb. delbrueckii* subsp. *bulgaricus* (PrtB) belongs to the sub-family of cysteine subtilisins and has a predicted molecular mass of 212 kDa, with 1946 residues (Gilbert et al. 1996). This enzyme is a metallo-enzyme requiring Zn^{+2} for its activity. Its proteolytic activity is optimum at 45–55°C and pH 5.2–5.8. The PrtB is more effective on β-casein and α-casein than whey proteins (Laloi 1989). β-casein is more readily hydrolyzed by yogurt bacteria than other casein fractions (Kalanzopoulous et al. 1990, Khalid et al. 1991).

On the other hand, some strains of *Lb. delbrueckii* subsp. *bulgaricus* (e.g., CRL 656) are able to degrade highly allergenic β-lactoglobulin (β-LG) fraction of whey proteins, making this protein fraction less immune-reactive, and these strains have potential to be used in the development of hypoallergenic dairy products (Pescuma et al. 2011). The activity of proteases from *Lb. delbrueckii* subsp. *bulgaricus* increases with the decrease in milk pH (pH <5.0). Therefore, these enzymes are more active in the later stage of fermentation (Stefanitsi and Garrel 1997). The proteolytic action of *S. thermophilus* on caseins during fermentation is rather weak (Meyer et al. 1989). The cell-wall envelope proteinase from *S. thermophilus* (PrtS), a member of subtilase family, is tightly anchored to the cell-wall via a mechanism involving the typical sortase A (SrtA) and initiates the breakdown of casein into small oligopeptides. PrtS activity is enhanced in the presence of Ca^{+2}, Mn^{+2} and Mg^{+2} (Fernandez-Espla et al. 2000) and it is a key enzyme of milk acidification by *S. thermophilus* (Dandoy et al. 2011).

The absence of active PrtS and PrtB in $PrtS^-$ and $PrtB^-$ mutants of *S. thermophilus* and *Lb. delbrueckii* subsp. *bulgaricus*, respectively, reduces the growth of these bacteria in growth medium (Courtin et al. 2002, Galia et al. 2009).

Peptide utilization by yogurt starters

Oligopeptides released by proteases from *Lb. delbrueckii* subsp. *bulgaricus* are further degraded to small molecular weight peptides and amino acids by endo- and exopeptidases, aminopeptidases, dipeptidases, tripeptidases and peptidyl peptidases (Kunji et al. 1996), and these nitrogenous compounds are utilized by *S. thermophilus* for their growth (Bockelman et al. 1992). Exopeptidases of *S. thermophilus* are more effective on peptides than those synthesized by *Lb. delbrueckii* subsp. *bulgaricus* (Bianchi-Salvadori et al. 1995). On the other hand, endopeptidase activity in *Lb. delbrueckii* subsp. *bulgaricus* is more pronounced and endopeptidases are primarily responsible for the hydrolysis of casein fractions (Bertrand-Harb et al. 2003). *S. thermophilus* have higher peptidase and aminopeptidase activities than its proteinase activity (Shankar 1977). Dipeptidases released from *S. thermophilus* are primarily effective on valily, leuciyl, alanyl and arginyl. Most strains of *S. thermophilus* have leucine-aminopeptidase activity (Bouillanne and Desmazeaud 1980). Some strains also possess aminopeptidase N (Pep N), aminopeptidase C (Pep C) and arginine-aminopeptidase activities. *Lb. delbrueckii* subsp. *bulgaricus* have four cell-wall associated aminopeptidases (AP 1-IV) (Laloi et al. 1991). These enzymes show optimum activity at 30°C and pH 5.5 (Ezzat et al. 1987).

Almost all strains of *Lb. delbrueckii* subsp. *bulgaricus* and *S. thermophilus* have X-propyl-dipeptidylamino-peptidase (X-pro-DPAP) activity (Meyer and Jordi 1987, Bockelmann et al. 1991). X-pro-DPAP is a metallo-enzyme and requires Ca^{+2} and Mn^{+2} for its activity. This enzyme is rapidly inhibited in the presence of Cu^{+2}, Fe^{+2}, Zn^{+2} or Hg^{+2} (Bockelmann et al. 1991). X-pro-DPAP is a key enzyme of casein hydrolyzation since it uses proline as a substrate and casein is a rich source of proline (Meyer and Jordi 1987). X-pro-DPAP is also primarily responsible for the formation of dipeptide in yogurt (Atlan et al. 1990, Laloi 1989). Proteolytic enzymes synthesized by yogurt starter bacteria are presented in Table 8.4. In general, the proteolytic activity is maximum in the logarithmic growth phase of yogurt bacteria and declines during the stationary phase of bacterial growth. Some free fatty acids (e.g., capric acid) have a negative effect on the proteolyitc activity of yogurt bacteria.

Lipolytic activity

Although yogurt starter bacteria have a limited lipolytic capacity, products of lipolysis contribute to the aroma/flavor development in yogurt. Triacylglycerol lipase from *S. thermophilus* hydrolyses tributyrin and triolein, but this enzyme shows fairly weak activity on milk fat (Tamime and Deeth 1980). *Lb. delbrueckii* subsp. *bulgaricus* have an intracellular esterase which

Table 8.4 Some properties of proteolytic enzymes synthesized by yogurt starter bacteria.

Enzyme	Origin	Enzyme type	Molecular weight (kDa)	Optimum pH
Pep N	*Lb. delbrueckii* subsp. *bulgaricus*			
	B-14	M	95	7.0
	S. thermophilus			
	ACA-DC 114	-	89	6.5
	CNRZ 302	M	97	7.0
	NCDO 537	-	96	-
Pep C	*Lb. delbrueckii* subsp. *bulgaricus*			
	B14	T	54	7.0
X-pro-DPAP	*Lb. delbrueckii* subsp. *bulgaricus*			
	B14	S	95	6.5
	CNRZ 397	S	82	7.0
	LBU 47	S	90	6.5
	S. thermophilus	S	165	>6.5
Proline *iminopeptidase*	*Lb. delbrueckii* subsp. *bulgaricus*			
	CNRZ 397	S	3	6.5
Dipeptidase (PEP D) (PEP V)	*Lb. delbrueckii* subsp. *bulgaricus*			
	B14	M	51	7.5

M: Metalloenzyme, **T:** Triolpeptidase, **S:** Serin-protease
After: Tamime and Robinson (2007)

acts on *orto-* and *para-*nitrophenyl (El-Soda et al. 1986). *Para-*nitrophenyl is hydrolyzed faster than *orto-*nitrophenyl (Khalid et al. 1990). Esterases from both yogurt starter bacteria show optimum activity at 40–50°C and around pH 7.

Urease activity

S. thermophilus is capable of producing CO_2 through a metabolic pathway different from homolactic fermentation, and urease activity plays a key role in this mechanism (Juillard et al. 1988). CO_2 is released during conversion of urea to ammonia by urease. Accumulation of ammonia in yogurt slows down the development of acidity which eventually affects the fermentation kinetics. In order to overcome this problem, urease-negative (urease$^-$) strains of *S. thermophilus* should be used in the manufacture of yogurt (Monnet et al. 2004, Mora et al. 2004). Metabolism of urea also affects the kinetics of growth of *S. thermophilus* (Pernoud et al. 2004). Urease activity in *S. thermophilus* decreases significantly during the stationary phase of growth (Juillard et al. 1988).

Oxygen metabolism

Lb. delbrueckii subsp. *bulgaricus* is capable of producing hydrogen peroxide (H_2O_2) in the presence of oxygen at high concentrations, and the H_2O_2 produced by this bacterium activates the lactoperoxidase system (LP-system) of milk. As a result of activation of the LP-system, the growth of *Lb. delbrueckii* subsp. *bulgaricus* is partly inhibited (Özer et al. 2003). Unlike *Lb. delbrueckii* subsp. *bulgaricus, S. thermophilus* cannot produce H_2O_2 at levels high enough to activate the LP-system in milk (Guirguis and Hickey 1987). Most strains of *S. thermophilus* do not possess NADH-oxidase activity, an enzyme plays active role in O_2 and H_2O_2 metabolism.

Vitamin metabolism

During yogurt fermentation, *S. thermophilus* produces folic acid which is further used by *Lb. delbrueckii* subsp. *bulgaricus* as a growth factor. The concentration of folic acid decreases at the stage of fermentation where the growth of *Lb. delbrueckii* subsp. *bulgaricus* is at logarithmic phase (Kaneko et al. 1987). Although it is not as important as folic acid, vit-B_{12} is also used by *Lb. delbrueckii* subsp. *bulgaricus* as a growth factor (Rao et al. 1984). Both yogurt bacteria can synthesize niacin and vit-B_6 during fermentation of yogurt.

8.3 Kefir

Kefir is a self-carbonated fermented milk product with white or greenish color. The raw material and the microflora of kefir are the determinative factors for the chemical and microbiological characteristics of kefir. The microflora and chemical composition of kefir are subject to regional variations (Güzel-Seydim et al. 2011). During fermentation, lactose-fermenting yeasts produce alcohol (ethanol) and CO_2, lactic acid bacteria convert lactose to lactic acid and a limited degree of proteolysis occurs in milk. The lactic acid and ethanol contents of kefir vary between 0.8–1.0% and 0.035–2.0%, respectively. Kefir has a balanced and yeasty aroma. It has an acidic but pleasant taste and the texture of kefir is thick but not gluey with an elastic consistency. Kefir grains range in size from 0.3 to 2.0 cm or more in diameter, and are characterized by forming an irregular, folded or uneven surface; the grains resemble cauliflower florets in shape and color. During fermentation period, the biomass of kefir grains increases continuously and the newly formed grains carry the characteristics of the grains seeded initially.

The total solids level of a fresh kefir grain is around 10–16% which is mainly constructed by proteins (30% of dry mass) and carbohydrates including polysaccharides (25–50% of dry mass). The exopolysaccharide material produced by kefir microflora is called "kefiran". Kefiran consists of monosaccharides including glucose, galactose and mannose in varying ratios, and the level of kefiran is affected by fermentation temperature but not fermentation time (Zajsek et al. 2011). Apart from its well-known texture-enhancing properties, kefiran has a health-promoting effects including immunomodulatory, anti-mutagenic, anti-ulceric, anti-allergic and anti-tumor activities, and acts as a prebiotic substance. Güzel-Seydim et al. (2006) found that the mutagenicity induced by aflatoxin B1, sodium azide, and methyl methanosulfate was significantly reduced by kefir extracts. Additionally, kefir contains high levels of conjugated linoleic acid (CLA) which has anti-carcinogenic, hypocholesterolemic and anti-atherogenic effects (Ip et al. 1994). Early studies revealed that a water soluble polysaccharide isolated from kefir grains inhibited the growth of Ehrilch carcinoma and Sarcoma 180 to a great extent (Shiomi et al. 1982, Murofushi et al. 1983). More recently, de Moreno de LeBlanc et al. (2006) demonstrated that breast cancer tumors were reduced and IgA(+) cells were increased in mice treated with kefir and cell-free extracts of kefir for 27 days. Cenesiz et al. (2008) showed an anti-oxidative effect of kefir in mice with colonic abnormal crypt formation. On the contrary, Hlastan-Ribic et al. (2005) failed to find any protective effect of kefir against colorectal epithelial tumors in Wistar rats. *Lactobacillus kefiranofaciens* M1 isolated from natural kefir grains was shown to have a potential to be applied in fermented dairy products as an alternative therapy for intestinal disorders (Chen et al. 2012).

Cholesterol-reducing effect of kefir has long been known (Yoon et al. 1998). The origin and fermentation conditions of kefir are the determinative factors for the cholesterol-assimilation activity in kefir microflora. On the other hand, both cell *in vitro* test and animal studies have yielded contradictory results on cholesterol- or lipid peroxidation reducing effects of kefir. St-Onge et al. (2002), for example, detected no reduction in lipid peroxidation level in 13 hypocholesterolemic men upon consumption of kefir. On the contrary, Güven et al. (2003) demonstrated the protective effects of kefir against lipid peroxidation in mice. Cell-wall materials such as α-mannan and β-glucan from *Kluyveromyces marxianus* YIT 8292 were reported to be responsible in hypocholesterolemic effect in rats (Yoshida et al. 2005). Urdaneta et al. (2007) observed that the levels of triacylglycerol, cholesterol and HDL increased in Wistar rats fed with kefir supplemented diet.

Few studies are available on the ACE-inhibitory action of kefir in the literature. Two peptides (PYVRYL and LVYPT) detected in kefir made from caprine milk were reported to have ACE-inhibitory activity (Quiros et al. 2005). Maeda et al. (2004) established a relationship between the consumption of kefir and reduction in blood pressure in rats.

Animal studies have revealed that kefir consumption stimulates the immune system significantly (Thoreux and Schmucker 2001). The immunomodulatory activity of kefiran depends on kefir dose and numbers of cells in kefir grains (Vinderola et al. 2005).

Kefir contains a number of anti-bacterial agents including organic acids, hydrogen peroxide, acetaldehyde, CO_2 and bacteriocins (Powell et al. 2007). Ulusoy et al. (2007) found that kefir made by using a commercial starter blend (PROBAT K3 from Danisco) had a strong anti-pathogenic effect against *Salmonella enteritidis* (ATCC 13076), *Listeria monocytogenes* (ATCC 7644), *Staphylococcus aureus* (ATCC 29213), *Bacillus cereus* (ATCC 11778) and *Eschericia coli* (ATCC 8739). *Salmonella* Typhimurium was also reported to be inhibited by *Lactobacillus* spp. isolated from kefir (Santos et al. 2003). The anti-microbial effect of kefir on Gram(+) bacteria was reported to be far more pronounced than that on Gram(–) bacteria or yeasts (Garrote et al. 2000). The anti-microbial effect of kefir is comparable with common antibiotics such as ampicillin and gentamycin. Reduction in lactose content of kefir makes it more suitable for people suffering from lactose-intolerance. Kefir consumption caused a reduction in the severity of flatulence by 1%. Health-promoting and anti-bacterial effects of kefir were well-documented in elsewhere (Güzel-Seydim et al. 2010, 2011, Sarkar 2008).

8.3.1 Microbiology of kefir grains

Lactic acid bacteria (~10^8–10^9 cfu ml^{-1}), yeasts (~10^5–10^6 cfu ml^{-1}), acetic acid bacteria (~10^5–10^6 cfu ml^{-1}) and possibly a mold are the major microbial groups of a kefir grain (Simova et al. 2002, Garrote et al. 2001, Güzel-Seydim et al. 2010). Depending on the origin of kefir grains, the ratio between these microbial groups shows variations. For example, lactobacilli were reported to be the dominant microbial group (80%) in Polish origin kefir grain, followed by yeasts (12%) and lactococci (8%) (Anonymous 2002). On the contrary, Kroger (1993) found that European origin kefir grains contained streptococci at much higher levels (10^8–10^{10} cfu ml^{-1}) than the other microbial groups (10^5 cfu ml^{-1} yeast, 10^5 cfu ml^{-1} thermophilic lactobacilli and 10^2–10^3 cfu ml^{-1} mesophilic lactobacilli). Table 8.5 presents the microorganisms isolated from different kefir grains. Variability in the population of lactic streptococci and yeasts in kefir grains are much more pronounced than that of lactobacilli (Ninane et al. 2005, Witthuhn et al. 2004). Wide variation in microflora of kefir grains makes it difficult to obtain an optimal and uniform starter culture necessary for obtaining a quality kefir beverage (Sarkar 2008).

Lactobacillus kefiranofaciens is the most prevalent bacteria of kefir grains. The carbohydrate metabolism in *Lb. kefiranofaciens* is somehow different from other homofermentative species of the genus *Lactobacillus*

Table 8.5 Microflora of kefir from different origins.

Lactococci	Lactobacilli	Leuconostocs	Yeasts and moulds	Acetic acid bacteria	Others
L. lactis subsp. *lactis*	*Lb. kefir*	*Leu. mesenteroides* subsp. *mesenteroides*	*Bretannomyces anomalus*	*Acetobacter aceti*	*Streptococcus thermophilus*
L. lactis subsp. *cremoris*	*Lb. kefiranofaciens*	*Leu. mesenteroides* subsp. *cremoris*	*Candida colliculosa*	*A. rasens*	*Enterococcus durans*
L. lactis subsp. *lactis* biovar. *diacetylactis*	*Lb. kefirgranum* (recently *Lb. kefiranofaciens* var. *kefirgranum*)	*Leu. mesenteroides* subsp. *dextranicum*	*C. friedrichii*	*A. pasteurianus*	*E. faecium*
L. filant	*Lb. parakefir*		*C. kefyr*		*Bifidobacterium animalis* subsp. *animalis*
	Lb. acidophilus		*C. holmii*		
	Lb. brevis		*C. valida*		
	Lb. casei		*C. inconspicua*		
	Lb. helveticus		*C. maris*		
	Lb. delbrueckii		*C. pseudotropilactis*		
	Lb. rhamnosus		*C. tenuis*		
	Lb. fermentum		*C. albicans*		
	Lb. gasseri		*Cryptococcus kefyr*		
	Lb. paracasei subsp. *pseudoplantarum*		*Debaryomyces hansenii*		
	Lb. paracasei subsp. *tolurans*		*Issatchenkia orientalis*		
	Lb. paracasei subsp. *paracasei*		*Kluyveromyces marxianus* var. *fragilis*		

Table 8.5 contd....

Table 8.5 contd.

Lactococci	**Lactobacilli**	**Leuconostocs**	**Yeasts and moulds**	**Acetic acid bacteria**	**Others**
L. filant	*Lb. delbrueckii* subsp. *bulgaricus*		*K. marxianus* var. *lactis*		
	Lb. cellobiosus		*K. marxianus* var. *marxianus*		
	Lb. pracasei subsp. *alactosus*		*Mycotorula lactis*		
	Lb. plantarum		*M. lactosa*		
	Lb. viridescens		*Pichia fermentans*		
	Lb. fructivorans		*Saccharomyces cerevisiae*		
	Lb. hilgardii		*S. dairenis*		
	Lb. paraplantarum		*S. exigus*		
			S. unisporus		
			S. turicensis		
			Torulaspora delbrus		
			T. delbrueckii		
			Zygosaccharomyces florentinus		
			Geotrichum candidum (possibly contaminant)		

(Güzel-Seydim et al. 2010). *Lb. kefiranofaciens* and *Lb. kefirgranum* have the same 16S rDNA sequence and hence the latter organism was re-classified as *Lactobacillus kefiranofaciens* subsp. *kefirgranum* (Vancanneyt et al. 2004). *Lb. kefir, Lb. brevis, Lb. paracasei, Lb. plantarum* and *Lb. acidophilus* are the most frequently isolated species of the genus *Lactobacillus* from kefir grain. Santos et al. (2003) isolated totally 53 strains of six different *Lactobacillus* species including *Lb. kefir, Lb. brevis, Lb. paracasei, Lb. plantarum, Lb. acidophilus* and *Lb. kefiranofaciens. Leuconostoc mesenteroides* subsp. *mesenteroides, Leu. mesenteroides* subsp. *cremoris* and *Leu. mesenteroides* subsp. *dextranicum* also constitute the dominant flora in kefir grains of different origins. The microstructure of a kefir grain as studied by electron microscopy (both SEM and TEM) revealed that kefir grain has a spongy fibrillar structure with a reticular lamellar matrix and a fiber mass especially in the center of the grain (Güzel-Seydim et al. 2005, Bottazzi and Bianchi 1980). A sheet-like structure and scroll-like forms in kefir were also reported by Marshall et al. (1984). The outer layer of the kefir grain generally contains the long rod shaped lactic acid bacteria (including *Lb. kefir*) (Toba et al. 1990). The yeasts are usually located in the inner part (core) of the kefir grain (Güzel-Seydim et al. 2005, Lin et al. 1999). Curved rods are only present in the interior section of the grain matrix.

There is a balance between yeasts and bacteria at the intermediate zone of the grain and a progressive change according to the distance from the core have been reported. The varieties of *Kluyveromyces marxianus* including var. *marxianus, fragilis* and *lactis* are frequently isolated from kefir grain. *K. marxianus* is primarily responsible for the development of characteristic yeasty aroma of kefir drink as well as high level of ethanol production. *Saccharomyces* spp. (including *cerevisiae, unisporus, exiguous, humaticus, turicensis* and *delbrueckii*) which is widely present in kefir grain is capable of fermenting D-glucose, D-galactose and sucrose but not lactose (Angulo et al. 1993, Rohm et al. 1992). Other yeast species isolated from kefir grain are *Torulopsis holmii, Candida* spp. (e.g., *holmii* and *friedrichii*) (Wang et al. 2008, Angulo et al. 1993), *Torulaspora delbrueckii* (Angulo et al. 1993), *Issatchenkia orientalis* (Lattore-Garcia et al. 2007), and *Pichia fermentans* (Rohm et al. 1992). There is a symbiotic relationship between yeasts and lactic acid bacteria in the kefir grain (Sarkar 2008). Yeasts provide an environment which is favorable for the growth of lactic acid bacteria (possibly by providing growth stimulants). Recently, Wang et al. (2012) showed that *Lb. kefiranofaciens* and *S. turicensis* possess strong auto-aggregation ability and *Lb. kefir* shows significant biofilm formation properties in kefir grain. According to same authors, the formation of kefir grains begins with the self aggregation of *Lb. kefiranofaciens* and *S. turicensis* to form small aggregates. Then, *Lb. kefir*-a biofilm-producer with a high hydrophobisity at pH 4.2 and high net positive cell-surface charge, attach to this newly formed granules.

Since the kefir grain contains microorganisms of various genus, it is rather difficult to lay out the whole microbiological profile of kefir grains by using conventional culture-dependent methods. Therefore, more sensitive methods such as denaturing gradient gel electrophoresis (DGGE) of partially amplified 16S rDNA, in monitoring the microbial flora of kefir grain are needed (Leite et al. 2012, Garbers et al. 2004, Lattore-Garcia et al. 2007, Wang et al. 2008). For example, conventional isolation revealed the presence of *Lb. helveticus*, *Lb. kefir* and *Acetobacter syzygii*; however, these bacteria were not among the sequenced DGGE bands. On the contrary, conventional culture-dependent method failed to isolate *Lb. satsumensis*, *Lb. uvarum* and *Gluconobacter japonicus* that were sequenced by PCR-DGGE method. Kök-Taş et al. (2012) demonstrated the presence of *Bifidobacterium bifidum* in kefir grain for the first time by using PCR.

8.3.2 Biochemistry of kefir

The unique physical, chemical and sensory characteristics of kefir are determined by biochemical metabolites formed during fermentation. Lactic acid and flavor compounds including acetaldehyde, acetoin, diacetyl, ethanol, acetic acid and CO_2 are the major biochemical products produced by mainly lactic acid bacteria (Beshkova et al. 2003). A slow increase in the level of lactic acid is observed during the early stages of fermentation, followed by a rapid increase at the later periods of fermentation. As with lactic acid, the formation of acetaldehyde and acetoin takes place at the later stages of fermentation (Güzel-Seydim et al. 2000). On the contrary, the levels of orotic, citric and pyruvic acids slightly decrease as the fermentation continues. The metabolic activities of kefir flora continue during the cold storage of kefir. The concentrations of carbonyl compounds, CO_2 and ethanol of kefir increase during storage period. Kefir starter produces more carbonyl compounds than kefir grains. While some vitamins including vit-B_1, vit-B_2, folic acid, vit-K and vit-P (riboflavenoid) are produced during fermentation of kefir, other vitamins are utilized by the microflora. When *Propionibacterium freudenreichii* is incorporated into kefir starter, the concentrations of vit-B_{12} and folate are increased. The presence of macro-minerals including K, Ca, Mg, P and micro-elements such as Cu, Zn, Fe, Mn, Co and Mo in kefir have been reported.

Limited levels of lipolysis and proteolysis occur during and/or after fermentation of kefir. As a result of lipolysis by mainly yeasts including *Torulopsis* spp. and *Candida* spp., methyl ketones (i.e., 2-nonanone and 2-heptanone), alcohols, esters and lactones are formed. The yeasts are also able to degrade caseins to small molecular weight peptides and free amino acids (Wszolek et al. 2006, Kahala et al. 1993). Free amino acids are further converted to alcohols, aldehydes, volatile acids, esters and sulphur-containing compounds. α-lactalbumin (α-LA) and β-casein (β-CN) are more easily hydrolyzed than α-CN by kefir microflora (Ferreira et al. 2010). The degree of protein hydrolysis depends on grain to milk ratio and fermentation temperature. Ferreira et al. (2010) showed that β-lactoglobulin (β-LG) withstood hydrolysis by kefir flora during fermentation. The threonine, serine, alanine, lysine and ammonia contents of Turkish kefir were found to be significantly higher than yogurt (Güzel-Seydim et al. 2003). While the glutamic acid level of kefir was decreased, the concentration of gamma aminobutyric acid (GABA) which is effective in lowering blood pressure was increased (Gronnevik et al. 2011). The formation of biogenic amines including tyramine, putrescine, cadaverine and spermidine during fermentation of kefir was reported. Metabolic pathways of yeast fermentation in kefir have been well-documented by Walker (1998).

8.4 Koumiss

Koumiss is a traditional drink of nomad cattle breeders with ancient origins, and is common in Eastern Europe and Central Asia (Wang et al. 2008, Özer and Özer 1999). It is traditionally produced from mare's milk by a combined fermentation to lactic acid and alcohol, and its highly nutritive and curative characteristics are well known (Mu et al. 2012). In Mongolia, camel's milk is also used in the preparation of traditional koumiss. A koumiss-like fermented milk product is manufactured from cow's milk (skimmed or whole) in some European countries and the USA. In traditional koumiss production, freshly drawn mare's milk is seeded with previous day's product in a smoked horsehide called saba or turdusk-burdusk (Danova et al. 2005, Özer and Özer 1999). This smoked horsehide contains koumiss microflora from the previous season. Koumiss fermentation usually takes 3 to 8 hr. Traditionally, koumiss production is started early summer and ceased in late autumn, and the koumiss starter is kept in glass bottle until re-use. Re-activation of koumiss starter microflora is achieved by keeping the koumiss starter at room temperature for about 24 hr. The re-activated starter is mixed with fresh mare's milk (or camel's milk in some cases). This mixture is vigorously stirred for about an hour to incorporate air into the mixture to stimulate the growth of yeasts which are responsible for the generation of alcohol.

In industrial koumiss production, cow's milk is modified by means of membrane technology (UF, MF or NF) to adjust the chemical composition of cow's milk to mare's milk (Kücükcetin et al. 2003). Alternatively, cow's milk is diluted with potable water to desired casein level (Malacarne et al. 2002). Whey protein powder or concentrate is added to the diluted milk and glucose, sucrose or hydrolyzed lactose is added to stimulate the growth of koumiss microflora (especially yeasts). Commercial starter cultures including *Kluyveromyces marxianus* var. *lactis*, *Lb. delbrueckii* subsp. *bulgaricus* and *Lb. acidophilus* are used in the inoculation of industrial koumiss (Viljoen 2001). The starter is added to processed milk (e.g., fat adjustment, heat treatment, etc.) by continuously stirring the milk. Fermentation is allowed at 25–26°C until the acidity reaches ~0.50% lactic acid, which normally takes about 1 hr. Then the fermenting milk is homogenized, cooled (~20°C) and packaged. The packaged product is further incubated at 18–20°C for about 1.5–2 hr and then is stored at 4–6°C for 12–24 hr before dispatch.

8.4.1 Microbiology of koumiss

The basic microflora of koumiss is constructed by the lactobacilli (*Lb. delbrueckii* subsp. *bulgaricus*, *Lb. casei*, *Lb. leichmanii*, *Lb. plantarum*, *Lb. helveticus*, *Lb. fermentum*, *Lb. buchneri* and *Lb. acidophilus*) and lactose-fermenting yeasts (*Saccharomyces* spp., *Kluyveromyces* spp. and *Candida koumiss*, *Torula lactis*, *T. koumiss*) (Montanari et al. 1997). Non-lactose-fermenting yeasts (e.g., *Saccharomyces cartilaginosus*) and non-carbohydrate-fermenting yeasts (e.g., *Mycoderma* spp.) may also be present in traditional koumiss. Wang et al. (2008) isolated and identified 12 strains of *Lactobacillus* spp. from homemade Chinese origin koumiss made from mare's milk, with *Lb. helveticus*, *Lb. casei*, *Lb. fermentum* and *Lb. plantarum* being dominant species. Sedláček et al. (2010) identified six *Lb. helveticus* strains isolated from traditional koumiss made from mare's milk by rep-PCR fingerprinting with $(GTG)_5$ primer. The authors concluded that the isolates analyzed represented a new ecovar, *Lb. helveticus* ecovar. Koumiss. To date, the most comprehensive study on the characterization of dominant yeast flora of koumiss was conducted by Mu et al. (2012). The authors detected twelve different yeast species belonging to nine genera including *Candida pararugosa*, *Dekkera anomala*, *Geotrichum* spp., *Issatchenkia orientalis*, *Kazachstania unispora*, *Kluyveromyces marxianus*, *Pichia deserticola*, *P. fermentans*, *P. membranoefaciens*, *Peronospora manshurica*, *S. cerevisiae* and *Torulaspora delbrueckii*. *K. marxianus*, *Kazachstania unispora* and *S. cerevisiae* were the dominant species present in traditional koumiss.

As with kefir, the microflora of koumiss is determined by the origin of the product and climatic conditions of the regions where koumiss is produced. For example, while *S. unisporus*—a galactose-fermenting yeast—was prevalent in Kazakhstan-origin koumiss, *Lactobacillus* spp. (e.g., *rhamnosus*, *paracasei* subsp. *paracasei*, *paracasei* subsp. *tolerans*, *curvatus*) were reported to be the dominant group in koumiss from inner Mongolia and China. The average numbers of bacteria and yeasts in koumiss are 5×10^7 cfu ml^{-1} and 1–2×10^7 cfu ml^{-1}, respectively. A number of metabolites are generated during the fermentation of koumiss milk including alcohol, glycerol, succinic acid and acetic acid. As a result of the accumulation of lactic acid and ethanol, the population of bacteria and yeasts decline gradually during the cold storage period. Most yeast strains from koumiss have moderate to strong proteolytic and lipolytic activities, which might allow them to contribute to the final sensorial profiles of the product (Mu et al. 2012). *Streptococcus* spp. may be present in koumiss microflora but their contribution to aroma and flavor of koumiss is fairly limited. *Acetobacter* spp. are also of only minor importance. The other yeasts reported to be isolated from traditional koumiss products belong to *Rhodotorula* spp.

Despite the lack of strong scientific evidence regarding the health-promoting effects of koumiss, it has long been believed to cure some illnesses including tuberculosis, disorders of the stomach and colon, and hepatitis. On the other hand, the total serum cholesterol and triglycerides levels in artificially-induced hyperlipidemial mice were reported to be reduced upon feeding with koumiss for 14 days (Pan et al. 2011). The presence of free amino acids in koumiss at high levels contributes to the nutritive value of this product. Chen et al. (2010) achieved to identify four ACE-inhibitory peptides from koumiss made from mare's milk, indicating a potential of this traditional product on improvement of cardiovascular health of individuals. Similarly, positive results were obtained by Sun et al. (2009) who investigated the role of *Lb. helveticus* isolated from Xinjiang koumiss in China on management of anti-hypertension. *In vitro* studies demonstrated that koumiss has an anti-bacterial effect against *E. coli*, *S. aureus* and species of *Mycobacterium*, *Bacillus*, *Serratia* and *Shigella*, stemming probably from the organic acids in the product. While *Lb. fermentum* SM-7 isolated from koumiss showed a strong anti-bacterial effect against *E. coli* and *S. aureus* (Pan et al. 2011), *Lb. plantarum* IMAU10116 isolated from Chinese koumiss had a great anti-fungal activity against *Penicillium roqueforti* (Wang et al. 2011).

In order to improve the health-promoting effect of koumiss, blending koumiss starter with probiotic bacteria (e.g., *Lb. rhamnosus* or *Lb. acidophilus*) is recommended. Many studies revealed that traditional koumiss microflora contains potential probiotic strains of lactobacilli, especially *Lb. casei* Zhang (Wang et al. 2010, Wu et al. 2009a,b, Bilige et al. 2009). The full genomic sequencing of *Lb. casei* Zhang has been completed by Zhang et al. (2010).

References

Abdi, R., S.M. Zeinoddin and Z.S. Soleimanian. 2012. Efficiency of modified skim milk base media to achieve high exopolysaccharide/cell ratios by *Lactobacillus delbrueckii* subsp. *bulgaricus* SZ2 in optimise conditions defined by the response surface methodology. Int. J. Food Sci. Technol. 47: 768–775.

Adams, R.M., S. Yoast, S.E. Mainzer, K. Moon, A.L. Palombella, D.A. Estell, S.D. Power and B.K. Schmidt. 1994. Characterization of two cold-sensitive mutants of the beta-galactosidase from *Lactobacillus delbrueckii* subsp. *bulgaricus*. J. Biol. Chem. 269: 5666–5672.

Alm, L. 1982. Effect of fermentation on L(+) and D(–) lactic acid in milk. J. Dairy Sci. 65: 515–520.

Almiron-Roig, E., F. Mulholland, M.J. Gasson and A.M. Griffin. 2000. The complete *cps* gene cluster from *Streptococcus thermophilus* NCFB 2393 involved in the biosynthesis of a new exopolysaccharide. Microbiol.-UK 146: 2793–2802.

Angulo, L., E. Lopez and C. Lema. 1993. Microflora present in kefir grains of the Galician region (North-West of Spain). J. Dairy Res. 60: 263–267.

Anonymous. 2002. Milk and Milk Product—Fermented Milks. PN-A-86061, Polski Komitet Normalizacji, Miar i Jako ci. Wydawnictwo Normalizacyjne, Warszawa.

Ascon-Reyes, D.B., M.A. Ascon-Cabrera, N. Cochet and J.M. Lebeault. 1995. Indirect conductance for measurements of carbon dioxide produced by *Streptococcus salivarius* ssp. *thermophilus* TJ 160 in pure and mixed cultures. J. Dairy Sci. 78: 8–16.

Atlan, D., P. Laloi and R. Portalier. 1990. X-propyl-dipeptidyl aminopeptidase of *Lactobacillus delbrueckii* subsp. *bulgaricus*: Characterization of the enzyme and isolation of deficient mutants. Applied Environ. Microbiol. July: 2174–2179.

Axelsson, L. 1998. Lactic acid bacteria: Classification and physiology. pp. 1–72. *In:* S. Salminen and A. von Wright [eds.]. Lactic Acid Bacteria: Microbiology and Functional Aspects. Marcel Dekker Inc., New York, USA.

Baranowska, M. 2006. Intensification of the synthesis of flavour compounds in yogurt by milk enrichment with their precursors. Polish J. Food Nutr. Sci. 15(Suppl. 1): 5–11.

Beal, C. and G. Corrieu. 1991. Influence of pH, temperature and inoculum composition on mixed cultures of *Streptococcus thermophilus* 404 and *Lactobacillus bulgaricus* 398. Biotechnol. Bioeng. 38: 90–98.

Benthin, S. and J. Villadsen. 1995. Different inhibition of *Lactobacillus delbrueckii* subsp. *bulgaricus* by D-lactic acid and L-lactic acid—Effects on lag phase, growth rate and cell yield. J. Appl. Bacteriol. 78: 647–654.

Bertrand-Harb, C., I.V. Ivanova, M. Dalgalarrondo and T. Haertllé. 2003. Evolution of β-lactoglobulin and α-lactalbumin content during yogurt fermentation. Int. Dairy J. 13: 39–45.

Beshkova, D.M., E.D. Simova, G.I. Frengova, Z.I. Simov and Z.P. Dimitrov. 2003. Production of volatile aroma compounds by kefir starter cultures. Int. Dairy J. 13: 529–535.

Bianchi-Salvadori, B., P. Camaschella and S. Cislaghi. 1995. Rapid enzymatic method for biotyping and control of lactic acid bacteria used in the production of yogurt and some cheeses. Int. J. Food Microbiol. 27: 253–261.

Bianchi-Salvadori, B. 1997. Lactic acid bacteria: Biochemical characteristics affecting the texture of fermented milk products. pp. 48–62. *In:* IDF Proceedings of Texture of Fermented Dairy Products and Dairy Desserts. International Dairy Federation, Vicenza, Italy.

Bilige, M., W.J. Liu, W. Rina, L.P. Wang, T.S. Sun, J.G. Wang, H.P. Li and H.P. Zhang. 2009. Evaluation of potential probiotics properties of the screened *Lactobacilli* isolated from home-made koumiss in Mongolia. Annal. Microbiol. 59: 493–498.

Black, B.A., V.S.Y. Lee, Y.Y. Zhao, Y. Hu, J.M. Curtis and M.G. Gänzle. 2012. Structural identification of novel oligosaccharides produced by *Lactobacillus bulgaricus* and *Lactobacillus plantarum*. J. Agric. Food Chem. 60: 4886–4894.

Bockelmann, W., M.B. Fobker and M. Teuber. 1991. Purification and characterisation of the X-propyl-dipeptidyl aminopeptidase from *Streptococcus cremoris* AM2. J. Dairy Res. 1: 51–66.

Bockelmann, W., Y. Schulz and M. Teuber. 1992. Purification and characterization of an aminopeptidase from *Lactobacillus delbrueckii* subsp. *bulgaricus*. Int. Dairy J. 2: 95–107.

Bongers, R.S., M.H. Hoefnagel and M. Kleerebezem. 2005. High-level acetaldehyde production in *Lactococcus lactis* by metabolic engineering. Appl. Environ. Microbiol. 71: 1109–1113.

Bottazzi, V. and F. Bianchi. 1980. A note on scanning electron microscopy of microorganisms associated with the kefir granule. J. Appl. Bacteriol. 48: 265–268.

Bouillanne, C. and M.J. Desmazeaud. 1980. A statistical method for grouping some *Streptococcus thermophilus* strains according to their technological or enzymological characteristics. Lait 60: 458–473.

Bourgoin, F., A. Pluvinet, B. Gintz, B. Decaris and G. Guedon. 1999. Are horizontal transfers involved in the evolution of the *Streptococcus thermophilus* exopolysaccharide synthesis loci? Gene 223: 151–161.

Bouzar, F., J. Cerning and M. Desmazeaud. 1997. Exopolysaccharide production and texture-promoting abilities of mixed-strain starter cultures in yogurt production. J. Dairy Sci. 80: 2310–2317.

Broadbent, J.R., D.J. McMahon, D.L. Welker, C.J. Oberg and S. Moineau. 2003. Biochemistry, genetics, and applications of exopolysaccharide production in *Streptococcus thermophilus*: A review. J. Dairy Sci. 86: 407–423.

Bubb, W.A., T. Urashima, R. Fujiwara, T. Shinnai and H. Ariga. 1997. Structural characterization of the exocellular polysaccharide produced by *Streptococcus thermophilus* OR 901. Carbohydrate Res. 301: 41–50.

Burrus, V.C., C. Bontemps, B. Decaris and G. Guedon. 2001. Characterization of a novel type II restriction-modification system, Sth368I, encoded by the integrative element ICESt1 of *Streptococcus thermophilus* CNRZ368. Appl. Environ. Microbiol. 67: 1522–1528.

Cenesiz, S., A.K. Devrim, U. Kamber and M. Sozmen. 2008. The effect of kefir on glutathione (GSH), malondialdehyde (MDA) and nitric oxide (NO) levels in mice with colonic abnormal crypt formation (ACF) induced by azoxymethane (AOM). Deut. Tierarzt. Wochensch. (DTW) 115: 15–19.

Cerning, J., C. Bouillanne, M.J. Desmazeaud and M. Landon. 1986. Isolation and characterisation of exocellular polysaccharide produced by *Lactobacillus bulgaricus*. Biotechnol. Lett. 8: 625–628.

Cerning, J. 1988. Exocellular polysaccharide production by *Streptococcus thermophilus*. Biotechnol. Lett. 10: 255–260.

Cerning, J., C. Bouillanne, M. Landon and M.J. Desmazeaud. 1990. Comparison of exocellular polysaccharide production by thermophilic lactic acid bacteria. Sci. Aliment. 10: 443–451.

Chaves, A.C.S.D., M. Fernandez, A.L.S. Lerayer, I. Mierau, M. Kleerebezem and J. Hugenholtz. 2002. Metabolic engineering of acetaldehyde production by *Streptococcus thermophilus*. Appl. Environ. Microbiol. 68: 5656–5662.

Chen, Y., Z. Wang, X. Chen, Y. Liu, H. Zhang and T. Sun. 2010. Identification of angiotensin I-converting enzyme inhibitory peptides from koumiss, a traditional fermented mare's milk. J. Dairy Sci. 93: 884–892.

Chen, Y.P., P.J. Hsiao, W.S. Hong, T.Y. Dai and M.J. Chen. 2012. *Lactobacillus kefiranofaciens* M1 isolated from kefir grains ameliorates experimental colitis *in vitro* and *in vivo*. J. Dairy Sci. 95: 63–74.

Cheng, H. 2010. Volatile flavour compounds in yogurt: A review. Critical Rev. Food Sci. 50: 938–950.

Cochu, A., C. Vadeboncoeur, S. Moineau and M. Frenette. 2003. Genetic and biochemical characterization of the phosphoenolpyruvate:glucose/mannose phosphotransferase system of *Streptococcus thermophilus*. Appl. Environ. Microbiol. 69: 5423–5432.

Courtin, P., V. Monnet and F. Rul. 2002. Cell-wall proteinases PrtS and PrtB have a different role in *Streptococcus thermophilus*/*Lactobacillus bulgaricus* mixed cultures in milk. Microbiol. 148: 3413–3421.

Dal Bello, F.D., J. Walter, C. Hertel and W.P. Hammes. 2001. *In vitro* study of prebiotic properties of levan-type exopolysaccharides from *Lactobacilli* and non-digestible carbohydrates using denaturing gradient gel electrophoresis. Syst. Appl. Microbiol. 24: 232–237.

Dandoy, D., C. Fremaux, M.H. de Frahan, P. Horvarth, P. Boyaval and L. Fontaine. 2011. The fast milk acidifying phenotype of *Streptococcus thermophilus* can be acquired by natural transformation of the genomic island encoding the cell-envelope proteinase PrtS. Microbial Cell Fact. 10(Suppl. 1): S21.

Danova, S., K. Petrov, P. Pavlov and P. Petrova. 2005. Isolation and characterization of *Lactobacillus* strains involved in koumiss fermentation. Soc. Dairy Technol. 2: 100–105.

Degeest, B., F. Vaningelgem, A.P. Laws and L. de Vuyst. 2001. UDP-N-acetylglucosamine 4-epimerase activity indicates the presence of N-acetylgalactosamine in exopolysaccharides of *Streptococcus thermophilus* strains. Appl. Environ. Microbiol. 67: 3976.

Delley, M., B. Mollet and H. Hottingher. 1990. DNA probe for *Lactobacillus delbrueckii*. Appl. Environ. Microbiol. 56: 1967–1970.

de Moreno de LeBlanc, A., C. Matar, E. Farnworth and G. Perdigon. 2006. Study of immune cells involved in the antitumor effect of kefir in a murine breast cancer model. J. Dairy Sci. 90: 1920–1928.

de Souza Oliveira, P.R., B.R. Torres, P. Perego, M.N. de Oliveira and A. Converti. 2012. Co-metabolic models of *Streptococcus thermophilus* in co-culture with *Lactobacillus bulgaricus* or *Lactobacillus acidophilus*. Biochem. Eng. J. 62: 62–69.

de Vin, F., P. Radström, L. Herman and L. de Vuyst. 2005. Molecular and biochemical analysis of the galactose phenotype of dairy *Streptococcus thermophilus* strains reveals four different fermentation profiles. Appl. Environ. Microbiol. 71: 3659–3667.

de Vos, W.M. 1996. Metabolic engineering of sugar metabolism in lactic acid bacteria. pp. 127–146. *In:* Proceedings of the Fifth Symposium on Lactic Acid Bacteria: Genetics, Metabolism and Applications. Federation of European Microbiological Societies and Netherland Institute for Microbiology, Veldhooven, The Netherlands.

de Vos, W.M. and J. Hugenholtz. 2004. Engineering metabolic highways in *Lactococci* and other lactic acid bacteria. Trends Biotechnol. 22: 72–79.

de Vuyst, L. and B. Degeest. 1999. Heteropolysaccharides from lactic acid bacteria. FEMS Microbiol. Rev. 23: 153–177.

de Vuyst, L., F. de Vin, F. Vaningelgem and B. Degeest. 2001. Recent developments in the biosynthesis and applications of heteropolysaccharides from lactic acid bacteria. Int. Dairy J. 11: 687–707.

Doco, T., J.M. Wieruszeski, B. Fournet, D. Carcano, P. Ramos and A. Loones. 1990. Structure of an exocellular polysaccharide produced by *Streptococcus thermophilus*. Carbohydrate Res. 198: 313–321.

Driessen, F.M. 1981. Protocooperation of yogurt bacteria in continuous culture. pp. 99–120. *In:* M.E. Blushell and J.H. Slater [eds.]. Mixed Culture Fermentations. Academic Press, London, UK.

Duboc, P. and B. Mollet. 2001. Applications of exopolysaccharides in the dairy industry. Int. Dairy J. 11: 759–768.

El-Soda, M., M. Korayem and N. Ezzat. 1986. The esterolytic and lipolytic activities of lactobacilli. III. Detection and characterization of the lipase system. Milchwiss. 41: 353–355.

Ezzat, N., C. Zevaco, M. El-Soda and J.C. Gripon. 1987. Partial purification and characterization of a cell-wall associated proteinase from *Lactobacillus bulgaricus*. Milchwiss. 42: 95–97.

Faber, E.J., P. Zoon, J.P. Kamerling and J.F.G. Vliegenthart. 1998. The exoploysaccharides produced by *Streptococcus thermophilus* Rs and Sts have the same repeating unit but differ in viscosity of their milk cultures. Carbohydrate Res. 310: 269–276.

Faber, E.J., M.J. van den Haak, J.P. Kamerling and J.F.G. Vliegenthart. 2001a. Structure of the exopolysaccharide produced by *Streptococcus thermophilus* S3. Carbohydrate Res. 331: 173–182.

Faber, E.J., J.P. Kamerling and J.F.G. Vliegenthart. 2001b. Structure of the extracellular polysaccharide produced by *Lactobacillus delbrueckii* subsp. *bulgaricus* 291. Carbohydrate Res. 331: 183–194.

Fajardo-Lira, C., M. Garcia-Garibay, C. Wacher-Rodarte, A. Farres and V.M.E. Marshall. 1997. Influence of water activity on the fermentation of yogurt made with extracellular polysaccharide-producing or non-producing starters. Int. Dairy J. 7: 279–281.

Farrow, J.A.E. and M.D. Collins. 1984. DNA base composition, DNA-DNA homology and long-chain fatty acid studies on *Streptococcus thermophilus* and *Streptococcus salivarius*. J. Genetics Microbiol. 130: 357–362.

Fernandez-Espla, M., P. Garault, V. Monnet and F. Rul. 2000. *Streptococcus thermophilus* cell wall-anchored proteinase: Release, purification, and biochemical and genetic characterization. Appl. Environ. Microbiol. 66: 4772–4778.

Ferreira, I.M.P.L.V.O., O. Pinho, D. Monterio, S. Faria, S. Cruz, A. Perreira, A.C. Rogue and P. Tavares. 2010. Effect of kefir grains on proteolysis of major milk proteins. J. Dairy Sci. 93: 27–31.

Fira, D., M. Kojic, A. Banina, I. Spasojevic, I. Strahinic and L. Topisirovic. 2001. Characterization of cell envelope-associated proteinases of thermophilic lactobacili. J. Appl. Microbiol. 90: 123–130.

Fonden, R., K. Leporanta and U. Svensson. 2006. Nordic/Scandinavian fermented milk products. pp. 156–173. *In:* A.Y. Tamime [ed.]. Fermented Milks. Blackwell Publishing, Oxford, UK.

Foucaud, C. and B. Poolman. 1992. Lactose transport system of *Streptococcus thermophilus*. Functional reconstitution of the protein and characterization of the kinetic mechanism of transport. J. Biol. Chem. 267: 22078–22094.

Galia, W., C. Perrin, M. Genay and A. Dary. 2009. Variability and molecular typing of *Streptococcus thermophilus* strains displaying different proteolytic and acidifying properties. Int. Dairy J. 19: 89–95.

Garbers, I.M., T.J. Britz and R.C. Witthuhn. 2004. PCR-based DGGE typification and identification of the microbial concortium present in kefir grains. World J. Microbiol. Biotechnol. 20: 687–693.

Garrote, G.L., A.G. Abraham and G.L. de Antoni. 2000. Inhibitory power of kefir: The role of organic acids. J. Food Protect. 63: 364–369.

Garrote, G.L., A.G. Abraham and G.L. de Antoni. 2001. Chemical and microbiological characterization of kefir grains. J. Dairy Res. 65: 149–154.

Germond, J.E., L. Lapierre, M. Delley and B. Mollet. 1995. A new genetic element in *Lactobacillus delbrueckii* subsp. *bulgaricus*. Mol. Gen. Genetics 248: 407–416.

Germond, J.E., M. Delley, N. D'Amico and S.F.J. Vincent. 2001. Heterologous expression and characterization of the exopolysaccharide from *Streptococcus thermophilus* Sfi39. Eur. J. Biochem. 268: 5149–5156.

Germond, J.E., L. Lapierre, M. Delley, B. Mollet, G.E. Felis and F. Dellaglio. 2003. Evolution of the bacterial species *Lactobacillus delbrueckii*: A partial genomic study with reflections on prokaryotic species concept. Mol. Biol. Evol. 20: 93–104.

Gilbert, C., D. Atlan, B. Blanc, R. Portalier, J.E. Germond, L. Lapierre and B. Mollet. 1996. A new cell surface proteinase: Sequencing and analysis of the prtB from *Lactobacillus delbrueckii* subsp. *bulgaricus*. J. Bacteriol. 178: 3059–3065.

Griffin, A.M., V.J. Morris and M.J. Gasson. 1996. The cpsABCDE genes involved in polysaccharide production in *Streptococcus salivarius* ssp. *thermophilus* strain NCBF 2393. Gene 183: 23–27.

Grobben, G.J., M.R. Smith, J. Sikkema and J.A.M. De Bont. 1996. Influence of fructose and glucose on the production of exopolysaccharides and the activities of enzymes involved in the sugar metabolism and the synthesis of sugar nucleotides in *Lactobacillus delbrueckii* subsp. *bulgaricus* NCFB 2772 grown in continuous culture on glucose and fructose. Appl. Microbiol. Biotechnol. 48: 516–521.

Grobben, G.J., I.C. Boels, J. Sikkema and J.A.M. de Bont. 2000. Influence of ions on growth and production of exopolysaccharides by *Lactobacillus delbrueckii* subsp. *bulgaricus* NCFB 2772. J. Dairy Sci. 67: 131–135.

Gronnevik, H., M. Falstad and J.A. Narvhus. 2011. Microbiological and chemical properties of Norwegian kefir during storage. Int. Dairy J. 21: 601–606.

Gruter, M., B.R. Leeflang, J. Kuiper, J.P. Kamerling and J.F.G. Vliegenthart. 1993. Structural characterization of the exopolysaccharide produced by *Lactobacillus delbrueckii* subsp. *bulgaricus* RR grown in skim milk. Carbohydrate Res. 239: 209–226.

Guirguis, N. and M.W. Hickey. 1987. Factors affecting the performance of thermophilic starters. 2. Sensitivity to the lactoperoxidase system. Aus. J. Dairy Technol. 42: 14–16.

Güven, A., A. Güven and M. Gülmez. 2003. The effect of kefir on the activities of GSH-Px, GST, CAT, GSH and LPO levels in carbontetrachloride-induced mice tissues. J. Vet. Med. B50: 412–416.

Güzel-Seydim, Z.B, A.C. Seydim and A. Greene. 2000. Organic acids and volatile flavor compounds evolved during refrigerated storage of kefir. J. Dairy Sci. 83: 275–277.

Güzel-Seydim, Z.B, A.C. Seydim and A. Greene. 2003. Comparison of amino acid profiles of milk, yogurt and Turkish kefir. Milchwiss. 58: 158–161.

Güzel-Seydim, Z.B., J. Wyffels, A.C. Seydim and A. Greene. 2005. Turkish kefir and kefir grains: Microbial enumeration and electron microscopy observations. Int. J. Dairy Technol. 58: 25–29.

Güzel-Seydim, Z.B., A.C. Seydim, A. Greene and T. Taş. 2006. Determination of anti-mutagenic properties of some fermented milks including changes in the total fatty acid profiles including CLA. Int. J. Dairy Technol. 59: 209–215.

Güzel-Seydim, Z.B., T. Kök-Taş and A. Greene. 2010. Kefir and koumiss: Microbiology and technology. pp. 143–163. *In:* F. Yildiz [ed.]. Development and Manufacture of Yogurt and Other Functional Dairy Products. CRC Press, Boca Raton, FL, USA.

Güzel-Seydim, Z.B., T. Kök-Taş, A. Greene and A.C. Seydim. 2011. Functional properties of kefir. Critical Rev. Food Sci. Nutr. 51: 261–268.

Hammes, W.P. and R.F. Vogel. 1995. The genus *Lactobacillus*. pp. 19–54. *In:* B.J.B. Wood and W.H. Holzapfel [eds.]. The Genera of Lactic Acid Bacteria. Chapman & Hall, London, UK.

Hao, P., H. Zheng, Y. Yu, G. Ding, W. Gu, Z. Yu, S. Ren, M. Oda, T. Konno, S. Wang, X. Li, S.Z. Ji and G. Zhao. 2011. Complete sequencing and pan-genomic analysis of *Lactobacillus delbrueckii* subsp. *bulgaricus* reveal its genetic basis for industrial yogurt production. PLoS ONE 6: 1–9.

Hardie, J.M. and R.A. Whilley. 1995. The genus *Streptococcus*. pp. 55–124. *In:* B.J.B. Wood and W.H. Holzapfel [eds.]. The Genera of Lactic Acid Bacteria. Vol. 2. Blackie Academic and Professional, London, UK.

Hassan, A.N., J.F. Frank, K.A. Schmidt and S.I. Shalabi. 1996. Textural properties of yogurt made with encapsulated nonropy bacteria. J. Dairy Sci. 79: 2098–2103.

Heller, K.J. 2001. Probiotic bacteria in fermented foods: Product characteristics and starter organisms. Am. J. Clin. Nutr. 73: 374S–379S.

Herrero, A.M., T. Requena, A.J. Reviejo and J.M. Pingarron. 2004. Determination of L-lactic acid in yogurt by a bienzyme amperometric graphite-Teflon composite biosensor. Eur. Food Res. Technol. 219: 556–559.

Herve-Jimenez, L., I. Guillouard, E. Guedon, S. Boudebbouze, P. Hols, V. Monnet, E. Maquin and F. Rul. 2009. Postgenomic analysis of *Streptococcus thermophilus* co-cultivated in milk with *Lactobacillus delbrueckii* subsp. *bulgaricus*: Involvement of nitrogen, purine, and iron metabolism. Appl. Environ. Microbiol. 75: 2062–2073.

Hess, S.J., R.F. Roberts and G.R. Ziegler. 1997. Rheological properties of nonfat yogurt stabilized using *Lactobacillus delbrueckii* subsp. *bulgaricus* producing exopolysaccharide or using commercial stabilizer systems. J. Dairy Sci. 80: 252–263.

Hiratsuka, Y. and G. Li. 1992. Alcohol and eye diseases: A review of epidemiologic studies. J. Study Alcohol Drug 62: 372–402.

Hlastan-Ribic, C., A. Cerar, D. Pokorn, M. Perse and A. Zebic. 2005. Effects of kefir containing various levels of fat on chemically induced colorectal epithelial tumors in Wistar rats. Nutr. Res. 25: 55–63.

Hols, P., F. Hancy, L. Fontaine, B. Grossiord, D. Prozzi, N. Leblond-Bourget, B. Decaris, A. Bolotin, C. Delorme, S.D. Ehrlich, E. Guedon, W. Monnet, P. Renault and M. Kleerebezem. 2005. New insight in the molecular biology and physiology of *Streptococcus thermophilus* revealed by comparative genomics. FEMS Mirobiol. Rev. 29: 435–463.

Horiuchi, H. and Y. Sasaki. 2012. Effect of oxygen on symbiosis between *Lactobacillus bulgaricus* and *Streptococcus thermophilus*. J. Dairy Sci. 95: 2904–2909.

Hugenholtz, J. 1993. Citrate metabolism in lactic acid bacteria. FEMS Microbiol. Rev. 12: 165–178.

Hutkins, R.W., H.A. Morris and L.L. McKay. 1985. Galactose transport in *Streptococcus thermophilus*. Appl. Environ. Microbiol. 50: 772–776.

Hutkins, R.W. and H.A. Morris. 1987. Carbohydrate metabolism by *Streptococcus thermophilus*: A review. J. Food Protect. 50: 876–884.

Ip, C., M. Singh, H.J. Thompson and J.A. Scimeca. 1994. Conjugated linoleic acid suppresses mammary carcinogenesis and proliferative activity of the mammary gland in the rat. Cancer Res. 54: 1212–1215.

Juillard, V., M.J. Desmazeaud and H.E. Spinnler. 1988. Mise en évidence d'une activité uréasique chez *Streptococcus thermophilus*. Can. J. Microbiol. 34: 818–822.

Kahala, M., E. Pahkala and A. Pihlanto-Leppälä. 1993. Peptides in fermented Finnish milk products. Agric. Sci. Finland 2: 379–386.

Kalantzopoulos, G., E. Tsakalidou and E. Manolopoulou. 1990. Proteinase, peptidase and esterase activities of cell-free extracts from wild strains of *Lactobacillus delbrueckii* subsp. *bulgaricus* and *Streptococcus salivarius* subsp. *thermophilus* isolated from traditional Greek yogurt. J. Dairy Res. 57: 593–601.

Kaneko, T., H. Suzuki and T. Takahahsi. 1987. Studies on biosynthesis and utilization of folic acid and vitamin B12 by lactic acid bacteria. Dairy Sci. Abst. 49: 201.

Khalid, N.M., M. El-Soda and E. Marth. 1990. Esterases of *Lactobacillus helveticus* and *Lactobacillus delbrueckii* ssp. *bulgaricus*. J. Dairy Sci. 73: 2711.

Khalid, N.M., M. El-Soda and E. Marth. 1991. Peptide hydrolyses of *Lactobacillus helveticus* and *Lactobacillus delbrueckii* ssp. *bulgaricus*. J. Dairy Sci. 74: 29–45.

Kikuchi, M., K. Yamaguchi and Y. Matsui. 1984. The effect of heating skim milk on symbiosis between *Lactobacillus bulgaricus* and *Streptococcus thermophilus*. J. College Dairy. Natural Sci. 10: 349–357.

Kitazawa, H., T. Harata, E. Uemura, T. Saito, T. Kaneko and T. Itoh. 1998. Phosphate group requirement for mitogenic activation of lymphocytes by an extracellular phosphopolysaccharide from *Lactobacillus delbrueckii* ssp. *bulgaricus*. Int. J. Food Microbiol. 40: 169–175.

Kök-Taş, T., Y.F. Ekinci and Z.B. Güzel-Seydim. 2012. Identification of microbial flora in kefir grains produced in Turkey using PCR. Int. J. Dairy Technol. 65: 126–131.

Kroger, M. 1993. Kefir. Cult. Dairy J. 28: 26–29.

Kunji, E.R.S., I. Mierau, A. Haqting, B. Poolman and W.N. Konings. 1996. The proteolytic systems of lactic acid bacteria, Antonie Leeuwen. Int. J. Gen. Mol. Microbiol. 70: 187–221.

Kurultay, Ş., Ö. Öksüz and B. Kaptan. 2005. Determination of the effects of different amino acids, sodium formate and their combinations on some growth characteristics of mixed and single cell cultures of yogurt bacteria. J. Tekirdag Agric. Fac. 2: 153–160.

Kücükcetin, A., H. Yaygin, J. Hinrichs and U. Kulozik. 2003. Adaptation of bovine milk towards mares' milk composition by means of membrane technology for koumiss manufacture. Int. Dairy J. 13: 945–951.

Laloi, P. 1989. Biochemical, physiological and genetic analysis of caseinolytic aminopeptidases from *Lactobacillus bulgaricus*. Ph.D. Thesis. University of Lyon, Lyon, France.

Laloi, P., B. Atlan, B. Blanc, C. Gilbert and R. Portalier. 1991. Cell-wall associated proteinase of *Lactobacillus delbrueckii* subsp. *bulgaricus* CNRZ 397: Differential extraction, purification and properties of the enzyme. Appl. Microbiol. Biotechnol. 36: 196–202.

Lankes, H., B.H. Özer and R.K. Robinson. 1998. The effect of elevated milk solids and incubation temperature on the physical properties of natural yogurt. Milchwiss. 53: 510–513.

Lattore-Garcia, L., L. del Castillo-Agudo and J. Polaina. 2007. Taxonomical classification of yeasts isolated from kefir based on the sequence of their ribosomal RNA genes. World J. Microbiol. Biotechnol. 23: 785–791.

Laws, A.P. and V.M. Marshall. 2001. The relevance of exopolysaccharides to the rheological properties in milk fermented with ropy strains of lactic acid bacteria. Int. Dairy J. 11: 709–721.

Le Bras, G. and J.R. Garel. 1991. Properties of D-lactate dehydrogenase from *Lactobacillus bulgaricus*: A possible different evolutionary origin for the D- and L-lactate dehydrogenases. FEMS Microbiol. Lett. 79: 89–94.

Lee, K.W., K.Y. Shin and S.C. Baick. 1998. D(–) and L(+) lactic acid determination *of Lactobacillus acidophilus* during fermentation and storage period. Korean J. Food Sci. Technol. 30: 168–174.

Leeflang, B.R., E.J. Faber, P. Eerbel and J.F.G. Vliegenhart. 2000. Structure elucidation of glycoprotein glycans and of polysaccharides by NMR spectroscopy. J. Biotechnol. 77: 115–122.

Lees, G.J. and G.R. Jago. 1976a. Acetaldehyde: An intermediate in the formation of ethanol from glucose by lactic acid bacteria. J. Dairy Res. 43: 75–83.

Lees, G.J. and G.R. Jago. 1976b. Formation of acetaldehyde from threonine by lactic acid bacteria. J. Dairy Res. 43: 75–83.

Lees, G.J. and G.R. Jago. 1977. Formation of acetaldehyde from 2-deoxy-D-ribose-5-phosphate in lactic acid bacteria. J. Dairy Res. 44: 139–144.

Leite, A.M.O., B. Mayo, C.T.C.C. Rachid, R.S. Peixoto, J.T. Silva, V.F.M. Paschoalin and S. Delgado. 2012. Assessment of the microbial diversity of Brazilian kefir grains by PCR-DGGE and pyrosequencing analysis. Food Microbiol. 31: 215–221.

Lemoine, J., F. Chirat, J.M. Wieruszeski, G. Strecker, N. Favre and J.R. Neeser. 1997. Structural characterization of the exocellular polysaccharides produced by *Streptococcus thermophilus* SFi39 and SFi12. Appl. Environ. Microbiol. 63: 3512–3518.

Letort, C. and V. Juillard. 2001. Development of a minimal chemical defined medium for the exponential growth of *Streptococcus thermophilus*. J. Appl. Microbiol. 91: 1–7.

Levander, F. and P. Radström. 2001. Requirement for phosphoglucomutase in exopolysaccharide biosynthesis in glucose- and lactose-utilizing *Streptococcus thermophilus*. Appl. Environ. Microbiol. 67: 2734–2738.

Levander, F., M. Svensson and P. Radström. 2002. Enhanced exopolysaccharide production by metabolic engineering of *Streptococcus thermophilus*. Appl. Environ. Microbiol. 68: 784–790.

Lin, C.W., H.L. Chen and J.R. Liu. 1999. Identification and characterization of lactic acid bacteria and yeasts isolated from kefir grains in Taiwan. Aus. J. Dairy Technol. 54: 14–18.

Loones, A. 1989. Changes in milk composition during yogurt production. Fermented Milks: Curr. Res. 129–137.

Low, D., J.A. Ahlgren, D. Horne, D.J. McMahon, C.J. Oberg and J.R. Broadbent. 1998. Role of *Streptococcus thermophilus* MR-1C capsular exopolysaccharide in cheese moisture retention. Appl. Environ. Microbiol. 64: 2147.

Lyck, S., L.-E. Nilsson and A.Y. Tamime. 2006. Miscellaneous fermented milk products. pp. 217–236. *In:* A.Y. Tamime [ed.]. Fermented Milks. Blackwell Publishing, Oxford, UK.

Maeda, H., X. Zhu, K. Suzuki and S. Kitamura. 2004. Structural characterization and biological activities of an exopolysaccharide kefiran produced by *Lactobacillus kefiranofaciens* WT-2B(T). J. Agri. Food Chem. 52: 5533–5538.

Makino, S., S. Ikegama, H. Kano, S. Sashihara, H. Sugano, T. Horiuchi, M. Saito and T. Oda. 2006. Immunomodulatory effects of polysaccharides produced by *Lactobacillus delbrueckii* subsp. *bulgaricus* OLL1073R-1. J. Dairy Sci. 89: 2873–2881.

Malacarne, M., F. Martuzzi, A. Summer and P. Mariani. 2002. Protein and fat composition of mare's milk: Some nutritional remarks with reference to human and cow's milk. Int. Dairy J. 12: 869–877.

Manca de Nadra, M.C., R.R. Raya, A.A. Pesce de Ruiz Holgado and G. Olivier. 1987. Isolation and properties of threonine aldolase of *Lactobacillus bulgaricus* YOP12. Milchwiss. 42: 92–94.

Marranzini, R.M., R.H. Schmidt, R.B. Shireman, M.R. Marshall and J.A. Cornell. 1989. Effect of threonine and glycine concentrations on threonine aldolase activity of yogurt microorganisms during growth in modified milk prepared by ultrafiltration. J. Dairy Res. 72: 1142–1148.

Marshall, V.M.E., W. Cole and L.A. Mabbitt. 1982. Yogurt made from single starter organisms using heat- or enzyme-treated milk or milk to which casein hydrolysate or sodium formate is added. J. Dairy Res. 49: 147–152.
Marshall, V., W.M. Cole and B.E. Brooker. 1984. Observations on the structure of kefir grains and the distribution of the microflora. J. Appl. Bacteriol. 57: 491–497.
Marshall, V.M.E. and A.Y. Tamime. 1997. Starter cultures employed in the manufacture of biofermented milks. Int. J. Dairy Technol. 50: 35–41.
Marshall, V.M., H. Dunn, M. Elvin, N. McLay, Y. Gu and A.P. Laws. 2001a. Structural characterization of the exopolysaccharide produced by *Streptococcus thermophilus* EU20. Carbohydrate Res. 331: 413–422.
Marshall, V.M., A.P. Laws, Y. Gu, F. Levander, P. Radström, L. de Vuyst, B. Degeest, F. Vaningelgem, H. Dunn and M. Elvin. 2001b. Exopolysaccharide-producing strains of thermophilic lactic acid bacteria cluster into groups according to their EPS structure. Lett. Appl. Microbiol. 32: 433–437.
Martin, F., R. Cachon, K. Pernin, J. de Coninck, P. Gervais, E. Guichard and N. Cayot. 2011. Effect of oxidoreduction potential on aroma biosynthesis by lactic acid bacteria in non-fat yogurt. J. Dairy Sci. 94: 614–622.
Mende, S., L. Krzyanowski, J. Weber, D. Jaros and H. Rohm. 2012. Growth and exopolysaccharide yield of *Lactobacillus delbrueckii* ssp. *bulgaricus* DSM 20081 in batch and continuous bioreactor experiments at constant pH. J. Biosci. Bioeng. 113: 185–191.
Meyer, J. and R. Jordi. 1987. Purification and characterization of X-prolyl-dipeptidyl-aminopeptidase from *Lactobacillus lactis* and from *Streptococcus thermophilus*. J. Dairy Sci. 70: 738–745.
Meyer, J., D. Howald, R. Jordi and M. Fürst. 1989. Location of proteolytic enzymes in *Lactobacillus lactis* and *Streptococcus thermophilus* and their influence on cheese ripening. Milchwiss. 44: 678–681.
Monnet, V., S. Condon, T.M. Cogan and J.C. Gripon. 1996. Metabolism of starter cultures. pp. 47–100. *In:* T.M. Cogan and J.P. Accolas [eds.]. Dairy Starter Cultures. VCH Publishers, New York, USA.
Monnet, C., S. Pernoud, A. Sepulchre, C. Fremaux and G. Corrieu. 2004. Selection and properties of *Streptococcus thermophilus* mutants deficient in urease. J. Dairy Sci. 87: 1634–1640.
Montanari, G., C. Zambonelli and G. Fiori. 1997. The koumiss, milk beverage. Ind. Aliment. 36: 5–9.
Mora, D., E. Maguin, M. Masiero, C. Parini, G. Ricci, P.L. Manachini and D. Daffonchio. 2004. Characterization of urease genes cluster of *Streptococcus thermophilus*. J. Appl. Microbiol. 96: 209–219.
Moreira, M., A. Abraham and G. de Antoni. 2000. Technological properties of milks fermented with thermophilic lactic acid bacteria at suboptimal temperature. J. Dairy Sci. 83: 395–400.
Mu, Z., X.Y. Yang and H. Yuan. 2012. Detection and identification of wild yeast in Koumiss. Food Microbiol. 31: 301–308.
Murofushi, M., M. Shiomi and K. Aibara. 1983. Effect of orally administered polysaccharide from kefir grain on delayed-type hypertensivity and tumor growth in mice. Japan. J. Medical Sci. Biol. 36: 49–54.
Nakamura, T., Y. Syukunobe, R. Doki, K. Shimoda, T. Yoshida and M. Kuwazuru. 1991. Stimulating effect of milk casein hydrolysates on the growth of *Streptococcus thermophilus*. Nippon Shok. Kogyo Gakkaishi 38: 858.
Navarini, L., A. Abatangelo, C. Bertocchi, E. Conti, M. Bosco and F. Picotti. 2001. Isolation and characterization of the exopolysaccharide produced by *Streptococcus thermophilus* SFi20. Int. J. Biol. Macromol. 28: 219–226.
Ninane, V., G. Berben, J.M. Romnee and R. Oger. 2005. Variability of the microbial abundance of a kefir grain starter cultivated in partially controlled conditions. Biotechnol. Agron. Soc. Environ. 9: 191–194.

Nishimura-Uemura, J., H. Kitazawa, Y. Kawai, Y. Itoh, M. Oda and T. Saito. 2003. Functional alteration of murine macrophages stimulated with extracellular polysaccharides from *Lactobacillus delbrueckii* ssp. *bulgaricus* OLL1073R-1. Food Microbiol. 20: 267–273.

Ott, A., J.E. Germond and A. Chaintreau. 2000. Origin of acetaldehyde during milk fermentation using ^{13}C-labelled precursors. J. Agric. Food Chem. 48: 1512–1517.

Özer, B.H. and R.K. Robinson. 1999. The behaviour of starter cultures in concentrated yogurt (Labneh) produced by different techniques. Lebensmitt-Wiss. und-Technol. 32: 391–395.

Özer, D. and H.B. Özer. 1999. Fermented products of Eastern Europe and Asia: koumiss, kefir and tarhana. pp. 798–805. *In:* R.K. Robinson, C. Batt and P. Patel [eds.]. Encyclopedia of Food Microbiology. Academic Press, London, UK.

Özer, B.H. and A.F. Atasoy. 2002. Effect of addition of aminoacids, treatment with β-galactosidase and use of heath-shocked cultures on the acetaldehyde level in yogurt. Int. J. Dairy Technol. 55: 166–170.

Özer, B.H., A.S. Grandison, R.K. Robinson and M. Atamer. 2003. Effects of lactoperoxidase and hydrogen peroxide on the rheological properties of yogurt. J. Dairy Res. 70: 227–232.

Pan, D.D., X.Q. Zeng and Y.T. Yan. 2011. Characterization of *Lactobacillus fermentum* SM-7 isolated from koumiss, a potential probiotic bacterium with cholesterol-lowering effects. J. Sci. Food Agric. 91: 512–518.

Pearce, L. and S. Flint. 2002. *Streptococcus thermophilus*. pp. 2577–2582. *In:* H.J. Roginski and P. Fuquay [eds.]. Encyclopedia of Dairy Science. Academic Press, London, UK.

Pébay, M., C. Colmin, C. Guédon, J.M. Simonet and B. Decaris. 1993. Chromosomal genetic instability in *Streptococcus thermophilus.* Lait 73: 181–190.

Pernoud, S., C. Fremaux, A. Sepulchre, G. Corrieu and C. Monnet. 2004. Effect of the metabolism of urea on the acidifying activity of *Streptococcus thermophilus*. J. Dairy Sci. 87: 550–555.

Pescuma, M., E.M. Hebert, H. Rabesona, M. Drouet, Y. Choiset, T. Haertle, F. Mozzi, G.F. de Valdez and M.J. Chobert. 2011. Proteolytic action of *Lactobacillus delbrueckii* subsp. *bulgaricus* CRL 656 reduces antigenic response to bovine beta-lactoglobulin. Food Chem. 127: 487–492.

Petry, S., S. Furlan, M.J. Crepeau, J. Cerning and M. Desmazeaud. 2000. Factors affecting exocellular polysaccharide production by *Lactobacillus delbrueckii* subsp. *bulgaricus* grown in a chemically defined medium. Appl. Environ. Microbiol. 66: 3427–3431.

Pigeon, R.M., E.P. Cuesta and S.E. Gilliland. 2002. Binding of free bile acids by cells of yogurt starter culture bacteria. J. Dairy Sci. 85: 2705.

Poolman, B., R. Modderman and J. Reizer. 1992. Lactose transport system of *Streptococcus thermophilus*—role of histidine residues. J. Biol. Chem. 267: 9150–9157.

Poolman, B. 1993. Energy transduction in lactic acid bacteria. FEMS Microbiol. Rev. 12: 125–147.

Powell, J.E., R.C. Witthuhn, S.D. Todorov and L.M.T. Dicks. 2007. Characterization of bacteriocin ST8KF produced by a kefir isolate *Lactobacillus plantarum* ST8KF. Int. Dairy J. 17: 190–198.

Quirós, A., B. Hernández-Ledesma, M. Ramos, L. Amigo and I. Recio. 2005. Angiotensin converting enzyme inhibitory activity of peptides derived from caprine kefir. J. Dairy Sci. 88: 3480–3487.

Radke-Mitchell, L. and W.E. Sandine. 1986. Influence of temperature on associative growth of *Streptococcus thermophilus* and *Lactobacillus bulgaricus*. J. Dairy Sci. 69: 2558–2568.

Rantsiou, K., T. Phister, L. McKay, G. Dunny and D. Mills. 1999. Broad host range mobilization of plasmid derivatives by the lactococcal conjugal element pRS01. pp. E13. *In:* Proceedings of the Sixth Symposium on Lactic Acid Bacteria. Federation of European Microbiological Societies and Netherland Institute for Microbiology, Veldhooven, The Netherlands.

Rao, D.R., A.V. Reddy, S.R. Pulusani and P.E. Cornwell. 1984. Biosynthesis and utilization of folic acid and vitamin B12 by lactic cultures in skim milk. J. Dairy Sci. 67: 1169–1174.

Raya, R.R., M.C. Manca de Nadra, A. Pesce de Ruiz Holgado and G. Oliver. 1986. Acetaldehyde metabolism in lactic acid bacteria. Milchwiss. 41: 397–399.

Rodriguez-Serrano, G.M., G. Perez-Hernandez, F. Gallardo, L. Gomez-Ruiz and M. Garcia-Garibay. 2002. Performance of yogurt cultures during the fermentation of whey concentrated by ultrafiltration for the elaboration of a fermented beverage. Milchwiss. 57: 540–543.

Rogout, A., A.P. Holgado, G. Oliver and F. Sineriz. 1989. Presence of an L(+)-lactate dehydrogenase in cells of *Lactobacillus delbrueckii* subsp. *bulgaricus*. Biochimie 71: 639–644.

Rohm, H., F. Eliskases-Lechner and M. Braver. 1992. Diversity of yeasts in selected dairy products. J. Appl. Bacteriol. 72: 370–376.

Ruas-Mediedo, P. and C.G. de los Reyes-Gavilian. 2005. Methods for the screening, isolation, and characterization of exopolysaccharides produced by lactic acid bacteria. J. Dairy Sci. 88: 843–856.

Rysstad, G., W.J. Knutsen and R.K. Abrahamsen. 1990. Effect of threonine and glycine on acetaldehyde formation in goats' milk yogurt. J. Dairy Res. 57: 401–410.

Santos, A., M. San Mauro, A. Sanchez, J.M. Torres and D. Marquina. 2003. The antimicrobial properties of different strains of *Lactobacillus* spp. isolated from kefir. Syst. Appl. Microbiol. 26: 434–437.

Sagui, F., P. Conti, G. Roda, R. Contestabile and S. Riva. 2008. Enzymatic synthesis of ω-carboxy-β-hydroxy-(L)-α-amino acids. Tetrahedron 64: 5079–5084.

Sarkar, S. 2008. Biotechnological innovations in kefir production: Review. British Food J. 110: 283–295.

Schmidt, B.F., R.M. Adams, C. Requadt, S. Power and S.E. Mainzer. 1989. Expression and nucleotide sequence of the *Lactobacillus bulgaricus* β-D-galactosidase gene cloned in *Eschericia coli*. J. Bacteriol. 171: 625–635.

Sedláček, I., D. Nováková and P. Švec. 2010. Ribotyping and biotyping of *Lactobacillus helveticus* from the koumiss. Eur. Food Res. Technol. 230: 753–758.

Serror, P., T. Sasaki, S.D. Ehrlich and E. Maguin. 2002. Electrotransformation of *Lactobacillus delbrueckii* subsp. *bulgaricus* and *L. delbrueckii* subsp. *lactis* with various plasmids. Appl. Environ. Microbiol. 68: 46–52.

Shankar, P.A. 1977. Interrelationships of *Streptococcus thermophilus* and *Lactobacillus bulgaricus* in yogurt culture. Ph.D. Thesis, University of Reading, Reading, UK.

Shiomi, M., K. Sasaki, M. Murofushi and K. Aibara. 1982. Anti-tumor activity in mice of orally-administered polysaccharide from kefir grain. Japan. J. Medical Sci. Biol. 35: 75–80.

Simova, E., D. Beshkova, A. Angelov, T. Hristozova, G. Frengova and Z. Spasov. 2002. Lactic acid bacteria and yeast in kefir grains and kefir made from them. J. Ind. Microbiol. Biotechnol. 28: 1–6.

St-Onge, M.P., E.R. Farnworth, T. Savard, D. Chabolt, A. Mafu and P.J.H. Jones. 2002. Kefir consumption does not alter plasma lipid levels or cholesterol fractional synthesis rates relative to milk in hyperlipidemic men. BMC Compl. Alter. Med. 2: 1–7.

Stefanitsi, D. and J.R. Garrel. 1997. A zinc-dependent proteinase from the cell wall of *Lactobacillus delbrueckii* subsp. *bulgaricus*. Lett. Appl. Microbiol. 24: 180–184.

Stingele, F., J.R. Neeser and B. Mollet. 1996. Identification and characterisation of the *eps* (exopolysaccharide) gene cluster from *Streptococcus thermophilus* Sfi6. J. Bacteriol. 178: 1680–1690.

Stingele, F., J.W. Newell and J.R. Neeser. 1999. Unraveling the function of glycosyltransferases in *Streptococcus thermophilus* Sfi6. J. Bacteriol. 181: 6354–6360.

Sun, T.S., S.P. Zhao, H.K. Wang, C.K. Cai, Y.F. Chen and H.P. Zhang. 2009. ACE-inhibitory activity and gamma-aminobutyric acid content of fermented skim milk by *Lactobacillus helveticus* isolated from Xinjiang koumiss in China. Eur. Food Res. Technol. 228: 607–612.

Svensson, M., E. Lohmeier-Vogel, E. Waak, U. Svensson and P. Radström. 2007. Altered nucleotide sugar metabolism in *Streptococcus thermophilus* interferes with nitrogen metabolism. Int. J. Food Microbiol. 113: 195–200.

Svensson, M., E. Waak, U. Svensson and P. Radström. 2005. Metabolically improved exopolysaccharide production by *Streptococcus thermophilus* and its influence on the rheological properties of fermented milk. Appl. Environ. Microbiol. 71: 6398–6400.

Tamime, A.Y. and H.C. Deeth. 1980. Yogurt: Technology and biochemistry. J. Food Protect. 43: 939–977.

Tamime, A.Y. 2002. Fermented milks: A historical food with modern applications. A review. Eur. J. Clin. Nutr. 56: S2–S15.

Tamime, A.Y. and R.K. Robinson. 2007. Tamime and Robinson's Yogurt Science and Technology. CRC Press, Boca Raton, USA.

Thoreux, K. and D.L. Schmucker. 2001. Kefir milk enhances intestinal immunity in young but not old rats. J. Nutr. 1341: 807–812.

Tinson, W., A.J. Hillier and G.R. Jago. 1982. Metabolism of *Streptococcus thermophilus*. I. Utilization of lactose, glucose and galactose. Aus. J. Dairy Technol. 37: 8–13.

Tiwari, P. and A.K. Misra. 2007. Synthesis of a pentasaccharide repeating unit of the extracellular polysaccharide produced by *Lactobacillus delbrueckii* subsp. *bulgaricus* 291. J. Carbohydrate Chem. 26: 239–248.

Toba, T., K. Arihara and S. Adachi. 1990. Distribution of microorganisms with particular reference to encapsulated bacteria in kefir grains. Int. J. Food Microbiol. 10: 219–224.

Tong, D., C. Xia, B.Q. Hua, L.W. Jun, Z.J. Chao and Z.H. Ping. 2012. Effect of L-threonine concentration on acetaldehyde production and glyA in fermented milk by *Streptococcus thermophilus*. Food Biotechnol. 26: 280–292.

Tuinier, R., E. ten Grotenhuis, C. Holt, P.A. Timmins and C.G. Kruif. 1999. Depletion interaction of casein micelles and an exocellular polysaccharide. Phys. Rev. E60: 848–856.

Ulusoy, B.H., H. Colak, H. Hampikyan and M.E. Erkan. 2007. An *in vitro* study on the antibacterial effect of kefir against some food-borne pathogens. Turkish Microbiol. Soc. 37: 103–107.

Urdaneta, E., J. Barrenetxe, P. Aranguren, A. Irigoyen, F. Marzo and F. Ibáñez. 2007. Intestinal beneficial effect of kefir supplemented diet in rats. Nutr. Res. 27: 653–658.

Ustok, F.I., C. Tarı and S. Harsa. 2007. Effect of symbiotic relationship of *Lactobacillus bulgaricus* 77 and *Streptococcus thermophilus* 95/2 on betagalactosidase and lactic acid production. J. Biotechnol. 131(suppl.): S224.

Vancanneyt, M., J. Mengaud, I. Cleenwerck, K. Vanhonacker, B. Hoste, P. Dawyndt, M.C. Degivry, D. Ringuet, D. Janssens and J. Swings. 2004. Reclassification of *Lactobacillus kefirgranum* Takizawa et al. 1994 as *Lactobacillus kefiranofaciens* subsp. *kefirgranum* subsp. nov. and emended description of *L-kefiranofaciens* Fujisawa et al. 1988. Int. J. Syst. Evol. Microbiol. 54: 551–556.

van den Bogaard, P.T.C. 2002. Catabolite control of sugar metabolism in *Streptococcus thermophilus*. Ph.D. Thesis. Wageningen University, Wageningen, The Netherlands.

van de Gutche, M., S. Penaud, C. Grimaldi, V. Barbe, K. Bryson, P. Nicolas, C. Robert, S. Oztas, S. Mangenot, A. Couloux, V. Loux, R. Dervyn, R. Bossy, A. Bolotin, J.M. Batto, T. Walunas, J.F. Gibrat, J. Weissenbach, S.D. Ehrlich and E. Maguin. 2006. The complete genome sequence of *Lactobacillus bulgaricus* reveals extensive and ongoing reductive evolution. PNAS-Proc. National Academy Sci. United States of America 103: 9274–9279.

Vaningelgem, F., R.V. Meulen, M. Zamfir, T. Adriany, A.P. Laws and L. de Vuyst. 2004. *Streptococcus thermophilus* ST 111 produces a stable high-molecular-mass exopolysaccharide in milk-based medium. Int. Dairy J. 14: 857–864.

van Kranenburg, R., M. Kleerebezem and W.M. de Vos. 2000. Nucleotide sequence analysis of the lactococcal EPS plasmid pNZ4000. Plasmid 43: 130–136.

Varga, F. 1998. The origins, production and control of acetaldehyde in yogurt. Ph.D. Thesis. The University of Reading, Reading, UK.

Vasilijevic, T., N.P. Shah and P. Jelen. 2005. Growth characteristics of *Lactobacillus delbrueckii* subsp. *bulgaricus* ATCC 11842 as affected by different neutralisers. Aus. J. Dairy Technol. 60: 3–9.

Vaughan, E.E., P.T.C. van den Bogaard, P. Catzeddu, O.P. Kuipers and W.M. de Vos. 2001. Activation of silent gal genes in the lac-gal regulon of *Streptococcus thermophilus*. J. Bacteriol. 183: 1184–1194.

Vescovo, M., G.L. Scolari and V. Bottazzi. 1989. Plasmid-encoded ropiness production in *Lactobacillus casei* ssp. *casei*. Biotechnol. Lett. 10: 709–712.

Vinals, C., X. Debolle, E. Depiereux and E. Feytmans. 1995. Knowledge-based modelling of the D-lactate dehydrogenase 3-dimensional structure. Prot.-Struc. Func. Gene. 21: 307–318.

Viljoen, B.C. 2001. The interaction between yeasts and bacteria in dairy environments. Int. J. Food Microbiol. 69: 37–44.

Vinderola, C.G., J. Duarte, D. Thangavel, G. Perdigon, E. Farnworth and C. Matar. 2005. Immunomodulating capacity of kefir. J. Dairy Res. 72: 195–202.

Walker, G.M. 1998. Yeast metabolism. pp. 203–264. *In:* G.M. Walker [ed.]. Yeast Physiology and Biotechnology. John Wiley & Sons, Chichester, UK.

Wang, J., X. Chen, W. Liu, M. Yang, Airidengcaicike and H. Zhang. 2008. Identification of *Lactobacillus* from koumiss by conventional and molecular methods. Eur. Food Res. Technol. 227: 1555–1561.

Wang, H.K., C. Dong, Y.F. Chen, L.M. Cui and H.P. Zhang. 2010. A new probiotic cheddar cheese with high ACE-inhibitory activity and gamma-aminobutyric acid content produced with koumiss-derived *Lactobacillus casei* Zhang. Food Technol. Biotechnol. 48: 62–70.

Wang, H.K., J. Shi, H.P. Zhang and W. Qi. 2011. A survey of some antifungal properties of lactic acid bacteria isolates from koumiss in China. Int. J. Dairy Technol. 64: 585–590.

Wang, S.Y., K.N. Chen, Y.M. Lo, M.L. Chiang, H.C. Chen, J.R. Liu and M.J. Chen. 2012. Investigation of microorganisms involved in biosynthesis of the kefir grain. Food Microbiol. 32: 274–285.

Wilkins, D.W., R.H. Schmidt and L.B. Kennedy. 1986. Threonine aldolase activity in yogurt bacteria as determined by headspace gas chromatography. J. Agri. Food Chem. 34: 150–152.

Witthuhn, R.C., T. Schoeman and T.J. Britz. 2004. Isolation and characterization of the microbial population of different South African kefir grains. Int. J. Dairy Technol. 57: 33–37.

Wszolek, M., B. Kupiec-Teahan, H. Skov Guldager and A.Y. Tamime. 2006. Production of kefir, koumiss and other related products. pp. 174–216. *In:* A.Y. Tamime [ed.]. Fermented Milks. Blackwell Publishing, Oxford, UK.

Wu, R., W. Wang, J.C. Wang, H.P. Li, B. Menghe, J.R. Wu, M.R. Guo and H.P. Zhang. 2009a. Proteomics analysis of *Lactobacillus casei* Zhang: A new probiotic bacterium isolated from traditional homemade Koumiss in Inner Mongolia. Mol. Cell. Prot. 8: 2321–2338.

Wu, R., L.P. Wang, J.C. Wang, H.P. Li, B. Menghe, J.R. Wu, M.R. Guo and H.P. Zhang. 2009b. Isolation and preliminary probiotic selection of lactobacilli from koumiss in Inner Mongolia. J. Basic Microbiol. 49: 318–326.

Yoon, Y.H., J.K. Cho, Y.J. Baek and C.S. Huh. 1998. Antimutagenic activity of *Lactobacillus* spp. isolated from kefir and yogurt and non-starter strains. Korean J. Animal Sci. 41: 39–44.

Yoshida, Y., W. Yokoi, K. Ohishi, M. Ito, E. Naito and H. Sawada. 2005. Effect of cell-wall of *Kluyveromyces marxianus* YIT 8292 on the plasma cholesterol and fecal sterol excretion in rats fed on a high-cholesterol diet. Biosci. Biotechnol. Biochem. 69: 714–723.

Zhang, W., D. Yu, Z. Sun, R. Wu, X. Chen, W. Chen, H. Meng, S. Hu and H. Zhang. 2010. Complete genome sequence of *Lactobacillus casei* Zhang: A new probiotic strain isolated from traditional homemade koumiss in Inner Mongolia, China. J. Bacteriol. 192: 5268–5269.

Zajsek, K., M. Kolar and A. Gorsek. 2011. Characterization of the exopolysaccharide kefiran produced by lactic acid bacteria entrapped within natural kefir grains. Int. J. Dairy Technol. 64: 544–548.

Zourari, A., J.P. Accolas and M.J. Desmazeud. 1992. Metabolism and biochemical characteristics of yogurt bacteria. A review. Lait 72: 1–34.

CHAPTER 9

Development of Fermented Milk Products Containing Probiotics

Claude P. Champagne

9.1 Introduction

In North-America, up to 93% of consumers believe certain foods have health benefits that may reduce the risk of disease (Clydesdale 2005). This has fostered the development of functional foods (FF). Global sales of probiotic ingredients, supplements and foods amounted to $21.6 billion in 2010 (BCC Research 2011) and between $24.2 billion (Transparency Market Research 2013) and 27.9 billion (marketsandmarkets.com 2013) in 2011. In North America alone, the probiotics market for human nutrition has revenues of US$1.3 billion, with about 150 companies operating in this market (Rajagopal 2012). Probiotics are arguably the main bioactive components of these fermented FF. In accordance, the main health attributes of such FF is gut health and digestive health still constitutes a key trend for innovation in the FF market (Mellentin 2012).

These data show the importance of probiotic-containing yogurt and explains why dairy processors are still developing novel yogurts. Nevertheless, it is notable that in this "strict definition" market, one-shot dose delivery products, now more widely known as active health drinks, now outsell more traditional "spoonable" functional yogurt lines. Products such as phytosterol margarines cost up to three times the price of standard lines, while probiotic drinks are also very expensive in comparison with yogurt drinks, although they, as unique products, have no actual basis for comparison (Leatherhead Food Int. 2006). It is not surprising that an analysis of FF product launches between January 2005 and April 2006 thus highlighted well over 200 introductions over the period (Leatherhead Food

Int. 2006). These market data show that strong commercial activities remain in the development of probiotic-containing yogurts, but that novel specialty milk products are on the increase and offer interesting market opportunities. Additionally, frozen yogurts and cheese are increasingly considered for enrichment by probiotics. Thus, fermented milks can generate a high number of products on the market.

A wide-ranging study of yogurt production with 14 commercial cultures revealed that counts in probiotic bacteria could vary between 4.0×10^5 and 7.7×10^8 cfu ml^{-1} in fresh products (Kneifel et al. 1993). Such data point to a thousand-fold variation in probiotic content, as a function of strain and production practices. This chapter will therefore address the technological issues which surround the development of probiotic bacteria in yogurts as well as in specialized high probiotic density fermented milks.

From a technological standpoint, there are many challenges in the development of a probiotic-containing dairy product:

1. Selection of the strain(s)
2. Determination of as well as preparation of the populations to add
3. Enable growth and survival during processing
4. Enable viability and functionality during storage
5. Assess the viable counts of the probiotic strain(s), particularly when multiple probiotic strains are added and when there are also starter cultures added
6. Manage the effects on sensory properties.

In this chapter, the focus will be given on three of these challenges: strain selection, processing and storage issues. The main dairy products this review will address are yogurt, cheese, fermented milks which do not contain yogurt or cheese cultures and frozen-yogurt.

9.2 The "ideal" product-strain selection and identification

When developing a new probiotic-based dairy product, a manufacturer should strive for the "ideal" product. Arguably, the ideal product would have the following attributes:

1. Species and strain(s) clearly identified on the label, for example, *Bifidobacterium longum* R0175
2. Viable counts at the "best before" date, for example 1 billion per portion of 100 ml
3. Clinical data proving a health attribute.

Many products only state on the label that it contains probiotic bacteria and provides the species added, for example *Lactobacillus acidophilus*. This carries an assumption that all *Lb. acidophilus* strains are the same with respect to generating a given health effect, which is like saying that all humans (*Homo sapiens*) can be good surgeons. Strains differ, and the health effects could be linked to them (Reid 2008). Therefore, strain identity should be given. There are a certain number of strains which already have a body of evidence (Shah 2007), and a critical assessment of some strains is available (Sanders and Klaenhammer 2001, Reid 2008). In the past, potential probiotic strains were selected for addition into yogurt on the basis of their technological properties. These technological properties would be from two perspectives: (i) the commercial producers of cultures and (ii) the food manufacturers (Champagne et al. 2005). Today, however, the ability of the cultures to provide substantiated health effects is paramount. These probiotic strains typically have the ability to survive and possibly grow in the gastro-intestinal tract (GIT), but might not have the abilities to survive the rigors of food processing and storage. Consequently, technological adaptations may be required, which is the subject of this chapter.

It is believed that the quantity of cells ingested influences the clinical effect, which explains why it is stated in the definition of a probiotic that an "adequate amount" must be given (Araya et al. 2002). Very few manufacturers state the guaranteed populations on the label itself, often limiting this information to the product's web site. Although this approach is understandable, in light of the challenges linked to providing a given viable population, the absence of such information on the label itself might eventually undermine consumer confidence. Serious manufacturers will therefore need to carry out stability tests of their products to establish the evolution of viability during processing and storage. In practice, this often results in the need to inoculate a significantly higher number of cells in the product than is claimed on the label in order to provide for viability losses, particularly during storage.

Very few products carry an approved specific health claim on the label. Companies obviously wish that their products do indeed exert a health effect. Many suppliers can provide valuable data on the mode of action of their cultures. The bioactive compounds specifically derived from the probiotic bacteria's metabolism, which can be termed "probioactive" (Farnworth and Champagne 2009), are increasingly being identified (enzymes, peptides, exopolysaccharides, cell-wall fractions, gamma amino butyric acid, etc.). Knowing what "probioactive" is involved and the quantity required could facilitate obtaining the health claim, although there are still foreseeable problems (Table 9.1). If the health benefits in the clinical trials were clearly linked to the presence of cells, and not the matrix consumed, then viable cell counts would presumably be critical in providing the effect from the food. A number of clinical trials have been carried out with the probiotic being delivered through caplets or powders. Presumably, the small amount of a food-grade non-medicinal filling in a caplet (about 0.3 g) would not

Table 9.1 Properties of the probiotic culture which has resulted in a positive clinical trial using a non-dairy matrix (caplet, rehydrated powder, cereal, etc.), which could enhance the probability of transferring the health effect to a dairy-based food system.

Source of the health effect	Property to look for	Reason	Foreseeable problem
Known compound (peptide, EPS,[1] enzyme or other)	1. A defined metabolite generated by the probiotic bacteria during fermentation has been identified (probioactive).	Increases the probability that a given health effect can be transferred to a different food matrix because viability issues of the cells can be irrelevant.	The dairy food matrix could still affect the stability and the activity of the probioactive in the GIT.[3]
	2. The concentration of the "probioactive" required in the portion consumed is known.	Requires the assurance that the probioactive compound can be analyzed and its concentration in the FF[2] guaranteed.	
Cells—undefined bioactive compound	Select a health effect that was clearly linked to "cells" (high bacterial density caplet?) rather than a compound in the non-dairy product. The quantity of cells required must be known.	If the health effect is linked to cell numbers, the chance of being able to transfer the claim to a food could be linked to the viable population in the food.	The food matrix will definitely affect (1) viability during production and storage and (2) the subsequent survival to GI[4] conditions.
	In a cell-based effect, the state of the cells in the dairy food must be the same as those given in the clinical trial.	Encapsulation will strongly affect the survival of probiotic bacteria in the GI system and potentially enhance their functionality. If the dairy product is planned to contain free, unprotected cells, then the clinical study should also have been carried out with such cultures.	If the clinical train was carried out with encapsulated cultures, one could wish to also use encapsulated cultures in the dairy food. This could affect texture.
	The dairy food must provide at least as good a protection to upper GI conditions (stomach, bile, pancreatic juice) than the non-dairy matrix used in the clinical trial.	In a cell-based effect, the population of viable probiotic bacteria in the lower intestine and colon is presumably linked to the health effect.	1. *In vitro* tests must be carefully chosen to provide a good picture of survival to gastric conditions. 2. The food matrix could not only affect survival to the upper GIT, but also growth in the lower intestines.

[1]**EPS**: Exopolysaccharide, [2]**FF**: Functional food, [3]**GIT**: Gastro-intestinal tract, [4]**GI**: Gastro-intestinal.

contribute much to the health effect and it can be hypothesized that the benefit obtained through a caplet is "cell linked". Unfortunately, simply providing identical cell populations in the food as were provided by the nutraceuticals, might not be enough to guarantee the "transferability" of the health effect to the dairy matrix. This would require clinical data with the dairy matrix itself, which is expensive to obtain. It can be foreseen that legislators will eventually demand such studies. The food industry arguably has much fewer resources than the pharmaceutical sector for human clinical studies.

So how does a dairy processor select a strain in order to improve the chances of effectively obtaining a positive response in a clinical trial? Some suggestions are presented in Table 9.1. Although applying this selection process will not guarantee a positive clinical trial, the dairy industry nevertheless has reason to be optimistic. Some data show that dairy products constitute a much better delivery system than powders containing free, non-encapsulated cells (Table 9.2). It could turn out that some dairy formulations provide much better delivery systems of probiotic bacteria to the GIT than the powders or uncoated caplets used in some previous clinical trials.

Table 9.2 Recovery of *Lactobacillus rhamnosus* GG in stools, as a function of the delivery matrix.

Food product	Population ingested in serving (cfu portion^{-1})	cfu in stools (cfu g^{-1}, wet mass)
Powder	2×10^{10}	1×10^{6}
Fermented milk (200 ml)	1×10^{10}	3×10^{7}
Milk (200 ml)	1×10^{8}	1×10^{7}
Cheese	3×10^{8}	7×10^{7}
Fruit juice (200 ml)	2×10^{9}	1×10^{6}

Table prepared from the data of Saxelin et al. (2003).

9.3 Inoculation of probiotics into the dairy matrix

Two methods serve to inoculate the dairy matrix with lactic cultures: (1) direct vat inoculation (DVI), from concentrated frozen or dried cultures and (2) liquid cultures prepared in bulk starter tanks. DVI has many advantages with respect to quality control and processing flexibility. However, DVI is approximately 20% more expensive than bulk starter preparation. Therefore, in large yogurt and cheese processing plants, bulk starters are still frequently prepared. In bulk starter preparation, a growth medium, which typically contains dairy ingredients as well as minerals and peptones, is inoculated with a relatively small amount of cells, and the medium is incubated in appropriate temperature and time parameters to propagate the cells. This bulk starter is then typically added at 1 to 2% (v/v) of the processing milk.

However, inoculation with fresh bulk liquid starters is not seen with probiotic bacteria. DVI is the norm. Various reasons explain this situation:

1. In cheese and yogurt manufacturing, the starters are expected to carry out an immediate acidifying activity. At identical population levels, fresh liquid starter cultures have shorter lag times before acidification initiates, which reduce processing time. This explains why large yogurt and cheese processors prepare fresh cultures. However, probiotic bacteria are not required to contribute extensively to acidification, and there is no need for immediate, high, specific acidifying activity. Thus, inoculating a frozen or dried form, which generally results in slow initial activity, does not create a processing problem.
2. Specific probiotic population levels are required, and this is easier to standardize with DVI.
3. No need to invest in specific equipment (fermentation tanks) for probiotics. Although one could attempt to grow the probiotics in the same tank as the bulk starter, it is unfortunately difficult to obtain at high probiotic population levels in mixed cultures, and to control starter:probiotic ratios. Preparation of probiotic bulk cultures would probably require the dairy plant to add fermentation vats/systems specifically for the probiotic bacteria.
4. Probiotic bacteria are more fastidious organisms to grow than starter cultures.

Two DVI formats are mainly available: frozen pellets and freeze-dried powders. Technological issues which surround DVI practices are linked to preservation of the cultures prior to use, and thawing/rehydration procedures. The "golden rule" with respect to storing and handling commercial cultures is to "follow the manufacturers' instructions".

The first point to consider is the storage conditions of the unopened commercial products. It is recommended to store frozen cultures at –40°C or below. Consumer freezers which typically maintain –20°C are inappropriate for extended storage of the frozen pellets. When thawed, they must be used very rapidly. Freeze-dried cultures can be stored at higher temperatures, and a refrigerator at 4°C is considered adequate. In such conditions, probiotic and lactic cultures in practice will lose approximately 0.2 log in viable counts over one year (Champagne et al. 1996). At 22°C, mortality rates are 10 times higher.

The second technological point on DVI practices is with respect to opened products (Champagne and Møllgaard 2008). The inoculation requirements are often such that only a fraction of a 1 kg culture package is needed. Therefore, situations occur where a commercial product is opened, a sample taken and the rest is kept for ulterior use. With the frozen cultures, the container must be immediately placed back in the freezer when the required amount of frozen cell pellets has been taken. A frozen product must not be thawed before taking the required amount. A thawed product cannot be re-frozen with satisfactory results with equipment found in dairy plants. Specialized suppliers of probiotic bacteria typically use rapid-freezing systems, often based on liquid nitrogen, which are critical to survival of the cultures to the freezing process. With dried products, moisture absorption by the powder will increase the water activity in the product. If the relative humidity of the room is of 30% and the powder is exposed to this environment long enough to the point that absorption stops, then the water activity (a_w) of the powder will be at 0.3. Although the product will still have the appearance of a powder, this a_w will be highly detrimental to the stability of the cultures. Data show that an increase from 0.1 to 0.3 in a_w of a milk-based product will result in only a 2% increase in moisture, but that the stability during storage will be 10 times lower (Ishibashi et al. 1985). Therefore, when a fraction of a package containing a freeze-dried powder is taken, the sachet must be closed as rapidly as possible and placed back at 4°C. The "best-before" date becomes voided.

The third point is the re-suspension medium. In yogurt and cheese processing, many plants add the frozen pellets or the powder directly into the pasteurized processing milk, at the beginning of fabrication, where they dissolve. Little data are available for frozen cultures, but it has been shown that milk is a good medium for rehydration of lactic cultures (de Valdez et al. 1985a). Because freeze-dried commercial products can contain over 10^{11} cfu g^{-1}, in some situations the inoculation rate can be of only 0.01%. Adding such small amounts of powder to the milk product is sometimes a problem and some companies may want to prepare a liquid cell suspension for inoculation. In this instance, water is not recommendable for rehydration of freeze-dried cultures (Sinha et al. 1982), and nor are fruit juices. Indeed, rehydration of lactic cultures in a high acid environment typical of fruits is very detrimental to viability (Sinha et al. 1982). Good results are obtained with milk-based solutions having between 10 and 20% total solids. This being said, carrying out DVI of probiotics in the processing milk is not as much a concern as with starters with respect to uneven distribution of the cells in milk. The main goal is to have the required cell count at the "portion" level of the dairy product.

The fourth point is rehydration temperature. One would assume that rehydrating at 4°C would be ideal but this is not the case. Therefore, when inoculating ice cream products, it is not advisable to add powders directly to the cool dairy blend just prior to freezing. In fact, temperatures between

30 and 37°C are best for post-hydration viabilities (Table 9.3). Care must be taken, however, not to go above 45°C, because this results in denaturation of enzymes with subsequent viability losses. In the case of yogurt production, where processing milk is between 37 and 45°C, DVI directly into the processing milk is appropriate from the temperature perspective.

The fifth element is the powder to liquid ratios during hydration. Data of de Valdez et al. (1985b) show that rehydrating at high powder to liquid ratios favors recovery of the cells, presumably because of smaller osmotic shocks when high solids levels are present. This would suggest that rehydration in processing milk would not be ideal, but the fact that yogurt processing milk contains above 10% milk solids may prevent extensive problems. When concentrated cell suspensions are needed, providing hydration conditions where medium solids are around 20% seem best (taking into account solids brought by the culture itself) (de Valdez et al. 1985b).

The last point is rehydration time. One could assume that extending the time of hydration prior to inoculation into the processing milk would result in growth initiation and higher populations in the rehydrated cell suspension. Data from de Valdez et al. (1985b) suggest that, when very concentrated cell suspensions are involved, viability losses occur after 30 minutes or rehydration time. Presumably, in such high cell densities, acidification of the medium could be detrimental to viability.

Whatever the reasons for viability losses due to rehydration temperature, medium, solids levels or time, these data point to the need to standardize the rehydration-inoculation procedure in DVI. To avoid variability in viable counts, it seems wise to define appropriate procedures and try to keep them constant. One must also keep in mind that there are wide differences between species of probiotic bacteria as well as between strains of the same species with respect to behavior during rehydration. It could well be found that conditions favorable to one strain may not apply to a different culture.

Table 9.3 Effect of rehydration temperature on viable counts of *Lactobacillus delbrueckii* subsp. *bulgaricus*.

Temperature (°C)	Viable count (cfu g^{-1})
4	2.5×10^{9}
10	3.5×10^{9}
15	3.5×10^{9}
20	5.0×10^{9}
25	1.1×10^{10}
30	1.7×10^{10}
37	1.6×10^{10}

Table prepared from data of Mille et al. (2004).

All this being said, there are evidently little data published on the effects of changes in, or even mistakes during, rehydration procedures on the viability of probiotics in the finished products. Only a handful of studies are published, often in laboratory media. More information would be warranted on this deceptively simple task.

9.4 Yogurt

9.4.1 Milk as a growth medium for probiotic bacteria

Probiotic cultures can grow in milks of many species (Abu-Taraboush 1994, Abu-Taraboush et al. 1998, Murad et al. 1997, Gomes and Malcata 1998) but, most of the time, rather slowly (Klaver et al. 1993). Although some studies report differences in growth as a function of milk animal source (Abu-Taraboush et al. 1998, Gomes and Malcata 1998, Al Saleh 2000), growth of probiotic bacteria in milk is very much strain-related (Klaver et al. 1993, Saxelin et al. 1999). These wide variations between strains with respect to their ability to grow rapidly in milk are not limited to probiotic bacteria. The same is observed with yogurt (Shihata and Shah 2002) and cheese (Heap and Richardson 1985) starter cultures. Therefore, no clear conclusion can be reached as to which milk source is better suited for the growth of probiotic cultures. Most studies found in the literature were carried on cow milk and, unless otherwise stated, the term "milk" will refer to cow milk.

Probiotic cultures such as *Lb. acidophilus* or *Bifidobacterium* spp. generally grow faster on synthetic media than on milk (Misra and Kuila 1991, Gaudreau et al. 2005). The slow growth of bifidobacteria in milk appears partially due to low lactase (Desjardins and Roy 1991) which was also the case for *Lactobacillus rhamnosus* (Gaudreau et al. 2005). This has also been seen with cheese lactococci starter cultures as well (Lawrence et al. 1976, Sinha 1990).

Milk supplementation by minerals, amino acids and nucleotides have all improved the growth of *Lactobacillus johnsonii* (Elli et al. 1999) and amino acids are frequently seen as good supplements for bifidobacteria (Klaver et al. 1993). This would suggest that high proteolytic activities would enhance growth rates and levels, but it appears that this is not always the limiting factor with bifidobacteria (Desjardins et al. 1991). Indeed, some *Lb. acidophilus* and *Bifidobacterium* cultures have proteolytic activities as high as those of yogurt starters (Shihata and Shah 2000), but still do not multiply as fast in milk. In some instances, mixing a non-proteolytic *Bifidobacterium* strain with a highly proteolytic *Lb. acidophilus* culture will be helpful (Klaver et al. 1993), but if the starter lactic acid bacteria (LAB) grow too fast the fermentation time is shortened and counts in probiotic bacteria can even be lowered (Shihata and Shah 2002).

Numerous studies show that enriching milk with supplements enhances the growth of probiotics. This constitutes an opportunity for innovation. Indeed, examples of potential supplements are extracts or juices of the following: yeast (Kim et al. 1995), citrus (Sendra et al. 2008), ginseng (Goh et al. 1993), tomato (Babu et al. 1992), peanut (Murad et al. 1997), soy (Yajima et al. 1992), cereals (Kyung and Young 1993, Vasiljevic et al. 2007), honey (Ustunol and Gandhi 2001), berries (Kailasapathy et al. 2008), mango (Kailasapathy et al. 2008), herbs (Ray-Chowdhury et al. 2008) and whey (Christopher et al. 2006). A noteworthy trend is the addition of prebiotics, which include: fructooligosaccharides (FOS) (Bruno et al. 2002) such as chicory's inulin (Aryana et al. 2007, Juhkam et al. 2007). Prebiotic "candidates" such as lactulose (Bruno et al. 2002), oat and barley glucans (Vasiljevic et al. 2007), galactooligosaccharides (GOS) (Shin et al. 2000), starch/maltodextrins (Bruno et al. 2002) and raffinose (Martinez-Villaluenga and Gomez 2007) have also been used. In view of the fact that many probiotics are poor users of lactose, the addition of a more readily available carbohydrate can selectively enhance the growth of probiotics during the fermentation. Thus, addition of 5% lactulose, GOS or inulin reduced the doubling time of bifidobacteria in milk by 25 to 50% (Shin et al. 2000, Bruno et al. 2002).

In addition to the above supplements, manufacturing processes often include artificial and natural flavors, sugar or sweeteners, and preservatives. Sucrose at less than 10% is generally not inhibitory to probiotic bacteria (Shah and Ravula 2000, Vinderola et al. 2002). Aspartame and acesulfame sweeteners are not toxic at concentrations used in industrial practice (Vinderola et al. 2002). Some fruit and vanilla flavors are detrimental to growth of probiotics, but this is quite variable (Vinderola et al. 2002). Preservatives natamycin and lysozyme are not a problem at concentrations used commercially, but data show that nisin inhibits a significant proportion of probiotic cultures (Vinderola et al. 2002).

Little is known on fat metabolism of bifidobacteria (Tamime et al. 1995) or of probiotic lactobacilli. Some data suggest that the presence of fat slightly improves milk fermentation and cell populations (Elli et al. 1999, Lanciotti et al. 2004). Fatty acids are growth factors at low levels (Partanen et al. 2001). Unless lipolysis has occurred, there are only small amounts of free fatty acids in milk (Lawrence et al. 1976). Fatty acids are mostly in triglycerides and are largely unavailable to the cultures. Therefore, it is not clear if the small benefits of milk fat on growth are linked to a nutritional component of the fat fraction or if the presence of fat affects milk's nutritional value for probiotic bacteria in another fashion, particularly during heating and homogenization.

Variations of growth rates are also function of the "age" of milk. Milk is generally collected at the farm within 48 hr, but it can be stored at the processing plant for up to three additional days. There are chemical and microbiological differences between fresh milk and one which has been stored for five days, even at 4°C. Such extended storage occurs when the plant has limited processing activities during the weekends. These milks tend to have high counts in psychrotrophic bacteria. This microbiota will often possess substantial proteolytic and lipolytic activities. A high level of proteolysis will stimulate the growth of probiotic bacteria (Srivinas et al. 1997). Very limited lipolysis might be stimulatory, but extensive lipolysis has the opposite effect. It cannot be recommended to use old milks because off-flavors and texture problems occur in the yogurt. However, one must consider the age of milk as a parameter which will affect the development of the cultures.

It must be stated, however, that the low growth rates of some probiotic bacteria in milk might erroneously be attributed to the nutritional aspects of milk. The redox level could be the limiting factor for some cultures. This is particularly the case with bifidobacteria, where deaeration favored the growth of many strains (Klaver et al. 1993). How milk is prepared is therefore critical to growth of starter and probiotic bacteria.

9.4.2 Milk preparation

Besides ingredient composition, milk preparation for yogurt manufacture varies by two main parameters: solids level as well as the heating/homogenization process.

Typically, the milk blend used for yogurt manufacture will have between 10 and 15% non-fat milk solids, be heated at 80–95°C for 1 to 30 min, and homogenized at 50–70°C at pressures of 100–200 kg cm^{-2} (Tamime and Robinson 1985). This is much different from typical cheeses, where milk is not concentrated, heated at only 73°C for 15 s (pasteurization) and is not homogenized. All these parameters affect the growth of lactic cultures.

In many cases, growth of lactic starter cultures is better in heated milks (Feldstein and Westhoff 1979, Mortazavian et al. 2006) because anti-bacterial components are destroyed, oxygen is removed and a lower redox level is reached, limited proteolysis occurs and formic acid is produced. When the growth of yogurt cultures was examined on milks heated at 85°C for 15 min or at 95°C for 5 min, the streptococci preferred in the lesser-heated milk, while lactobacilli developed much better in milk previously heated at 95°C for 5 min (Kurultay et al. 2006).

Non-fat milk solids (NFMS) influence the numbers of cells in the yogurt products as well as strain ratios (Al-Dabbagh and Allan 1989, Torriani et al. 1996, Alvarez et al. 1998). The effect of NFMS is obviously linked to the quantity of nutrients but also to buffering ability. Therefore, increases of NFMS from 8.5% to 12–14% generally enhance viable counts in the fermented products. However, high NFMS increase osmotic conditions which become detrimental to growth (Al-Dabbagh and Allan 1989). Lowering of milk a_w with sugar favors the cocci (Larsen and Añón 1989, Shah and Ravula 2000). Little is known unfortunately of the effect of milk solids on ratios of probiotic bacteria in yogurt cultures. When sugar is added to the blend, *Lb. acidophilus* and bifidobacteria had 1 log reductions in populations in the yogurts with 12% sucrose, as compared to supplemented with 8% (Shah and Ravula 2000). However, the probiotic bacteria were similarly affected by the addition of sucrose at concentrations above 8% as were the starter cultures. If the probiotic lactobacilli do indeed react similarly as *Lactobacillus delbrueckii* subsp. *bulgaricus*, then adjustments of milk solids above 14% might not be advisable.

In summary, there are at least 10 parameters in the formulation of the milk blend which will affect the development of probiotics in yogurt (Table 9.4). In many cases, this means that they will need to be standardized in order to have reproducible results.

Table 9.4 Parameters which need to be considered in the development of a probiotic-containing yogurt.

Milk blend	Fermentation	Storage
• Animal source • Pre-processing storage time of raw milk • Non-fat solids • Fat content • Growth supplements • Sugar level • Flavors and fruits • Preservatives • Heating parameters • Redox level	• Compatible starter • Form of starter or probiotic (liquid, DVI) • If dried DVI, rehydration parameters (solids, temperature, time) • Inoculation level of starter or probiotic (cfu ml^{-1}) • Moment of inoculation of probiotic • Fermentation temperature • Fermentation time	• pH (yogurt and after fruit addition) • Moment of inoculation of the probiotic • *Lb. delbrueckii* subsp. *bulgaricus* content and activity (H_2O_2, over-acidification) • Redox level. Addition of anti-oxidants? • Packaging, particularly with respect to oxygen permeability • Encapsulation

9.4.3 Competition with yogurt starters

When milk is inoculated with probiotic cultures alone, fermentation times can be up to 24 hr (Table 9.5). With these pure cultures, populations attained typically range between 3 and 10×10^8 cfu ml^{-1} of fermented milk. If a yogurt starter culture is blended with the probiotic bacteria, the probiotic counts are typically between 3 and 10 times lower than if they were grown in pure cultures.

Table 9.5 Effect of fermentation time, probiotic or streptococci cultures as well as strain pairing on the viable counts of fermented milks.

Strains		Initial (cfu ml^{-1})	Fermented milk (cfu ml^{-1})	Fermentation time (hr)
Pure R0083	*S. thermophilus*	2.0×10^7	1.2×10^9	8
Pure ST5	*S. thermophilus*	2.3×10^7	8.7×10^8	24
Pure R0175	*B. longum*	2.3×10^7	3.6×10^8	24
Pure R0052	*Lb. helveticus*	1.9×10^7	2.8×10^8	24
Mixed R0083 + R0175	*S. thermophilus*	1.0×10^7	1.1×10^9	8
	B. longum	9.1×10^6	5.4×10^7	8
	Ratio S/B	1	20	
Mixed ST5 + R0175	*S. thermophilus*	9.6×10^6	7.4×10^8	20
	B. longum	9.1×10^6	6.8×10^7	20
	Ratio S/B	1	11	
Mixed R0083 + R0052	*S. thermophilus*	1.0×10^7	1.4×10^9	8
	Lb. helveticus	1.2×10^7	2.6×10^8	8
	Ratio S/L	1	5	

Source: Champagne et al. (2009)

With a starter, fermentation time is generally much shorter, but this varies with the starter strain (Table 9.5). An important property of fast-acidifying yogurt cultures is often their proteolytic activity (Shihata and Shah 2002). The first reason fermented milks obtained from mixed cultures of probiotic and starter bacteria are lower in probiotic bacteria would therefore appear to be the fermentation time. In this scenario, the shorter fermentation time limits the extended growth of the probiotic bacteria. One study in which a slower *Streptococcus thermophilus* culture was used to carry out the fermentation helped slightly in improving the *B. longum* population in the fermented product (Table 9.5). This gives some support to the effect of fermentation time. However, even with the slow starter, the probiotic population was still much lower in the mixed culture than when the probiotic was grown as a pure culture (Table 9.5). Therefore, additional factors contribute to the detrimental effect of yogurt cultures on probiotic populations.

A partial explanation is that some bifidobacteria grow rapidly during the first hours of incubation, but multiplication stops at low population levels (Samona and Robinson 1994). This pattern is typical of protease-negative lactococci (Juillard et al. 1995). Therefore, nitrogen metabolism becomes critical in the relationship between the starter and the probiotic culture. When starters are not highly proteolytic, it can be feared that competition would occur for the assimilation of the limited amount of free peptides and amino acids in milk as noted for cheese cultures (Juillard et al. 1987). Hence, the proteolytic activities of the various strains in the blends will influence the strain ratios (Klaver et al. 1993).

Starter cultures produce additional inhibitory products to lactic acid. Since some probiotic bacteria are strongly affected by the redox level (Klaver et al. 1993), the production of hydrogen peroxide (H_2O_2) is particularly to be feared. *Lactobacillus delbrueckii* is a recognized producer of H_2O_2 (Villegas and Gilliland 1998). H_2O_2 strongly influences the storage stability of probiotic cultures in yogurt (Lankaputhra et al. 1996) but it remains to be determined if H_2O_2 production during fermentation is also detrimental. Bifidobacteria which synthesize peroxidases are less susceptible to this occurrence (Shimamura et al. 1992).

Cheese starter cultures synthesize bacteriocins (Carr et al. 2002). Bifidobacteria are sensitive to nisin (Kheadr et al. 2004), but this bacteriocin is produced by lactococci in cheese starters. Many lactobacilli also produce bacteriocins which raises a concern. However, the inhibition of the growth of probiotic bacteria by bacteriocins of yogurt starters is undocumented. It is known that *Lb. delbrueckii* subsp. *bulgaricus* inhibits probiotics to a greater extent than *S. thermophilus* (Sarkar 2008), but the link towards bacteriocins is unknown.

Fresh liquid cultures will initiate growth and acidification much faster than cells which are added to the processing milk frozen or dried. Therefore, another reason probiotic bacteria do not seem to compete well with starters is simply linked to the form of inoculation. Yogurt cultures are often added in the fresh liquid form, while probiotics are inoculated in the frozen or dried state. Even if thawing or rehydration conditions of the probiotics are acceptable, the growth of the probiotic bacteria will be delayed as compared to that of a fresh liquid starter.

These inoculation practices and interactions between yogurt starters on probiotics have an important impact on the inoculation level of probiotics. Since manufacturers claim given cell counts, it is recommended that during the manufacture of dairy products with bifidobacteria, the milk should be inoculated with the final number of bifidobacteria required for the product (Roy 2005).

In addition to providing a milk-based medium favorable to the development of probiotic bacteria (sections 9.4.1 and 9.4.2 above) a number of techniques can be used to attempt to improve the growth of probiotic cultures in a mixed starter-probiotic fermentation: (1) selection of a compatible yogurt starter, (2) reduce the *Lb. delbrueckii* subsp. *bulgaricus* content of the starter, (3) use an oxygen-consuming *S. thermophilus* culture, (4) deaeration of milk, (5) inoculate the starter later in the fermentation, (6) inoculate with a sonicated yogurt culture, (7) lower the starter inoculation level, and (8) lower the fermentation temperature from 42 to 37°C. The first three are related to the composition of the yogurt starter. In the past, the probiotic culture was the one being tested for compatibility with the current yogurt starter. The opposite may now need to be carried out. It is extremely costly to carry out clinical tests with probiotic bacteria and a given strain might be selected for its demonstrated health benefits. In this instance, it is the yogurt starter which must now be subject to compatibility testing with the probiotic culture. Starters which produce inhibitory compounds towards the probiotic bacteria are obviously to be avoided. This is why some authors recommend to reduce or even eliminate the *Lb. delbrueckii* subsp. *bulgaricus* strains in the yogurt culture (Rybka and Kailasapathy 1995). Indeed, some *Lb. delbrueckii* subsp. *bulgaricus* strains are undesirable for two reasons: production of H_2O_2 and over-acidification of milk. However, one must be careful, *Lb. delbrueckii* subsp. *bulgaricus* contributes greatly to the fermentation, particularly by its proteolysis (Tamime and Robinson 1985, Juillard et al. 1987) and acetaldehyde production, and removing this culture might affect the acidification rate as well as the flavor of the product.

Some probiotic cultures are very sensitive to oxygen, and growth will not occur even if the medium is adequate for growth. It was previously mentioned that milk could be supplemented with anti-oxidants to achieve the goal of reducing the redox level. To influence this redox level, two other approaches are also possible: the selection on an oxygen-scavenging strain of *S. thermophilus* (Ishibashi and Shimamura 1993) and deaeration of milk (Tamime et al. 1995). Deaeration might not be necessary if care is simply taken not to re-introduce oxygen in the fermentation tank following pasteurization. Indeed, heating of the yogurt milk blend is generally above 85°C for many minutes which strongly reduces the oxygen level. To prevent oxygen re-introduction in the milk blend, pumping and agitation must be kept at a minimum, and gas flushing of the headspace of the fermentation vat can be considered. If medium composition and redox levels are adequate, then playing with the starter inoculation procedures can be successful. Inoculating with less viable cells of the starter, for example at 10^6 cfu ml^{-1} while maintaining the probiotic inoculation at 10^7 cfu ml^{-1}, will favor the development of the probiotic bacteria for two reasons: (1) longer fermentation time and (2) it takes longer for the pH to drop to acid values and the extended incubation time at pH values above 5.5 helps the growth of probiotic bacteria. In one application based on this strategy, ruptured cells are used (Shah et al. 1997). Presumably, the liberation of lactase from the ruptured starter cells add to the two factors just mentioned. Indeed, lactose assimilation is often a limiting factor in the growth of probiotic bacteria in milk, and this is an elegant method of combining inoculation and addition of lactase.

Finally, lowering the fermentation temperature to 37°C will promote the selective growth of the *Lb. acidophilus* and bifidobacteria (Mortazavian et al. 2006). Presumably this would also be the case for *Lb. casei*, *Lb. rhamnosus* and *Lb. plantarum* cultures of which many have the ability to grow at 15°C (Carr et al. 2002). Not surprisingly, Kneifel et al. (1993) report that some processes use a 32°C incubation temperature. Traditionally, yogurt fermentation is carried out between 40 and 44°C (Tamime and Robinson 1985). Therefore using temperatures under 40°C is an adaptation. The optimal growth temperature for *Lb. delbrueckii* subsp. *bulgaricus* is close to 43°C, while that of *S. thermophilus* is around 40°C (Beal et al. 1989). Therefore, lowering the incubation temperature to 37°C not only affects the level of probiotic in the product but also the strain ratios of the yogurt starters. Again, it must be kept in mind that such changes will affect the acidification rate (consequently the fermentation time), flavor and acidification during storage. This being said, lower-than-traditional incubation temperatures are increasingly used in yogurt production because exopolysaccharide product is sometimes enhanced under these conditions (Farnworth et al. 2007).

In summary, there are at least seven parameters in the selection and use of yogurt starter which will affect the development of probiotics in yogurt (Table 9.4) and which require standardization.

9.4.4 Storage

Many studies have examined the stability of probiotic bacteria in yogurt during storage and a few are presented in Table 9.6. Data show wide variations in stability. In cheese, there is potential for growth of probiotics during storage (Stanton et al. 1998) but this is not the case for yogurt. As a rule, probiotic bacteria die during storage. On an average, a 1 log reduction in viable counts will occur during a month of storage at 4°C, but values can reach 6 log. Stability is very much dependant on strain. In the past, stability during storage was an important criterion for the selection of bifidobacteria and *Lb. acidophilus* strains. It still is, but the limited number of strains having demonstrated health effects reduces the strain choices. Therefore, it might be necessary to adapt production and storage parameters to enhance the stability of the probiotic bacteria during storage of yogurts.

Acidity is probably the main detrimental factor on viability during storage. The problem of sensitivity to acidity of the probiotic cultures is compounded by (1) the addition of fruits and (2) continued acidification (over-acidification). One means of addressing the problem of fruit acidity is to package the yogurt and the fruits separately (Koch and Carnio 2001). To prevent a continued drop of pH during storage, it appears important

Table 9.6 Effect of some processing and storage conditions on the growth or viability of probiotic cultures in yogurt or specialty fermented milks.

Processing/storage condition	Effect	Reference
Sugar addition (up to 16%)	Decrease in *Bifidobacterium* and *Lb. acidophilus* during fermentation as sugar concentration increases	Shah and Ravula (2000)
Addition of fruit or flavors	Variable. Rarely stimulatory, more often inhibitory	Vinderola et al. (2002)
Addition of preservatives	Natamycin not a problem, but nisin inhibits many strains	Vinderola et al. (2002)
Addition of starter	Less growth of the probiotics	Roy et al. (1997)
Presence of oxygen in milk	Less growth of the probiotics, particularly bifidobacteria	Klaver et al. (1993)
Freezing	Bifidobacteria show 1 log drop in low acid product (pH 5.85) but at least 2 log drop in cfu when pH is 4.47. Strain dependant. *Lb. acidophilus* appears to have better resistance	Laroia and Martin (1991), Modler and Villa-Garcia (1993), Mitchell and Gilliland (1983), Brashears and Gilliland (1995)
Drying	6 to 150 fold reduction of viable populations after drying. *Bifidobacterium* more sensitive than *Lb. acidophilus*	Rybka and Kailasapathy (1995, 1997)
Refrigerated storage (typically 4°C)	Bifidobacteria losses in cfu counts of up to 6 logs over 40 days, typically around 1 log. Cultures of *Lb. acidophilus* seem more stable. Highly affected by strain and pH	Micanel et al. (1997), Rybka and Kailasapathy (1995), Shah et al. (1995), Roy et al. (1997), Dave and Shah (1996), Nighswonger et al. (1996), Medina and Jordano (1994)
Carbonation	No effect overall on probiotic populations during manufacture. CO_2 stimulated growth of *Lb. acidophilus* but production time was shorter	Vinderola et al. (2000)

to select lactobacilli that have weak over-acidification properties, or to reduce and even exclude *Lb. delbrueckii* subsp. *bulgaricus* from the starter (Kailasapathy and Rybka 1997).

Oxygen in the medium can lead to H_2O_2 production by either the probiotic culture itself or by lactobacilli from the starter (Villegas and Gilliland 1998). Certain probiotic cultures are very sensitive to oxygen (Dave and Shah 1997a,b, Meile et al. 1997), and this is presumably linked to variable levels of peroxidases (Talwalkar and Kailasapathy 2003b) which eliminate H_2O_2. The ability of *Lb. delbrueckii* subsp. *bulgaricus* to produce H_2O_2 partially explains why its removal from fermented milk starter cultures has had success in improving the survival of *Lb. acidophilus* in fermented milks (Rybka and Kailasapathy 1995). The problem of oxygen sensitivity is compounded by the occurrence of a synergistic inhibition of bifidobacteria by acid and H_2O_2 (Lankaputhra et al. 1996). This being said, if a probiotic culture does have the ability to synthesize peroxidases, it is better that the culture be added into the yogurt at the beginning of the fermentation (Hull et al. 1984), rather than at the end with, for example, the fruits. Presumably, the synthesis of the peroxidases occurs during the fermentation, which helps protect the probiotic cells during the subsequent storage.

Many studies have focused on the way to prevent the detrimental effects of oxygen on the probiotic cultures. In addition to adjustments in *Lb. delbrueckii* subsp. *bulgaricus* content in starters, the most common are the addition of anti-oxidants, such as ascorbic acid (Dave and Shah 1997a) and packaging with materials impermeable to oxygen (Talwalkar et al. 2004). Other ingredients, such as prebiotics (Martinez-Villaluenga et al. 2006) are also helpful in improving the stability of probiotics during storage, although they may not act on the redox level.

Of the various methods used to enhance the survival of cells to acidic environments, encapsulation seems promising (Doleyres and Lacroix 2005). Microentrapment (ME) in gel beads was shown to be very effective for increased stability during storage in yogurt (Sultana et al. 2000). It was first believed that ME protected against acidity, but there is evidence that the benefit of ME to stability in yogurt is rather due to protection from oxygen (Talwalkar and Kailasapathy 2003a).

9.4.5 Other yogurt-type fermented milks

There are many types of fermented milks other than yogurt (Farnworth 2004). Many, such as Kefir, use different starter cultures, which can be either mesophilic or thermophilic. Obviously, fermented milks based on mesophilic starters will require adaptations to incubation temperature, or selection of probiotic cultures more adapted to this parameter. Fermented milks based on mesophilic cultures are typically fermented between 25 and 30°C. An increase in incubation temperature is warranted if the fermented milk contains *Lb. acidophilus* or *Bifidobacterium* cultures. If it is imperative to keep the incubation temperatures below 30°C, presumably for the production of flavor compounds, then probiotic strains of *Lb. plantarum*, *Lb. casei* or *Lb. rhamnosus* should be considered for the product. The problems, challenges and strategies which were described above for yogurt, and summarized in Table 9.4, apply in most cases.

9.5 Fermented milks having high probiotic bacterial densities

Recent yogurt products contain one billion probiotic cells per portion (90 to 150 ml) at the end of storage, which means there will ultimately be approximately 10^7 cfu ml^{-1}. Since the total bacterial population in yogurt (starter + probiotic cells) is typically 20 to 100 times that number, probiotic bacteria therefore only represent a small fraction of the microbiota. There are novel products which contain as much as 50 times more probiotic bacteria per portion than traditional yogurt and which, in fact, may not contain starter cultures at all. In many cases, the product only contains probiotic bacteria. For the purpose of this chapter, fermented milks having more than 10 billion cells per portion will be termed "high probiotic density" (HPD) products. This represents approximately 10^8 cfu ml^{-1}, which is typically 10 times more than in yogurt.

There are excellent reviews of specialty HPD products containing bifidobacteria (Roy 2005), which even contain specific manufacturing processes of commercial products (Tamime et al. 1995). Therefore, this section will focus on general principles required to obtain HPD milks.

There are basically three options to prepare products having HPD: (1) DVI, (2) fermentation without pH control and (3) fermentation with pH control and subsequent blending with fruit juice or with other beverages.

To our knowledge, DVI is not practiced in the HPD line because of costs. Assuming there is a 1 log loss in viability during storage, ensuring a 10^8 cfu ml^{-1} population at the end of storage requires a 10^9 cfu ml^{-1} inoculation level. Many dried commercial probiotic products contain 1 to 5×10^{11} cfu g^{-1}, which means that the inoculation level may be as high as 1%. This cost can be seen as being prohibitive.

The most widely used technique is probably fermentation without pH control using only probiotic cultures. The population in the fermented milk will depend on the buffering capacity of the milk as well as on the addition of growth supplements. Examples of HPD products appear in Table 9.5 (data on the four pure cultures). In milk enriched at 4.5% protein as for yogurt, but otherwise without growth supplements, the population indeed gets above the 10^8 cfu ml^{-1} threshold, but there is little room for loss of viability during storage. In order to obtain populations above 10^9 cfu ml^{-1}, growth factors must be added. If one wishes this growth supplement to be of dairy origin, casein hydrolysates and lactose-based oligosaccharides seem the best choice. If non-dairy ingredients are permitted, then adding yeast, soy or fruit extracts are possible. Various peptones are also available. Obviously these supplements will affect flavor and this must be kept in mind.

The third option is to carry out fermentation with external pH control. When the growth medium is appropriate, a bifidobacteria population of 10^{10} cfu ml^{-1} can be achieved (Misra and Kuila 1991). In this case, the probiotic bacterial concentrate can be blended with a variety of juices and beverages, and the latter could constitute 90% or more of the product. This provides great latitude for innovation in HPD beverages. Obviously prebiotics can be added at this stage in order to prepare symbiotic blends.

9.6 Cheese

Very good reviews have recently been published on probiotics in cheese (Karimi et al. 2011, da Cruz et al. 2009, Grattepanche et al. 2008). In these manuscripts the evolution of probiotic bacteria in many cheese varieties is described. Readers will see that wide variations occur in the stability of probiotics in cheese. Therefore, this section will examine some reasons for these observations and attempt to propose strategies to enhance the viability of probiotics during cheese-making.

9.6.1 Milk preparation

There are much fewer variations in milk preparation for cheese than there are for yogurt. With respect to standardization, the level of fat is the main component which is adjusted. As mentioned with yogurt, triglycerides in their native form do not have an important effect on the growth of probiotics during processing. Potential exceptions in cheese are Feta, Parmesan and Blue-type cheeses. In the two former, lipases are added in order to develop the characteristic flavors. Free fatty acids have variable effects on probiotics. Since some fatty acids are essential growth factors, low levels will stimulate growth, but high concentrations become detrimental (Carvalho et al. 2004, Kankaanpaa et al. 2004, Talon et al. 2000). There are little data on the stability of probiotics in such cheeses. In Feta cheese, there was up to 3 log loss in the viable counts of *Lb. acidophilus* and *B. lactis* over a seven-week period (Kailasapathy and Masondole 2005). Microencapsulation was not helpful to prevent this viability loss. It is unknown to what extent lipolysis specifically affected the viability of the cultures and if adaptation or selection of the cultures is required.

In modern plants, the protein level could also be standardized using ultrafiltration (UF). Growth rates of *Lactococcus* cultures are either slower in UF retentate than in milk or are unaffected and wide variations in acid production and generation times between cultures were noted (Orme et al. 1994). It is unknown how probiotic bacteria fare in these UF-concentrated milk substrates.

In cheese-making, the heat treatment which is typically applied is limited to pasteurization, generally at 73°C for 15 s. The reason is that heating negatively affects the rennet coagulation of milk. Therefore, cheese-makers heat the least possible to obtain their sanitary goals. Consequently, the effect of heat treatment of the growth of probiotics during cheese processing is much less than in yogurt manufacture.

The situation with respect to other ingredients added to milk, such as flavors, fruits, preservatives and sugars, is quite different from yogurt. Cheese milk is almost never supplemented in this fashion.

9.6.2 Processing and competition with cheese starters

With respect to processing itself, the most critical point to address is the moment of inoculation. In Cheddar, it was found that addition of *B. longum* directly in milk before renneting enabled the highest viable counts of bifidobacteria in the cheese (Fortin et al. 2011a,b). This practice was found better than inoculation just before cheddaring or at the salting step.

As in yogurt, probiotics do not compete well with starters. During Cheddar cheese-making, the quantity of cells which are found in the cheese and whey at the end of fermentation is very similar to that which is inoculated in the milk (Table 9.7). The reasons for limited growth in Cheddar cheese-making are: (1) fermentation time generally limited to 3.5 to 5 hr, (2) most of the processing occurs at a sub-optimal temperature range of 32–35°C, (3) acidification occurs during processing, and (4) cheese-making steps often require agitation, which incorporates oxygen in the medium.

Many cheese-making processes have a "cooking" step in order to enhance whey expulsion from the grains. In Cheddar, it will vary between 37 and 40°C, which is actually good for probiotics. In theory, processes which

Table 9.7 Effect of oxygen level and strain on viable counts of probiotics in curds and whey in laboratory-scale Cheddar cheese manufacturing.

Strain	Condition	Total population variation (log cfu ml^{-1})	Distribution (% in curds)
Bifidobacterium lactis BB-12	Control[a]	– 0.09	90
	Low DO[b]	– 0.08	92
Bifidobacterium longum R0175	Control	– 0.03	81
	Low DO	– 0.02	84
Lactobacillus helveticus R0052	Control	+ 0.01	90
	Low DO	+ 0.14	89
Bifidobacterium longum 15708	Control	+ 0.02	78
	Low DO	+ 0.23	79
Bifidobacterium infantis 15697	Control	– 0.01	84
	Low DO	+ 0.04	82

Source: Fortin et al. (2011a)

[a] Control had 7.4 mg l^{-1} of oxygen in milk

[b] Low DO: Low dissolved oxygen, which represented 2.0 mg l^{-1}

include such a step (Cheddar) should allow better growth of the probiotics than those which do not (Camembert, Brie, fromage frais). However some Swiss-type cheeses, such as Emmenthaler, include a cooking step that can reach 50°C in extreme cases. Probiotic cultures may suffer from exposure at 46°C and above (Champagne et al. 2007). Therefore, in such cheeses, the cheese-maker must avoid or limit the use of these high temperatures, and select a strain that can tolerate them. Cottage cheese also includes a cooking step, but temperatures reached are quite high and can cause loss of viability. In this case, it is better to add the probiotic bacteria in the dressing cream (Blanchette et al. 1996).

Even though the cheese curd mass obtained represents between 8 and 15% of the mass of milk used originally, between 78 and 92% of the cells are found in it (Table 9.7). Therefore, milk coagulation enables a strong retention of the cells, but the recovery rate varies as a function of strain. Retention of bacteria in the rennet curd might even be a selection parameter of the culture.

In selecting a probiotic strain, bacteriocin production might also be a concern. Some *Lactococcus lactis* cultures produce bacteriocins, the most notable is nisin. Data show that some probiotic strains are sensitive to nisin (Vinderola et al. 2002). Therefore in selecting a probiotic culture for cheese, one might want to examine if a bacteriocin is produced by the starter and if the probiotic culture is sensitive to it.

9.6.3 Storage

Very rarely do we see increases in viable counts of probiotics in yogurt during storage. However, this is not the case for cheese. Some lactobacilli grow readily during storage (Stanton et al. 1998). One reason is the pH of cheese which, at pH 5.0–5.4 for most fresh cheese, is typically one unit higher than that of yogurt. Another reason is that cheese ripening is carried out between 7 and 12°C at which temperatures some probiotics can slowly grow. Cultures of *Lb. rhamnosus* and *Lb. casei* should particularly be considered in this perspective.

Another difference between cheese and yogurt is salting. Salting can negatively influence the stability of probiotics during the pressing and storage periods (Fortin et al. 2011a,b). Cheese are typically salted at 1.5 to 2.5% salt on a weight/weight basis, but this results in a salt-in-whey level of 3 to 8% as a function of the moisture level of the cheese. Cheese-makers can rarely adjust their process in this sense, because of the impact of salt on safety and flavor. Therefore selection of a proper probiotic culture with respect to tolerance to salt might be necessary.

In cheeses, gas production during storage is sometimes desirable (Gouda, Swiss-type) but in some instances it is not (Cheddar). With respect to

Cheddar, adjunct cultures of lactobacilli having a strictly homofermentative metabolism is desirable (Stanton et al. 1998). Inoculation with probiotics can potentially serve this purpose. Probiotic strains of the species *Lb. helveticus* or *Lb. casei* would seem recommendable.

As can be seen, there are many differences between cheese and yogurt, and probiotic cultures which are well suited for yogurt might not be for cheese. A summary of desirable strain traits for a probiotic destined for cheese-making is found in Table 9.8.

Table 9.8 Parameters which might be considered in the selection of a probiotic culture for incorporation into cheese.

Stressful condition or problem	Selection parameter
Lack of growth due to manufacturing temperatures at sub-optimum temperatures	Ability to grow at 32°C
Viability loss at cooking at 45–50°C (some Swiss-type cheeses)	Tolerance to these temperatures
Exposure to oxygen during processing due to agitation, or during storage of fresh cheeses which are not vacuum-packed	Low sensitivity to oxygen
Bacteriocin production by starter	Low sensitivity to nisin or other bacteriocins
Salting of curd or cheese	Tolerance to salt-in-water levels between 4 and 7%
Loss of cells in whey	Select strain which has higher recovery level in rennet curd
Extended ripening period in an acid environment	Ability to slowly grow at 7–12°C and at pH 5.0–5.4
Undesirable splits or gas openings	Use a strictly homofermentative culture

9.7 Frozen yogurt

When frozen yogurt is made from yogurt, rather than unfermented ice cream, the addition of probiotics is basically made as in the traditional process. Therefore, strategies presented in Section 9.4 above apply.

However, some manufacturers do not carry out cream fermentation and simply add yogurt cultures and probiotics to the product. In this situation, inoculation can be carried out at two levels: in the liquid cream blend prior to freezing or in the "soft" ice cream product which exits the freezing apparatus. Processes will obviously vary, but the temperature of the ice cream mix exiting the freezer is around –8°C and is in a semi-solid state which enables the addition and blending of various ingredients (cookies, fruits, and a powder of probiotic bacteria). Once the ingredients have been added, the ice cream is stored at –20°C or below and harden.

As was mentioned previously, addition of a powder of freeze-dried bacteria to a cold medium can result in viability losses. Therefore directly to the cold cream blend prior to freezing is not recommended. Rehydration in a small portion of warmed cream is advisable.

Although it would be assumed that probiotics would be stable in the frozen product during storage, this is not always the case. Losses of up to 4 log cfu g^{-1} can occur with bifidobacteria (Pinto et al. 2012). Microencapsulation of the cultures appears to be the best way of improving stability (Sheu et al. 1993, Pinto et al. 2012).

Unfortunately, most data on ice cream and frozen yogurt pertain to the stability during storage. Very little is known of the effect of inoculation practices or cell adaptation techniques on subsequent survival to freezing and storage.

9.8 Conclusion

Adding probiotics to fermented milks holds many challenges, but it also provides the opportunity for novel products. Thus, ingredients which are added to the fermented milk formulations can be selected not only for their beneficial effect on growth and stability of the probiotics in the fermented milks but also to provide additional health benefits. As an example, antioxidants could be used to protect the probiotics from the detrimental effect of oxygen during storage, but also for potential risk reduction of cancer.

The HPD products also provide new opportunities for flavors. The current populations of probiotic bacteria in yogurt are geared to try to prevent the production of off-flavors by probiotic cultures. Since bifidobacteria partially convert carbohydrates into acetic acid, these cultures can affect flavor (de Vuyst 2000). Furthermore, probiotics show different metabolic patterns than the starters with respect to secondary compounds. As an example, while *Lb. delbrueckii* subsp. *bulgaricus* produces acetaldehyde in yogurt, an important aroma compound, some *Lb. acidophilus* converts acetaldehyde to ethanol (de Vuyst 2000, Ostlie et al. 2005). There are variable abilities to assimilate citrate and produce aromatic compounds such as diacetyl (Ostlie et al. 2005). Therefore, in addition to their health effects, milks fermented with probiotic bacteria alone could constitute new products lines with respect to sensory properties.

9.9 Acknowledgment

The author thanks Edward Farnworth for constructive suggestions to improve some sections of this manuscript.

References

Abu-Taraboush, H.M. 1994. Growth behavior of *Lactobacillus acidophilus* and biochemical characteristics and acceptability of acidophilus milk made from camel milk. Milchwiss. 49: 379–382.

Abu-Taraboush, H.M., M.M. Al-Dagal and M.A. Al-Royli. 1998. Growth, viability and proteolytic activity of bifidobacteria in whole camel milk. J. Dairy Sci. 81: 354–361.

Al-Dabbagh, W.Y. and M.C. Allan. 1989. Influence of concentration of non-fat dried milk solids (NFDM) on growth and acid production of *Lactobacillus bulgaricus* and *Streptococcus thermophilus*. Asian J. Dairy Res. 8: 115–121.

Al-Saleh, A.A. 2000. Growth of *Bifidobacterium longum* NCFB 2716 in ultrafiltered skimmed milk retentates from cow, camel and sheep. Aus. J. Dairy Technol. 55: 145–147.

Alvarez, F., M. Arguello, M. Cabero, F.A. Riera, R. Alvarez, J.R. Iglesias and J. Granda. 1998. Fermentation of concentrated skim-milk: Effects of different protein. J. Sci. Food Agric. 76: 10–16.

Araya, M., L. Morelli, G. Reid, M.-E. Sanders, C. Stanton, M. Pineiro and P.B. Embarek. 2002. Guidelines for the evaluation of probiotics in food. Joint FAO/WHO Working Group Report on Drafting Guidelines for the Evaluation of Probiotics in Food, London, ON, Canada.

Aryana, K.J., S. Plauche, R.M. Rao, P. McGrew and N.P. Shah. 2007. Fat-free plain yogurt manufactured with inulins of various chain lengths and *Lactobacillus acidophilus*. J. Food Sci. 72: M79–M84.

BCC Research. 2011. The probiotics market: Ingredients, supplements, foods. Report code: FOD035C. http://www.bccresearch.com/report/probiotics-market-ingredients-foods-fod035c.html.

Babu, V., B.K. Mital and S.K. Garg. 1992. Effect of tomato juice addition on the growth and activity of *Lactobacillus acidophilus*. Int. J. Food Microbiol. 17: 67–70.

Beal, C., P. Louvet and G. Corrieu. 1989. Influence of controlled pH and temperature on the growth and acidification of pure cultures of *Streptococcus thermophilus* 404 and *Lactobacillus bulgaricus* 398. Appl. Microbiol. Biotechnol. 32: 148–154.

Blanchette, L., D. Roy, G. Bélanger and S.F. Gauthier. 1996. Production of cottage cheese using dressing fermented by bifidobacteria. J. Dairy Sci. 79: 8–15.

Brashears, M.M. and S.E. Gilliland. 1995. Survival during frozen and subsequent refrigerated storage of *Lactobacillus acidophilus* cells as influenced by the growth phase. J. Dairy Sci. 78: 2326–2335.

Bruno, F.A., W.E.V. Lankaputhra and N.P. Shah. 2002. Growth, viability and activity of *Bifidobacterium* spp. in skim milk containing prebiotics. J. Food Sci. 67: 2740–2744.

Carr, F.J., D. Chill and N. Maida. 2002. The lactic acid bacteria: A literature survey. Crit. Rev. Microbiol. 28: 281–370.

Carvalho, A.S., J. Silva, P. Ho, P. Teixeira, F.X. Malcata and P. Gibbs. 2004. Relevant factors for the preparation of freeze-dried lactic acid bacteria. Int. Dairy J. 14: 835–847.

Champagne, C.P., F. Mondou, Y. Raymond and D. Roy. 1996. Effect of polymers and storage temperature on the stability of freeze-dried lactic acid bacteria. Food Res. Int. 29: 555–562.

Champagne, C.P., N. Gardner and D. Roy. 2005. Challenges in the addition of probiotic cultures to foods. Crit. Rev. Food Sci. Nutr. 45: 61–84.

Champagne, C.P., N.J. Gardner and C. Lacroix. 2007. Fermentation technologies for the production of exopolysaccharide-synthesizing *Lactobacillus rhamnosus* concentrated cultures. Electr. J. Biotechnol. 10: 211–220.

Champagne, C.P. and H. Møllgaard. 2008. Production of probiotic cultures and their addition in fermented foods. pp. 513–532. *In:* E.R. Farnworth [ed.]. Handbook of Fermented Functional Foods, 2nd edition. CRC Press/Taylor & Francis, Boca Raton, FL, USA.

Champagne, C.P., J. Green-Johnson, Y. Raymond, J. Barrette and N. Buckley. 2009. Selection of probiotic bacteria for the fermentation of a soy beverage in combination with *Streptococcus thermophilus*. Food Res. Int. 42: 612–621.
Clydesdale, F. 2005. Functional foods: Opportunities and challenges. Food Technol. 58: 35–40.
Christopher, M.D., V. Padmanabha-Reddy and K. Venkateswarlu. 2006. Effect of whey protein concentrate supplementation on the viability of *Bifidobacterium bifidum* in probiotic yogurt. J. Food Sci. Technol. 43: 552–554.
da Cruz, A.G., F.C.A. Buriti, C.H.B. de Souza, J.A.F. Faria and M.I.I. Saad. 2009. Probiotic cheese: Health benefits, technological and stability aspects. Trends Food Sci. Technol. 20: 344–354.
Dave, R.I. and N.P. Shah. 1996. Evaluation of media for selective enumeration of *Streptococcus thermophilus, Lactobacillus bulgaricus, Lactobacillus acidophilus* and bifidobacteria. J. Dairy Sci. 79: 1529–1536.
Dave, R.I. and N.P. Shah. 1997a. Effectiveness of ascorbic acid as an oxygen scavenger in improving viability of probiotic bacteria in yogurts made with commercial starter cultures. Int. Dairy J. 7: 435–443.
Dave, R.I. and N.P. Shah. 1997b. Effect of cysteine on the viability of yogurt and probiotic bacteria in yogurts made with commercial starter cultures. Int. Dairy J. 7: 537–545.
Desjardins, M.L. and D. Roy. 1991. β-galactosidase and proteolytic activities of bifidobacteria in milk: A preliminary study. Milchwiss. 46: 11–13.
de Valdez, G.F., G.S. de Giori, A.P. de Ruiz-Holgado and G. Oliver. 1985a. Effect of the rehydration medium on the recovery of freeze-dried lactic acid bacteria. Appl. Environ. Microbiol. 50: 1339–1341.
de Valdez, G.F., G.S. de Giori, de Ruiz-Holgado and G. Oliver. 1985b. Rehydration conditions and viability of freeze-dried lactic acid bacteria. Cryobiol. 22: 574–577.
de Vuyst, L. 2000. Technology aspects related to the application of functional starter cultures. Food Technol. Biotechnol. 38: 105–112.
Doleyres, Y. and C. Lacroix. 2005. Technologies with free and immobilised cells for probiotic bifidobacteria production and protection. Int. Dairy J. 15: 973–988.
Elli, M., R. Zink, R. Reniero and L. Morelli. 1999. Growth requirements of *Lactobacillus johnsonii* in skim and UHT milk. Int. Dairy J. 9: 507–513.
Farnworth, E. 2004. The beneficial health effects of fermented foods—Potential probiotics around the world. J. Nutr. Funct. Medical Foods. 4: 93–117.
Farnworth, E.R., C.P. Champagne and M.R. van Calsteren. 2007. Exopolysaccharides from lactic acid bacteria: Food uses, production, chemical structures, and health effects. pp. 353–371. *In:* R.E.C. Wildman [ed.]. Handbook of Nutraceuticals and Functional Foods, 2nd edition. CRC Press, Boca Raton, FL, USA.
Farnworth, E.R. and C.P. Champagne. 2009. Delivery of viable probiotics by foods. pp. 3–17. *In:* R.R. Watson and V. Preedy [eds.]. Bioactive Foods in Promoting Health: Probiotics-Prebiotics. Elsevier/Academic Press, Oxford, UK.
Feldstein, F.J. and D.C. Westhoff. 1979. The influence of heat treatment of milk on starter activity. What about UHT? Cult. Dairy Prod. J. 14: 11–15.
Fortin, M.H., C.P. Champagne, D. St-Gelais, M. Britten, P. Fustier and M. Lacroix. 2011a. Effect of time of inoculation, starter addition, oxygen level and salting on the viability of probiotic cultures during Cheddar cheese production. Int. Dairy J. 21: 75–82.
Fortin, M.H., C.P. Champagne, D. St Gelais, M. Britten, P. Fustier and M. Lacroix. 2011b. Viability of *Bifidobacterium longum* in Cheddar cheese curd during manufacture and storage: Effect of microencapsulation and point of inoculation. Dairy Sci. Technol. 91: 599–614.
Gaudreau, H., C.P. Champagne and P. Jelen. 2005. The use of crude cellular extracts of *Lactobacillus delbrueckii* ssp. *bulgaricus* 11842 to stimulate growth of a probiotic *Lactobacillus rhamnosus* culture in milk. Enz. Microbial. Technol. 36: 83–90.
Goh, J.S., Y.S. Chae, C.G. Gang, I.K. Kwon, M. Choi, S.K. Lee and H. Park. 1993. Studies on the development of ginseng-yogurt and its health effect. I. Effect of ginseng extracts on

the acid production by lactic acid bacteria and the distribution of intestinal microflora of mouse. Korean J. Dairy Sci. 15: 216–225.

Gomes, A.M.P. and F.X. Malcata. 1998. Development of probiotic cheese manufactured from goat milk: Response surface analysis via technological manipulation. J. Dairy Sci. 81: 1492–1507.

Grattepanche, F., S. Miescher-Schwenninger, L. Meile and C. Lacroix. 2008. Recent developments in cheese cultures with protective and probiotic functionalities. Dairy Sci. Technol. 88: 421–444.

Heap, H.A. and G.H. Richardson. 1985. The proteolytic effect of fast-coagulating and slow-coagulating strains of *Streptococcus cremoris*. New Zealand J. Dairy Sci. Technol. 20: 155–161.

Hull, R.R., A.V. Roberts and J.J. Mayes. 1984. Survival of *Lactobacillus acidophilus* in yogurt. Aus. J. Dairy Technol. 39: 164–166.

Ishibashi, N., T. Tatematsu, S. Shimamura, M. Tomota and S. Okonogi. 1985. Effect of water activity on the viability of freeze-dried bifidobacteria and lactic acid bacteria. pp. 227–232. *In:* Fundamentals and Application of Freeze Drying to Biological Materials, Dryes and Foodstuffs. Conference proceedings of the Int. Inst. Refrigeration (Paris, France), Tokyo. http://www.iifiir.org/clientbookline/service/.

Ishibashi, N. and S. Shimamura. 1993. Bifidobacteria: Research and development in Japan. Food Technol. 47: 126, 129–134.

Juhkam, K., P. Elias, M. Roasto and T. Tamme. 2007. Viability of *Lactobacillus acidophilus* in yogurt containing inulin or oligofructose during refrigerated storage. Milchwiss. 62: 52–54.

Juillard, V., H.E. Spinnler, M. Desmazeaud and J. Boquien. 1987. Cooperation and inhibition between lactic acid bacteria used in dairy industry. Lait 67: 149–172.

Juillard, V., D. Le Bars, E.R.S. Kunji, W.N. Konings, J.C. Gripon and J. Richard. 1995. Oligopeptides are the main source of nitrogen for *Lactococcus lactis* during growth in milk. Appl. Environ. Microbiol. 61: 3024–3030.

Kailasapathy, K. and S. Rybka. 1997. *Lactobacillus acidophilus* and *Bifidobacterium* ssp.—Their therapeutic potential and survival in yogurt. Aus. J. Dairy Technol. 52: 28–35.

Kailasapathy, K. and L. Masondole. 2005. Survival of free and microencapsulated *Lactobacillus acidophilus* and *Bifidobacterium lactis* and their effect on texture of feta cheese. Aus. J. Dairy Technol. 60: 252–258.

Kailasapathy, K., I. Harmstorf and M. Phillips. 2008. Survival of *Lactobacillus acidophilus* and *Bifidobacterium animalis* ssp. *lactis* in stirred fruit yogurts. LWT- Food Sci. Technol. 41: 1317–1322.

Kankaanpaa, P., B. Yang, H. Kallio, E. Isolauri and S. Salminen. 2004. Effects of polyunsaturated fatty acids in growth medium on lipid composition and on physicochemical surface properties of lactobacilli. Appl. Environ. Microbiol. 70: 129–136.

Karimi, R., M.A. Mortazavian and A.G. da Cruz. 2011. Viability of probiotic microorganisms in cheese during production and storage: A review. Dairy Sci. Technol. 91: 283–308.

Kheadr, E., N. Bernoussi, C. Lacroix and I. Fliss. 2004. Comparison of the sensitivity of commercial strains and infant isolates of bifidobacteria to antibiotics and bacteriocins. Int. Dairy J. 14: 1041–1053.

Kim, Y.J., S.C. Baick and J.H. Yu. 1995. Effect of oligosaccharides and *Bifidobacterium* growth promoting materials in yogurt base made by *Bifidobacterium infantis* 420. Korean J. Dairy Sci. 17: 167–173.

Klaver, F.A.M., F. Kingma and A.H. Weerkamp. 1993. Growth and survival of bifidobacteria in milk. Neth. Milk Dairy J. 47: 151–164.

Kneifel, W., D. Jaros and F. Erhard. 1993. Microflora and acidification properties of yogurt and yogurt-related products fermented with commercially available starter cultures. Int. J. Food Microbiol. 18: 179–189.

Koch, N. and M.C. Carnio. 2001. Probiotics: Quo vadis? The industrial point of view. pp. 64–77. *In:* D. Roy [ed.]. Probiotics and Health—The intestinal microflora. Fondation des Gouverneurs-Edisem, St-Hyacinthe, QC, Canada.

Kurultay, S., Ö. Öksüz and B. Kaptan. 2006. Effects of different heat treatments of milk on some growth characteristics of mixed and single cell cultures of yogurt bacteria. Milchwiss. 61: 52–55.
Kyung, H.-K. and T.-K. Young. 1993. The preparation of yogurt from milk and cereals. Korean J. Food Sci. Technol. 25: 130–135.
Lanciotti, R., L. Vannini, P. Pittia and M.E. Guerzoni. 2004. Suitability of high-dynamic-pressure-treated milk for the production of yogurt. Food Microbiol. 21: 753–760.
Lankaputhra, W.E.V., N.P. Shah and M.L. Britz. 1996. Survival of bifidobacteria during refrigerated storage in the presence of acid and hydrogen peroxide. Milchwiss. 51: 65–69.
Laroia, S. and J.H. Martin. 1991. Effect of pH on survival of *Bifidobacterium bifidum* and *Lactobacillus acidophilus* in frozen fermented dairy desserts. Cult. Dairy Prod. J. 26: 13–24.
Larsen, R.F. and M.C. Añón. 1989. Effect of water activity a_w of milk on acid production by *Streptococcus thermophilus* and *Lactobacillus bulgaricus*. J. Food Sci. 54: 917–921.
Lawrence, R.C., T.D. Thomas and B.E. Terzaghi. 1976. Reviews of the progress of dairy science: Cheese starters. J. Dairy Res. 43: 141–193.
Leatherhead Food International. 2006. The International Markets for Functional Foods: Moving into the mainstream. Leatherhead Food Internationals, Leatherhead, Surrey, UK.
marketsandmarkets.com. 2013. Probiotics market by products (functional foods, dietary supplements, specialty nutrients, animal feed), applications (regular, therapeutic, preventive health care) and ingredients (lactobacilli, bifidobacteria, yeast): Global Trends & Forecasts to 2017. Report code: FB 1046. Available at: http://www.marketsandmarkets.com/Market-Reports/probiotic-market-advanced-technologies-and-global-market-69.html.
Martinez-Villaluenga, C., J. Frias, R. Gomez and C. Vidal-Valverde. 2006. Influence of addition of raffinose family oligosaccharides on probiotic survival in fermented milk during refrigerated storage. Int. Dairy J. 16: 768–774.
Martinez-Villaluenga, M. and R. Gomez. 2007. Characterization of bifidobacteria as starters in fermented milk containing raffinose family of oligosaccharides from lupin as prebiotic. Int. Dairy J. 17: 116–122.
Medina, L.M. and R. Jordano. 1994. Survival of constitutive microflora in commercially fermented milk containing bifidobacteria during refrigerated storage. J. Food Protect. 57: 731–733.
Meile, L., W. Ludwig, U. Rueger, C. Gut, P. Kaufman, G. Dasen, S. Wenger and M. Teuber. 1997. *Bifidobacterium lactis* sp. nov., a moderately oxygen tolerant species isolated from fermented milk. Syst. Appl. Microbiol. 20: 57–64.
Mellentin, J. 2012. Ten key trends in food, nutrition and health—2012. Report, New Nutrition Business, www.new–nutrition.com.
Micanel, N., I.N. Haynes and M.J. Playne. 1997. Viability of probiotic cultures in commercial Australian yogurts. Aus. J. Dairy Technol. 52: 24–27.
Mille, Y., J.P. Obert, L. Beney and P. Gervais. 2004. New drying process for lactic bacteria based on their dehydration behaviour in liquid medium. Biotechnol. Bioeng. 88: 71–76.
Misra, A.K. and R.K. Kuila. 1991. Intensified growth of *Bifidobacterium* and preparation of *Bifidobacterium bifidum* for a dietary adjunct. Cult. Dairy Prod. J. 26: 4–6.
Mitchell, S.L. and S.E. Gilliland. 1983. Pepsinized sweet whey medium for growing *Lactobacillus acidophilus* for frozen concentrated cultures. J. Dairy Sci. 66: 712–718.
Modler, H.W. and L. Villa-Garcia. 1993. The growth of *Bifidobacterium longum* in a whey-based medium and viability of this organism in frozen yogurt with low and high levels of developed acidity. Cult. Dairy Prod. J. 28: 4–8.
Mortazavian, A.M., M.R. Ehsani, S.M. Mousavi, S. Sohrabvandi and J.A. Reinheimer. 2006. Combined effects of temperature-related variables on the viability of probiotic micro-organisms in yogurt. Aus. J. Dairy Technol. 61: 248–252.

Murad, H.A., F.A. Fathy and S. Abdel-Ghani. 1997. Growth of *Bifidobacterium bifidum* in buffalo milk supplemented with peanut milk and some amino acids. Egyptian J. Dairy Sci. 25: 75–84.

Nighswonger, B.D., M.M. Brashears and S.E. Gilliland. 1996. Viability of *Lactobacillus acidophilus* and *Lactobacillus casei* in fermented milk products during refrigerated storage. J. Dairy Sci. 79: 212–219.

Orme, B.J., D.J. McMahon and R.K. Thunell. 1994. Variable growth and acid production of lactococci in whole milk concentrated by ultrafiltration and diafiltration. J. Dairy Sci. 77: 3454–3459.

Ostlie, H.M., J. Treimo, J.A. Narvhus and L. de Vuyst. 2005. Effect of temperature on growth and metabolism of probiotic bacteria in milk. Int. Dairy J. 15: 989–997.

Partanen, L., N. Marttinen and T. Alatossawa. 2001. Fats and fatty acids as growth factors for *Lactobacillus delbrueckii*. Syst. Appl. Microbiol. 24: 500–506.

Pinto, S.S., C.B. Fritzen-Freire, I.B. Muñoz, P.L.M. Barreto, E.S. Prudêncio and R.D.M.C. Amboni. 2012. Effects of the addition of microencapsulated *Bifidobacterium* BB-12 on the properties of frozen yogurt. J. Food Eng. 111: 563–569.

Rajagopal, N. 2012. The North American probiotics market. http://www.naturalproductsinsider.com/articles/2012/10/the-north-american-probiotics-market.aspx. Accessed on October 3, 2012.

Ray-Chowdhury, B., R. Chakraborty and U. Raychaudhuri. 2008. Study on beta-galactosidase enzymatic activity of herbal yogurt. Int. J. Food Sci. Nutr. 59: 116–122.

Reid, G. 2008. Probiotics and prebiotics—Progress and challenges. Int. Dairy J. 18: 969–975.

Roy, D., I. Mainville and F. Mondou. 1997. Bifidobacteria and their role in yogurt-related products. Microecology Therapy 26: 167–180.

Roy, D. 2005. Technological aspects related to the use of bifidobacteria in dairy products. Lait 85: 39–56.

Rybka, S. and K. Kailasapathy. 1995. The survival of culture bacteria in fresh and freeze-dried AB yogurts. Aus. J. Dairy Technol. 50: 51–57.

Rybka, S. and K. Kailasapathy. 1997. Effect of freeze drying and storage on the microbiological and physical properties of AB-yogurt. Milchwiss. 52: 390–394.

Samona, A. and R.K. Robinson. 1994. Effect of yogurt cultures on the survival of bifidobacteria in fermented milks. J. Soc. Dairy Technol. 47: 58–60.

Sanders, M.E. and T.R. Klaenhammer. 2001. The scientific basis of *Lactobacillus acidophilus* NCFM functionality as a probiotic. J. Dairy Sci. 84: 319–331.

Sarkar, S. 2008. Effect of probiotics on biotechnological properties of yogurt: A review. British Food J. 110: 717–740.

Saxelin, M., B. Grenov, U. Svensson, R. Fondén, R. Reniero and T. Mattila-Sandholm. 1999. The technology of probiotics. Trends Food Sci. Technol. 10: 387–392.

Saxelin, M., R. Korpela and A. Mayra-Makinen. 2003. Introduction: Classifying functional dairy products. pp. 1–15. *In:* T. Mattila-Sandholm and M. Saarela [eds.]. Functional Dairy Products. Vol. 1. CRC Press, Boca Raton/Woodhead Publishing Ltd., FL, USA.

Sendra, E., P. Fayos, Y. Lario, J.A. Fernandez-Lopez, E. Sayas-Barbera and J. Perez-Alvarez. 2008. Incorporation of citrus fibers in fermented milk containing probiotic bacteria. Food Microbiol. 25: 13–21.

Shah, N.P., W.E.V. Lankaputhra, M.L. Britz and W.S.A. Kyle. 1995. Survival of *Lactobacillus acidophilus* and *Bifidobacterium bifidum* in commercial yogurt during refrigerated storage. Int. Dairy J. 5: 515–521.

Shah, N.P., E. Warnakulsuriya and W.E.V. Lankaputhra. 1997. Improving viability of *Lactobacillus acidophilus* and *Bifidobacterium* spp. in yogurt. Int. Dairy J. 7: 349–356.

Shah, N.P. and R.R. Ravula. 2000. Influence of water activity on fermentation, organic acids production and viability of yogurt and probiotic bacteria. Aus. J. Dairy Technol. 55: 127–131.

Shah, N.P. 2007. Functional cultures and health benefits. Int. Dairy J. 17: 1262–1277.

Sheu, T.Y., R.T. Marshall and H. Heymann. 1993. Improving survival of culture bacteria in frozen desserts by microentrapment. J. Dairy Sci. 76: 1902–1907.

Shihata, A. and N.P. Shah. 2000. Proteolytic profiles of yogurt and probiotic bacteria. Int. Dairy J. 10: 401–408.

Shihata, A. and N.P. Shah. 2002. Influence of addition of proteolytic strains of *Lactobacillus delbrueckii* subsp. *bulgaricus* to commercial ABT starter cultures on texture of yogurt, exopolysaccharide production and survival of bacteria. Int. Dairy J. 12: 765–772.

Shimamura, S., F. Abe, N. Ishibashi, H. Miyakawa, T. Yaeshima, T. Araya and M. Tomita. 1992. Relationship between oxygen sensitivity and oxygen metabolism of *Bifidobacterium* species. J. Dairy Sci. 75: 3296–3306.

Shin, H.S., J.H. Lee, J.J. Pestka and Z. Ustonol. 2000. Growth and viability of commercial *Bifidobacterium* spp. in skim milk containing oligosaccharides and inulin. J. Food Sci. 65: 884–887.

Sinha, R.N., A.K. Shukla, M. Lal and B. Ranganathan. 1982. Rehydration of freeze-dried cultures of lactic streptococci. J. Food Sci. 47: 668–669.

Sinha, R.P. 1990. Effect of growth media and extended incubation on the appearance of lactose-nonfermenting variants in lactococci. J. Food Protect. 53: 583–587, 629.

Srivinas, H., H.R. Prabha and P.A. Shankar. 1997. Characteristics of cultured milks, yogurt and probiotic yogurts prepared from pre-refrigerated milks. J. Food Sci. Technol. 34: 162–164.

Stanton, C., G. Gardiner, P.B. Lynch, J.K. Collins, G. Fitzgerald and R.P. Ross. 1998. Probiotic cheese. Int. Dairy J. 8: 491–496.

Sultana, K., G. Godward, N. Reynolds, R. Arumugaswamy, P. Peiris and K. Kailasapathy. 2000. Encapsulation of probiotic bacteria with alginate-starch and evaluation of survival in simulated gastrointestinal conditions and in yogurt. Int. J. Food Microbiol. 62: 47–55.

Talon, R., D. Walter and M.C. Montel. 2000. Growth and effect of staphylococci and lactic acid bacteria on unsaturated free fatty acids. Meat Sci. 54: 41–47.

Talwalkar, A. and K. Kailasapathy. 2003a. Effect of microencapsulation on oxygen toxicity in probiotic bacteria. Aus. J. Dairy Technol. 58: 36–39.

Talwalkar, A. and K. Kailasapathy. 2003b. Metabolic and biochemical responses of probiotic bacteria to oxygen. J. Dairy Sci. 86: 2537–2546.

Talwalkar, A., C.W. Miller, K. Kailasapathy and M.H. Nguyen. 2004. Effect of packaging materials and dissolved oxygen on the survival of probiotic bacteria in yogurt. Int. J. Food Sci. Technol. 39: 606–611.

Tamime, A.Y. and R.K. Robinson. 1985. Yogurt Science and Technology. Pergamon Press, Oxford, UK.

Tamime, A.Y., V.M.E. Marshall and R.K. Robinson. 1995. Microbiological and technological aspects of milks fermented by bifidobacteria. J. Dairy Res. 62: 151–187.

Torriani, S., F. Gardini, M.E. Guerzoni and F. Dellagio. 1996. Use of response surface methodology to evaluate some variables affecting the growth and acidification characteristics of yogurt cultures. Int. Dairy J. 6: 625–636.

Transparency Market Research. 2013. Probiotics market (dietary supplements, animal feed, foods & beverages)—global industry analysis, market size, share, trends, analysis, growth and forecast, 2012–2018. Available at http://www.transparencymarketresearch.com.

Ustunol, Z. and H. Gandhi. 2001. Growth and viability of commercial *Bifidobacterium* ssp. in honey-sweetened skim milk. J. Food Protect. 64: 1775–1779.

Vasiljevic, T., T. Kealy and V.K. Mishra. 2007. Effects of beta-glucan addition to a probiotic containing yogurt. J. Food Sci. 72: C405–C411.

Villegas, E. and S.E. Gilliland. 1998. Hydrogen peroxide production by *Lactobacillus delbrueckii* subsp. *lactis* I at 5°C. J. Food Sci. 63: 1070–1074.

Vinderola, C.G., M. Gueimonde, T. Delgado, J.A. Reinheimer and C.G. de los Reyes-Gavilan. 2000. Characteristics of carbonated fermented milk and survival of probiotic bacteria. Int. Dairy J. 10: 213–220.

Vinderola, C.G., G.A. Costa, S. Regenhardt and J.A. Reinheimer. 2002. Influence of compounds associated with fermented dairy products on the growth of lactic acid starter and probiotic bacteria. Int. Dairy J. 12: 579–589.

Yajima, M., S. Hashimoto, T. Saita and K. Matsuzaki. 1992. Method of preparing milk-fermented food. KK Yakult Honsha. European Patent Application, EP 0 486 738 A1, EP 90-312757.

CHAPTER 10

Microbiology of Cream, Butter, Ice Cream and Related Products

Hamid Ghoddusi and *Barbaros Özer**

10.1 Cream

Cream is defined as the part of milk that is rich in fat which has been separated by skimming or otherwise (Anonymous 1995, 1996). Depending on the heat treatment applied, cream can be classified as unheated (raw), pasteurized (at 63°C for 30 min or 72°C for 15 s), sterilized (at 108°C for 45 min or equivalent) and ultra high temperature (UHT) sterilized (at >140°C for 2 s). Based on the fat content of the end product and production methods, cream may also be divided into various categories such as single, double or heavy, half or light, whipping or whipped, light whipping cream and clotted cream (Anonymous 1993, 1996, 1999). In farmhouse cream production, classical thermization conditions (i.e., at 63–65°C for <1 min) are not effective enough to reduce the initial bacterial counts of cream to safe limits. Unless the thermized cream is cooled immediately to <4°C and processed within a short period of time, the quality losses in the end product are inevitable. Development of sour, rancid, bitter, stale, cheesy, yeasty and putrid taste, flavor and odor by mainly *Pseudomonas*, *Micrococci* and yeasts are the common taints found in farm-produced cream (Wilbey 2002). Additionally, slime formation caused by coliforms and, less extensively, lactococci is frequently observed in farm-produced cream. Unless the cream is kept under improper conditions, the formation of gas by lactose-fermenting yeasts, and surface taints due to mold growth on the surface of the cream are rare cases. One of the major problems associated

*Corresponding author

with farm-produced cream is sweet curdling caused by proteolytic enzymes released from aerobic sporeformers.

The separation of cream from milk is usually achieved at 40–45°C. This temperature range is ideal for microbial growth. Therefore, increasing separation temperature to above 60°C (i.e., 62–63°C) may help keeping bacterial counts of cream as low as possible. During separation of cream, somatic cells and foreign matters including dirt or any others in the form of slime as well as some bacteria associated with slime are removed. Separation of the cream followed by homogenization is not considered as the critical technological treatment as far as cream microbiology is concerned. However, one should keep in mind that mechanical stress created during homogenization may split bacterial clumps causing faster growth of such bacteria in cream. Homogenization also reduces the oxidative stability of the cream.

Heat treatment is the most critical step of cream production in terms of microbiological safety. Within the EU, the minimum heat treatment of cream is set at 72°C for 15 s. In the USA, a slightly higher minimum temperature (i.e., 74.4°C for 15 s) is mandated for cream with fat level of >10%. Much earlier, International Dairy Federation (IDF) recommended a minimum time-temperature combinations of pasteurization for cream with fat levels of 18% and >35% as 75°C for 15 s and 80°C for 15 s, respectively. Under these thermal conditions, it is unlikely that heat-treated cream contains vegetative forms of spoilage or pathogenic microorganisms. Heat stable spores and/or post-pasteurization contamination are the most likely sources of microorganisms in cream. On the other hand, if the cream is separated from milk held at 4°C for more than 24 hr, there is a risk of growth of heat-stable extracellular lipases from *Pseudomonas* spp. or other organisms in cream, leading likely to lipolytic spoilage in the end product.

The time-temperature combination of cream pasteurization is decided based on the fat content of the cream. The lethal effect of heat treatment on the cream microflora is rather more problematic and complex than that on the raw milk flora. This is particularly due to the protective effect of milk fat on bacteria and presence of microflora with differing thermolability (Wilbey 2002). Higher viscosity of cream also results in slower heat transfer during pasteurization. All these factors may well stimulate the germination of spores surviving heat treatment. Surely, UHT treatment to cream enhances the shelf-life of the end product. However, the cooked flavor caused by high heat treatment and gelation triggered by proteases surviving UHT treatment are the major limiting factors of the shelf-life of UHT cream. The shelf-life of UHT cream is usually 3–6 months. The filling stage of the UHT cream production is of critical importance for microbial contamination, since homogenization is achieved after sterilization, and homogenized cream is sent directly to the filling point. Slightly higher temperatures of filling (e.g.,

75–80°C) are recommended to avoid microbial contamination. In this case, the changes in cream viscosity should be considered.

In fresh cream, the predominating organisms vary depending on the temperatures of storage. While *Pseudomonas, Alcaligenes, Acinetobacter, Aeromonas* and *Achromobacter* are dominant in cream kept at 5°C, *Corynebacterium, Bacillus, Micrococcus, Lactobacillus* and *Staphylococcus* constitute the majority of the cream microflora stored at 30°C (Rothwell and Wilbey 1990). Since milk is emulsified in cream, and, therefore, the surface area of cream is extremely large, cream easily absorbs the odors from the environment, leading to quality losses.

Whipped or whipping cream is more likely to be spoiled by the microorganisms than the liquid cream since the introduction of air into the cream during whipping creates a suitable environment for the growth of most microorganisms (Wilbey 2002). If nitrogen is used as whipping agent, this problem is largely avoided. Products prepared or filled with cream are more likely to cause outbreaks than the cream consumption.

10.2 Butter

10.2.1 Composition, types and consumption of butter

The growing unhealthy image of animal fat in general and butter in particular as well as emerge of cheaper options such as margarine were the major contributing factors towards the decline in consumption of butter during 1960s which continued until recent years. The Canadian Dairy Commission reported a steady decrease in per capita consumption of butter from 8.42 kg in 1964 to 2.84 kg in 1999. Also according to FAO, between 1985 and 1996 there was a 40 67% decline in consumption of butter in 10 EU states. This trend has now been almost reversed in many regions as more details are available indicating that the butter consumption per capita starting to grow. The global butter production has reached about 10 million tons in 2011. The fastest increase in the butter production was recorded in the USA (15.7%) and New Zealand (11.6%). The annual increase in butter production in the EU countries (EU-27) was 3.0%. European countries (especially France, Finland, Luxembourg, Germany, Austria, Czech Republic and Switzerland), New Zealand and Australia are the leading butter-producing and consuming countries with annual consumptions of >4 kg per capita. These countries are followed by the USA, India, Iran, Russia and former Soviet republics, Italy, UK, the Netherlands and Denmark with annual consumptions of 2–4 kg per capita (Eurostat 2013). Table 10.1 shows the changes in butter production figures in some countries between 2010 and 2011. Such increase may largely be due to consumers' desire to eat more natural products, their concerns over trans fatty acids in margarine

Table 10.1 Changes in butter production figures in some countries between 2010 and 2011.

	2011 (million tons)	2010/2011 annual growth rates (%)
India	4.3	3.8
EU-27	2.0	3.0
USA	0.8	15.7
Pakistan*	0.6	-
New Zealand	0.5	11.6
Russia	0.2	4.6

* estimated
Source: Anonymous (2013)

and possible effect of the introducing products that are easier to spread and lower in saturated fats such as butter and vegetable oil based dairy blends in the early 1990s.

Cream which is the starting material for making butter is an oil-in-water (o/w) emulsion. During butter-making, cream goes through an emulsion inversion phase and a water-in-oil (w/o) emulsion occurs. This emulsion phase inversion happens during churning of cream, and as a result, milk fat concentration in cream (ca. 30–35%) is increased to about 80% in butter.

Butter is typically composed of 80–81% milk fat, 16–17% moisture, 1% carbohydrates and protein, and 1.2–1.5% sodium chloride. Butter with no added salt contains 82–83% milk fat. With many developments in dairy industries including butter manufacturing to widen the choices available to the consumers, different types of spreadable butters are now available: dairy spread (40% fat), blended spread (40% fat) and blended spread with high fat (70–80% fat) (The Dairy Council n.d.).

Two typical types of butter are sweet butter (made from unfermented sweet cream) and cultured/fermented butter, a flavor intense type made from fermented/ripened cream by lactic acid bacteria (Kornacki et al. 2001). The fermentation process reduces the pH from typical 6.5 in sweet butter to around 5.0 (Kornacki et al. 2001).

10.2.2 Manufacture of butter and dairy spreads

It is not the intention of this chapter to go through details of butter production. Briefly, the raw milk is separated into skim milk and cream. The principle behind this step is to increase the efficiency of the production process by increasing the yield of butter and reducing the yield of buttermilk (Simpson 2012). The cream, after pasteurization and standardization, could be marketed as different consumer products such as single/double cream, whipping cream and/or coffee cream. The cream to be used in the production of the butter is pasteurized, with the purpose of eliminating spoilage and pathogenic microorganism, inactivating enzymes and to make the butter less susceptible to oxidative degradation. The pasteurized cream (with a fat content of at least 35%) is inoculated with 1–2% starter culture in the production of ripened cream butter to ferment the citrate in the cream to yield acetoin and diacetyl with the latter responsible for the characteristic flavor of the cream butter (see also Chapter 2 for citrate metabolism in lactic acid bacteria). The ripened butter starters usually consist of acid-producers (e.g., *Lactococcus lactis* subsp. *lactis* and *L. lactis* subsp. *cremoris*) and aroma-producers (e.g., *Leuconostoc mesenteroides* subsp. *cremoris* and *Lactococcus lactis* biovar. *diacetylactis*) (Fernandes 2009). Alternatively, concentrated diacetyl permeate and lactic starter permeate containing lactic acid and flavor-enhancing lactic acid bacteria are added to the butter after churning (NIZO process) (Kimenai 1986). Diacetyl is primarily responsible for the development of characteristic butter flavor and inhibits Gram-negative bacteria (Jay 2000). In the manufacture of sweet cream butter, the fat phase of the cream is crystallized while the pasteurized cream is tempered at a low temperature of around 10°C.

The next stage is the churning the tempered sweet cream or cream in the ripening stage, which is probably the most critical step determining the textural quality of the butter (Simpson 2012). The physical force induced during the churning process breaks the o/w emulsion of cream and forms a continuous matrix of fat globule surrounding the water molecules. The typical temperature at this stage is set at 10–11°C. Figure 10.1 illustrates all major steps involved in making butter. The sweet cream butter goes through the same process except for the addition of the starter culture. The resulting butter grains may then be washed to remove non-fat-solids and be finally worked or kneaded.

Dairy spreads are basically produced by mixing fat and oil blends plus lipid soluble additives with water and dry protein milk (4:1 ratio) (Fernandes 2009). Flavorings, colorings, vitamins, salt and preservatives are added to the mixture prior to emulsification at 45°C. The emulsified mixture is pasteurized at 80–85°C for 2–3 s and then left for crystallization at 4–5°C. Afterwards, the product is worked and packaged for retail or whole sales. The level of salt (1–1.5%) added to dairy spread mix is far too low to guarantee microbial safety in the end product. Therefore, preservatives such as sorbic or benzoic acid and their salts are used to enhance the microbial stability of dairy spreads.

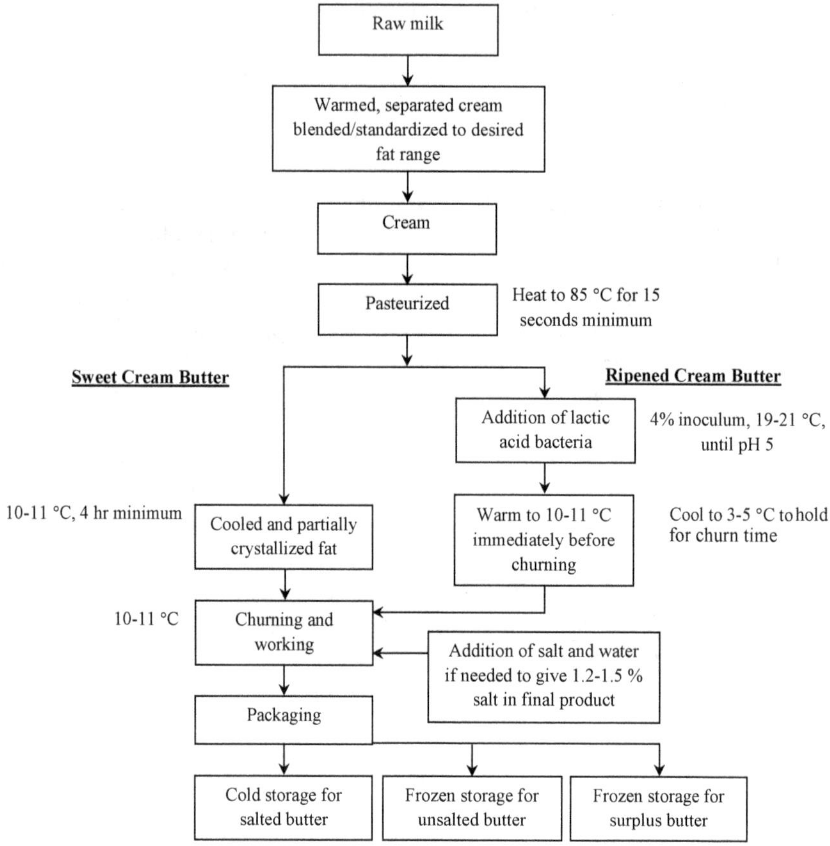

Figure 10.1 A typical flow chart for manufacture of sweet and cultured butter.
Reproduced with permission from CRC Press/ Taylor & Francis Group.
After: Kornacki et al. (2001)

10.2.3 Microbiological consideration in butter

As can be seen in Fig. 10.1, butter manufacturing process involves cream pasteurization, fermentation of cream (in cultured butter only), churning, working, salting, packaging, storage, and perhaps retail packaging, if needed. The microflora of the final products is affected by all these activities.

The physicochemical properties of butter, including pH, water activity (a_w) and other intrinsic factors such as salt concentration define the microbiological stability of the butter. Although moisture content is, to some extent, reflected in a_w value, the distribution, size and uniformity of water droplets all play a role in microbiological quality of the butter.

Like any other processed food, the microbial quality of butter is mainly affected by the raw material, the way in which they are prepared, heat treatments and storage conditions. So for any quality butter, apart from excellent sanitary processing conditions, the quality of starting material—cream—is critical as it is considered to be the main source of microorganisms in butter. The microbiological quality of the cream, in turn, is affected by the microflora of the raw milk. In that sense, the microbial quality of the butter starts at much earlier stages in the processing line and any quality control measure should begin much earlier down the processing line.

The raw milk may become contaminated with a wide range of pathogen and spoilage microorganisms during milking, milk handling equipment and storage (Jay et al. 2005). Although cream pasteurization inactivates the vegetative pathogens, the sporeforming bacteria such as psychrotrophic *Bacillus* and *Clostridium* spp. remain a threat.

A typical pasteurization of cream for butter manufacture is at or above 85°C for 15 s. While it is enough to inactivate all pathogenic microorganisms in vegetative form, many potential spoilage organisms and perhaps most of the lipases (which cause hydrolytic rancidity in butter) and many of the bacterial spores may survive this heat treatment. Higher pasteurization temperatures (e.g., 95–112°C) are commonly used to inactivate all these microorganisms (Schweizer 1986).

For the spores that survive heat treatment, pasteurization may play a heat-shock role in triggering spore activation (tyndallization). The subsequent activity of these microorganisms during cold storage may lead to production of lipolytic enzymes that are relatively heat resistant.

Manufacturing process impact on microbial load of butter and cream

Butter is generally considered a safe product from microbiological point of view, nevertheless, incidences such as listeriosis outbreak in Finland in 1988/9 where six people died proved otherwise and are alarming (Lyytikäinen et al. 2000).

The fact that microorganisms cannot grow in fatty phase of the emulsion type products such as butter and cream highlights the importance of water phase in such products. The distribution of water droplets in w/o emulsion (butter), therefore, is of paramount importance. An interesting observation by Hammer and Babel (1957) suggests that in 1 g of w/o emulsion of butter about 10–18 billion water droplets are dispersed. Considering the expected low total microbial counts of pasteurized cream ($<2\times10^4$ cfu ml^{-1}) (Jay et al. 2005), the majority of the droplets are expected to be free from microorganisms. In cream pasteurized with the right time/temperature combination, few microorganisms should survive and thus the microbial quality of butter depends primarily on processing stages as well as the water and brine quality used.

The size of the water droplets depends on the homogeneity of distribution, finer droplets indicating a better distribution of water molecules. Ideally, 10 μm is optimum size for water droplets (Varnam and Sutherland 1994), but they may be as big as 30 μm (Brunner 1976). Water droplet size rules the availability of nutrients or inhibitors, which are limited in finely dispersed water droplets (Foster et al. 1957). Therefore, the working stage during conventional (batch churn) butter manufacture is crucial (Hammer and Babel 1957) and may result in uneven distribution of droplets containing microorganisms causing variation in the microbiological risks of butter. On the other hand, most of the microorganisms associated with cream are lost with buttermilk during churning.

The working stage has also an impact on salt distribution in butter with added salt. Insufficient working results in an uneven distribution of salt in the water droplets (Milner 1995). The inhibitory effect of salt that can be added in finely ground or brine form is due to osmotic gradients. The growth of yeasts and molds on the surface of the butter appears to be less affected by the water or salt distribution in butter. Firstly, because the low level of salt used in butter does not greatly reduce the water activity to below minimum inhibitory level for these organisms. Secondly, the access to air and humid condition at the surface of the butter seems to be of more importance to their growth. This, however, is different for spoilage bacteria as they tend to grow in areas of butter where the water droplets are large and salt concentration is low, both a result of poor working. In a well-distributed water and salt in butter (e.g., containing 16% moisture and 1% salt), the bacteria can face an actual salt strength of up to 6.25% (Adams and Moss 2008).

Addition of salt to butter also has a lowering impact on the freezing point and, as a result, the psychrotrophic microorganisms present in butter may be able to grow at temperatures less than 0°C (Hammer and Babel 1957).

Although modern and continuous butter-making practices exclude washing step, in batch butter-making processes, the butter granules are washed and the wash water itself is a source of contamination (e.g., *Pseudomonas* spp.). The microbiological quality of water used at different stages of butter production is vital for the safety and stability of the finished product. The uses of chilled water for washing off the excess buttermilk, and for making brine are two examples. The total aerobic counts of water should be less than 10 to 100 cfu ml^{-1} at 22 and 37°C, respectively (Murphy 1990).

Another crucial stage in butter-making is the packaging, where butter is exposed to plant environment, worker's hand and ambient temperature of processing hall, all of which increase the risk of spoilage. Packaging butter into retail size immediately after churning and washing is preferred but not always feasible in batch operations. In such cases, the bulk packages of butter must be re-worked and re-packed to consumer size before being made available to consumers. This extra re-work stage increases the risk of contamination and spoilage of the end product (Milner 1995).

10.2.4 Survival and growth of spoilage and pathogenic microorganisms in butter

As mentioned earlier, butter is not an ideal medium for the growth of microorganisms, especially for the pathogens. This is well supported by rare reported incidence of food infection/poisoning associated with butter. This may be explained by the use of ingredients pasteurized well beyond usual time/temperature pasteurization norms used for milk. Hence, the greatest cause and source of butter contamination and spoilage is post-pasteurization related. Few intoxication outbreaks reported for staphylococci during 1970s in USA are just an example (CDC 1977).

Holliday et al. (2003) conducted a study to assess the viability of *Salmonella, Escherichia coli* 0157:H7 and *Listeria monocytogenes* in butter when subjected to temperature and physical abuse. Three different types of butter namely, sweet cream whipped salted butter (pH 6.4), sweet cream whipped unsalted butter (pH 4.51) and light salted butter (pH 4.58) were all inoculated with three pathogens to simulate post-processing contamination. The products which were stored at a temperature of 4.4°C were then held at 37°C under high relative humidity of 85% so as to induce condensation on the surface before it is stored again at 4.4 or 21°C. The results indicated that all three pathogens grew within 1–2 days in sweet cream whipped salted butter stored at 21°C, while at 4.4°C only *L. monocytogenes* was capable of growing for 7 to 14 days on the same product demonstrating the ability of *L. monocytogenes* to grow at refrigeration conditions. It was also observed that none of the pathogens grew on sweet cream whipped unsalted butter and light salted butter most likely due to the low pH which does not permit the growth of most microorganisms.

Voysey et al. (2009) suggested that the size of the water droplets and its dispersion throughout the butter has an effect on the growth rate and the maximum population of pathogens such as *L. monocytogenes.* They demonstrated that *L. monocytogenes* was able to grow in 'coarse' as compared to 'fine' grain butter. Salt content was also shown to affect the growth and interestingly its growth at 8°C was demonstrated to be comparable to 21°C (Voysey et al. 2009). Lindberg et al. (1998) detected *Enterobacteriaceae* in high numbers after storage at 7°C in 6% (26 of 428 retail samples) of retail packs of pasteurized milk and cream. Salt added to butter at moderate levels (i.e.,

<1.5%) is ineffective for inhibiting *Staphylococcus aureus*. The fate of *S. aureus* in the butter churned from the Cheddar cheese whey cream was studied by Halpin-Dohnalek and Marth (1989). The authors demonstrated that the viability of *S. aureus* was enhanced in lightly salted (1%) whey cream butter and in butter made from neutralized whey cream held at 4, 25 or 30°C for up to four weeks.

10.2.5 Microbial spoilage of butter

The cream quality and the sanitary conditions used in the butter processing are the major determinants of microbiological quality of butter. All three major groups of organisms—bacteria (e.g., *Pseudomonas* spp.), yeasts (e.g., *Candida* spp.), and molds (e.g., *Geotrichum*)—have been implicated in spoilage of butter on the surface causing flavor defects such as putridity, rancidity and/or fishy flavor as well as surface discoloration. The flavor defects in unsalted butter have been attributed to growth of coliforms, *Enterococcus* and *Pseudomonas* in water-phase of butter.

Microorganisms of concern in the spoilage of butter are mainly psychrotrophs, which are predominantly Gram-negative, rod shaped microorganisms. The main characteristic of psychrotrophs, which makes them important in the spoilage of butter, is their ability to survive at low temperatures (i.e., 3–7°C) and the production of enzymes such as lipase and protease which catalyze the hydrolysis of lipids and proteins in the butter, respectively (Ledenbach and Marshall 2009). The rich nutrient content of the butter also makes it susceptible to spoilage microorganisms other than *Pseudomonas* such as *Serratia, Acinetobacter* and *Moraxella*.

Two main types of butter spoilage are color change at the surface (surface taint) and rancidity. The major culprit for both is believed to be *Pseudomonas* spp. *Shewanella putrefaciens* (formerly *Alteromonas putrefaciens*) or *Flavobacterium* spp. which also play role in development of surface taints in butter. Some species such as *P. putrefaciens* are able to grow on butter surface and produce a putrid odor within relatively short period of time (7–10 days) at refrigeration temperature. The odor is suggested to be the result of releasing organic acids such as isovaleric acid (Jay et al. 2005). Black discoloration and shunk-like odor are also developed in butter by *Pseudomonas nigrificans* (Kornacki et al. 2001) and *Pseudomonas mephitica* (Kornacki et al. 2001, Jakson and Morgan 1954), respectively. *Pseudomonas fragi* and, in rare cases, *Pseudomonas fluorescens* as well as non-microbial lipases degrade milk fat into free fatty acids leading to hydrolytic rancidity in butter (Jay et al. 2005). Hydrolytic rancidity in butter can also be the result of activity of *Micrococcus* spp. (Boor and Fromm 2006) and molds such as *Rhizopus, Geotrichum, Penicillium* and *Cladosporium* (Irbe 1993). Formation of 3-methylbutanal by *Lactococcus lactis* var. *maltigenes* is the major cause of 'malty' flavor defect in butter (Kornacki et al. 2001). Microbial groups associated with spoilage of butter, cream and dairy spreads are presented in Table 10.2.

Table 10.2 Microorganisms associated with spoilage of cream, butter and dairy spreads.

Microorganisms	Defects
Cream	
Bacillus cereus	Bitty cream, sweet curdling
Bacillus licheniformis, B. coagulans, B. subtilis	Bitterness, thinning in sterilized cream
Geobacillus stearothermophilus	Spoilage in UHT cream
Pseudomonas fluorescens, P. fragi, P. putrefaciens, P. nigrifaciens	Spoilage in double cream, apple-like ester formation, putrid odor, blackish discoloration
Candida lipolyticum, Geotrichum candidum, Torula cremoris, Candida pseudotropicalis, Torulopsis sphaerica	Yeasty or fruity flavors, gas formation
Penicillium spp.	Surface taints (if the product is stored at 0–1°C)
Rhodotorula mucilaginosa	Bitterness
Proteus } in the case of defective cans or leaking seams	Bitterness and thinning
Coliforms } in the case of defective cans or leaking seams	Gas formation
Lactic acid bacteria } in the case of defective cans or leaking seams	Acid curdling
Butter	
Shewanella putrefaciens *Pseudomonas putrefaciens* *Flavobacterium* spp.	Surface taints
Pseudomonas nigrificans	Black discoloration
Pseudomonas mephitica	Shunk-like odor
Lactococcus lactis var. *maltigenes*	Malty flavor
Micrococcus spp.	Lipolytic spoilage
Molds (e.g., *Penicillium, Aspergillus, Mucor, Cladosporidium, Geotrichum, Alternia, Rhizopus* spp.)	Surface discoloration and taints
Yeasts (e.g., *Rhodotorula, Cryptococcus, Torulopsis, Candida lipolytica*)	Lipolytic spoilage
Dairy spreads	
Penicillium spp. *Trichoderma harzianum* *Yarrowia lipolytica* *Bacillus polymyxa* *Enterococcus faecium*	Tainting

Compiled from Anonymous (2005), Boor and Fromm (2006), Jackson and Morgan (1954), Kornacki et al. (2001), Phillips et al. (1981), Varnam and Sutherland (1994), van Zijl and Klapwijk (2000), Wilbey (2002).

Psychrotroph pseudomonads that grow during storage of raw milk or cream could result in spoilage of butter after manufacture. They are able to do so through their heat-resistant proteases and lipases that are not destroyed by pasteurization.

Suryanvanshi and Ghosh (2010) carried out a study to determine the optimum pH and temperature at which *Pseudomonas aeruginosa* NCIM 2036 can produce lipase in unsalted white butter. The results showed that *P. aeruginosa* produced lipase with significant activity at 5°C. Such activity at 25°C leading to fat hydrolysis of butter by the microbial enzymes was also observed. It was concluded that hydrolysis must be avoided as it can lead to severe economic loss following butter going rancid and not meeting standards. Good sanitation and hygiene practices as well as avoiding the use of unpasteurized milk are amongst the important measures to prevent such losses.

The main factors contributing to control of microorganisms in butter include:

(a) the moisture phase and its fine dispersion,
(b) low-temperature storage,
(c) uniform dispersion of salt in moisture (in salted butter) and
(d) use of lactic cultures (in ripened butter) (Hammer and Babel 1957, Olsen et al. 1988).

Freezing storage temperature has a different effect on the inhibition of microbial growth in salted and unsalted butter. In salted butter, the freezing point in the moisture phase is depressed due to the presence of salt and perhaps other solutes and, therefore, microbial activities are not completely stopped.

Addition of food preservatives is another way of controlling microbial growth in butter. Potassium sorbate (as low as 0.1%) was reported to inhibit the growth of coliforms and molds (Kaul et al. 1979), though this approach may raise concerns over the addition of chemical preservatives.

10.2.6 Examples of food-borne illness outbreaks associated with butter in recent years

L. monocytogenes is a Gram-positive pathogen, which is a facultative anaerobe and psychrotrophic in nature, meaning it is capable of surviving either in the presence or absence of oxygen and at temperatures as low as 0°C. The ability of this pathogen to survive under such conditions makes it significant to the butter industry. Its osmotolerant nature also gives it an added advantage of surviving in salted butter. *L. monocytogenes* is normally found in raw milk as a result of contamination during the milking process. When producing butter it is advisable to use cream that has been pasteurized, since pasteurization destroys all pathogens present in the cream. Butter made from pasteurized cream is not at risk of causing a disease outbreak if good hygienic and sanitation procedures are followed in the production. That is, workers must practice personal hygiene and also the machinery used in producing the butter must also be cleaned very well since *L. monocytogenes* is capable of adhering to the steel surface of the machinery. It is an opportunistic pathogen, having an incubation period of 1–7 days, causing listeriosis which can lead to severe illness with high hospitalization and death rates. At risk groups are pregnant women, newborns, elderly people and immunocompromised adults (Rogga et al. 2005).

L. monocytogenes was implicated in an outbreak of disease in the Huddersfield area in May 2003. Around the same time, it had been reported that a dairy in the area was having problems with low level of *L. monocytogenes* contamination, which was detected during a routine sampling by the local authorities in April, with further confirmation of its presence of up to $1.8{\times}10^2$ cfu g^{-1} in one batch and less than $0.2{\times}10^2$ cfu g^{-1} in another batch. The presence of *Listeria* in the butter led to a revision of the cleaning procedures at the dairy resulting in the absence of the pathogen in productions following the implementation. The use of 2 kg catering packs of butter in local sandwich outlets were sidelined as the likely cause of the outbreak instead of the consumption of contaminated products at home. The product was later re-called by the dairy.

L. monocytogenes serotype 3a was also found to be the cause of a food-borne disease outbreak in Finland between 1998 and 1999. According to Maijala et al. (2001), a single dose of 4.87 $\log_{10}$ cfu g^{-1} could have been sufficient to cause illness in this outbreak.

An outbreak of *E. coli* O157:H7 affected 202 pupils of an elementary school in 2001, with 11 hospitalized. Unpasteurized butter was reported to be the cause of this outbreak and children spread the *E. coli* O157:H7 to family members and school's staff. Shiga-toxin, but not *E. coli* O157, was detected in manure from the butter maker's milking cows.

Clostridium perfringens toxin was implicated in a food-borne disease outbreak in Pueblo, Colorado linked to the consumption of butter. The stool sample of some of the patients tested positive for *C. perfringens*. The Pueblo Health Department carried out an extensive inspection of their premises, with the results indicating that there were multiple critical violations. This led to the closing down of the premises.

Table 10.3 summarizes some of the food-borne diseases outbreaks associated with butter, cream and ice cream.

Table 10.3 Outbreak of illness from consumption of butter, cream and ice cream.

Product	Causative Organism	Year	Country	State/city
Butter	*Listeria monocytogenes*	1987	USA	
	Listeria monocytogenes (serotype 3a) (with six deaths)	1999	Finland	
	Listeria monocytogenes	2003	UK	
	Listeria monocytogenes	2011	Canada	
	Norovirus	2007	USA	Florida
	Norovirus	2008	USA	California
	E. coli 0157:H7	2001	USA	North California
	Staphylococcus aureus	1970	USA	
	Staphylococcal enterotoxin A	1977	USA	
Garlic butter	*Campylobacter jejuni* enteritis	1995	USA	
Butter-blend spread	*Staphylococcus intermedius*	1991	USA	
Ice cream	*E. coli* 0157:H7	2009	USA	Ohio
	E. coli 0157:H7	2008	USA	Vermont
	Salmonella hadar	2008	USA	California
	VTEC *E. coli* 0145 and 026	2007	Belgium	Antwerp
	Salmonella typhi	1947	Wales	
	Salmonella paratyphi	1970s	UK	North Devon
	Salmonella enteritidis	1992	UK	
	Salmonella enteritidis	1993	Wales	
	Salmonella enteritidis	1994	USA	Minnesota
	Norovirus	2007	USA	Pennsylvania
	Hepatitis A	2007	USA	Minnesota
	Listeria spp. (with four deaths)	1986	USA	
	Listeria monocytogenes	1989	Belgium	
	Between 1950–1955 11 outbreaks associated with ice cream and ice lollies were reported			
Cream	*Staphylococcus aureus*	2011	USA	Guam
	Salmonella enteritidis	1991	Spain	Navarra
	Salmonella enteritidis PT4	1992	Wales	
	Salmonella enteritidis (largest outbreak ever recorded affected 224,000 people)	1994	USA	
	Salmonella enteritidis	2009	USA	California
	Salmonella enterica	2008	USA	Missouri
	Salmonella Typhimurium DT104	1998	UK	
	Clostridium perfringens	2008	USA	Illinois
	Eschericia coli O157:H7	1998	UK	
	Hepatitis A	1975	Scotland	
Cream-filled profiterols	*Salmonella* Typhimurium DT40	1986	UK	

Data compiled from outbreakdatabase, Center for Disease Control (CDC), Anonymous (1970, 1977, 1998a, 1998b, 2008), CDR (2004), Chaudhuri et al. (1975), Evans et al. (1996), Hobbs and Gilbert (1978), Khambaty et al. (1994), Lyytikäinen et al. (2000), Mascola et al. (1998), Papademas and Bintis (2002), Rothwell (1990), Sesma et al. (1992), Zhao et al. (2000).

10.2.7 Advances in detection of pathogens in butter, cream and ice cream

The direct epifluorescent filter technique (DEFT) can be used to predict the keeping quality of pasteurized cream as long as undiluted cream is filtered (Kroll and Rodriguez 1986). Although not as efficient as DEFT or plate count method, the cytochrome *c* oxidase test can also be employed to monitor the keeping quality of pasteurized cream (Kroll 1985).

There is an increasing interest in use of molecular biology based methods for detection of pathogens in dairy products. Apart from the speed of identifying potential pathogens, most of these methods have the advantage of being growth-independent. Therefore, pathogens in food undergone certain processes such as freezing would still be detectable whereas the conventional culturing methods usually fail to pick up the contaminants in food stored, for example, at freezing temperature. Also the detection threshold of these DNA-based techniques is by far lower than standard microbiological techniques for enumeration of pathogens of concerns. Moreover such methods could reduce the enrichment time for organisms such as *Salmonella* dramatically to achieve the same results.

Culture based methods on the other hand are labor-intensive and time-consuming especially when handling many samples. In addition, poor specificity of the selective agar media and the presence of competing microorganisms could lead to false positive results.

A recent study by Kawasaki et al. (2009) illustrated the potential use of multiplex PCR for simultaneous detection of three major food pathogens namely, *Salmonella* spp., *L. monocytogenes*, and *E. coli* O157:H7 in foods and in food subjected to freezing including cultured butter, fresh cream and ice cream. Multiplex PCR with the presence of 16S rRNA and *nuc* gene is also effective in detecting *S. aureus* in butter and related products (Gucukoglu et al. 2012). Strachan et al. (1997) reported that an automated rapid immunosensor was capable of detecting *S. aureus* enterotoxin B in cream down to a level of 5 ng g^{-1} in approximately 10 min. Lateral flow immunochromatographic devices (LDFs) can be successfully exploited to detect staphylococcal enterotoxin B at concentrations below 0.25 ng ml^{-1} in milk products including butter.

10.3 Ice cream

Although frozen status of ice cream and frozen dairy desserts gives both manufacturers and consumers a positive image about the microbial safety of the finished product, it does not completely eliminate the risk of transmission of pathogenic microorganisms or toxins to humans via frozen products (Marshall and Arbuckle 2000). *Salmonella* spp. was reported to survive for seven years in ice cream. Therefore, before freezing, the ice cream mix should be subjected to adequate heat treatment to provide food safety (Georgala and Hurst 1963). *Bacillus cereus* is unlikely to survive pasteurization; however, *B. careus* spores can cause illness if the initial level of spores is too high and time and temperature control after pasteurization are not adequate (Fernandes 2009).

Ice cream is a frozen foam of which air cells, fat globules, caseins, stabilizing gums and ice crystals form the suspended phase (Marshall 1998). Viscous syrup which constitutes the continuous phase of ice cream contains microorganisms. Many factors including physical forces of ice crystals, pH and osmotic pressure affect the viability of microorganisms in ice cream. During freezing of ice cream mix, acids, salts and sugars are concentrated in the syrup. All these substances show detrimental effect against microorganisms and spores at varying levels. In general, bacterial spores and spores of molds and yeasts are more resistant against these substances than the vegetative cells. Gram-negative bacteria are generally more sensitive than Gram-positive bacteria and vegetative cells of molds and yeasts. Growing extracellular ice crystals is one of the major factors damaging the microbial cells. Therefore, any factor restricting growth of ice crystals such as fat globules and air cells may reduce the damage done to microbial cells by ice crystals (Marshall 1998).

In the manufacture of ice cream, a number of ingredients including milk, cream sweeteners, flavorings, nuts, confectionary products, eggs and egg products and coloring materials are used. The ingredients added to ice cream mix after pasteurization carry potential risk of transmitting contaminant microorganisms to ice cream. Majority of the outbreaks associated with ice cream have involved ice cream made from raw materials (i.e., raw milk, cream or eggs). Improper heating and/or post-production contamination are among the other sources of outbreaks. As long as a proper heat treatment is applied, milk is not considered a source of microbial contamination. However, sometimes raw milk is used in the manufacture of home-made ice creams. In addition, if the milk with poor hygienic quality or with high spore counts is used in ice cream production, spores may withstand pasteurization and spoil the finished product.

Sweeteners used in ice cream production are available in two different forms: crystalline and syrup forms. It is unlikely that bacteria grow in crystalline sugars (e.g., dextrose, fructose, sucrose or maltose), but yeasts may be present in such sugars. On the other hand, aerobic osmophilic fungi can grow in syrup. This group of microorganisms is usually present on the surface of ice cream. Therefore, surface treatments (e.g., exposure to UV light) or sealing of the containers may well inhibit the growth of yeasts, if present. However, in bulk productions, air is inevitably incorporated into the tank; thus the initial level of osmophilic aerobic yeasts must be as low as possible.

The microbiological stability of syrup with high sugar concentration (a_w 0.70) is higher than those with high a_w (i.e., 0.80). Corn syrup is widely used in ice cream productions. Comparing to the sugars containing glucose polymers, higher concentration of corn syrup is needed to inhibit microbial growth. It should be kept in mind that osmophilic yeasts (especially those belong to the genus *Zygosaccharomyces*) can grow in syrup with an a_w as low as 0.65. *Candida, Pichia, Schizosaccharomyces* and *Torula* spp. are among the other yeast flora likely present in corn syrup. The preparation of corn syrup is of critical importance for microbial growth since steeping step includes soaking of corn in water at 45–50°C for about 48 hr at pH 4.0 (Whistler and Paschall 1967). These conditions stimulate the growth of microorganisms. Addition of sulfur dioxide at a level of 0.1–0.2% may well inhibit the growth of microorganisms in corn syrup (Marshall 1998). Other sources of contaminants in ice cream manufacture are given in Table 10.4.

Frozen yogurt is another product of frozen dairy products segment. Frozen yogurt contains two yogurt bacteria (*Streptococcus thermophilus* and *Lactobacillus delbrueckii* subsp. *bulgaricus*). The results of scientific works on the resistance of yogurt bacteria to freezing are somehow contradictory. For example, Lopez et al. (1998) demonstrated that both yogurt bacteria were not affected by freezing to a great extent and the numbers of these bacteria remained almost stable in frozen yogurt stored at –45°C for 3–4 months. Similarly, lactase activity was also maintained in frozen yogurt during storage (Mashayekh and Brown 1992). On the contrary, Sheu and Marshall (1993) observed that the counts of two strains of *Lb. delbrueckii* subsp. *bulgaricus* decreased by 45% and 90% during freezing. A more moderate decrease (ca. 5%) in viable cell counts of yogurt bacteria during storage at –20°C for two weeks was also observed (Sheu and Marshall 1993).

There is a relationship between changes in the numbers of yogurt starters during freezing and the type of freezer. In case of use of classical freezers, the counts of bacteria remain almost unchanged; whereas, the use of a Taylor®-type freezer suitable for soft ice cream-making results in 1.5 or 2.0 log reductions in the counts of yogurt bacteria (Thompson and Mistry 1994). Microentrapment of bacterial cells protects the bacteria against freezing to some extent.

Ice cream and frozen yogurt may be considered as vehicles for probiotic uptake (Hekmat and McMahon 1992). Hagen and Narvhus (1999) evaluated the viability of *Lactobacillus reuteri, Lb. acidophilus, Lb. rhamnosus* GG and *Bifidobacterium bifidum* in ice cream stored at –20°C for 52 weeks. The product was demonstrated to contain probiotic bacteria at levels above therapeutic minimum (i.e., 10^6 cfu g^{-1}) throughout storage. Similar results were reported by Marshall (1998) who suggested that *B. bifidum* and *Lactobacillus* spp. could be added to ice cream and frozen dairy desserts as dietary adjunct. In order to improve and/or maintain the viability of probiotic bacteria

Table 10.4 Potential sources of contamination of various microbial groups to ice cream.

Source	Microbial groups	Remarks
Honey	Yeasts (e.g., *Zygosaccharomyces* spp.)	Yeasts are natural habitats of flowers.
Flavorings	Vegetative microbial cells Bacterial spores (<100 per g)	Extracts contain alcohol that sufficiently dehydrate and destroy vegetative cells. Cocoa and chocolate liquor consist bacterial spores when contaminated during or after pressing and grinding.
Fruits	Yeasts (e.g., *Saccharomyces* and *Cryptococcus*) Molds (e.g., *Alternia, Aspergillus, Botrytis, Fusarium, Geotrichum, Mucor, Penicillium, Rhizopus*) Bacteria (e.g., *Bacillus, Pseudomonas, Achromobacter*)	Relatively low pH of fruits favors the growth of yeasts and molds. Freezing destroys some microorganisms but does not guarantee complete inhibition. Adding hypochlorite to wash water is an effective way of reducing microbial counts of fruits before freezing. Candied or glaciéd fruits have a_w low enough to prevent microbial growth.
Nuts	Various microorganisms	Nuts consist of microorganisms originating from mainly soil. Treatments applied to separate nut meats and shell may reduce the microbial count. a_w of most nuts (i.e., 0.70) prevent microbial growth. Nuts (both tree nuts and ground nuts) are considered to be free of pathogens. However, mycotoxins carry a potential risk.
Egg yolks	*Salmonella* spp. and *Micrococci* in fresh eggs Gram-negative rods (e.g., *Pseudomonas, Proteus*)	Egg is usually contaminated during laying. Iron is essential for the growth of most bacteria. Glycoproteins present in shell membrane bind iron. Also, ovotransferrin chelates iron, making them unavailable for bacteria. Binding biotin and riboflavin by avidin and ovoflavoprotein, respectively, inhibits bacteria that need these two vitamins for their growth. Transovarian infection of *Salmonella* is a major concern. Pasteurization and freezing reduce the counts of bacteria to a great extent. *Bacillus, Micrococcus* and *Enterococcus* may withstand pasteurization and freezing.

in ice cream and frozen desserts, addition of prebiotics (i.e., 1% resistant starch) or microencapsulation of bacterial cells are the effective options. Homayouni et al. (2008) showed that both options increased the survival rate of probiotic bacteria in ice cream stored at –20°C for 180 days. Addition of inulin as prebiotic also stimulated the viability of *Lb. rhamnosus* and *Lb. casei* in ice cream (Di Criscio et al. 2010). In order to stimulate the growth and viability of probiotic strains in frozen products, addition of cystein of casein hydrolysates to frozen yogurt mix prior to fermentation is recommended (Ravula and Shah 1998, Shah and Ravula 2000).

From therapeutic point of view, gastrointestinal survival and adhesion ability of probiotics are of primary importance. Scientific evidences on probiotic ice cream are rather contradictory. Ranadheera et al. (2012) observed enhanced acid and bile tolerances of *Lb. acidophilus* LA-5, *Bifidobacterium animalis* subsp. *lactis* BB-12, and *Propionibacterium jensenii* 702 added as dietary adjunct into ice cream. On the other hand, while the resistance to bile in *Lb. rhamnosus* was not affected in frozen-thawed ice cream, the bile tolerance of *Lb. acidophilus* was more susceptible to the process and cold storage (Abghari et al. 2011). There is an adverse relationship between overrun level of ice cream and viability of *Lb. acidophilus*. Therefore, in order to maintain the probiotic status of ice cream, the overrun level should be around 45–50% in probiotic ice cream (Ferraz 2012).

As stated earlier, the post-pasteurization contamination is the major reason of presence of pathogens in ice cream. Chilling and freezing conditions (i.e., static or dynamic chilling and freezing) slightly affect the kinetic behavior of *L. monocytogenes*. Freezing conditions did not cause a severe stress in *L. monocytogenes* cells capable of leading to a significant additional lag phase during the subsequent growth of the pathogen at chilling conditions (Gougouli et al. 2008). Lactose and sucrose can act as cryoprotectants on *L. monocytogenes* by depressing freezing point of the suspension media in ice cream (Dean and Zottola 1996). Therefore, it is fair to assume that freezing/hardening or low temperature storage of ice cream are not enough to provide microbial safety in ice cream (Papademas and Bintis 2002). Therefore, strict hygienic measures should be practiced to minimize the pathogenic risks in ice cream. Although recent outbreaks reported to be associated with ice cream and/or frozen dairy desserts are rather rare, between 1986 and 1998 over 10,000,000 liters of ice cream contaminated with pathogens were re-called in the USA (Ryser 1999). Table 10.5 illustrates the predominant pathogenic microorganisms found in ice cream in retail sale in various countries. Lethal rates of pathogenic bacteria during freezing of soft served ice cream show differences. For example, *S. aureus* shows great resistance against freezing and spores of *Bacillus* are not affected. On the contrary, *E. coli*, *P. fluorescens* and *Saccharomyces lactis* show non-linear destruction rate curves of log counts versus time. Young cells of *E. coli* are more susceptible to destruction than older cells.

In order to provide microbial safety of ice cream, low dose irradiation (i.e., 1 kGy) may be effectively used with no sensory deterioration in the final product (Kamat et al. 2000). At higher irradiation doses (i.e., >5 kGy), *Listeria* spp., *Salmonella* spp. and *E. coli* are completely killed in ice cream (Lee et al. 2009). Jo et al. (2007) showed that for the inactivation of *Listeria ivanovii*, *E. coli* and *Salmonella* Thyphimurium in ice cream, irradiation doses of 3, 1 and 0.1 kGy were required, respectively.

Table 10.5 Predominant pathogenic microorganisms of ice cream in retail sale.

Predominant pathogen	Country	References
Salmonella enteritidis	Poland (survey period of 1961–1991)	Głośnicka and Kurnikoska (1994)
Listeria spp.	England and Wales	Greenwood et al. (1991)
Listeria monocytogenes	Finland (survey period of 1990–1997)	Miettinen et al. (1999)
Listeria monocytogenes *Listeria innocua*	California/USA (an incidence of 12% among 922 samples)	Walker et al. (1991)
Listeria monocytogenes	USA (survey period of 1997–1999 with six re-calls)	FDA Enforcement Report (www.fda.gov/Safety/Recalls/EnformenetReports/defaut.html)
Listeria monocytogenes	India (53% of packed and 100% of open ice creams were contaminated)	Warke et al. (2000)
Fecal coliforms	Turkey (15 of 46 samples contained coliforms above the legal limits)	Kocak et al. (1998)
Fecal coliforms *Staphylococcus aureus* *Salmonella* spp.	Italy (survey period of 1987–1988)	Massa et al. (1989)
Fecal coliforms	Pakistan (66% of the samples contained coliforms above the legal limits)	Masud (1989)
Enterobacteriaceae	Venezuella (77% of the samples failed to comply with legal requirements)	de Tamsut and Garcia (1989)
Eschericia coli O157:H7 (detected by RTi PCR)	India	Singh et al. (2009)
Verocytotoxin-producing *E. coli* O145 and O26	Belgium	Eurosurveilance (European Communicable Disease Bulletin 2008)

Biofilm in ice cream plants poses a potential risk of contamination of pathogenic and non-pathogenic microorganisms. Gunduz and Tuncel (2006) investigated the sites of biofilm formation in an ice cream plant. Most of the biofilm formation was detected on the conveyor belt of the packaging machine 8 hr after the beginning of the production. Most of the Gram-negative bacteria isolated from the biofilms belonged to *Enterobacteriaceae* family including *Proteus, Enterobacter, Citrobacter, Shigella, Eschericia* and *Edwardsiella* spp. *Aeromonas, Plesiomonas, Moraxella, Pseudomonas* and *Alcaligenes* spp. were also isolated from the biofilms. Gram-positive bacteria isolated from the biofilms were *Staphylococcus, Bacillus, Listeria* spp. and lactic acid bacteria including *Streptococcus, Leuconostoc* and *Pediococcus* spp.

Miettinen et al. (1999) characterized 41 *L. monocytogenes* isolates originated from an ice cream plant by serotyping and pulsed field gel electrophoresis (PFGE) during the years 1990–1997. Totally 12 different PFGE types were distinguished based on information obtained with three rare-cutting enzymes (*Apa*I, *Asc*I and *Sma*I). The dominant PFGE type was found to have persisted in the ice cream plant for seven years. *Cryptosporidium parvum* is a protozoan capable of causing massive water-borne outbreaks. The most possible way of transferring *C. parvum* oocyst into ice cream is the contaminated water via food contact surfaces. Deng and Cliver (1999) showed that although approximately 20% of oocyts were viable before hardening, none were viable after 24 hr at –20°C.

Impedance measurement technique is a feasible, rapid and accurate alternative to classical plating technique in the detection of microbial count. Detection of *Salmonella* spp. and *Listeria* spp. in ice cream by PCR requires pre-enrichment using an appropriate enrichment medium (Velazquez-Estrada et al. 2010). However, direct detection of pathogens in ice cream by PCR is also possible (i.e., detection of *Salmonella* spp. by PCR against *hilA* gene) (Marathe et al. 2012).

References

Abghari, A., M. Sheik-Zeinoddin and S. Solemanian-Zad. 2011. Nonfermented ice cream as a carrier for *Lactobacillus acidophilus* and *Lactobacillus rhamnosus*. Int. J. Food Sci. Technol. 46: 84–92.

Adams, M. and M.O. Moss. 2008. Food Microbiology. 3rd edition. The Royal Society of Chemistry, Cambridge, UK. 112 p.

Anonymous. 1970. Staphylococcal food poisoning traced to butter: Alabama. Morbidity and Mortality. Weekly Report 28: 129–130.

Anonymous. 1977. Presumed staphylococcal food poisoning associated with whipped butter. Morbidity and Mortality Weekly Report 26: 268.

Anonymous. 1993. Dairy Industry Regulation 1993, Section 5. State of Queensland, Australia.

Anonymous. 1995. UK Regulations. The Dairy Products (Hygiene) Regulations 1995, SI 1995, No. 1086, HMSO, London, UK.

Anonymous. 1996. UK Regulations. Food Labeling Regulations 1996, SI 1996, No. 1499, HMSO, London, UK.

Anonymous. 1998a. Defective pasteurisation linked to outbreak of *Salmonella typhimurium* definitive phage type 104 infection in Lancashire. CDR Weekly 8: 335, 338.

Anonymous. 1998b. Cases of *Escherichia coli* O157 infection associated with unpasteurised cream. CDR Weekly 8: 377.

Anonymous. 1999. South Carolina Code of Regulations Chapter 61. *In:* State Regulations 23 (10), South Carolina, FL, USA.

Anonymous. 2005. Oil- and fat-based foods. pp. 480–521. *In:* International Commission on Microbiological Specifications for Foods [ed.]. Microbial Ecology of Food Commodities. Vol. 6. Plenum Publishers, London, UK.

Anonymous. 2008. Euro Surveillance. European Communicable Disease Bulletin. Vol. 13. Date of electronic publication: 2008 February 14.

Anonymous. 2013. Dairy sector statistics in Turkey and in the World. National Dairy Council of Turkey, Ankara, Turkey.

Boor, K. and H. Fromm. 2006. Managing microbial spoilage in the dairy industry. pp. 171–193. *In:* C.W. Blackburn [ed.]. Food Spoilage Microorganisms. Woodhead Publishing Ltd., Cambridge, UK.

Brunner, J.R. 1976. Characteristics of edible fluids of animal origin: Milk. pp. 619–658. *In:* O.R. Fennema [ed.]. Principles of Food Science. I. Food Chemistry. Marcel Dekker, New York, NY, USA.

CDC. 1977. Presumed staphylococcal food poisoning associated with whipped butter. Morbid Mortal Weekly Report 19: 271.

CDR. 2004. *Listeria monocytogenes* infections in England and Wales in 2004. Com. Dis. Rep. Weekly 14: 37.

Chaudhuri, A.K.R., G. Cassie and M. Silver. 1975. Outbreak of foodborne type-A hepatitis in greater Glasgow. Lancet 2: 223–225.

Dean, J.P. and E.A. Zottola. 1996. Use of nisin in ice cream and effect on survival of *Listeria monocytogenes* as a function of milk components. J. Food Protect. 59: 476–480.

de Tamsut, L.S. and C.E. Garcia. 1989. Microbiological quality of vanilla ice cream manufactured in Caracas, Venezuela. Arch. Latinoam. Nutr. 39: 46–56.

Deng, M.Q. and D.O. Cliver. 1999. *Cryptosporidium parvum* studies with dairy products. Int. J. Food Microbiol. 46: 113–121.

Di Criscio, T., A. Fratianni, R. Mignogna, L. Cinquanta, R. Coppola, E. Sorrentino and G. Panfili. 2010. Production of functional probiotic, prebiotic and symbiotic ice creams. J. Dairy Sci. 93: 4555–4564.

Evans, M.R., J.P. Tromans, E.L.S. Dexter, C.D. Ribeiro and D. Gardner. 1996. Consecutive *Salmonella* outbreaks traced to the same bakery. Epidem. Infect. 116: 161–167.

Fernandes, R. 2009. Microbiology Handbook. Dairy Products. Leatherhead Food International Ltd., Surrey, UK.

Ferraz, J.L. 2012. Sensory acceptance and survival of probiotic bacteria in ice cream produced with different overrun levels. J. Food Sci. 77: S24–S28.

Foster, E.M., F.E. Nelson, M.L. Speck, R.N. Doetsch and J.C. Olson. 1957. Microbiology of Cream and Butter. Dairy Microbiology. Prentice Hall, Englewood Cliffs, NJ, USA.

Georgala, D.L. and A. Hurst. 1963. The survival of food poisoning bacteria in frozen foods. J. Appl. Bacteriol. 26: 346–358.

Głośnicka, R. and D. Kurnikoska. 1994. The epidemiological situation of *Salmonella enteritidis* in Poland. Int. J. Food Microbiol. 21: 21–30.

Gougouli, M., A.S. Angelidis and K. Koutsoumanis. 2008. A study on the kinetic behavior of *Listeria monocytogenes* in ice cream stored under static and dynamic chilling and freezing conditions. J. Dairy Sci. 91: 523–530.

Greenwood, M.H., D. Roberts and P. Burden. 1991. The occurrence of *Listeria* species in milk and dairy products: A national survey in England and Wales. Int. J. Food Microbiol. 12: 197–206.

Gucukoglu, A., T.O. Kevenk, T. Uyanik, O. Cadirci, G. Terzi and M. Alisarli. 2012. Detection of enterotoxigenic *Staphylococcus aureus* in raw milk and dairy products by multiplex PCR. J. Food Sci. 77: M620–623.

Gunduz, G.T. and G. Tuncel. 2006. Biofilm formation in an ice cream plant. Antonie van Leeuwen. Int. J. Gen. Mol. Microbiol. 89: 329–336.

Hagen, M. and J.A. Narvhus. 1999. Production of ice cream containing probiotic bacteria. Milchwiss. 54: 265–268.

Halpin-Dohnalek, M.I. and E.H. Marth. 1989. Fate of *Staphylococcus aureus* in whey, whey cream and whey cream butter. J. Dairy Sci. 72: 3149–3155.

Hammer, B.W. and F.J. Babel. 1957. Dairy Bacteriology. 4th edition. Chapman & Hall, New York, NY, USA.

Hekmat, S. and D.J. McMahon. 1992. Survival of *Lactobacillus acidophilus* and *Bifidobacterium bifidum* in ice cream for use as a probiotic food. J. Dairy Sci. 75: 1415–1422.

Hobbs, C.B. and R.J. Gilbert. 1978. Food Poisoning and Food Hygiene. Edward Arnold, London, UK.

Holliday, S.L., B.B. Adler and L.R. Beuchat. 2003. Viability of *Salmonella, Escherichia coli* O157:H7, and *Listeria monocytogenes* in butter, yellow fat spreads, and margarine as affected by temperature and physical abuse. Food Microbiol. 20: 159–168.

Homayouni, A., A. Azizi, M.R. Ehsani, M.S. Yarmand and S.H. Razavi. 2008. Effect of microencapsulation and resistant starch on the probiotic survival and sensory properties of symbiotic ice cream. Food Chem. 111: 20–55.

Irbe, R.J. 1993. Microbiology of reduced-fat foods: Science for the food industry of the 21st century. pp. 359–382. *In:* M. Yalpani [ed.]. Biotechnology, Supercritical Fluids and Other Advanced Technologies for Low Calorie, Healthy, Food Alternatives. ATL Press, Mount Prospect, IL, USA.

Jackson, H.W. and M.E. Morgan. 1954. Identity and origin of the malty aroma substance from milk cultures of *Streptococcus lactis* var. *maltigenes*. J. Dairy Sci. 37: 1316–1324.

Jay, J.M. 2000. Modern Food Microbiology. 6th edition. Chapman and Hall, New York, NY, USA.

Jay, J.M., M.J. Loessner and D.A. Golden. 2005. Modern Food Microbiology. Springer, New York, NY, USA.

Jo, C., H.-J. Kim., D.-H. Kim, W.-K. Lee, J.-S. Ham and M.-W. Byun. 2007. Radiation sensitivity of selected pathogens in ice cream. Food Cont. 18: 859–865.

Kamat, A., R. Warke, M. Kamat and P. Thomas. 2000. Low-dose irradiation as a measure to improve microbial quality of ice cream. Int. J. Food Microbiol. 62: 27–35.

Kaul, J., J. Singh and R.K. Kuila. 1979. Effect of potassium sorbate on the microbial quality of butter. J. Food Protect. 42: 656–657.

Kawasaki, S., P.M. Fratamico, N. Horikoshi, Y. Okada, K. Takeshita, T. Sameshima and S. Kawamoto. 2009. Evaluation of a multiplex PCR system for simultaneous detection of *Salmonella* spp., *Listeria monocytogenes*, and *Escherichia coli* O157:H7 in foods and in food subjected to freezing. Foodborne Path. Dis. 6: 81–89.

Khambaty, F.M., R.W. Bennett and D.B. Shah. 1994. Application of pulsed-field gel electrophoresis to the epidemiological characterisation of *Staphylococcus intermedius* implicated in a food-related outbreak. Epidem. Infect. 113: 75–81.

Kimenai, M.P. 1986. Cream crystallization. Section III-NIZO method. pp. 11–15. *In:* Continuous Butter Manufacture. IDF Bulletin No. 204. International Dairy Federation, Brussels, Belgium.

Kocak, C., M. Akan and H. Yardimci. 1998. Bacteriological quality of ice cream marketed in Ankara. Veteriner Fak. Der. Ankara Univ. 45: 131–134.

Kornacki, J., R. Flowers and R. Bradley, Jr. 2001. Microbiology of butter and related products. pp. 127–150. *In:* E. Marth and J. Steele [eds.]. Applied Dairy Microbiology. Marcel Dekker Inc., New York, NY, USA.

Kroll, R.G. 1985. Cytochrome *c* oxidase test for rapid detection of psychrotrophic bacteria in milk. J. Appl. Bacteriol. 59: 137–141.

Kroll, R.G. and U.M. Rodriguez. 1986. The direct epifluorescent filter technique, cytochrome *c* oxidase test and plate count method for predicting the keeping quality of pasteurized milk. Food Microbiol. 3: 185–194.

Ledenbach, L.H. and R.T. Marshall. 2009. Microbiological spoilage of dairy products. pp. 41–70. *In:* W.H. Sperber and M.P. Doyle [eds.]. Compendium of the Microbiological Spoilage of Foods and Beverages. Springer Publ., New York, NY, USA.

Lee, J.W., H.J. Kim., Y. Yoon, J.H. Kim., J.S. Ham, M.W. Byun, M. Baek, C. Jo and M.G. Shin. 2009. Manufacture of ice cream with improved microbiological safety by using gamma irradiation. Radiation Phys. Chem. 78: 593–595.

Lindberg, A.-M., Å. Ljungh, S.A. Ahrné, S. Löfdahl and G. Molin. 1998. Enterobacteriaceae found in high numbers in fish, minced meat and pasteurized milk or cream and the presence of toxin encoding genes. Int. J. Food Microbiol. 39: 11–17.

Lopez, M., L. Medina and R. Jordano. 1998. Survival of lactic acid bacteria in commercial frozen yogurt. J. Food Sci. 63: 706–708.

Lyytikäinen, O., T. Autio, R. Maijala, P. Ruutu, T. Honkanen-Buzalski, M. Miettinen, M. Hatakka, J. Mikkola, V.J. Anttila, T. Johansson, L. Rantala, T. Aalto, H. Korkeala and A. Siitonen. 2000. An outbreak of *Listeria monocytogenes* serotype 3a infections from butter in Finland. J. Infect. Dis. 181: 1838–1841.

Maijala, R., O. Lyytikäinen, T. Johansson, T. Autio, T. Aalto, L. Haavisto and T. Honkanen-Buzalski. 2001. Exposure of *Listeria monocytogenes* within an epidemic caused by butter in Finland. Int. J. Food Microbiol. 70: 97–109.

Marathe, S.A., R. Chowdhury, R. Bhattacharya, A. Govindsamay-Nagarajan and D. Chakravorty. 2012. Direct detection of Salmonella without pre-enrichment in milk, ice cream and fruit juice by PCR against *hil*A gene. Food Cont. 23: 559–563.

Marshall, R.T. 1998. Ice cream and frozen yogurt. pp. 81–108. *In:* E.H. Marth and J.L. Steele [eds.]. Applied Dairy Microbiology. Marcel Dekker Inc., New York, NY, USA.

Marshall, R.T. and W.S. Arbuckle. 2000. Ice Cream. 5th edition. Aspen Publishers, Gaithersburg, Maryland, USA.

Mascola, L., L. Chun, J. Thomas, W.F. Bibe, B. Schwartz, C. Salminen and P. Heseltine. 1998. A case-control study of a cluster of perinatal listeriosis identified by an active surveillance system in Los Angeles County. Proc. Society Ind. Microbiol.-Comprehensive Conference on *Listeria monocytogenes*. Rohnert Park, CA, USA.

Mashayekh, M. and R. Brown. 1992. Stability of *Lactobacillus delbrueckii* subsp. *bulgaricus*, *Streptococcus salivarius* subsp. *thermophilus* and 3-galactosidase activity in frozen cultured ice cream. Cult. Dairy Prod. J. 27: 4–8.

Massa, S., G. Poda, D. Cesaroni and L.D. Trovatelli. 1989. A bacteriological survey of retail ice cream. Food Microbiol. 6: 129–134.

Masud, T. 1989. Microbiological quality and public health significance of ice cream. J. Pak. Med. Assoc. 39: 102–104.

Miettinen, M.K., K.J. Björkroth and H.J. Korkeala. 1999. Characterization of *Listeria monocytogenes* from an ice cream plant by serotyping and pulsed field gel electrophoresis. Int. J. Food Microbiol. 46: 187–192.

Milner, J. 1995. Butter and low-fat dairy spreads. pp. B1–B5. *In:* LFRA Microbiology Handbook. Leatherhead Food RA, Leatherhead, Surrey, UK.

Murphy, M.F. 1990. Microbiology of butter. pp. 109–130. *In:* R.K. Robinson [ed.]. Dairy Microbiology. Vol. 2. The Microbiology of Milk Products. Elsevier, London, UK.

Olsen, J.A., A.E. Yousef and E.H. Marth. 1998. Growth and survival of *Listeria monocytogenes* during making and storage of butter. Milchwiss. 43: 487–489.

Papademas, P. and T. Bintis. 2002. Microbiology of ice cream and related products. pp. 213–260. *In:* R.K. Robinson [ed.]. Handbook of Dairy Microbiology: The Microbiology of Milk and Milk Products. John Wiley, New York, NY, USA.

Phillips, J.D., M.W. Griffiths and D.D. Muir. 1981. Growth and associated enzyme activity of spoilage bacteria in pasteurised double cream. J. Soc. Dairy Technol. 34: 113–118.

Ranadheera, C.S., C.A. Evans, M.C. Adams and S.K. Baines. 2012. *In vitro* analysis of gastrointestinal tolerance and intestinal cell adhesion of probiotics in goat's milk ice cream and yogurt. Food Res. Int. 49: 619–625.

Ravula, R. and N. Shah. 1998. Effect of acid casein hydrolysate and cysteine on the viability of yogurt and probiotics bacteria in fermented frozen dairy desserts. Aus. J. Dairy Technol. 53: 175–179.

Rogga, K.J., J. Samelis, A. Kakouri, M.C. Katsiari, I.N. Savvaidis and M.G. Kontominas. 2005. Survival of *Listeria monocytogenes* in Galotyri, a traditional Greek soft acid-curd cheese, stored aerobically at 4 degrees C and 12 degrees C. Int. Dairy J. 15: 59–67.

Rothwell, J. 1990. Microbiology of ice cream and related products. pp. 1–40. *In:* R.K. Robinson [ed.]. Dairy Microbiology. Elsevier Science Publishers Ltd., New York, NY, USA.

Rothwell, J. and R.A. Wilbey. 1990. Microbiology of cream and dairy desserts. pp. 41–108. *In:* R.K. Robinson [ed.]. Dairy Microbiology. Elsevier Science Publishers Ltd., New York, NY, USA.

Ryser, E.T. 1999. Incidence and behavior of *Listeria monocytogenes* in unfermented dairy products. pp. 359–410. *In:* E.T. Ryser and E.H. Marth [eds.]. *Listeria*, Listeriosis and Food Safety. 2nd edition. Marcel Dekker, New York, NY, USA.

Schweizer, M. 1986. Heat treatment. pp. 36. *In:* Continuous Butter Manufacture. Bulletin No. 204. International Dairy Federation (IDF), Brussels, Belgium.

Sesma, B., M. Moreno and J. Eguaras. 1992. Foodborne *Salmonella enteritidis* outbreak: A problem of hygiene or technology? An investigation by means of HACCP monitoring. pp. 1065–1068. *In:* Foodborne Infections and Intoxications. Proceedings of the 3rd World Congress. Vol. 2. FAO/WHO, Institute of Veterinary Medicine, Berlin, Germany.

Shah, N. and L.R. Ravula. 2000. Microencapsulation of probiotics bacteria and their survival in frozen fermented dairy desserts. Aust. J. Dairy Technol. 55: 139–144.

Sheu, J.-S. and R.T. Marshall. 1993. Improving survival of culture bacteria in frozen desserts by microentrapment. J. Dairy Sci. 76: 1902–1907.

Simpson, B.K. 2012. Food Biochemistry and Food Processing. Wiley Blackwell, Iowa, USA.

Singh, J., V.K. Batish and S. Grover. 2009. A scorpion probe-based real-time PCR assay for detection of *E. coli* O157:H7 in dairy products. Foodborne Pathog. Dis. 6: 395–400.

Strachan, N.J.C., P.G. John and I.G. Millar. 1997. Application of a rapid automated immunosensor for the detection of *Staphylococcus aureus* enterotoxin B in cream. Int. J. Food Microbiol. 35: 293–297.

Suryavanshi, M.V. and J.S. Ghosh. 2010. Spoilage of white unsalted butter by psychrophilic lipolysis of *Pseudomonas aeruginosa* NCIM 2036. British J. Dairy Sci. 26–29.

Thompson, L.D. and A.N. Mistry. 1994. Compositional changes in frozen yogurt during fermentation, frozen storage and soft serve freezing. Cult. Dairy Prod. J. 29: 12–16.

van Zijl, M.M. and P.M. Klapwijk. 2000. Yellow fat products (butter, margarine, dairy and non-dairy spreads). pp. 784–806. *In:* B.M. Lund, T.C. Baird-Parker and G.W. Gould [eds.]. The Microbiological Safety and Quality of Food. Vol. 1. Aspen Publisher, Gaithersburg, Maryland, USA.

Varnam, A.H. and J.P. Sutherland. 1994. Milk and Milk Products. Chapman & Hall, New York, NY, USA.

Velazquez-Estrada, R.M., M.M. Hernández-Herrero, T.J. López-Pedemonte, W.J. Briñez-Zambrano, B. Guamis-López and A.X. Roig-Sagués. 2010. Inactivation of *Listeria monocytogenes* and *Salmonella enterica* serovar. Senftenberg 775W inoculated into fruit juice by means of ultra high pressure homogenization. Food Cont. 22: 313–317.

Voysey, P.A., P.A. Anslow, K.J. Bridgwater, B. Lavender and L. Watson. 2009. The effects of butter characteristics on the growth of *Listeria monocytogenes*. Int. J. Dairy Technol. 62: 326–330.

Walker, R.L., L.H. Jensen, H. Kinde, A.V. Alexander and L.S. Owens. 1991. Environmental survey for *Listeria* species in frozen milk product plants in California. J. Food Protect. 54: 178–182.

Warke, R., A. Kamat, M. Kamat and P. Thomas. 2000. Incidence of pathogenic psychrotroph in ice cream sold in some retail in Mumbai, India. Food Cont. 11: 77–83.

Whistler, R.L. and E.F. Paschall. 1967. Starch: Chemistry and Technology. Vol. II. Academic Press, New York, NY, USA.

Wilbey, R.A. 2002. Microbiology of cream and butter. pp. 123–174. *In:* R.K. Robinson [ed.]. Dairy Microbiology Handbook: The Microbiology of Milk and Milk Products. John Wiley Inc., New York, NY, USA.

Zhao, T., M.P. Doyle and D.E. Berg. 2000. Fate of *Campylobacter* jejuni in butter. J. Food Protect. 63: 120–122.

Internet sources

CDC. Multistate foodborne outbreak investigations. Available at http://www.cdc.gov/outbreaknet/outbreaks.html. Accessed on April 30, 2013.

European Communicable Disease Bulletin. 2008. www.Eurosurveillance.org/Public/Articles/Archives.aspx.

Eurostat. 2013. http://epp.eurostat.ec.europa.eu/portal/page/portal/eurostat/home/. Accessed on June 15, 2013.

FAO. FAOSTATS nutrition data: food balance sheet. Available at http://faostat3.fao.org/home/indexhtml. Accessed on May 1, 2013.

FDA Enforcement Report. www.fda.gov/Safety/Recalls/EnformenetReports/defaut.html. Accessed on May 15, 2013.

Outbreakdatabase. Foodborne illness outbreak database. Available at http://www.outbreakdata.com. Accessed on April 30, 2013.

The Canadian Dairy Commission. *Per capita* consumption of cheese and butter. Available at www.cdc-ccl.gc.ca/CDC/index-eng.php?id=3800. Accessed on April 4, 2013.

The Dairy Council. The Nutritional Composition of Butter and Dairy Spreads. Available at www.milk.co.uk/resources/. Accessed on April 30, 2013.

CHAPTER 11

Microbiology of Evaporated, Condensed and Powdered Milk

*Ayse Demet Karaman** and *Valente B. Alvarez*

11.1 Evaporated milk

11.1.1 Product description and properties

Evaporated milk is the commercial name of sterilized, unsweetened condensed cow's milk from which a considerable portion of the water has been removed (Nieuwenhuijse 2011a). Thus, evaporated milk is a product with high solids content and its processing includes standardization, homogenization, concentration and sterilization. The FDA Standard of Identity 21 131.130 for evaporated milk specifies a minimum of 6.5% milk fat, 16.5% milk solids-non-fat, 23% total milk solids and indicates a specified quantity of added Vitamin D or Vitamin A per fluid ounce. Similar milk products are evaporated skimmed milk, evaporated low-fat milk, evaporated filled milk, and evaporated goat's milk (FDA 2011a). However, the Codex Alimentarius standard requires evaporated milk to have at least 7.5% milk fat, 25% total milk solids and 34% milk protein solids-not-fat. The Codex also has specifications for similar products including evaporated skimmed milk, evaporated partly skimmed milk, and evaporated high fat milk (Codex 2010a).

Evaporated milk products have a long shelf-life and they can be kept for several months without refrigeration, even at tropical temperatures. The 'use by' date of evaporated milk may be as long as 18 months after production, although 6–12 months of shelf-life is more common (Nieuwenhuijse 2011a). Evaporated milk is produced by removing about 60% of the water from whole milk. Evaporated milk also contains added

*Corresponding author

salts that are necessary to stabilize and maintain the optimum viscosity of the product after sterilization. Salts such as calcium chloride or sodium carbonates, phosphates, or citrates may be added. A thickening agent such as carrageenan can be added at a low concentration, e.g., 0.015%, to slow down creaming (Walstra et al. 1999). The fat source is normally milk fat but in some countries producers are allowed to make evaporated 'filled milk', where fats from other sources are used to replace milk fat (Fernandes 2009). Palm oil is used most frequently to replace milk fat for cost reduction purposes or local preferences. Fat stability to oxidation is of utmost importance when using edible oils as a milk fat substitute (Nieuwenhuijse 2011a).

Most evaporated milk is sold directly to consumers who use it in home cooking (Fernandes 2009). However, it can be used as fluid milk after reconstituting it with water. After dilution, flavor and nutritive value of the product are not greatly different from that of fresh milk (Walstra et al. 1999). In some countries with low milk production, especially in the tropics, evaporated milk is still a general-use milk product (Rowe and Donaghy 2011). In other markets, evaporated milk is used for specific purposes. For example, when it is used to whiten coffee or tea, evaporated milk is consumed after 1:1 dilution with (boiled) water (Nieuwenhuijse 2011a).

11.1.2 Technology of evaporated milk

The typical processing scheme to manufacture evaporated milk requires high-quality, fresh whole milk. Upon receiving, raw milk is tested and accepted, and then undergoes preliminary treatments such as clarification, fat separation or milk standardization. Milk is usually concentrated by vacuum evaporation at 45–60°C. Concentration can also be performed using reverse osmosis but it is not common. Subsequently, homogenization is done at two pressure stages: 15–25 MPa and 5–10 MPa at 65°C (Oliveira et al. 2009). Table 11.1 shows the processing conditions and the preferred ultra high temperature (UHT) treatment at 115–118°C for 1–6 min. However, a longer heat treatment (e.g., below 100°C for 20 min) could be applied.

Metal cans are still the traditional container for evaporated milk in sizes of 170 g (6 oz) or 411 g (14.5 oz). However, in some Western European countries, glass bottles with a twist-off cap are used for packaging evaporated milk that is used to whiten coffee (Nieuwenhuijse 2011a). Glass bottles can be sterilized batch wise in an autoclave or continuously. Glass bottles are thermally processed at 100–120°C over 15–20 min or at 140°C for 3 s. In other instances, concentrated milk could be sterilized by UHT treatment as an alternative process (Oliveira et al. 2009). Various types of aseptic packaging systems such as aluminum foil-lined milk cartons and aluminum or polystyrene portion cups can also be used, as long as the above mentioned criteria are fulfilled. Sterilized evaporated milk is stored

Table 11.1 Process parameters and microbiological incidence of evaporated, condensed and powdered milk.

Product	Processing conditions	Incidence of microorganisms	Microbial effects	References
Evaporated milk	Vacuum evaporation at 45–60°C Homogenization at 15–25 MPa and 5–10 MPa UHT processing at 115–118°C for 1–6 min Storage at 20°C	*Bacillus licheniformis* *B. coagulans*, *B. marcerans*, *B. subtilis*, *Geobacillus stearothermophilus*	Spoilage problems	Oliveira et al. (2009) Nieuwenhuijse (2011a)
		B. subtilis, *B. megaterium*	Brownish liquid with a bitter taste	Fernandes (2009)
		Clostridium spp.	Blowing in cans	
Condensed milk	HTST pasteurization at 71.6°C for 15 s or UHT processed at 138°C >4 s or 82.2°C–93.3°C for 15 min Concentration at 54.4–57.2°C	*Candida* spp.	Spoilage	Fernandes (2009) Oliveira et al. (2009) Modi (2009) Robinson and Itsaranuwat (2002)
		Aspergillus repens *A. glaucus*	Colored lumps Off-flavor	
Sweetened condensed milk	Homogenization at 2–5 MPa at 70°C Concentrated sucrose solution addition (62.5–64.5%) Vacuum evaporation up to 80°C Cooling at 30–32°C Lactose crystals addition (0.05%) Cooling at 15–18°C Resting for 12–14 hr and crystallization			

Table 11.1 contd....

Table 11.1 contd.

Product	Processing conditions	Incidence of microorganisms	Microbial effects	References
Dried milk	Heating levels: Low (74°C for 30 s) Medium (80–100°C for 1–2 min) High (120–135°C for 1 to 2 min)	*Pseudomonas* spp. *Bacillus* and *Clostridium* species *E. faecalis* *S. thermophilus*, *Micrococcus* spp.	Degradation in milk fat and protein	Fernandes (2009) Rowe and Donaghy (2011) Skanderby et al. (2009) Walstra et al. (1999)
	Homogenization Heating (85–95°C for several min) Evaporation to 50% total solids Drying	Food-borne pathogens such as *Salmonella* spp., (*S. worthington*) *S. aureus*, *Bacillus* spp. (*B. cereus, B. licheniformis*) *Cronobacter* spp. (*C. sakazakii*) *Clostridium* spp. (*C. botulinum*) *Enterobacter* spp. *L. monocytogenes* *A. flavithermus*	Food poisoning outbreaks	Fernandes (2009) Lepoutre et al. (2005) Asao et al. (2003) FAO/WHO (2007) Ruckert et al. (2004) Brett et al. (2005)

at 20°C. However, when stored at ambient temperatures, and especially above 25°C, Maillard reactions may occur leading to undesired color and flavor changes (Oliveira et al. 2009, Nieuwenhuijse 2011a).

11.1.3 Microbiology of evaporated milk

Processing operations influence the microflora of evaporated milk. The initial centrifugal clarification stage removes some bacteria, principally heavy spores (Rowe and Donaghy 2011). Pre-heating temperatures of 120–122°C for several minutes are applied to denature whey proteins, but these temperatures also destroy all of the non-sporeforming bacteria and many of the less resistant sporulating types. The temperatures employed are normally too high to allow the growth of even thermophilic bacteria and only a few bacterial spores are likely to survive such a process (Robinson and Itsaranuwat 2002, Fernandes 2009). During the condensing operation, whether it is batch or continuous, the temperature seldom exceeds 54.5°C, and it will not kill those bacteria that may have survived preheating.

After condensing, the product is homogenized and if the usual food safety precautions during homogenization are observed, contamination from this processing step is not a problem. The product is then cooled and placed in storage, where the final standardization of composition takes place. It is important to ensure that cooling is rapid and sufficient to minimize any microbial growth during procedures. Holding the product under appropriate refrigeration, until it is packed and sterilized, is essential because the product is not sterile and can spoil if microbial growth occurs (Robinson and Itsaranuwat 2002). The sterilization process should kill all vegetative organisms and spores, although spoilage problems may arise due to the presence of *Bacillus licheniformis*, *B. coagulans*, *B. marcerans*, *B. subtilis* and especially *Geobacillus stearothermophilus* that are likely to survive the process. The spores of these microorganisms do not germinate unless the cans are stored at high ambient temperatures (>43°C). Aseptically filled evaporated milk must be processed following a strict hygiene procedure equivalent to those used in filling UHT milk products to avoid the possibility of re-contamination during filling (Fernandes 2009).

Treating evaporated milk with a time-temperature combination that would provide 'absolute sterility' is possible, but such processing conditions would produce an unacceptable level of 'cooked' flavor, a dark color, and probably some modified physical characteristics. Therefore, evaporated milk is processed at the lowest temperature-time conditions to achieve 'commercial sterility' that reduces the change levels of flavor and physical characteristics. If processing and packaging are performed properly, then spoilage is most likely to be the result of the action of heat-stable extracellular enzymes derived from psychrotrophic bacteria growing in the original raw milk (Rowe and Donaghy 2011). Also, small numbers of non-pathogenic spores occasionally may survive the heat treatment and subsequently germinate (Modi 2009).

There is some risk of post-contamination of evaporated milk after sterilization due to microleaks in cans or during aseptic filling. Some processing equipment, especially evaporators, may promote the growth of thermophilic sporeforming bacteria such as *G. stearothermophilus*. Many evaporators contain large, stainless steel surfaces in sections where the product temperature is 45–65°C that may allow the growth and sporulation of bacteria. If the running time is long or cleaning of the equipment is inadequate, it may lead to an appreciable spore count in the concentrated milk. Residual spores can germinate and grow if the product is stored at temperatures above 40°C, as may occur in tropical countries (Nieuwenhuijse 2011a).

Spoilage of commercially sterile canned evaporated milk is uncommon, but it can happen and it is the result of either under-processing or post-process contamination as discussed earlier. Bacteria that survive the thermal processing can cause acid coagulation, a slight cheesy odor and flavor, and 'flat sour' spoilage of the milk. However, many strains are obligate thermophilic and are only a problem at elevated storage temperatures or if cooling was too slow. The presence of *B. subtilis* produces a non-acid curd that develops into a brownish liquid with a bitter taste. A similar problem is the survival of *B. megaterium* which is responsible for the formation of a coagulum, accompanied by cheesy odor and gas. Blown cans occur very occasionally and are associated with putrefactive spoilage caused by *Clostridium* spp. All these incidences of microorganisms are also shown in Table 11.1.

11.2 Condensed and sweetened condensed milk

11.2.1 Product description and properties

Most condensed milk is made by evaporation, and the degree of concentration is usually within the range (milk:water) of 2.5:1 to 4:1, depending on the usage (Fernandes 2009). Condensed milk may be made from whole, skimmed, or reduced fat milks, depending on the end use. Due to its composition and properties, condensed milk is an important source of milk solids in confectionary, bakery products, ice cream, concentrated yogurt and other products. When it is manufactured for these purposes, it is bulk packaged in large quantities. The typical composition of the condensed milk is a minimum of 7.5% fat and a minimum of 25% total milk solids. The FDA Standard of Identity 21 131.115 and 120 specifies that the milk fat and total milk solids of the concentrated milk are not less than 7.5 and

25.5%, respectively. The Council Directive 2001/114/EC (EU 2001) and the Condensed Milk and Dried Milk Regulation in England (FSA 2003) have specifications for different condensed products (condensed high-fat milk, condensed milk, condensed partly skimmed milk and condensed skimmed milk) and for sweetened condensed milk types (sweetened condensed milk, sweetened condensed partly skimmed milk and sweetened condensed skimmed milk).

Since no sterilization process is involved at any stage of manufacturing, the final product is not sterile. Although the water activity (a_w) is reduced, it is not sufficient to completely inhibit microbial growth. Thus, these products must be refrigerated as quickly as possible and refrigeration maintained during transport to their destination (Rowe and Donaghy 2011). The shelf-life varies from a few days to weeks for condensed milk, depending on the degree of contamination, the severity of the heat treatment applied, and the effectiveness of temperature control during cooling and storage (Fernandes 2009).

Sugar can be added to condensed milk to make what is known as sweetened condensed milk. The finished product may be made bulk as a food ingredient or in small cans or tubes for retail sale to the consumer. After evaporation, sufficient sugar is added, usually as sucrose or glucose, to prevent most microbial growth (Fernandes 2009). Sweetened condensed milk specifies a minimum of 8% fat and 28% total milk-solids (Chandan 2011, FDA 2011b,c). When it is made from skim milk, it contains at least 0.5% fat and 24% milk solids. The primary difference between condensed and sweetened condensed milks is the sugar content. Sweetened condensed milk is preserved by the addition of sugar, which increases the osmotic pressure to a point that inhibits the growth of most microorganisms. Therefore, sugar addition is primarily responsible for maintaining the quality of the product and for its fairly long shelf-life when compared to condensed milk. The increased milk solids content also increases the osmotic pressure (Modi 2009). The product is packed in hermetically sealed metal containers for retail trade, and milk cans, barrels, steel drums or bulk tanks for industrial purposes (Fernandes 2009). Sweetened condensed milk has multiple uses including cooking, confectionary chocolate bars, and to enrich tea or coffee. It is also used as a milk drink after dilution (Rowe and Donaghy 2011).

11.2.2 Technology of condensed and sweetened condensed milk

In the manufacturing of bulk condensed milk, milk is separated and standardization of fat content is often carried out after concentration, if necessary. Unless skimmed milk is used, homogenization is also usually carried out at this stage. To manufacture condensed milk, a pre-heating process is applied using a continuous heater or a 'hot well'. After pre-heating, milk can be HTST pasteurized at 71.6°C (161°F) for 15 s or UHT processed at 138°C for >4 s. Both thermal processes decrease the bacteria level and make the milk more stable. Also, higher temperatures of 82.2–93.3°C for as much as 15 min may be applied to obtain a higher viscosity with desirable characteristics. The preheated and processed milk is then concentrated in a vacuum-pan or in a multiple-effect evaporator at a temperature range of 54.4–57.2°C (Fernandes 2009). Products processed under these conditions are not considered to be commercially sterile and they are immediately cooled and continuously held at temperatures below 7°C (Modi 2009). Creaming and coalescence do not often occur and homogenization is not always done. However, when producing a low viscous sweetened condensed milk, homogenization is required and is performed at 2–6 MPa at 70°C. Concentrated sucrose solution (62.5–64.5%) is heat treated and added to the concentrated milk at the end of the evaporation stage. Milk is usually concentrated by vacuum evaporation at relatively high temperatures up to 80°C. Then, cooling is performed to 30–32°C and the product is transferred to a tank where fine lactose crystals are added (0.05%). The product should be vigorously agitated for a period of 1 hr. Condensed milk must be cooled to a temperature at which the lactose is supersaturated so that the lactose seeds do not dissolve. Cooling should be at 15–18°C. The product is kept in the tank for 12–14 hr so that the crystallization is completed (Table 11.1) (Oliveira et al. 2009). Product for retail sale is mainly packed in tinned cans. Small (20–40 g) portion packs made of aluminum foil and plastic laminate, metal or plastic tubes, and aluminum-lined 'milk' cartons are also commonly used (Nieuwenhuijse 2011b).

11.2.3 Microbiology of condensed and sweetened condensed milk

Milk used in the manufacturing of condensed milk has many microbial hazards. Since the product is not sterilized during or after condensing and the concentration of milk solids is not high enough to inhibit microbial development, condensed milk may contain a number of viable microorganisms.

Sometimes the equipment and the heater, but primarily the 'hot well' or surge tanks can serve as incubators for thermophilic bacteria under the processing conditions for condensed milk. Furthermore, when such equipment is operated for long periods without intermediate clean-up, or when the milk supply contains excessive numbers of thermophilic bacteria, the numbers may build up to a point where acid and unclean flavors result. Therefore, proper sanitation and control in the preceding phases of the operation are essential (Robinson and Itsarauwat 2002). The preheating conditions of milk and the evaporator temperatures between 54.4°C and 57.2°C are very suitable for the growth of thermophilic bacteria. Because the product does contain microorganisms that have survived processing, and usually contains microorganisms that have contaminated the product subsequent to heat treatment, the time over which the product can be held without serious microbial spoilage is frequently very limited, especially at ambient temperatures (Robinson and Itsaranuwat 2002). Psychrotrophic bacteria, yeasts, or molds may cause spoilage if the product is held for long periods or under improper storage conditions (Modi 2009). Deterioration usually occurs by osmophilic yeasts, most of which belong to the genus *Candida* spp. The yeasts do not start growth easily, especially if the sugar concentration is high. Additionally, some micrococci may grow in sweetened condensed milk, though slowly, especially if water activity and temperature are high. Similarly, some strains of molds such as *Aspergillus repens* and *A. glaucus* can grow as long as oxygen is present. These microorganisms produce fairly firm colored lumps and off-flavor development in the product (Table 11.1).

The natural enzymes of milk are also inactivated during processing. However, the proteolytic and lipolytic enzymes resulting from excessive microbial growth may not be affected and may cause problems in the finished product. Heat stable proteolytic and lipolytic enzymes elaborated by psychrotrophic bacteria in the raw milk can also cause spoilage (Walstra et al. 1999, Rowe and Donaghy 2011).

Sugar is normally an unimportant source of microorganisms, but when it is made under unsanitary conditions, it might be contaminated with mold spores, osmophilic yeasts, or bacteria that will produce acid and gas. The sugar may be added to the forewarmed milk prior to entry to the vacuum pan, or late in the condensing operation as a solution at about 65% (w/v) concentration. Addition of sugar before forewarming does reduce the microbiological effectiveness of the heating, but addition to the forewarmed milk may contribute to age thickening, so the latter procedure is used primarily for bulk product to be used quickly (Robinson and Itsaranuwat 2002).

Sweetened condensed milk is normally packed in cans that are first sterilized by flaming. The cans should be filled accurately to avoid any air in the head space, as this could permit the growth of osmophilic yeast (Oliveria et al. 2009).

11.3 Dried milk

11.3.1 Product description and properties

Milk can be preserved by dehydrating it to various degrees. The dehydrated milk has a longer shelf-life than milk preserved by pasteurization and refrigeration. Milk from which all water has been withdrawn is called dried milk or milk powder (Modi 2009). Dried milk can be found commercially as full fat, reduced fat and skim-milk powder. Dried milk powders are used in dairy products, confectionary, infant formulas, and manufactured-food industries. They are also used in coatings, soups, sauces, and ready-to-eat meals (Rowe and Donaghy 2011). The function of milk powder in these products is related to the components of the powder, primarily protein, fat and lactose (Deeb et al. 2010).

The FDA Standard of Identity 21 131.127 for non-fat dry milk fortified with Vitamins A and D specifies that each quart of the reconstituted product should contain 2000 and 400 International Units (IU) of Vitamin A and D, respectively. Additionally, dry whole milk contains 26% or more but less than 40% by weight of milk fat and it contains not more than 5% by weight of moisture excluding fats on a milk solids-not-fat basis (FDA 2011d,e). The Codex Standard for Milk Powders and Cream Powder specifies in detail the internationally agreed definitions for the origin, composition and permitted additives for milk powder (Codex 2010b). The standard specifies that only milk and cream are permitted as raw materials for milk powders. However, the protein content of the powder may be adjusted by addition of skimmed or partially skimmed milk or by addition of the retentate or permeate obtained from the treatment of milk by ultrafiltration (UF). According to this standard, whole milk powder should have 34% milk protein in milk solids-non-fat and milk fat content between 26–42%. Partly skimmed milk should have the same protein level as whole milk powder but different fat levels of 1.5–26%. Skimmed milk powder requires the same protein levels as the previous products but a maximum of 1.5% fat level. High fat content in whole milk powder (WMP) makes it very susceptible to oxidation, which can result in off-flavors. The US industry standard for shelf-life of WMP is 6 to 9 mon under storage at <27°C and <65% relative humidity, although recent research demonstrated flavor changes by 3 mon when stored at ambient temperature (Llyod et al. 2009, Carunchia-Whetstine and Drake 2007).

11.3.2 Technology of dried milk

The steps in milk powder production are milk clarification, cooling, standardization, evaporation, homogenization (optional), drying, and packaging (Table 11.1). Milk is heated and then concentrated before evaporation and drying. Skimmed milk may also be subjected to low, medium and high heat processes to give varying degrees of protein denaturation as required. Typically, a low heat process is 74°C for 30 s, a medium heat process 80–100°C for 1–2 min, and a high heat process at 120–135°C for 1 to 2 min. The high heat process may be equivalent to an UHT treatment. Milk is heated at 85–95°C for several minutes to manufacture dried whole milk powders. The heat-treated milk is then evaporated to about 50% total solids before drying (Fernandes 2009, Rowe and Donaghy 2011).

The milk stream is generally thermally concentrated but it can also be concentrated through membranes (ultrafiltration or infiltration). Residence time and temperature exposure for the thermal process are generally less than 60 s and lower than 72°C, respectively. This thermal concentration process should contribute to further enhance the microbiological quality of the milk powder. In some evaporation plants the milk stream is heated at temperatures between 140 and 150°C that generates a product with high microbiological quality. Dried milk products made under these conditions are used as ingredients in baby food manufacturing (Rowe and Donaghy 2011).

Dry milks have been produced by basically two processes, roller (drum) and spray dry techniques. During roller processing, two large rollers, usually heated internally and located adjacent and parallel to each other, revolve at a desired speed. The rollers revolve in opposite directions, contacting a reservoir of either pasteurized fluid or condensed milk. During rotation, the fluid milk product dries on the hot roller surface. After approximately ¾ of a revolution, a carefully positioned, sharp stationary knife detaches the milk product, now in the form of a thin dry sheet. Then, milk is conveyed by an auger to a hammer mill where it undergoes a physical treatment to convert it into uniformly fine particles which are packaged, usually in 50 or 100 lb bulk packages (Fernandes 2009). There are two systems of spray drying that are used to make milk powder: jet or nozzle dryers and rotary atomizer dryers. Each of these dryers produces powders with specific characteristics. In the case of jet dryer systems, milk is fed to a high-pressure pump, and then pumped under pressure to a series of jets or nozzles within the drying chamber. The impact of hot air on the milk causes the liquid to break up into fine droplets that form milk powder (Modi 2009). In the rotary atomizer, the milk enters the drying chamber through slots or holes located at the periphery of a circular disc fixed to a rotating shaft. Modern spray dryers usually consist of several stages and incorporate a final drying stage using a fluidized bed dryer. Spray dryers operate by mixing pre-heated atomized milk droplets with heated air at an inlet temperature of 180–220°C. The air is cooled to an exit temperature of 70–95°C, and moisture is removed from the milk. This gives very rapid drying at a vaporization temperature of about 50°C, and the temperature of the dried particles remains below that of the cooled air. The dried particles are then removed, cooled, and packaged (Fernandes 2009, Modi 2009). The powder may be packed into multilayer paper bags with a polyethylene inner layer. Whole milk powder, however, is often packed in tins or in plastic containers to minimize oxygen uptake (Walstra et al. 1999).

11.3.3 Microbiology of dried milk

The microbial quality of milk powder is determined by the quality of the raw material, the nature and extent of processing and the extent of post-production contamination (Skanderby et al. 2009). Most of the bacteria and food-borne disease outbreaks associated with dried milk products have usually been the result of post-pasteurization contamination by pathogens. However, there have been cases that microorganisms originate in raw milk and they spread through the plant during processing (Fernandes 2009). Therefore, it is important to have low microbiological loads in raw milk in order to obtain a high quality finished product in milk powdered plants (Oliveira et al. 2000).

The psychrotrophic bacteria found in cold stored milk comprise mainly of *Pseudomonas* spp. and, although not pathogenic in nature, they commonly produce extracellular lipase and protease that degrade the milk fat and protein, respectively. Although most psychrotrophic bacteria are readily destroyed by pasteurization, the extracellular enzymes they produce are remarkably heat-resistant and cause problems when they are present in the powder that is incorporated in finished goods. Counts of psychrotrophic organisms in milk not exceeding 10^6 cfu ml^{-1} at the point of heat treatment to manufacture milk powder can minimize the problem of high resistant enzymes in the finished product (Skanderby et al. 2009).

The microorganism contamination can also occur during processing and manufacturing of dry milk. The occurrence can take place when milk is concentrated in an evaporator before spray drying at a temperature range of 45–55°C. Some heat-resistant bacteria can grow under these conditions and can increase disproportionally the bacterial count of the concentrate. However, pasteurization of milk at 72°C for 15 s directly before the evaporation reduces the bacterial count in the final powder. Thus, the higher the temperature and the longer the holding time, the more efficient is the process to inactivate microorganisms. It has been found that sporeforming thermophilic bacteria such as *Bacillus* and *Clostridium* species in milk powder can increase exponentially after only 12–16 hr of operation of a plant. As illustrated in Table 11.1, other bacteria that can be present in milk powder are heat-resistant *Enterococcus faecalis, Lactobacillus brevis, Streptococcus thermophilus* and *Micrococcus* spp. These bacteria may grow slowly at refrigeration temperature. Therefore, their presence at high counts is indicative of deficiencies in the cold storage chain.

Powder is produced by atomizing the concentrated milk in a stream of very hot air. However, during the evaporative cooling process of dry milk, the temperature of the atomized milk droplet remains low and may favor microbial growth. Consequently, the bacterial load of the concentrate largely would determine that of the powder. Also, the heat-resistance of the bacteria increases significantly with the dry matter content. A study reported that about 70% of *E. faecalis* and *E. faecium* survived during drying, whereas survival of *Staphylococcus aureus* varied widely. Approximately 10^{-4} to 10^{-5} of the initial count of *Salmonella* spp. and *Escherichia coli* survived the drying process (Walstra et al. 1999). Licari and Potter (1970) demonstrated that *Salmonella* spp. are not completely eradicated from milk powder by spray drying and were only reduced slightly but not eliminated during storage in milk powder at 25°C and 35°C. Another condition that can deteriorate the microbial quality of milk powder is the environmental contamination from air or surfaces. Damaged surfaces of equipment such as cracks in spray dryer walls can become reservoirs of bacteria that may develop within the insulating material of the spray dryer. Such reservoirs of bacteria are resistant to normal cleaning and disinfection procedures and can harbor pathogenic or spoilage microorganisms (Skanderby et al. 2009).

The principal food-borne pathogens associated with dried milk powders that have caused significant outbreaks are *Salmonella* spp., *S. aureus, Bacillus* spp., *Cronobacter sakazakii* and *Clostridium* spp. Among them, salmonella contamination is regarded as a serious potential hazard in these products in several countries including the USA and UK. A nationwide outbreak in the USA was associated with non-fat milk powder produced in one plant between 1964 and 1965. The problem was caused by a widespread contamination in both finished product and environmental samples. This problem made necessary significant improvements in the hygiene, sanitation and process control programs in the plant. Another outbreak with non-fat dried milk contaminated with *Salmonella* Typhimurium and *S. agona* occurred in Oregon in 1979. There was a major salmonellosis outbreak associated with dried milk powder-based infant formula that was contaminated with *S. ealing* in the UK in 1986. The problem was linked to raw milk that was contaminated and then spread through the plant (Fernandes 2009). A similar outbreak that involved elderly patients was caused by the consumption of contaminated milk powder with serotype *Salmonella worthington* in France in 2005 (Lepoutre et al. 2005). Salmonella outbreaks have also been traced to milk powder worldwide including countries such as Spain, Korea and Australia (Forsyth et al. 2003, Park et al. 2004, Soler et al. 2008).

Contamination of dried milk powders with staphylococcal enterotoxins was a significant problem in the 1950s. Several outbreaks were caused by staphylococci food-poisoning due to ingestion of dried milk powder. The outbreaks were reported in England in 1953, Japan in 1955, and Puerto Rico in 1956 (Rowe and Donaghy 2011). A recent large outbreak of staphylococcal food-poisoning that affected over 13,420 people was reported in Japan in 2000. Consumption of semi-skimmed liquid milk made from skimmed powder that was contaminated with heat-stable toxin from *S. aureus* was the cause of the outbreak (Asao et al. 2003).

There are other types of bacteria that can be found in dried milk products. *Cronobacter* and *Enterobacter* spp. are not normally regarded as food-borne pathogens; however, there have been a number of sporadic outbreaks of neonatal meningitis that were caused by *C. sakazakii* in several countries. The outbreaks were associated with the consumption of dried milk. The incidence rate of this microorganism in powdered infant milk formula was reported to range from 3% up to 25% and the fatality rate was high (FAO/WHO 2007). The causes of these incidences in most cases are related to gross temperature abuse, or poor hygiene in the manufacturing process or during preparation of the formulas (Iversen and Forsythe 2003). The most prevalent sources of contamination are external roofs above spray driers, air in treatment areas and where high foot traffic occurs (Craven et al. 2010).

The presence of *B. cereus*, other *Bacillus* spp., *Clostridium botulinum* and *Listeria monocytogenes* in dried milk products, especially infant formula is also a concern (Rowe and Donaghy 2011). Various levels of *B. cereus* contamination in dried whole milk and non-fat dry milk have been reported in USA (62.5%), Germany (13–43%), UK (17%) and Brazil (28%). Twenty-eight milk powders from 18 different countries were examined for the number and type of contaminating thermophilic bacilli. The dominant isolate was *Anoxybacillus flavithermus* strain C followed by *B. licheniformis* strain F (Rueckert et al. 2005). Infant botulism that results from the ingestion of spores of *C. botulinum* is rare; however, there was a case in the UK where an infant formulated milk powder was the possible cause (Brett et al. 2005).

Doyle et al. (1985) studied survival of *L. monocytogenes* during the manufacture and storage of non-fat dry milk using inoculated milk that was not pasteurized before drying. Reductions of 1 to 1.5 $\log_{10}$ *L. monocytogenes* g^{-1} occurred during the spray drying process and the organism progressively died during storage. Although the incidence of listeriosis is very low, less than 10 cases per million individuals, the mortality rate is very high (20–40%) (Rola et al. 2006). All incidences of microorganisms in dried milk are given in Table 11.1.

Aflatoxin M1 has occasionally been found in dried milk and, if present, significant amounts of them are able to withstand the processing and storage conditions of finished product for long periods (Fernandes 2009).

FDA Standard of Identity 21 does not have microbial specifications for evaporated, condensed and powdered milk. However, the US Standards for Grades of non-fat dry milk (spray process) specifies a maximum standard plate count of 10,000 per gram for U.S. extra grade and 75,000 per gram for U.S. standard grade. For instant non-fat dry milk the maximum standard plate count is 10,000 per gram and a maximum coliform count of 10 per gram. The Codex standard for evaporated milks, sweetened condensed milks, and milk powders and cream powder (Codex 2010a,b,c) specifies that products should comply with any microbiological criteria established in accordance with the Principles for the Establishment and Application of Microbiological Criteria for Foods (CAC/GL 21-1997). Although there are certain microorganisms (e.g., *C. perfringens, S. aureus* and *Vibrio parahaemolyticus*) that can cause food-borne illness, this regulation does not specify maximum or minimum limits for these types of microorganisms (Codex 2010d).

References

Asao, T., Y. Kumeda, T. Kawaaii, T. Shibata, H. Oda, K. Haruki, H. Nakazawa and S. Kozaki. 2003. An extensive outbreak of staphylococcal food poisoning due to low-fat milk in Japan: Estimation of Enterotoxin A in the incriminated milk and powdered skim milk. Epidemiol. Infect. 130: 33–40.

Brett, M.M., J. McLauchlin, A. Harris, S. O'Brien, N. Black, R.J. Forsyth, D. Roberts and F.J. Bolton. 2005. A case of infant botulism with a possible link to infant formula milk powder: Evidence for the presence of more than one strain of *Clostridium botulinum* in clinical specimens and food. Medical Microbiol. 54: 769–776.

CAC/GL 21. 1997. Principles for the establishment and application of microbiological criteria for foods. http://www.fao.org/DOCREP/005/Y1579E/y1579e04.html.

Carunchia-Whetstine, M.E. and M.A. Drake. 2007. The flavor and flavor stability of skim and whole milk powders. pp. 217–251. *In:* K.R. Cadwallader, M.A. Drake and R. McGorrin [eds.]. Flavor of Dairy Products. ACS Publishing, Washington, DC, USA.

Chandan, R.C. 2011. Dairy ingredients for food processing: An overview. pp. 1–33. *In:* R.C. Chandan and A. Kilara [eds.]. Dairy Ingredients for Food Processing. Blackwell Publishing Ltd., Ames, Iowa, USA.

Codex. 2010a. Codex standard for evaporated milks. Codex Stan 281-1971. Formerly CODEX STAN A-3-1971. Adopted in 1971, Revision 1999, Amendment 2010.

Codex. 2010b. Codex standard for sweetened condensed milks. Codex Stan 282-1971. Adopted in 1971, Revision 1999, Amendment 2010.

Codex. 2010c. Codex standard for milk powders and cream powder. Codex Stan 207-1999. Adopted in 1999, Amendment 2010.

Codex. 2010d. Codex general standard for contaminants and toxins in food and feed. Adopted 1995, Revised 1997, 2006, 2008, 2009, Amended 2010.

Craven, H.M., C.M. McAuley, L.L. Duffy and N. Fegan. 2010. Distribution, prevalence and persistence of *Cronobacter* (*Enterobacter sakazakii*) in the nonprocessing and processing environments of five milk powder factories. J. Appl. Microbiol. 109: 1044–1052.

Deeb, A.M.M., I.I. Al-Hawary, I.M. Aman and D.M.H.A. Shahine. 2010. Bacteriological investigation on milk powder in the Egyptian market with emphasis on its safety. Global Vet. 4: 424–433.

Doyle, M.P., L.M. Meske and E.H. Marth. 1985. Survival of *Listeria monocytogenes* during the manufacture and storage of nonfat dry milk. J. Food Protect. 48: 740–742.

EU. 2001. Council Directive 2001/114/EC of 20 December 2001 relating to certain partly or wholly dehydrated preserved milk for human consumption. Official Journal of the European Commission L15: 19–23.

FAO/WHO (Food and Agriculture Organisation, World Health Organisation). 2007. *Enterobacter sakazakii* and *Salmonella* in powdered infant formula: Meeting report, Rome.

In: Microbiological Risk Assessment Series, No. 10. Food and Agriculture Organisation/ World Health Organisation, Geneva.

FDA. 2011a. The FDA Standard of Identify 21 131.130 for evaporated milk. Code of Federal Regulations, Title 21. Vol. 2. Revised as of April 1, 2011, CITE: 21CFR131.130. http://www.accessdata.fda.gov/scripts/cdrh/cfdocs/cfcfr/CFRSearch.cfm?fr=131.130.

FDA. 2011b. Concentrated milk. The FDA Standard of Identify 21 131.115. Code of Federal Regulations, Title 21. Vol. 2. Revised as of April 1, 2011, CITE: 21CFR131.115. http://www.accessdata.fda.gov/scripts/cdrh/cfdocs/cfcfr/CFRSearch.cfm?fr=131.115.

FDA. 2011c. Sweetened condensed milk. The FDA Standard of Identify 21 131.120. Code of Federal Regulations, Title 21. Vol. 2. Revised as of April 1, 2011, CITE: 21CFR131.120. http://www.accessdata.fda.gov/scripts/cdrh/cfdocs/cfcfr/CFRSearch.cfm?fr=131.120.

FDA. 2011d. Nonfat dry milk. The FDA Standard of Identify 21 131.127. Code of Federal Regulations, Title 21. Vol. 2. Revised as of April 1, 2011, CITE: 21CFR131.127. http://www.accessdata.fda.gov/scripts/cdrh/cfdocs/cfcfr/CFRSearch.cfm?fr=131.125.

FDA. 2011e. Nonfat dry milk fortified with vitamins A and D. The FDA Standard of Identify 21 131.127. Code of Federal Regulations, Title 21. Vol. 2. Revised as of April 1, 2011, CITE: 21CFR131.127. http://www.accessdata.fda.gov/scripts/cdrh/cfdocs/cfcfr/CFRSearch.cfm?fr=131.127.

Fernandes, R. 2009. Concentrated and Dried Milk Products. pp. 21–33. *In:* R. Fernandes [ed.]. Microbiology Handbook: Dairy Products. Leatherhead Food International Ltd., Leatherhead, Surrey, UK.

Forsyth, J.R., N.M. Bennett, S. Hogben, E.M. Hutchinson, G. Rouch, A. Tan and J. Taplin. 2003. The year of the Salmonella seekers-1977. Aus. N. Z. J. Public Health 27: 385–389.

FSA. 2003. Food Standards Agency. The condensed milk and dried milk regulations 2003 (as amended). Version 2, 12 pages. http://www.food.gov.uk/multimedia/pdfs/condensedguid03.pdf.

Iversen, C. and S. Forsythe. 2003. Risk profile of *Enterobacter sakazakii*, an emergent pathogen associated with infant milk formula. Trends Food Sci. Technol. 14: 443–454.

Lepoutre, A., E. Espie, V. Vailant, A. Guignard, S. Larrieu, M. Bordes-Couecou, E. Girou, A. Lemire, M. Crolard, C. Dummartin, I. Poujol, H. de Vlk, A.M. Navel, C. Rousseau, L. Filleul, V. Schwoebel and F.-X. Weill. 2005. Outbreak of *Salmonella worthington* infection in elderly people due to contaminated milk powder, France, January–July 2005. Eurosurveillance 10: 29.

Licari, J.J. and N.N. Potter. 1970. Salmonella survival during spray drying and subsequent handling of skim milk powder. III. Effects of storage temperature on Salmonella and dried milk properties. J. Dairy Sci. 53: 877–882.

Lloyd, M.A., S.J. Hess and M.A. Drake. 2009. Effect of nitrogen flushing and storage temperature on flavor and shelf-life of whole milk powder. J. Dairy Sci. 92: 2409–2422.

Modi, H.A. 2009. Concentrated milks, dried milks and dry whey products. pp. 45–61. *In:* H.A. Modi [ed.]. Dairy Microbiology. Aavishkar Publishers, Jaipur, India.

Nieuwenhuijse, J.A. 2011a. Evaporated milk. pp. 862–869. *In:* J.W. Fuquay, P.F. Fox and P.L.H. McSweeney [eds.]. Encyclopedia of Dairy Sciences. Elsevier Ltd., San Diego, CA, USA.

Nieuwenhuijse, J.A. 2011b. Sweetened condensed milk. pp. 869–873. *In:* J.W. Fuquay, P.F. Fox and P.L.H. McSweeney [eds.]. Encyclopedia of Dairy Sciences. Elsevier Ltd., San Diego, CA, USA.

Oliveira, C.A.F., L. Mestieri, M.V. Santos, J.F.G. Moreno, A. Spers and P.M.L. Germano. 2000. Effect of microbiological characteristics of raw milk on the quality of whole milk powder. Brazilian J. Microbiol. 31: 95–98.

Oliveira, M.N., A.L.B. Penna and G.H. Nevarez. 2009. Production of evaporated milk, sweetened condensed milk and 'Dulce de Leche'. pp. 149–179. *In:* A.Y. Tamime [ed.]. Dairy Powders and Concentrated Products. Blackwell Publishing Ltd., Oxford, UK.

Park, J.K., W.S. Seok, B.J. Choi, H.M. Kim, B.K. Lim, S.S. Yoon, S. Kim, Y.S. Kim and J.Y. Park. 2004. *Salmonella enterica serovar* London infections associated with consumption of infant formula. Yonsei Med. J. 45: 43–48.

Robinson, R.K. and P. Itsaranuwat. 2002. The microbiology of concentrated and dried milks. pp. 175–211. *In:* R.K. Robinson [ed.]. Dairy Microbiology Handbook. 3rd edition. John Wiley and Sons Inc., New York, NY, USA.
Rola, J.G., W. Korpysa-Dzirba and B. Wojton. 2006. Detection of *Listeria monocytogenes* in powdered milk by Vidas LMO 2-Collaborative study. J. Rapid Meth. Aut. Mic. 14: 362–368.
Rowe, M. and J. Donaghy. 2011. Microbiological aspects of dairy ingredients. pp. 59–103. *In:* R.C. Chandan and A. Kilara [eds.]. Dairy Ingredients for Food Processing. Blackwell Publishing Ltd., Ames, Iowa, USA.
Rueckert, A., R.S. Ronimus and H.W. Morgan. 2005. Development of a rapid detection and enumeration method for thermophilic bacilli in milk powders. J. Microbiol. Meth. 60: 155–167.
Skanderby, M., V. Westergaard, A. Partridge and D.D. Muir. 2009. Dried milk products. pp. 180–234. *In:* A.Y. Tamime [ed.]. Dairy Powders and Concentrated Products. Blackwell Publishing Ltd., Oxford, UK.
Soler, P., S. Herrera, J. Rodríguez, J. Cascante, R. Cabral, A. Echeita-Sarriondia and S. Mateo. 2008. Nationwide outbreak of *Salmonella enterica* serotype Kedougou infection in infants linked to infant formula milk, Spain. Eurosurveillance 13: 1–2.
Walstra, P., T.J. Geurts, A. Noomen, A. Jellema and M.A.J.S. van Boekel. 1999. Concentrated Milks. pp. 425–470. *In:* P. Walstra, T.J. Geurts, A. Noomen, A. Jellema and M.A.J.S. van Boekel [eds.]. Dairy Technology: Principles of Milk Properties and Processes. Marcel Dekker, Inc., New York, NY, USA.

CHAPTER 12

Functional Dairy Ingredients

Ana Raquel Madureira, Ana Gomes and *Manuela Pintado**

12.1 Introduction

The consumers at present are aware that choices on food will affect their health positively or negatively according to their composition and amount consumed. World food industry is conscious of this phenomenon, and during the last two decades it has invested all its efforts in research and development of healthier and more nutritious foods. Around the world, there are several types of products available on the markets named by the main academic and industrial research community as "functional foods". Nevertheless, on both sides of the Atlantic Ocean, there are different perspectives on this concept. In the USA, for example, the term of functional foods is replaced by "nutraceuticals". In general, these foods can be defined as *"foods/ingredients that specifically and actively contribute (have a function) to a state of good health"*.

The European Functional Food Science Programme, funded by the European Union (EU) and led by the International Life Sciences Institute (ILSI) states that *"A food can be regarded as functional if it is satisfactorily demonstrated to affect beneficially one or more target functions in the body beyond adequate nutritional effects, in a way that is relevant to either an improved state of health and well-being and/or a reduction of risk of disease"* and, two of the possible functions that are attributed, include physiological enhancement and reducing the risk of disease. In the USA, functional foods are defined as *"naturally derived bioactive compounds, including live active cultures, that have health-promoting, disease-preventing properties, and that can be delivered in a number of different ways"*. There are big differences between Europe and

*Corresponding author

the USA in terms of definitions, regulations, and approaches related to the development of functional foods. In the USA it is clearly distinguished the ingestion of a dietary supplement (vitamin, capsule, pill, etc.), from the whole food such as yogurt and cheese as functional food and the ingestion of food product, which was intentionally fortified with a nutraceutical (probiotic yogurt, calcium fortified milk, etc.). As this chapter will more focus on European legislation, the term of "functional food" will be used hereafter, although in some cases the US classes will also be highlighted.

The dairy industry is the largest producer of functional foods or ingredients, using milk as the main base for production of such products. The recognition that certain dairy components can contribute to functionality in foods was achieved by Japanese researchers in 1980s. In general, we can find products such as milk, cheese, whey, fermented milks, whey protein concentrates and isolates, with identified functionality related with their components such as proteins, peptides, lipids, carbohydrates, minerals and vitamins. Their biological activities are mainly related to the cardiovascular system (anti-hypertensive, anti-thrombotic, cholesterol reduction) and immune system, or exert a specific role, such as delaying osteoporosis, anti-microbial effect, improving gut microflora balance, and anti-cancer, anti-oxidant activities. Furthermore, milk is considered an excellent vector for other functional ingredients such as probiotics, prebiotics, vitamins and minerals, due to technological feasibility as well as conferring protective environments for their stability and functionality.

This chapter will focus on the most common and popular functional dairy products such as probiotic and prebiotic dairy foods as well as bioactive compounds extracted from whey, casein (caseinophosphopeptides) and other ingredients naturally found in dairy products such as conjugated linoleic acids, milk fat globule membrane and many others. Readers are recommended to refer to Chapter 9 for more information about development of probiotic dairy products.

12.2 Probiotic and prebiotic dairy foods

12.2.1 Probiotics

The word *probiotic* is derived from the Greek language—meaning *"for life"*, and was originally used by Kollath in 1953, to denote all organic and inorganic food complexes which are able to restore health of patients suffering from a form of malnutrition resulting from consuming too much highly refined foods—in contrast to (harmful) antibiotics. Among other definitions reported in the meantime, more recently, FAO/WHO has defined probiotics as *"live microorganisms which, when consumed in adequate amounts, confer a health benefit on the host"* (FAO/WHO 2001).

Most bacteria bearing probiotic features belong to the genera *Lactobacillus* or *Bifidobacterium*, which are common but non-dominant members of the indigenous microbiota of the gastro-intestinal tract (GIT) (Walter et al. 2001). They are mostly lactic acid bacteria (LAB) that ferment milk sugar (lactose) into lactic acid and/or acetic acid, and also originate flavor compounds such as acetaldehyde and diacetyl, besides CO_2 (Marshall 1987). However, in order to be considered as probiotics, microorganisms should simultaneously comply with the requirement of being safe for human use, i.e., qualified presumption of safety (QPS) concept used in Europe or generally recognized as safe (GRAS) concept used in the USA. In this regard, recent criteria published by the European Food Safety Authority (EFSA) ought to be used as a proper reference. In any case (and not withstanding any further criterion introduced by EFSA), probiotics should not be vectors of any acquired or transmissible antibiotic resistance and virulence factor; be active and vital in the GIT; be viable at high levels so as to justify a possible measurable beneficial effect; be able to live and multiply in the intestine; and be able to confer a physiological benefit.

There is a growing body of experimental evidence in support of the hypothesis that maintenance of a healthy gut microbiota prevents GIT infections, inflammatory bowel diseases and even cancer (Salminen et al. 2005, Oelschlaeger 2010). In Table 12.1, the main probiotic health effects of probiotics are presented.

Table 12.1 Health benefits associated with probiotic bacteria.

Benefits of probiotic bacteria
Protection against infection
Reduced incidence of diarrhea
Reduced levels of cold and influenza-like symptom in children
Anti-microbial activity
Competitive exclusion of pathogens
Immune tolerance
Reduction in colorectal cancer biomarkers
Return to pre-antibiotic baseline flora
Epithelial barrier function
Increased cellular immunity
Increased humoral response
Lowering of blood cholesterol levels
Reduction in irritable bowel disease symptoms
Delivery of therapeutics

Adapted from O'Flaherty and Klaenhammer (2010).

The health-promoting activities of probiotic bacteria have been gradually characterized and confirmed as deeper and more accurate research studies have meanwhile been carried out. Metchnikoff (1907) was the first who showed that fermented milks, especially yogurt, had a favorable impact on longevity owing to their LAB flora, which may reduce formation of toxic products in the colon (Rettger and Cheplin 1921). Later, LAB of yogurt were associated to the lactose fermentation, which made this dairy product consumable by people with lactose malabsorption (Kolars et al. 1984, Marteau et al. 1990).

Probiotic dairy products

During the last two decades, probiotic bacteria have been incorporated in a range of dairy products including yogurts, soft-, semi-hard and hard cheeses, ice cream, milk powders and frozen dairy desserts, as well as a few non-dairy products (Table 12.2). For successful incorporation of the probiotic culture in any food, typically viable numbers of a minimum of 10^6–10^7 cfu

Table 12.2 Food products fermented with probiotics.

Matrix	**Probiotic culture**
Fermented milks/yogurt	*Bifidobacterium* spp. *Lb. acidophilus, Lb. rhamnosus, P. jensenii*
Ice-cream	*Bifidobacterium* spp. *Lb. johnsonii*
Cheese	*Lactobacillus* spp., *Bifidobacterium* spp.
Soya frozen dessert	*Lb. acidophilus, Lb. paracasei,* *B. animalis* subsp. *lactis, Lb. rhamnosus, S. boulardii*
Soya milk	*B. breve*
Oat bars	*B. animalis* subsp. *lactis*
Maize or rice pudding	*B. animalis, Lb. acidophilus, Lb. rhamnosus*
Oat meal gruel mixed with fruit drink	*Lb. plantarum*
Dry sausages	*Lb. rhamnosus*
Dried fruits	*Lb. rhamnosus*
African beverages	*B. animalis* subsp. *lactis*
Beet (juice)	*Lb. plantarum, Lb. acidophilus, Lb. casei, Lb. delbrueckii*
Blackcurrant (juice)	*Lb. plantarum*
Tomato (juice)	*Lb. plantarum, Lb. acidophilus, Lb. casei, Lb. delbrueckii*
Orange (juice)	*Lb. rhamnosus*
Maple sap	*B. animalis* subsp. *lactis* and *Lb. rhamnosus*
Table olives	*Lb. rhamnosus, Lb. paracasei, B. bifidum, B. longum*
Artichokes	*Lb. plantarum, Lb. paracasei*

Adapted from Madureira et al. (2010a).

g^{-1} in the final product are accepted as the therapeutic minimum level that quantifies for probiotic cultures processed at the moment of consumption (Talwalkar et al. 2004)—which means that 10^8–10^9 cfu g^{-1} should typically be provided, and daily consumption of 100 g of cheese or other food product also should take place (Jayamanne and Adams 2006). Furthermore, the food at stake should maintain its characteristic sensory attributes; otherwise the final product will hardly be marketable. Nevertheless, the existence of products containing non-viable probiotic strains is also a reality and one of the products available on the market is not a food but is an anti-diarrhea supplement called Lacteol™ (Salazar-Lindo et al. 2007).

Probiotic cheese

Various cheese types have received the attention of researchers with regard to incorporation of probiotics (see Table 12.3). When developing a probiotic cheese, the microorganisms should be culturable to high cell densities for eventual inoculation into the cheese vat and be able to retain viability and maintain the desirable characteristics throughout the whole manufacturing process, including the ripening period and retail chain and still be able to survive in the GIT (Sanz 2007). Cheese is one of the most versatile dairy foods, as it is also appealing to most people, and suitable for essentially all age groups, and offers a number of advantages over other products, especially in terms of delivery of viable probiotics (relatively high pH, fat content and mechanical consistency, coupled with its oxygen level); these features provide, in combination, an extra protection to probiotic strains during storage prior to ingestion, and even later on throughout the GIT (Ross et al. 2002, Bergamini et al. 2005). In their extensive review covering the development of fresh cheeses containing well-known potential probiotic cultures, Cruz et al. (2009) showed that a positive influence of probiotics on texture and sensory properties of the cheeses was evident, and that adequate viable cell numbers were maintained as storage time elapsed.

Conjugated linoleic acid produced by bifidobacteria

Conjugated linoleic acid (CLA) occurs naturally in red meat and dairy products, as a by-product of ruminal hydrogenation. It is a group of positional and geometric fatty acid isomers derived from linoleic acid (C18:2) with double conjugated bonds, in which milk fat is the richest dietary source (Parodi 1998). This compound can be produced by bio-hydrogenation pathway of linoleic acid by ruminal bacteria and via mammary delta-9 desaturase activity in *trans*-vaccenic acid (TVA) C18:1t, an isomer of oleic acid (*cis*-C18:1n-9) (Bergamo et al. 2003, Bisig et al. 2007, Hayes 2006).

Table 12.3 Cheeses formulated with probiotics.

Product	Probiotic
Fresh and soft cheeses	
Cottage	*B. infantis* ATCC 279206
Argentinean Fresco cheese	*Bifidobacterium* spp., *Lb. acidophilus, Lb. casei* and *Lactococcus lactis* subsp.
	Lb. paracasei A13, *B. bifidum* A1 and *Lb. acidophilus* A3
Estonian Pikante cheese	*Lb. fermentum* ME-3
Minas Fresh Cheese	*Lb. paracasei* subsp. *paracasei* LBC 82
Whey cheese (Requeijão)	LAFTI cultures (L26, L10 and B94), *B. animalis* BB12, *B. animalis* Bo
Crescenza	*B. bifidum, B. infantis* and *B. longum*
	Lb. paracasei A13 and *Lb. acidophilus* H5
Flor di latte	*Lb. casei, Lb. delbrueckii* subsp. *bulgaricus, Lb. paracasei, Lb. plantarum, Lb. rhamnosus* and *Lb. reuteri*
Caprine Brazilian "coalho" cheese	*Lb. acidophilus* LA5
Semi-hard and hard cheeses	
Gouda	*Lb. acidophilus* Ki and *B. lactis* Bo
Arzua-Ulloa	*Lactobacillus* spp.
Canestrato Pugliese	*B. bifidum* Bb02 and *B. longum* Bb46
Tallaga	*B. animalis* Bb12
Turkish White cheese	*L. lactis* subsp. *lactis, L. lactis* subsp. *cremoris* and *Lb. acidophilus* 593 N *Lb. fermentum* (AB5-18 and AK4-120) and *Lb. plantarum* (AB16-65 and AC18-82) *B. animalis* Bb-12 and *Lb. acidophilus* LA-5
Pategrás Argentino	*Lb. acidophilus, Lb. paracasei* subsp. *paracasei* and *B. animalis* subsp. *lactis* (human intestinal origin)
Cheddar	*B. bifidum*
	Lb. acidophilus LAFTI®L10, *B. lactis* LAFTI®B94 and *Lb. paracasei* LAFTI®L26, *Lb. acidophilus* La5, *Lb. casei* Lc1 and *B. lactis* Bb-12; *Bifidobacterium* spp. (HOWARU) and *Lb. rhamnosus* (HOWARU)
	B. longum 1941, *B. animalis* subsp. *lactis* LAFTI®B94, *Lb. casei* 279, *Lb. casei* LAFTI®L26, *Lb. acidophilus* 4962 and *Lb. acidophilus* LAFTI®L10
Caprine Kefalotyri hard cheeses	*Lb. rhamnosus* LC 705 and/or *Lb. paracasei* subsp. *paracasei* DC 412
Washed-curd cheese	*Lb. delbrueckii* subsp. *lactis* UO 004
Iranian white cheese	*B. animalis* (ATCC 25527) and *Lb. rhamnosus* (ATCC 7469)

Adapted from Madureira et al. (2010a).

TVA is considered the main trans fatty acid isomer found in the fat of ruminants and in dairy products (Santora et al. 2000). The trans fatty acids content in milk represents about 2% of total fatty acids, which can be increased to 4–10% of total fatty acids by enhancing dietary unsaturated oils content in the cow's diet. C18:2, *cis*9, *trans*11, the major isomer of CLA, has been linked to beneficial health effects, including decreasing risk of coronary heart disease and reducing incidence of some tumoral cancer types (Butler et al. 2011). Finally, α-linolenic acid (ALA), the major omega-3 fatty acid in milk, and the dietary precursor for three long-chain omega-3 polyunsaturated fatty acids (LC-PUFA) synthesis: eicosapentaenoic acid (EPA), docosapentaenoic acid (DPA), and docosahexaenoic acid (DHA) (Brenna et al. 2009), has been related to anti-arrhythmic effects in the heart, a positive impact on neurological function (by limiting central nervous system injury) and protection against coronary heart disease (Barceló-Coblijn and Murphy 2009). Since this compound can be produced by bio-hydrogenation pathway of linoleic acid by ruminal bacteria, recently, a number of dairy cultures and probiotic strains with the ability to convert linoleic acid to the *c*9, *t*11 and *t*9, *t*11 CLA isomers have been identified, including a number of bifidobacteria (Barrett et al. 2007, Rodríguez-Alcalá et al. 2011). This offers the potential for the development of novel CLA enriched dairy products (Xu et al. 2005) or alternatively the establishment of probiotic cultures in the GIT, capable of producing CLA from dietary linoleic acid (Ewaschuk et al. 2006).

Probiotic whey products

The carbohydrate reservoir of whey (in the form of lactose), coupled with the presence of other nutrients essential for microbial growth, makes whey a suitable raw material for production of various bio-products (Panesar et al. 2007). These include, among others, potable and fuel-grade alcohol (Dragone et al. 2008), lactic acid (Kourkoutas et al. 2005), single cell protein for livestock feed (Plessas et al. 2008), baker's yeast (Plessas et al. 2004) and fermented whey drinks (Paraskevopoulou et al. 2003).

In the special case of whey, the fermentation of such by-product is a means to modify its functional properties, often in attempts to add value thereto (Yang and Silva 1995). During fermentation, complex whey nutrients are converted to simpler forms, which can be more readily assimilated in the intestine. In particular, fermentation of whey by probiotic bacteria increases the amount of easily digested amino acids (Hitchins and McDonough 1989). Dietary calcium, in the form of calcium phosphate, also favors selective growth of intestinal lactobacilli, and even decreases the severity of *Salmonella* infections in rats (Bovee-Oudenhoven et al. 1997). The metabolic activity of probiotic microorganisms leads to release of flavoring compounds (e.g., certain organic acids), which make fermented whey products more palatable when compared with their non-fermented counterparts.

Partial replacement of milk by whey in the manufacture of fermented milk beverages constitutes a rational alternative—in terms of nutritional profile—to which the food industry has adhered more and more frequently. Fermented lactic beverages have been operationally defined as fermented milk. They are produced by mixing milk and whey, and contain a lactic culture and possibly other dairy ingredients (Gallardo-Escamilla et al. 2007). Such lactic beverages have been considered a particularly suitable vehicle for the uptake of probiotics (Almeida et al. 2008). Production of probiotic milk beverages containing whey in their formulation, has indeed gained a large market expression—especially owing to the ease of the underlying process, the use of the same equipment that are used for milk processing, and the low acidity level (and hence the sensory appeal) of the products, as well as the acceptable consistency and the associated healthy image of products (Castro et al. 2009). The incorporation of probiotic strains results by itself, in products with a pleasant taste and differentiable flavors (Macedo 1997). Several formulations developed are shown in Table 12.4.

The use of different whey sources, with or without milk addition, and addition of prebiotics such as fructooligosaccharides (FOS), inulin and oligofructose has been tested for their suitability for development of probiotic dairy-based beverages (Thamer and Penna 2005, 2006). An increase

Table 12.4 Formulations of some whey based beverages.

Raw material	Probiotic strains
35% (v/v) buffalo whey, 30% (v/v) soymilk or 35% (v/v) cow milk	*Lb. casei* Shirota and *B. adolescentis*
Whey (45%, 50% and 55%) + fructooligosaccharides (6%, 7% and 8%)	*S. thermophilus*, *Lb. delbrueckii* subsp. *bulgaricus*, *Bifidobacterium* spp. and *Lb. acidophilus*
Reconstituted whey, with or without addition of inulin	*Lb. acidophilus*, *Lb. casei* and *B. bifidum*
Whey and fruit juice	*Lb. rhamnosus* VTT E-97800 (E800) or *Lb. rhamnosus* Lc705 and *P. freudenreichii* subsp. *shermanii* JS
Whey	*Lb. reuteri* and 0.5% *B. bifidum*
Whey fortified with iron-amino acid chelate	*Lb. acidophilus*
40% (v/v) whey with 5% (v/v) açai pulp	*Lb. delbrueckii* subsp. *bulgaricus* LB 340 and *S. thermophilus* TA 40, *B. longum* BL 05 and *Lb. acidophilus* La14
Whey (20, 30, 35 or 40 ml 100 ml^{-1}) and oligofructose (2, 3, 3.5 and 5 g 100 ml^{-1})	*Lb. acidophilus* LA-5, *B. animalis* Bb-12 and *S. thermophilus*

Adapted from Madureira et al. (2010a).

in the health-promoting features can also be achieved by addition of fruit pulps, such as açai pulp, which is rich in anti-oxidant compounds, such as anthocyanins (400 mg 100 g^{-1} of fruit pulp) to whey (Zoellner et al. 2009).

12.2.2 Prebiotics

Another category of functional foods includes prebiotics, which are non-digestible food ingredients that affect the host by stimulating the growth and/or activity of beneficial bifidobacteria in the colon that can improve host's health (Gibson et al. 2004). Such compounds include several peptides, proteins, and certain lipids, and especially non-digestible carbohydrates (in particular, non-digestible oligosaccharides).

Oligosaccharides

Oligosaccharides are non-digestible compounds composed of 2–20 monosaccharides units with structures predominantly linear or branched and produced from carbohydrates. Human milk contains a complex mixture of more than 130 different oligosaccharides comprising a total concentration of 15–23 g l^{-1} in colostrum and 8–12 g l^{-1} in transitional and mature milk (Kunz et al. 2000). These oligosaccharides are resistant to enzymatic digestion in the upper GIT (Engfer 2000) and thus, reaching the colon intact, they may serve as substrates for colonic microflora fermentation. It has been shown that human milk oligosaccharides induce an increase in the number of bifidobacteria in the colonic flora, along with a reduction in the number of potentially pathogenic bacteria (Vandenplas 2002, Chierici et al. 2003).

Oligosaccharides include FOS found in fruits and vegetables, and galacto-oligosaccharides (GOS) found in milk besides other classes of complex carbohydrates. They can be produced by using bacterial strains and have been used in the production of functional fermented milks (Yadav et al. 2007). Examples of commercially available oligosaccharides include inulin/oligofructose, FOS, GOS, lactulose and isomalto-oligosaccharides (Özer et al. 2005) and have been used to develop a variety of functional dairy products (Aryana and McGrew 2007). GOS are composed of galactose units and are mainly obtained from lactose (mostly from whey or permeate) by the action of β-galactosidases. The major discoveries made on these compounds were recently and extensively reviewed by Gänzle (2012). Galactosidases catalyze the hydrolysis of lactose into glucose and galactose, and also the transgalactosylation reactions with lactose as acceptor of galactose units giving rise to GOS of different glycosidic linkages and molecular weights (Otieno 2010). GOS glycosidic bonds depend on the source of the enzyme. Glycosidic bonds between galactose units are β-(1–4) bonds when β-galactosidases are derived from *Bacillus circulans* (Mozaffar et al. 1984) or *Cryptococcus laurentii* (Ozawa and Ontsuka-Uchidat 1989), and are β-(1–6) bonds when enzymes derived from *Aspergillus oryzae* or *Streptococcus thermophilus* are used (Ito et al. 1990).

These types of oligosaccharides can also be obtained from lactulose (β-(1→4)-galactosyl-fructose), a synthetic disaccharide, produced by isomerization of lactose in basic media or enzyme-catalysed synthesis, with a significant impact on human digestion. In this case, they may be produced, e.g., by transgalactosylation reactions of prebiotic lactulose, by using β-galactosidases of different fungal origin (*A. oryzae, A. aculeatus* and *Kluyveromyces lactis*) (Martinez-Villaluenga et al. 2008, Hernández-Hernández et al. 2011). Their physiological action on the colonic motility pattern (Schumann 2002) and their ability to promote the selective growth of healthy intestinal bacteria, mainly bifidobacteria and lactobacilli populations, in the human gut have been extensively reported (de Preter et al. 2006). In fact, it has been reported that GOS from lactulose have the ability to promote the growth of bifidobacteria using *in vitro* fermentation systems with human fecal cultures (Cardelle-Cobas et al. 2009).

In vivo studies also were performed in order to assess the effect of GOS on the intestinal microbiota (Tiihonen et al. 2008, Vulevic et al. 2008, Westerbeek et al. 2010). Significant increases in the bifidobacteria population were observed in several of these studies. These compounds were shown to promote a good balance of intestinal microflora, decrease gastro-intestinal infections, reduce the level of cholesterol in serum, promote the colon cancer prevention, and enhance calcium absorption (Tuohy et al. 2005, Qiang et al. 2009).

Additionally, the combination of prebiotic and probiotic, known as symbiotic, have been shown to improve probiotic proliferation in the intestine (Holzapfel and Schillinger 2002) and modify gut bacterial community structure (Bartosch et al. 2005). Also, GOS from lactulose used as carbon source for lactobacilli species improve the resistance to simulated GIT conditions (Hernández-Hernández et al. 2012). The use of oligosaccharides, including GOS, as cryoprotectants of lactic acid bacteria has also been suggested (Tymczyszyn et al. 2011).

Whey/milk prebiotic properties

Research on how to improve the viability of bifidobacteria in various food matrices has indeed unfolded a promising relationship between probiotic strains and whey proteins. However the "bifidogenic effect", a particular contribution of human milk to support growth of probiotic bacteria, is not related to a single growth-promoting substance, but rather entails a complex set of interacting factors. In particular, a probiotic-type effect has been ascribed to the low concentration of proteins and phosphates, as well as the presence of lactoferrin, lactose, nucleotides and oligosaccharides. The actual prebiotic role of each of these substances is yet to be clearly defined, with the exception of oligosaccharides which promote the growth of *Bifidobacterium*-dominant microflora.

The prevalence of whey proteins over caseins has also been considered to play a positive role upon development of newborn's intestinal microflora. Hence, most formulae in use today throughout Europe have a protein composition in which whey proteins predominate; studies encompassing newborns fed on such formulae have showed a microflora different from those of breastfed infants (Coppa et al. 2006). Marshall et al. (1982), for instance, fortified milk with whey proteins and threonine, in attempts to make a more nutritious medium bearing a low redox potential for lipido-bacteria. Furthermore, Dave and Shah (1998) studied the effects of whey, whey protein concentrate (WPC) and acid casein hydrolyzates (ACH) on viability of *Streptococcus thermophilus, Lactobacillus delbrueckii* subsp. *bulgaricus, Lactobacillus acidophilus* and bifidobacteria in yogurt; addition of WPC and ACH improved viability of bifidobacteria to variable extents, depending on the degree of availability of the additional nitrogen supplied.

Finally, note that whey proteins are rich in sulphur-containing amino acid residues which, upon release by heat or enzymatic treatments, contribute to lower the redox potential. Caseinomacropeptide (CMP) has been recognized for several years as a bifidus growth-promoting factor; CMP isolated from sweet cow's whey, as well as WPC exhibited a bifidogenic effect on *Bifidobacterium animalis* subsp. *lactis* in milk (Janer et al. 2004). Therefore, addition of extra CMP (or WPC) to milk was recommended to complement milk in attempts to increase the viable counts of bifidobacteria in probiotic fermented milks. Addition of WPC also has been found to significantly influence the viability of *Bifidobacterium* strain Bb-12 in fermented goats' milk, during refrigerated storage (Ranadheera et al. 2010).

The production of oligosaccharides from whey components also has been researched. GOS can be produced from whey permeate by enzymatic hydrolysis using β-D-galactosidase. The enzyme can be used in two ways: in batch or in an immobilized form. Continuous production by hydrolysis from Maxilact 2000 L (*Kluyveromyces lactis* β-D-galacosidase) already has been made (Foda and Lopez-Leiva 2000). Moreno et al. (2008) have, in turn, proved that glycosylation of bovine β-lactoglobulin (β-LG) with GOS forms stable glycosylated peptides that can resist *in vitro* gastrointestinal digestion.

12.3 Milk fat globule membrane

There is a growing interest towards nutritional and technological properties of milk fat globule membrane (MFGM). The membrane which surrounds milk fat globule is known as MFGM. The milk fat globules gain the outer bilayer coat from the apical plasma membrane of secretory cells during secretion phase (Keenan and Mather 2006). This membrane is a true biological membrane, composed mainly of polar lipids—phospholipids (PL), triglycerides, cholesterol and unique membrane specific proteins (glycoproteins, enzymes) (Walstra et al. 2006). Among the main MFGM proteins are xanthine oxidase, and several glycoproteins like butyrophilin and mucin, and represent 0.4 g l^{-1} of milk proteins. Technologically, in milk, MFGM acts as a natural emulsifying agent that protects the fat globules against coalescence (protection against enzymes activity). The phospholipids are amphiphilic in nature, and are, therefore, considered as good emulsifiers (Dewettinck et al. 2008). There is growing evidence that the composition of the MFGM varies depending on the milk source, lactation season and feeding profile. It has been recently shown that, with advancing lactation, in sheep milk, the medium chain fatty acids increase in the core of MFGM and the long chain fatty acids, as well as the saturated fatty acids, on the other hand decrease (Martini et al. 2013).

Buttermilk is a by-product from butter manufacture and contains a mixture of skim milk-derived components and MFGM (Zanabria-Eyzaguirre and Corredig 2011). Buttermilk and butter serum are rich sources of MFGM. During dairy processing such as churning and agitation, the MFGM is broken into fragments and distributed in the serum phase (Rombaut et al. 2005). These fragments can be separated from the dairy constituents by using membrane filtration (Morin et al. 2007).

The proteins from MFGM are revealing to be the most important fragments from all remaining in membrane, and despite extensive efforts, their physiological functions are still not well-understood (Spitsberg 2005, Dewettinck et al. 2008). Most of the studies concerning these substances are made with a bovine source, but studies devoted to the colloidal and soluble protein fractions of mare milk have also emerged (caseins and whey proteins). Lactadherin variants were found in mare's MFGM by liquid chromatography coupled with tandem mass spectrometry analyses. Mare's milk origin MFGM was found to be much more complex than that from bovine milk; therefore raising questions regarding the precise function of these different isoforms, if any, in the equine mammary gland were evident (Cebo et al. 2010). Very few studies on MFGM have included goat's milk and up to date, two caprine MFGM proteins have been characterized, i.e., MUC1 (Sacchi et al. 2004) and xanthine oxidase (XO) (Atmani et al. 2004). MFGM from other milk sources (e.g., human milk) have also been explored owing to the progresses and advances in proteomics, i.e., mass spectrometry, and genetic techniques, i.e., molecular cloning and computer-assisted sequence analysis (Cavaletto et al. 2004). Nevertheless, these MFGM products have lower economic value compared with the mainstream products, but this value can be increased by isolating new added-value ingredients with higher or new functional properties (Riccio 2004).

12.3.1 Biological properties of MFGM constituents

In recent years, different factors with health beneficial properties (e.g., cholesterolemia-lowering factor, inhibition of cancer cell growth, and anti-bacterial effect) have been detected in bovine MFGM components (Dewettinck et al. 2008, Spitsberg 2005). The controversy about the use of MFGM (benefits and risks) is well-documented in the reviews of Dewettinck et al. (2008), Michalski and Januel (2006), Riccio (2004), Rutter (2006) and Spitsberg (2005).

Interaction with bacteria

Phospholipids and proteins from MFGM have been associated to anti-microbial activity (van Hooijdonk et al. 2000). Several *in vitro* studies showed that MFGM hydrolyzates exhibited variable anti-bacterial activity against test food pathogens (i.e., *Escherichia coli* O157:H7, *Listeria monocytogenes*, *Salmonella enterica* Typhimurium, *Pseudomonas fluorescens*, *Bacillus cereus*, *Lactobacillus acidophilus* and *Lactobacillus gasseri*) and this variation was not attributable to xanthine oxidase activity. The probiotic microorganisms were mostly resilient to these anti-bacterial fractions (Clare et al. 2008). Actually, the specific effect of MFGM against *E. coli* O157:H7 strain is on expression of Shiga-toxin genes as well as genes involved in the motility of *E. coli* O157:H7 (Tellez et al. 2012). In addition, owing to its unique biochemical composition, MFGM possesses anti-adhesive effects toward pathogens. Milk fat globules were extracted from bovine and goat milk and incubated with HT-29 human adenocarcinoma cells to assess the attachment and internalization of *Salmonella enteritidis*. Both bovine and caprine milk markedly reduced bacterial binding of *S. enteritidis* to HT-29 human adenocarcinoma cells (Guri et al. 2012). MFGM contains considerable amounts of various proteins predominantly sialylated, glycoproteins including mucins such as MUC1, which have been proposed as the basis of their therapeutic activity, and has been associated to the therapeutic and/or bioactivity against *Helicobacter pylori* (Gustafsson et al. 2006). Not all the constituents of MFGM are anti-microbial; instead they can serve as a source of energy for LAB present in cheese during ripening (Fox et al. 1994).

These constituents are the carbohydrates of the MFGM glycoconjugate side chains, which include *N*-acetyl-D-neuraminic acid (NANA), D-galactose (Gal), L-fucose (Fuc), D-mannose (Man), *N*-acetyl-D-glucosamine (GlcNAc) and *N*-acetyl-D-galactosamine (GalNAc). Non-starter lactic acid bacteria (NSLAB) can utilize sugars originating from glycoproteins and glycolipids found in the MFGM (Adamberg et al. 2005). Moe et al. (2012) demonstrated that sugars from glycoproteins and glycolipids were important components for survival and growth of lactobacilli involved in cheese ripening.

Anti-carcinogenic activity

Several biological properties of milk phospholipids have been identified, including protection against tumor growth (Parodi 1998). Milk sphingolipids have been reported to show an important role in preventing colorectal cancer (Schmelz et al. 2000). A native MFGM isolate obtained from raw milk was assessed for its anti-carcinogenic capacity using a colon cancer cell line (HT-29). The toxic effect of the MFGM extracts on HT-29 cells was confirmed by the significant reduction in lactate dehydrogenase enzyme (LDH) by the residual viable cells. An increase of caspase-3 activity (up to 26%) led to the conclusion that MFGM has an apoptotic effect on HT-29 cancer cells (Zanabria-Eyzaguirre and Corredig 2011).

Other relevant biological effects

MFGM plays an important role in infant nutrition and some of its components such as proteins and polar lipids have health-enhancing functions (Spitsberg 2005). However, there is no essential nutritional requirement for milk PL and other class of lipids, such as those containing a backbone of sphingoid bases, shingolipids (SL). Research is revealing evidence suggesting a relationship between dietary consumption of these lipids and health promotion (Spitsberg 2005). However, there has been relatively less information on the nutritional activity of milk polar lipids as compared with other bioactive milk lipid fractions such as CLA as well as short- and medium-chain fatty acids. Moreover, a little information is available on the effect of milk polar lipids on lipid metabolism related to cardiovascular heart diseases risk. Furthermore, MFGM phospholipids have been associated with memory improvement (McDaniel et al. 2003). In contrast, there are controversial findings on the relationship between MFGM phospholipids and multiple sclerosis (Riccio et al. 2011).

12.3.2 MFGM as delivery systems for nutraceutical compounds

MFGM, including its phospholipids and proteins, has been suggested as a promising raw material for production of liposomes for delivery of bioactives. It has been shown that a relatively simple and upgradable microfluidizing process created liposomes, which were roughly spherical with an outer layer surrounding an internal aqueous space observed in many of them. Their size distribution was bimodal, with a large number around 40 nm interspersed with larger particles (100–200 nm). It would be interesting to see this co-assembly of phospholipids and proteins being utilized in the future for the development of novel systems, possibly for simultaneous delivery of hydrophilic and lipophilic bioactives. Moreover, it is a fact that some of the MFGM proteins are surface active enzymes suggesting the possibility to form micro-reactors for enzymatic activity in the internal water phase (Singh 2006). Oil-in-water emulsions were prepared with α-tocopherol and lycopene and extracted MFGM fraction and conventional milk protein concentrates (i.e., whey proteins, caseinate), showing that MFGM-enriched ingredients can be used as delivery systems of liposoluble nutrients in food formulations (Bezelgues et al. 2009). Milk proteins are widely available, inexpensive, natural and GRAS (generally recognized as safe) raw materials with high nutritional value and good sensory properties, and they have many structural properties and functionalities, which make them highly suitable as vehicles, or as components for the construction of vehicles for delivering various bioactives.

12.4 Proteins and bioactive peptides

In the last two decades, evolution of separation technologies, viz. those relying on selective, porous membranes, has permitted a number of whey protein components to become widespread additives in foods. Whey may indeed be subject to several treatments, thus various whey products with specific qualitative and quantitative profiles of proteins, minerals, lipids and sugars are obtained. The aforementioned membrane-based separation technologies include ultrafiltration (UF) to concentrate proteins, or diafiltration (DF) to remove most lactose, minerals and low molecular weight components and consequently lead to whey protein concentrates (WPC). The protein concentration of WPC varies depending on the requirements in food processing (i.e., 35, 50, 65, 80%, w/v). When the threshold protein level of 90% (w/w) is reached, the resulting product is called "whey protein isolate (WPI)" which is a protein concentrate with high-quality and purity. Both types of whey products are used as vectors to promote many biological properties, following addition to specific foods.

Whey-derived growth factors—a recent category of dairy ingredients aimed at new health-promoting product concepts (e.g., prevention, or even treatment of certain gastro-intestinal diseases)—may serve as reliable replacers or as supplement to (fetal) bovine serum. Preliminary results of animal and human trials showed that the whey growth factor extract obtained from Cheddar cheese whey was effective upon oral mucositis and chronic ulcers (Michaelidou and Steijns 2006). There are various modes of release of peptides (with biological activity) from milk proteins, or of synthesis thereof from simpler molecules. Besides being susceptible to inorganic (acid or alkaline) catalysis, milk proteins can be hydrolyzed via gastric, pancreatic and microbial proteases, thus generating peptides that may play physiological roles. These peptides can trigger physiological effects: (i) on the nervous system, via their opiate and ileum-contracting activities; (ii) on the cardiovascular system, via their anti-thrombotic and anti-hypertensive activities; (iii) on the immune system, via their anti-microbial, anti-viral activities, and (iv) on the nutrition system. These roles were tackled in a review article recently published by our group (Madureira et al. 2010b). One such important physiological role is the angiotensin converting enzyme (ACE)-inhibitory activity, which will be described below in more detail.

12.4.1 ACE-inhibitory activity

The ACE-inhibitory activity depends on the protein substrate, and on the proteolytic enzymes used for its breakdown. Trypsin has been the most widely used enzyme to produce hydrolyzates with reasonable ACE-inhibitory activity. Characterization of hydrolysates of whey proteins, including the amino acid sequences of peptides therein that exhibit *in vitro* ACE-inhibitory activity and/or *in vivo* anti-hypertensive effects, is provided in Table 12.5. Digestion of α-lactalbumin (α-LA) and β-LG by enzymes (viz. pepsin, α-chymotrypsin, pancreatin, elastase, or carboxypeptidases A and B) reveals in fact that trypsin is normally required to produce high ACE-inhibitory activity from whey proteins (Pihlanto Leppäla et al. 2000); e.g., the peptides f(104–108) and f(142–148), released from α-LA and β-LG via trypsin, possessing ACE-inhibitory activities of 77 and 43 μM, respectively. Conversely, elastase—the gastrointestinal protease—is associated with a poor yield of ACE-inhibitory peptides from α-LA and β-LG (Mullally et al. 1997, Pihlanto-Leppäla et al. 2000). Also, peptides from whey protein concentrates were obtained by using enzymes from a vegetal source—*Cynara curdunculus*. These peptides were found to possess significant bioactivity, such as ACE-inhibitory activity under *in vitro* and *in vivo* conditions (Tavares et al. 2012). One of the major risk factors for cardiovascular disease is elevated blood pressure. Angiotensin I-converting enzyme plays a crucial

Table 12.5 Primary structural characteristics of whey peptides with ACE-inhibitory activity, and vectors of generation thereof.

Source protein	Amino acid sequence	Identity	Bioactivity (IC_{50} µM)
α-Lactalbumin	f(50–52)	YGL	409
	f(50–53)	YGLF	733.3
	f(52–53)	LF	349.1
	f(50–51)	YG	1522.6
	f(99–108)	VGNYWLAHK	327
	f(104–108)	WLAHK YGGVSLPEW	77 16
	f(105–110)	LAHKAL	621
		LKGYGVSLPEW	83
		GVSLPEW	30
		VSLPEW	57
		WLAHK	56
β-Lactoglobulin	f(15–20)	VAGTWY	210
	f(9–14)	GLDIQK	580
	f(102–105)	YLLF	172
	f(142–148)	ALPMHIR	42.6
	f(36–42)	IIAE SAPLRVY	28 8
Lactoferrin		Ac-RKWHFW-NH2	210.7
	f(17–31)	Ac-FKCRRWQWRMKKLGA-NH2	25.5
	f(20–25)	LRPVAA	4.14
	f(266–270)	LIWKL	0.47
	f(133–136)	RPYL	56.5
	f(232–238)	LNNSRAP	105.3
	f(221–222)	FP	315
	f(208–216)	ALKAWSVAR	3

Adapted from Madureira et al. (2010b).

role in the regulation of blood pressure, because it promotes conversion of angiotensin I to the potent vasoconstrictor angiotensin II and inactivates the vasodilator bradykinin (Fig. 12.1).

Synthetic ACE inhibitors block these processes and have long been used as anti-hypertensive agents. Milk proteins were identified as sources of ACE inhibitory peptides, which are currently the best-known class of bioactive peptides.

ACE-inhibiting peptides derived from casein are termed casokinins, whereas those derived from whey (viz. α-LA and β-LG) are called

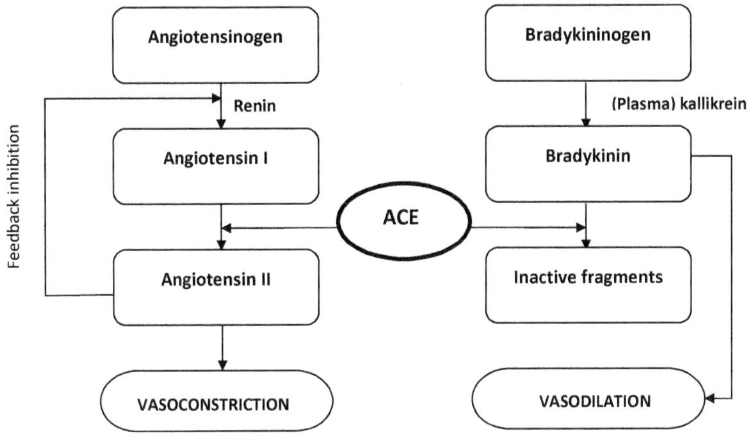

Figure 12.1 Biochemical pathways involving action of ACE in the regulation of blood pressure. Adapted from Madureira et al. (2010b).

lactokinins. ACE inhibition is classically measured as the concentration of compound needed to inhibit 50% of the original ACE activity (IC_{50}) (Gerdes et al. 2002).

Starter and non-starter lactic acid bacteria are commonly used in the manufacture of dairy products taking advantage namely of their proteolytic system, which contains at least 16 already characterized different peptidases. Some of these bacteria have strains with probiotic properties (i.e., *Lactobacillus helveticus*, *Lactococcus lactis* subsp. *cremoris* and *Lb. delbrueckii* subsp. *bulgaricus*) with their well-established proteolytic systems (Law and Haandrikman 1997), which consist of proteinase and peptidases (Christensen et al. 1999).

Several fermented milks and yogurts were found to have ACE-inhibitory (ACE-I) activity or possess peptides with this activity. *Lb. helveticus*, which is able to release ACE-I peptides, is widely used in the manufacture of such fermented milk products (two commercial examples are Calpis® and Evolus®). The best known ACE-I peptides, Valine-Proline-Proline (VPP) and Isoleucine-Proline-Proline (IPP), have been identified in milk fermented with *Lb. helveticus* strains (Seppo et al. 2003). Gobbetti et al. (2000) also provided evidence for the production of ACE-I peptides in milk fermented with *Lb. delbrueckii* subsp. *bulgaricus* or *L. lactis* subsp. *cremoris* for 72 hr.

Active ACE-inhibitory fractions in fermented milks contained mainly β-casein-derived peptides, with inhibitory concentration (IC_{50}) values ranging from 8.0 to 11.2 μg ml^{-1} (Korhonen and Pihlanto 2006). Papadimitriou et al. (2007) found peptides with ACE-inhibitory activity in probiotic sheep yogurt inoculated with *Lb. delbrueckii* subsp. *bulgaricus*, *S. thermophilus* and

Lactobacillus paracasei subsp. *paracasei* DC412, as mixed or single cultures. The former was the most proteolytic, followed closely by *Lb. paracasei* subsp. *paracasei* DC412 (with essentially similar proteolytic profiles). Assessment of the probiotic yogurts for anti-hypertensive peptides revealed the presence of β-CN f(114–121) which exhibited both anti-hypertensive and opiate-like activity. Ashar and Chand (2004) also identified an ACE-I peptide in milk fermented with *Lb. delbrueckii* subsp. *bulgaricus*; this peptide, bearing the sequence of Ser-Lys-Val-Tyr-Pro-Phe-Pro-Gly-Pro-Ile and derived from β-CN, had an IC_{50} value of 1.7 mg ml^{-1}. Fuglsang et al. (2003) studied a total of 26 wild-type LAB strains, belonging mainly to the *L. lactis* and *Lb. helveticus* species, for their ability to produce milk fermentates with ACE-I activity; most LAB produced ACE inhibitors but to varying extents. Recently, *L. lactis* strains isolated from artisanal dairy products and commercial starter cultures showed satisfactory potential for the production of fermented dairy products with ACE-I properties, and the lowest IC_{50} value reported was 13 μg ml^{-1} (Rodríguez-Figueroa et al. 2010). Cheeses were also screened for their ACE-inhibitory activities (Table 12.6).

Ripened-type cheeses contain numerous peptides that originate mainly from casein as a result of proteolysis during ripening. The presence of these bioactive peptides is dependent on the ripening stage of the cheese (Meisel

Table 12.6 Bioactive peptides identified in various cheeses.

Cheese type	Peptides
Italian varieties: Mozzarella, Crescenza, Italico, Gorgonzola	β-CN f(58–72)
Gouda	α_{s1}-CN f(1–9), β-CN f(60–68)
Festivo	α_{s1}-CN f(1–9), f(1–7), f(1–6)
Manchego	Ovine α_{s1}-, α_{s2}- and β-CN fragments
Australian varieties: Cheddar, Edam, Swiss, Feta, Camembert, Blue vein	NI
Finnish varieties: Edam, Emmental, Turunmaa, Cheddar	NI
Low fat Cheddar cheese	α_{s1}-CN (f 1–9), (f 1–7), (f 1–6)
Emmental	NI
Swiss cheeses	NI
Asiago d'allevo	α_{s1}-CN f(24–30), f(27–30), f(24–32), β-CN f(7–14), f(47–52), f(193–209)
Mexican fresco cheese	β-CN f(193–205), f(193–207), f(193–209) α_{S1}-CN f(1–15), f(1–22), f(14–23), f(24–34)
Portuguese whey cheese (Requeijão)	NI

NI: No identification of peptides was made.
Adapted from Madureira et al. (2010a).

1998, Saito et al. 2000). Incorporation of probiotic cultures as adjunct starters to improve proteolysis and enhance final cheese flavor has been attempted by several researchers (Gardiner et al. 1998, Bergamini et al. 2006). Probiotic bacteria strains including *B. longum* 1941, *Lb. casei* 279, *Lb. acidophilus* 4962, *B. animalis* subsp. *lactis* LAFTI®B94, *Lb. casei* LAFTI®L26 or *Lb. acidophilus* LAFTI®L10 were evaluated for their suitability in the manufacture of Cheddar cheese and several ACE-I peptides were identified in the resulting products (Ong et al. 2007). A fermented, low-fat hard cheese containing probiotic LAB (*Lb. acidophilus* and *Bifidobacterium* spp.) was found to produce high levels of ACE-inhibitory peptides derived from α_{s1}-CN (α_{s1}-CN (f1–9) with a level of 13.4 μM) (Saito et al. 1994).

12.4.2 Caseinophosphopeptides (CPPs)

Caseins are milk proteins which represent ca. 80% of the total milk protein. During milk biosynthesis after completion of polypeptide chain, a biological process occurs, which is known as phosphorylation. This process can occur at Ser, Thr, Glu, Asp, His, Lys and Tyr residues and it is performed by enzymes that will phosphorylate proteins with ATP. These enzymes are called kinases, and are membrane-bound in the smooth endoplasmic reticulum and Golgi apparatus present in the mammary gland (FitzGerald et al. 2004). The genetic polymorphism of casein determines the number of serine/threonine phosphate groups (Table 12.7), generating different levels of phosphorylation in the individual caseins.

The phosphorylation of caseins may originate peptides which are known as caseinophosphopeptides (CPPs). CPPs are rich in clusters of phosphorylated seryl (and occasionally threonine) residues and represent ca. 10% (w/w) of the primary sequence of the caseins (Cross et al. 2006). These peptides are encrypted within the primary sequence of caseins and must be released in order to become active (FitzGerald 1998). Similar to other peptides, the release of CPPs is a result of enzymatic hydrolysis of casein fractions. Enzymes from mammalians, plant or microbial sources

Table 12.7 Physicochemical properties of the four main groups of the phosphoproteins caseins.

Casein designation	Molecular weight (Daltons)	Number of residues (number of proline residues)	Number of phosphate groups
α_{s1}-Casein	23000	199 (17)	8–9
α_{s2}-Casein	25000	207 (10)	10–13
β-Casein	24000	209 (35)	5
κ-Casein	19000	169 (20)	1–2

such as pancreatic endopeptidases (McDonagh and FitzGerald 1998) and trypsin (FitzGerald 1998) are able to breakdown caseins to release clusters of phosphorylated seryl residues. Other methods used in the hydrolysis of caseins include physical and/or chemical processes such as ultrasonic, microwave and chemical treatments.

Natural action of the bacterial proteases during manufacture of cheese or other fermented dairy products can also contribute to the release of CPPs. Such CPPs that are accumulated during cheese ripening are ingested as food components (Ellegård et al. 1999). Hence, when dairy products reach the gut, the CPPs already present in the food product are digested into smaller molecules, while new CPPs may also occur during digestion of α_{s1}-, α_{s2}- and β-CN (FitzGerald 1998).

Regarding *in vivo* production, this may also occur by the action of proteases from intestinal microflora or by the digestive enzymes such as gastric and pancreatic hydrolyses after oral ingestion of such CPPs. Actually, CPPs have been found in the distal small intestine of the rat, and in the stomach, duodenum and distal ileum of human subjects following the consumption of dairy ingredients and products (Chabance et al. 1998, Meisel and FitzGerald 2003, Hartmann and Meisel 2007). Most of the commercially available CPP preparations are released from casein using the enzymatic hydrolysis.

Functional properties of CPPs

Functional properties of CPPs include mineral binding capacity, immunomodulatory, anti-oxidative and anti-carcinogenic activities.

Mineral binding property

The main property that has been widely described for CPPs is the capability of binding minerals. Peptides released from α_{s1}-CN and β-CN are the most studied and include α_{s1}-CN f(59–79)5P and β-CN f(1–25)4P, which contain a specific sequence known as the 'acidic motif', which is a 3 Ser phosphate group followed by two glutamic acid residues. This motif is responsible for the binding of these peptides to divalent minerals including Ca^{2+}, Zn^{2+}, Fe^{2+}, Mn^{2+} and Se^{2+} (Rose 2000, Kitts 2006). Phosphorylated Ser residues in close proximity to the peptide chain create a polar, acidic domain that is favorable for sequestering. Nevertheless, carboxylate groups also have been attributed as binding groups to calcium (Mekmene and Gaucheron 2011). The presence of SerP-SerP-SerP-Glu-Glu embedded in the bioactive peptide is a distinctive feature for all functional CPPs derived from whole β-CN-4P (1–25), α_{s1}-CN-5P (59–79), α_{s2}-CN-4P (1–21) and α_{s2}-CN-4P (46–70)

(Kitts 2006, Scholz-Ahrens and Schrezenmeir 2000). The CPP/minerals complexes formed have high electrical charges, making them resistant to further hydrolysis (Silva and Malcata 2005).

An increase in intercellular calcium in intestinal HT-29 cells in the presence of a commercial CPP mixture, in addition to pure β-CN (1–25)4P and to a lesser extent α_{s1}-CN (59–79)5P was shown, caused by uptake of extracellular calcium ions, with no involvement of the intracellular calcium stores. The influx of calcium into HT-29 cells was caused by β-CN (1–25)4P and was dependent on the phosphorylated 'acidic motif' and the preceding N-terminal region (Ferraretto et al. 2001, 2003). Furthermore, CPPs are able to aggregate to polymers and favor the flux of extracellular calcium into HT-29 cells (Gravaghi et al. 2007).

In addition, CPPs may bind and solubilize other trace elements, such as zinc, barium, selenium, nickel, cobalt and chromium (FitzGerald 1998). The addition of CPP at a level of 14 µmol to Caco-2 cells had a positive effect on zinc binding and uptake from phytate-containing solutions, whereas higher levels (i.e., 36 µmol and 72 µmol) inhibited zinc binding and uptake (Hansen et al. 1997). The zinc levels, the type of calcium salt and the CPP dose also were proved to be important factors as far as zinc bio-availability in Caco-2 cells was concerned (Etcheverry et al. 2004).

Improved dietary bioavailability of divalent cations by the CPPs is important in modulating mineral uptake and bone formation, and has been the subject of much research (FitzGerald 1998, Guéguen and Pointillart 2000, Tulipano et al. 2010). Improvement in oral health has also been associated with CPPs, notably due to their role in promoting remineralization of dental enamel, thereby combating the development of dental caries (FitzGerald 1998, Elsayad et al. 2009, Rehder Neto et al. 2009). During the last 30 years, the interest in the potential applications of CPPs in the field of dentistry has increased. With recent developments in minimally invasive treatments, the interest in natural remineralizing agents such as CPPs are becoming more relevant to dentistry (Borges et al. 2011, Gupta and Prakash 2011), and the discovery of new CPP sequences or the isolation and development of new CPP applications in the area of oral hygiene and bone health is becoming a reality (Reynolds et al. 2009, Reynolds 2010).

Anti-oxidant property

Researches on anti-oxidants derived from natural sources have largely been triggered by safety concerns and consumer demands (Korhonen and Pihlanto 2006). Several food protein hydrolyzates have been found to exhibit anti-oxidant activity (Saiga et al. 2003). CPPs have been shown to prevent oxidative reactions (Kitts 2005). Casein hydrolyzates and CPPs that retain the amino acid domain, or sequence with anti-oxidant activity, could be a better option to prevent rancidity in foods without affecting other food quality parameters. The bioactivity of milk peptides, including anti-oxidant activity, is dependent on a number of factors including amino acid sequences of protein substrates, pre-treatment, enzyme specificity, hydrolysis conditions such as pH, temperature, degree of hydrolysis, methods used in enzyme inactivation, separation and purification (Korhonen and Pihlanto 2006), as well as aggregation. Understanding the relationship between peptide composition and anti-oxidant activity could lead to the development of a new class of extremely effective, multi-functional anti-oxidants that could be used in many food applications such as development of functional foods fortified with unstable, unsaturated fatty acids (Elias et al. 2008).

Finally, since CPPs are effective cation chelators that form complexes with minerals and metals, they may control lipid oxidation in food emulsions. Nevertheless, the use of caseins as metal chelators in foods could be problematic because protein denaturation alters the metal-binding properties of proteins and can cause loss of protein solubility and formation of protein aggregates that alter food quality parameters. The CPPs could provide a more practical source of natural protein-based chelators having an activity that is not influenced by denaturation. Lipid oxidation in oil-in-water emulsions was inhibited more by casein hydrolyzates (Díaz et al. 2003) than that by enriched CPP preparations at equal phosphorus content. Later, the anti-oxidant activity of CPPs and casein hydrolyzates in a cell membrane model system of phosphatidylcholine liposomes and in ground beef was compared (Díaz and Decker 2004), and it was reported that iron-induced lipid oxidation was inhibited by both of them, with casein hydrolyzates being less effective than CPPs.

Other relevant properties

CPPs, among the other functions mentioned, also stimulate immunoglobulin A (IgA) production in mice (Ferraretto et al. 2001, Otani et al. 2003). Otani et al. (2003) reported that enhancement of intestinal IgA levels by CPPs-III was found to be involved in IL-6 production. This highlights the fact that the individual activities involved in immune function are interlinked. The CPPs enhanced mRNA expression of IL-6 and tumor necrosis factor-α (TNF-α), and this expression was stronger in the presence of both CPPs and peptidoglycan from *Lb. acidophilus* and lipopolysaccharide (LPS) from S. Typhimurium (Kawahara and Otani 2004). The CPPs used in these studies had enhanced calcium and phosphopeptide content compared with the other peptides. This reflects the increased calcium-solubilizing capacity of CPPs required for intestinal epithelial IL-6 secretion. *In vitro* results suggest that casein hydrolyzates mainly affect the production of Th1 cytokines and, hence, have a role in cellular immunity.

In addition, CPPs are related with reduction of glutathione (GSH) concentration and increased GSH-reductase activity in Caco-2 cells (Laparra et al. 2008). On the other hand, Hartmann and Meisel (2007) observed no deleterious cytochemical consequences (apoptotic, anti-proliferative or general cytotoxic effects) when they added a range of CPP preparations to human cancer cell lines (HL-60 and Caco-2 cells) and non-malignant polymorphonuclear leukocytes.

12.4.3 Commercial applications of CPPs

The ability of CPPs to both increase food stability and promote health makes them potential multifunctional food ingredients. Several research works and patents concerning the manufacture of CPPs have been developed in the last years (Nongonierma and FitzGerald 2012). These patents are relatively similar and based on modifications of the sequence of steps involved in the extraction protocol, namely, modification of the size of the CPP aggregates at different stages of the extraction procedure. Several of these patents concern the production of phosphopeptides salts for human dietetic use, mainly for application in enteral formulas (Brule et al. 1982, 1985, 1989, 1991, 1993a,b) as well as non-phosphorylated peptides for nutrition formulas (Brule et al. 1994a,b). Furthermore, there are references to the production of CPPs for use as ingredients or functional foods. These are targeted for preventing osteoporosis or for the treatment of anemic conditions (increment of calcium and iron adsorption), as well as compositions bearing CPPs for the management of lactating animals therapeutic conditions (Koide et al. 1994, Naito et al. 1995, Iscovich et al. 2011). Most applications use whole bovine milk or other substrates such as sodium caseinate, as a source of CPPs (FitzGerald 1998). Apart from the pancreatic enzymes or trypsin, enzymes from plant sources such as papain proteinase are also used in the production of CPPs. Extraction of CPPs is made, in general, by ultrafiltration with a wide range of molecular weight cut-off membranes or by ion exchange chromatography (Lihme et al. 1998). Crosslinking of CPPs to chitosan beads (Koide et al. 1994), or production of chitosan-CPPs nanocomplexes which involves electrostatic interactions (Hu et al. 2011) also has been described.

12.5 Concluding remarks

Large numbers of compounds and ingredients with bioactive properties are available in functional dairy matrices. The major representative group is probiotics and despite intensive research works on probiotics, there are still several concerns at present, related to the abusive use of this term/

concept in the commercial trades and scientific community research. Years of research showed us that probiotic bacteria cannot be treated as regular lactic bacteria. Actually, EFSA has recently declared that the term "probiotic" will no longer be allowed on food labels in the EU starting December 14, 2012, because it is not convinced that the strength of the science behind probiotic products meets its efficacy standards. On the other hand, the production of bioactive compounds such as CLA or bioactive peptides attracts the scientific community's attention more than ever. Besides this microbial source, biopeptides also can be produced from milk, casein or whey proteins by hydrolysis using proteolytic enzymes from different sources including animal, microbial or vegetable, such as *Cynara curdunculus*, as recently explored by our research group. In fact, regarding all the research work performed in functional ingredients, the production and properties of bioactive peptides have been the major area being explored in the last two decades, together with probiotics. Additionally, attention has been focused now on milk as a source of new functional lipids such as MFGM or oligosaccharides with prebiotic activity.

All these ingredients must be stable in each of functional dairy matrices as well as throughout storage. Actually, there is a serious concern in maintaining these functional ingredients intact during GIT passage to reach the intestine in a bioactive form. So, the encapsulation and the generation of new particles at nanoscale is a tendency that allows the protection of these functional ingredients and increases their bioavailability in the intestine and will allow more stable and active functional dairy products in the near future.

References

Adamberg, K., M. Antonsson, F.K. Vogensen, E.W. Nielsen, S. Kask, P.-L. Møller and Y. Ardö. 2005. Fermentation of carbohydrates from cheese sources by non-starter lactic acid bacteria isolated from semi-hard Danish cheese. Int. Dairy J. 15: 873–882.

Almeida, K.E., A.Y. Tamime and M.N. Oliveira. 2008. Acidification rates of probiotic bacteria in Minas Frescal cheese whey. Food Sci. Technol. 41: 311–316.

Aryana, K.J. and P. McGrew. 2007. Quality attributes of yogurt with *Lactobacillus casei* and various prebiotics. LWT-Food Sci. and Technol. 40: 1808–1814.

Ashar, M.N. and R. Chand. 2004. Fermented milk containing ACE-inhibitory peptides reduces blood pressure in middle aged hypertensive subjects. Milchwiss. 59: 363–369.

Atmani, D., M. Benboubetra and R. Harrison. 2004. Goats' milk xanthine oxidoreductase is grossly deficient in molybdenum. J. Dairy Res. 71: 7–13.

Barceló-Coblijn, G. and E.J. Murphy. 2009. Alpha-linolenic acid and its conversion to longer chain n-3 fatty acids: Benefits for human health and a role in maintaining tissue n-3 fatty acid levels. Prog. Lipid Res. 48: 355–374.

Barrett, E., R.P. Ross, G.F. Fitzgerald and C. Stanton. 2007. Rapid screening method for analyzing the conjugated linoleic acid production capabilities of bacterial cultures. App. Environ. Microbiol. 73: 2333–2337.

Bartosch, S., E.J. Woodmansey, J.C. Paterson, M.E. McMurdo and G.T. Macfarlane. 2005. Microbiological effects of consuming a synbiotic containing *Bifidobacterium bifidum*, *Bifidobacterium lactis*, and oligofructose in elderly persons, determined by real-time polymerase chain reaction and counting of viable bacteria. Clin. Infect. Dis. 40: 28–37.

Bergamini, C.V., E.R. Hynes, A. Quiberoni, V.B. Suárez and C.A. Zalazar. 2005. Probiotic bacteria as adjunct starters: Influence of the addition methodology on their survival in a semi-hard Argentinean cheese. Food Res. Int. 38: 597–604.

Bergamini, C.V., E.R. Hynes and C.A. Zalazar. 2006. Influence of probiotic bacteria on the proteolysis profile of a semi-hard cheese. Int. Dairy J. 16: 856–862.

Bergamo, P., E. Fedeleand, L. Iannibelli and G. Marzillo. 2003. Fat soluble vitamin contents and fatty acid composition in organic and conventional Italian dairy products. Food Chem. 82: 625–631.

Bezelgues, J.-B., F. Morgan, G. Palomo, L. Crosset-Perrotin and P. Ducret. 2009. Milk fat globule membrane as a potential delivery system for liposoluble nutrients. J. Dairy Sci. 92: 2524–2528.

Bisig, W., P. Eberhard, M. Collomb and R. Rehberger. 2007. Influence of processing on the fatty acid composition and the content of conjugated linoleic acid in organic and conventional dairy products—A review. Lait 87: 1–19.

Borges, B.C., J. de Souza Borges, L.S. de Araújo, C.T. Machado, A.J. dos Santos and I.V. de Assunçao Pinheiro. 2011. Update on nonsurgical, ultraconservative approaches to treat effectively noncavitated caries lesions in permanent teeth. Eur. J. Dent. 5: 229–236.

Bovee-Oudenhoven, I.M., D.S.M.L. Termont, P.J. Heidtand and R. van der Meer. 1997. Increasing the intestinal resistance of rats to the invasive pathogen *Salmonella enteritidis*: An additive effect of dietary lactulose and calcium. Gut 40: 497–504.

Brenna, J.T., N. Salem, Jr., A.J. Sinclair and S.C. Cunnane. 2009. Alpha-linolenic acid supplementation and conversion to n-3 long-chain polyunsaturated fatty acids in humans. Prost. Leukot. Essent. Fatty Acid. 80: 85–91.

Brule, G., L. Roger, J. Fauquant and M. Piot. 1982. Phosphopeptides from casein-based material. USPTO US4361587.

Brule, G., L. Roger, J. Fauquant and M. Piot. 1985. Phosphopeptides from casein-based raw material. USPTO US4495176.

Brule, G., L. Roger, J. Fauquant and M. Piot. 1989. Casein phosphopeptides composition. USPTO US4816398.

Brule, G., L. Roger, J. Fauquant and M. Piot. 1991. Casein phosphopeptide salts. USPTO US5028589.

Brule, G., L. Roger, J. Fauquant and M. Piot. 1993a. Non-phosphorylated peptides from casein-based material. USPTO US5219735.

Brule, G., L. Roger, J. Fauquant and M. Piot. 1993b. Non-phosphorylated peptides from casein-based material. USPTO US5216123.

Brule, G., L. Roger, J. Fauquant and M. Piot. 1994a. Nutrient composition containing non-phosphorylated peptides from casein-based material. USPTO US5352476.

Brule, G., L. Roger, J. Fauquant and M. Piot. 1994b. Nutrient composition containing non-phosphorylated peptides from casein-based material. USPTO US5334408.

Butler, G., S. Stergiadis, C. Seal, M. Eyre and C. Leifert. 2011. Fat composition of organic and conventional retail milk in northwest England. J. Dairy Sci. 94: 24–36.

Cardelle-Cobas, A., M. Fernández, N. Salazar, C. Martínez-Villaluenga, M. Villamiel, P. Ruas-Madiedo and C.G. de los Reyes-Gavilán. 2009. Bifidogenic effect and stimulation of short chain fatty acid production in human faecal slurry cultures by oligosaccharides derived from lactose and lactulose. J. Dairy Res. 76: 317–325.

Castro, F., T.M. Cunha, P.J. Ogliari, R.F. Teófilo, M.M.C. Ferreira and E.S. Prudêncio. 2009. Influence of different content of cheese whey and oligofructose on the properties of fermented lactic beverages: Study using response surface methodology. LWT-Food Sci. Technol. 42: 993–997.

Cavaletto, M., M.G. Giuffrida and A. Conti. 2004. The proteomic approach to analysis of human milk fat globule membrane. Clin. Chim. Acta 347: 41–48.

Cebo, C., H. Caillat, F. Bouvier and P. Martin. 2010. Major proteins of the goat milk fat globule membrane. J. Dairy Sci. 93: 868–876.

Chabance, B., P. Marteau, J.C. Rambaud, D. Migliore-Samour, M. Boynard, P. Perrotin, R. Guillet, P. Jollès and A.M. Fiat. 1998. Casein peptide release and passage to the blood in humans during digestion of milk or yogurt. Biochimie 80: 155–165.

Chierici, R., S. Fanaro, D. Saccomandi and V. Vigi. 2003. Advances in the modulation of the microbial ecology of the gut in early infancy. Acta Paediatr. Suppl. 91: 56–63.

Christensen, J.E., E.G. Dudley, J.A. Pedersen and J. Steele. 1999. Peptidases and amino acid catabolism in lactic acid bacteria. Antonie van Leeuwen. 76: 217–246.

Clare, D.A., Z. Zheng, H.M. Hassan, H.E. Swaisgood and G.L. Catignani. 2008. Antimicrobial properties of milkfat globule membrane fractions. J. Food Protect. 71: 126–133.

Coppa, G.V., L. Zampini, T. Galeazzi and O. Gabrielli. 2006. Prebiotics in human milk: A review. Dig. Liver Dis. 38: S291–S294.

Cross, K.J., N.L. Huq and E.C. Reynolds. 2006. Anticariogenic peptides. pp. 335–351. *In:* Y. Mine and F. Shahidi [eds.]. Nutraceutical Protein and Peptides in Health and Disease. CRC Press, New York, NY, USA.

Cruz, A.G., F.C.A. Buriti, C.H.B. Souza and S.M.I. Saad. 2009. Probiotic cheese: Health benefits, technological and stability aspects. Trends Food Sci. Technol. 20: 344–354.

Dave, R.I. and N.P. Shah. 1998. Ingredients supplementation effects on viability of probiotic bacteria in yogurt. J. Dairy Sci. 81: 2804–2816.

de Preter, V., T. Vanhoutte, G. Huys, J. Swings, P. Rutgeerts and K. Verbeke. 2006. Baseline microbiota activity and initial bifidobacteria counts influence responses to prebiotic dosing in healthy subjects. Aliment. Pharma. Ther. 15: 504–516.

Dewettinck, K., R. Rombaut, N. Thienpont, T.T. Le, K. Messens and J. van Camp. 2008. Nutritional and technological aspects of milk fat globule membrane material. Int. Dairy J. 18: 436–457.

Díaz, M., C.M. Dunn, J. McClements and E.A. Decker. 2003. Use of caseinophosphopeptides as natural antioxidants in oil-in-water emulsions. J. Agric. Food Chem. 51: 2365–2370.

Díaz, M. and E.A. Decker. 2004. Antioxidant mechanisms of caseinophosphopeptides and casein hydrolysates and their application in ground beef. J. Agric. Food Chem. 52: 8208–8213.

Dragone, G.M., S.I. Silva, J.B. Almeida and J.A. Teixeira. 2008. Optimization of ethanol production from cheese whey powder by *Kluyveromyces fragilis* using factorial design and response surface methodology. *In:* E.C. Ferreira and M. Mota [eds.]. Proceedings of the 10th International Chemical and Biological Engineering Conference—CHEMPOR 2008, September 4–6, Braga, Portugal.

Elias, R.J., S.S. Kellerby and E.A. Decker. 2008. Antioxidant activity of proteins and peptides. Crit. Rev. Food Sci. Nutr. 48: 430–441.

Ellegård, K.H., C. Gammelgård-Larsen, E.S. Sørensen and S. Fedosov. 1999. Process scale chromatographic isolation, characterization and identification of tryptic bioactive casein phosphopeptides. Int. Dairy J. 9: 639–652.

Elsayad, I., A. Sakr and Y. Badr. 2009. Combining casein phosphopeptides-amorphous calcium phosphate with fluoride: Synergistic remineralization potential of artificially demineralized enamel or not? J. Biomed. Opt. 14: article ID 044039.

Engfer, M.B. 2000. Human milk oligosaccharides are resistant to enzymatic hydrolysis in the upper gastrointestinal tract. Am. J. Clin. Nutr. 71: 1589–1596.

Etcheverry, P., J.C. Wallingford, D.D. Miller and R.P. Glahn. 2004. Calcium, zinc, and iron bioavailabilities from a commercial human milk fortifier: A comparison study. J. Dairy Sci. 87: 3629–3637.

Ewaschuk, J.B., J.W. Walker, H. Diaz and K.L. Madsen. 2006. Bioproduction of conjugated linoleic acid by probiotic bacteria occurs *in vitro* and *in vivo* in mice. J. Nutr. 136: 1483–1487.

FAO/WHO. 2001. Food and Agriculture Organization and World Health Organization. Evaluation of health and nutritional properties of probiotics in food including powder milk with live lactic acid bacteria: Report of a Joint FAO/WHO Expert Consultation, Córdoba, Argentina [cited Dec 15, 2005]. Available from: ftp://ftp.fao.org/es/esn/food/probio_ report_en.pdf.

Ferraretto, A., A. Signorile, C. Gravaghi and A. Fiorilli. 2001. Casein phosphopeptides influence calcium uptake by cultured human intestinal HT-29 tumor cells. J. Nutr. 131: 1655–1661.

Ferraretto, A., C. Gravaghi, A. Fiorilli and G. Tettamanti. 2003. Casein derived bioactive phosphopeptides: The role of phosphorylation and primary structure in promoting calcium uptake by HT-29 tumor cells. FEBS Letters 551: 92–98.

FitzGerald, R.J. 1998. Potential uses of caseinophosphopeptides. Int. Dairy J. 8: 451–457.

FitzGerald, R.J., B. Murray and D.J. Walsh. 2004. Hypotensive peptides from milk proteins. J. Nutr. 134: 980S–988S.

Foda, M.I. and M. Lopez-Leiva. 2000. Continuous production of oligosaccharides from whey using a membrane reactor. Process Biochem. 35: 581–587.

Fox, P.F., T.K. Singh and P.L.H. McSweeney. 1994. Proteolysis in cheese during ripening. pp. 1–31. *In:* A.T. Andrews and J. Varley [eds.]. Biochemistry of Milk Products. Royal Society of Chemistry, Cambridge, UK.

Fuglsang, A., D. Nilsson and N.C.B. Nyborg. 2003. Characterization of new milk-derived inhibitors of angiotensin converting enzyme *in vitro* and *in vivo*. J. Enzyme Inhib. Med. Chem. 18: 407–412.

Gallardo-Escamilla, F.J., A.L. Kelly and C.M. Delahunty. 2007. Mouthfeel and flavor of fermented whey with added hydrocolloids. Int. Dairy J. 17: 308–315.

Gänzle, M.G. 2012. Enzymatic synthesis of galacto-oligosaccharides and other lactose derivatives (hetero-oligosaccharides) from lactose. Int. Dairy J. 22: 116–122.

Gardiner, G., R.P. Ross, J.K. Collins and G.C. Fitzgerald. 1998. Development of a probiotic Cheddar cheese containing human-derived *Lactobacillus paracasei* strains. Appl. Environ. Microbiol. 64: 2192–2199.

Gerdes, S.K., J.W. Harper and G. Miller. 2002. Applications monograph cardiovascular health bioactive components of whey and cardiovascular health. pp. 1–8. *In:* The American Dairy Products Institute Monograph. US Dairy Export Council, Chicago IL, USA.

Gibson, G.R., H.M. Probert, J. van Loo, R.A. Rastall and M. Roberfroid. 2004. Dietary modulation of the human colonic microbiota: Updating the concept of prebiotics. Nutr. Res. Rev. 17: 259–275.

Gobbetti, M., P. Ferranti, E. Smacchi and F. Goffredi. 2000. Production of angiotensin I-converting enzyme inhibitory peptides in fermented milks started by *Lactobacillus delbruekii* ssp. *bulgaricus* SS1 and *Lactococcus lactis* ssp. *cremoris* FT5. Appl. Env. Microbiol. 66: 3898–3904.

Gravaghi, C., E. del Favero, L. Cantu, E. Donetti, M. Bedoni, A. Fiorilli, G. Tettamanti and A. Ferraretto. 2007. Casein phosphopeptide promotion of calcium uptake in HT-29 cells—Relationship between biological activity and supramolecular structure. FEBS J. 274: 4999–5011.

Guéguen, L. and A. Pointillart. 2000. The bioavailability of dietary calcium. J. Am. Coll. Nutr. 19: 119–136.

Gupta, R. and V. Prakash. 2011. CPP-ACP complex as a new adjunctive agent for remineralisation: A review. Oral Health Prevent. Dent. 9: 151–165.

Guri, A.M., M. Griffiths, C.M. Khursigara and M. Corredig. 2012. The effect of milk fat globules on adherence and internalization of *Salmonella enteritidis* to HT-29 cells. J. Dairy Sci. 95: 6937–6945.

Gustafsson, A., A. Hultberg, R. Sjöström, I. Kacskovics, M.E. Breimer, T. Borén, L. Hammarström and J. Holgersson. 2006. Carbohydrate-dependent inhibition of *Helicobacter pylori* colonization using porcine milk. Glycobiol. 16: 1–10.

Hansen, M., B. Sandstrom, M. Jensen and S.S. Sorensen. 1997. Effect of casein phosphopeptides on zinc and calcium absorption from bread meals. J. Trace Elem. Med. Biol. 11: 143–149.

Hartmann, R. and H. Meisel. 2007. Food-derived peptides with biological activity: From research to food applications. Curr. Opin. Biotechnol. 18: 163–169.

Hayes, D.G. 2006. Effect of temperature programming on the performance of urea inclusion compound-based free fatty acid fractionation. J. Am. Oil Chem. Soc. 83: 253–259.

Hernández-Hernández, O., F. Montañez, A. Clemente, F.J. Moreno and M.L. Sanz. 2011. Characterization of galactooligosaccharides derived from lactulose. J. Chrom. A. 1218: 7691–7696.

Hernández-Hernández, O., A. Muthaiyan, F.J. Moreno, A. Montilla, M.L. Sanz and S.C. Ricke. 2012. Effect of prebiotic carbohydrates on the growth and tolerance of Lactobacillus. Food Microbiol. 30: 355–361.

Hitchins, A.D. and F.E. McDonough. 1989. Prophylactic and therapeutic aspects of fermented milk. Am. J. Clin. Nutr. 49: 675–684.

Holzapfel, W.H. and U. Schillinger. 2002. Introduction to pre- and probiotics. Food Res. Int. 35: 109–116.

Hu, B., S.S. Wang, J. Li, X.X. Zeng and Q.R. Huiang. 2011. Assembly of bioactive peptide-chitosan nanocomplexes. J. Phys. Chem. 115: 7515–7523.

Iscovich, J.M., N. Silanikove and J. Iscovich. 2011. Pharmaceutical compositions comprising casein derived peptides and methods of use thereof. USPTO US7968513 B2.

Ito, M., Y. Deguchi, A. Myamori, K. Matsumoto, H. Kikuchi and K. Matsumoto. 1990. Effects of administration of galactooligosaccharides on the human fecal microflora, stool weight and abdominal sensation. Microbiol. Ecol. Health Dis. 3: 285–292.

Janer, C., C. Peláez and T. Requena. 2004. Caseinomacropeptide and whey protein concentrate enhance *Bifidobacterium lactis* growth in milk. Food Chem. 86: 263–267.

Jayamanne, V.S. and M.R. Adams. 2006. Determination of survival, identity and stress resistance of probiotic bifidobacteria in bio-yogurts. Lett. Appl. Microbiol. 42: 189–194.

Kawahara, T. and H. Otani. 2004. Stimulatory effects of casein phosphopeptide (CPP-III) on mRNA expression of cytokines in Caco-2 cells. Biosci. Biotechnol. Biochem. 68: 1779–1781.

Keenan, T.W. and I.H. Mather. 2006. Intracellular origin of milk fat globules and the nature of the milk fat globule membrane. pp. 137–171. *In:* P.F. Fox and P.L.H. McSweeney [eds.]. Advanced Dairy Chemistry: Lipids. 3rd ed. Springer, New York, NY, USA.

Kitts, D.D. 2005. Antioxidant properties of casein-phosphopeptides. Trends Food Sci. Tech. 16: 549–554.

Kitts, D.D. 2006. Calcium binding peptides. pp. 11–27. *In:* Y. Mine and F. Shahidi [eds.]. Nutraceutical Proteins and Peptides in Health and Disease. Nutraceutical Science and Technology. CRC Press, New York, NY, USA.

Koide, K., T. Fukushima, T. Kuwata, K. Itoyama and F. Miyazawa. 1994. Method for separation and concentration of phosphopeptides. USPTO US5290685.

Kolars, J.C., M.D. Levitt, M. Aouji and D.A. Savaiano. 1984. Yogurt, an auto digesting source of lactose. New England J. Med. 310: 1–3.

Korhonen, H.J.T. and A. Pihlanto. 2006. Bioactive peptides: Production and functionality. Int. Dairy J. 16: 945–949.

Kourkoutas, Y., V. Xolias, M. Kallis, E. Bezirtzoglou and M. Kanellaki. 2005. *Lactobacillus casei* cell immobilization on fruit pieces for probiotic additive, fermented milk and lactic acid production. Process Biochem. 40: 411–416.

Kunz, C., S. Rudloff, W. Baier, N. Klein and S. Strobel. 2000. Oligosaccharides in human milk: Structure, functional, and metabolic aspects. Ann. Rev. Nutr. 20: 699–722.

Laparra, J.M., A. Alegría, R. Barberá and R. Farré. 2008. Antioxidant effect of casein phosphopeptides compared with fruit beverages supplemented with skimmed milk against H_2O_2-induced oxidative stress in Caco-2 cells. Food Res. Int. 41: 773–779.

Law, J. and A. Haandrikman. 1997. Proteolytic enzymes of lactic acid bacteria. Int. Dairy J. 7: 1–11.

Lihme, A.O.F., M.I. Aagesen, C. Gammelgard-Larsen and K.H. Ellegard. 1998. Method of isolating biomolecules by ion exchange. USPTO US5780593.

Macedo, R.E.F. 1997. Development of a dairy beverage with soy water extract and buffalo whey fermented by a mixture culture of *Lactobacillus casei* Shirota and *Bifidobacterium adolescentis*. MSc Thesis, UFPR, Brazil.

Madureira, A.R., M.E. Pintado, A.M.P. Gomes and F.X. Malcata. 2010a. Use of whey protein matrices as food vectors for probiotic bacteria. PhD Thesis. Universidade Católica Portuguesa, Porto, Portugal.

Madureira, A.R., T.S. Tavares, M.E. Pintado, A.M.P. Gomes and F.X. Malcata. 2010b. Physiological properties of bovine whey peptides. J. Dairy Sci. 93: 1–6.

Marshall, V.M., W.M. Cole and L.A. Mabbit. 1982. Fermentation of specially formulated milk with single strains of Bifidobacteria. J. Soc. Dairy Technol. 35: 143–154.

Marshall, V.M. 1987. Fermented milks and their future trends. I: Microbiological aspects. J. Dairy Res. 54: 534–537.

Marteau, P., P. Pochart, B. Flourie, P. Pellier, L. Santos, J.F. Desjeux and J.C. Rambaud. 1990. Effect of chronic ingestion of a fermented dairy product containing *Lactobacillus acidophilus* and *Bifidobacterium bifidum* on metabolic activities of the colonic flora in humans. Am. J. Clin. Nutr. 52: 685–688.

Martinez-Villaluenga, C., A. Cardelle-Cobas, A. Olano, N. Corzo, M. Villamiel and M.L. Jimeno. 2008. Enzymatic synthesis and identification of two trisaccharides produced from lactulose by transgalactosylation. J. Agric. Food. Chem. 56: 557–563.

Martini, M., I. Altomonte and F. Salari. 2013. Evaluation of the fatty acid profile from the core and membrane of fat globules in ewe's milk during lactation. LWT-Food Sci. Technol. 50: 253–258.

McDaniel, M.A., S.F. Maier and G.O. Einstein. 2003. "Brainspecific" nutrients: A memory cure? Nutrition 19: 955–956.

McDonagh, D. and R.J. FitzGerald. 1998. Production of caseinophosphopeptides (CPPs) from sodium caseinate using a range of commercial protease preparations. Int. Dairy J. 8: 39–45.

Meisel, H. 1998. Overview on milk protein-derived peptides. Int. Dairy J. 8: 363–369.

Meisel, H. and R.J. FitzGerald. 2003. Biofunctional peptides from milk proteins: Mineral binding and cytomodulatory effects. Curr. Pharma. Design 9: 1289–1295.

Mekmene, O. and F. Gaucheron. 2011. Determination of calcium-binding constants of caseins. Food Chem. 127: 676–682.

Metchnikoff, E. 1907. Essais optimistes. Paris. The prolongation of life. Optimistic studies. Translated and edited by P. Chalmers Mitchell. Heinemann, London, UK.

Michaelidou, A. and J. Steijns. 2006. Nutritional and technological aspects of minor bioactive components in milk and whey: Growth factors, vitamins and nucleotides. Int. Dairy J. 16: 1421–1428.

Michalski, M. and C. Januel. 2006. Does homogenization affect the human health properties of cow's milk? Trends Food Sci. Technol. 17: 423–437.

Moe, K.M., D. Porcellato and S. Skeie. 2012. Metabolism of milk fat globule membrane components by nonstarter lactic acid bacteria isolated from cheese. J. Dairy Sci. 96: 727–739.

Moreno, F.J., J. Quintanilla-Lopez, R. Lebron-Aguilar, A. Olano and M. Sanz. 2008. Mass spectrometric characterization of glycated β-lactoglobulin peptides derived from galacto-oligosaccharides surviving the *in vitro* gastrointestinal digestion. J. Am. Soc. Mass Spectrom. 19: 927–937.

Morin, P., M. Britten, R. Jimenez-Flores and Y. Pouliot. 2007. Microfiltration of buttermilk and washed cream buttermilk for concentration of milk fat globule membrane components. J. Dairy Sci. 90: 2132–2140.

Mozaffar, Z., K. Nakanishi, R. Matsuno and T. Kamikubo. 1984. Purification and properties of β-galactosidase from *Bacillus circulans*. Agric. Biol. Chem. 48: 3053–3061.

Mullally, M.M., H. Meisel and R.J. FitzGerald. 1997. Angiotensin-I-converting enzyme inhibitory activities of gastric and pancreatic proteinase digests of whey proteins. Int. Dairy J. 7: 299–306.

Naito, H., T. Noguchi, R. Sato, K. Tsuji and H. Hidaka. 1995. Transparent acid drink containing acid soluble casein phosphopeptide. USPTO US5405756.

Nongonierma, A.B. and R.J. FitzGerald. 2012. Biofunctional properties of caseinophosphopeptides in the oral cavity. Caries Res. 46: 234–267.

Oelschlaeger, T.A. 2010. Mechanisms of probiotic actions—A review. Int. J. Medical Microbiol. 300: 57–63.

O'Flaherty, S. and T.R. Klaenhammer. 2010. The role and potential of probiotic bacteria in the gut, and the communication between gut microflora and gut/host. Int. Dairy J. 20: 262–272.

Ong, L., A. Henriksson and N.P. Shah. 2007. Angiotensin converting enzyme-inhibitory activity in Cheddar cheeses made with the addition of probiotic *Lactobacillus casei* sp. Lait 87: 149–165.

Otani, H., K. Nakano and T. Kawahara. 2003. Stimulatory effect of a dietary casein phosphopeptide preparation on the mucosal IgA response of mice to orally ingested lipopolysaccharide from *Salmonella typhimurium*. Biosci. Biotechnol. Biochem. 67: 729–735.

Otieno, D.O. 2010. Synthesis of β-galactooligosaccharides from lactose using microbial β-galactosidases. Comp. Rev. Food Sci. Food Safety 9: 471–482.

Ozawa, O. and K. Ontsuka Uchidat. 1989. Production of 4'-galactosyllactose by intact and immobilized cells of *Cryptococcus laurentii* OKN-4. J. Japan. Soc. Food Sci. Technol. 36: 325–328.

Özer, D., S. Akin and B. Özer. 2005. Effect of inulin and lactulose on survival of *Lactobacillus acidophilus* LA-5 and *Bifidobacterium bifidum* BB-02 in Acidophilus-Bifidus yogurt. Food Sci. Technol. Int. 11: 19–24.

Panesar, P.S., J.F. Kennedy, D.N. Gandhi and K. Bunko. 2007. Bioutilisation of whey for lactic acid production. Food Chem. 105: 11–17.

Papadimitriou, C.G., A. Vafopoulou-Mastrojiannaki, S.V. Silva, A.M. Gomes, F.X. Malcata and E. Alichanidis. 2007. Identification of peptides in traditional and probiotic sheep milk yogurt with angiotensin I-converting enzyme (ACE)-inhibitory activity. Food Chem. 105: 647–656.

Paraskevopoulou, A., I. Athanasiadis, M. Kanellaki, A. Bekatorou, G. Blekas and V. Kiosseoglou. 2003. Functional properties of single cell protein produced by kefir microflora. Food Res. Int. 36: 431–438.

Parodi, P.W. 1998. A role for milk proteins in cancer prevention. Aus. J. Dairy Sci. 53: 37–46.

Pihlanto-Leppälä, A., P. Koskinen, K. Piilola, T. Tupasela and H. Korhonen. 2000. Angiotensin I-converting enzyme inhibitory properties of whey protein digests: Concentration and characterization of active peptides. J. Dairy Res. 67: 53–59.

Plessas, S., L. Pherson, A. Bekatorou, P. Nigam and A.A. Koutinas. 2004. Bread making using kefir grains as baker's yeast. Food Chem. 93: 585–592.

Plessas, S., D. Koliopoulos, Y. Kourkoutas, C. Psarianos, A. Alexopoulos, R. Marchant, I.M. Banatand and A.A. Koutinas. 2008. Upgrading of discarded oranges through fermentation using kefir in food industry. Food Chem. 106: 40–48.

Qiang, X., C. YongLie and W. QianBing. 2009. Health benefit application of functional oligosaccharides. Carbohydr. Polymers 77: 435–441.

Ranadheera, C., S.K. Baines and M.C. Adams. 2010. Importance of food in probiotic efficacy. Food Res. Int. 43: 1–8.

Rehder Neto, F.C., F.A. Maeda, C.P. Turssi and M.C. Serra. 2009. Potential agents to control. Acta Odontol. Scand. 67: 321–332.

Rettger, L.F. and H.A. Cheplin. 1921. A treatise on the transformation of the intestinal flora with special reference to the implantation of *Bacillus acidophilus*. Yale University Press, London, UK.

Reynolds, E.C., S.G. Dashper, N.M. O'Brien-Simpson, G.H. Talbo and M. Malkoski. 2009. Antimicrobial peptides. USPTO US20030195150 A1.

Reynolds, E.C. 2010. Ionic complexes. USPTO US0215593 A1.

Riccio, P. 2004. The proteins of the milk fat globule membrane in the balance. Trends Food Sci. Technol. 15: 458–461.

Riccio, P., R. Rossano and G.M. Liuzzi. 2011. May diet and dietary supplements improve the wellness of multiple sclerosis patients? A molecular approach. Autoimmune Dis. 2010: ID 249842.

Rodríguez-Alcalá, L.M., T. Braga, F.X. Malcata, A. Gomes and J. Fontecha. 2011. Quantitative and qualitative determination of CLA produced by Bifidobacterium and lactic acid bacteria by combining spectrophotometric and Ag^+-HPLC techniques. Food Chem. 125: 1373–1378.

Rodríguez-Figueroa, J.C., R. Reyes-Díaz, A.F. González-Córdova, R. Troncoso-Rojas, I. Vargas-Arispuro and B. Vallejo-Cordoba. 2010. Angiotensin-converting enzyme inhibitory activity of milk fermented by wild and industrial *Lactococcus lactis* strains. J. Dairy Sci. 93: 5032–5038.

Rombaut, R., J. van Camp and K. Dewettinck. 2005. Analysis of phospho- and sphingolipids in dairy products by a new HPLC method. J. Dairy Sci. 88: 482–488.

Rose, R.K. 2000. Binding characteristics of *Streptococcus mutans* for calcium and casein phosphopeptide. Caries Res. 34: 427–431.

Ross, R.P., G. FitzGerald, K. Collins and C. Stanton. 2002. Cheese delivering biocultures: Probiotic cheese. Aus. J. Dairy Technol. 57: 71–78.

Rutter, E.R.F. 2006. Multiple sclerosis and milk: To drink or not to drink? Int. J. Dairy Technol. 59: 223–228.

Sacchi, P., A. Caroli, E. Cauvin, S. Maione, S. Sartore, D. Soglia and R. Rasero. 2004. Analysis of the Muc1 gene and its polymorphism in *Capra hircus*. J. Dairy Sci. 87: 3017–3021.

Saiga, A., S. Tanabe and T. Nishimura. 2003. Antioxidant activity of peptides obtained from porcine myofibrillar proteins by protease treatment. J. Agric. Food Chem. 51: 3661–3667.

Saito, Y., K.N. Wanezaki and A. Kawato. 1994. Structure and activity of angiotensin I converting enzyme inhibitory peptides from sake and sake lees. Biosci. Biotechnol. Biochem. 58: 1767–1772.

Saito, T., T. Nakamura, H. Kitazawa, Y. Kawai and T. Itoh. 2000. Isolation and structural analysis of antihypertensive peptides that exist naturally in Gouda cheese. J. Dairy Sci. 83: 1434–1440.

Salazar-Lindo, E., D. Figueroa-Quintanilla, M.I. Caciano, V. Reto-Valiente, G. Chauviere and P. Colin. 2007. Effectiveness and safety of *Lactobacillus* LB in the treatment of mild acute diarrhea in children. J. Pediatric Gastroent. Nutr. 44: 571–576.

Salminen, S.J., M. Gueimonde and E. Isolauri. 2005. Probiotics that modify disease risk. J. Nutr. 135: 1294–1301.

Santora, J.E., D.L. Palmquist and K.L. Roehrig. 2000. Vaccenic acid is desaturated to conjugated linoleic acid in mice. J. Nutr. 130: 208–215.

Sanz, Y. 2007. Ecological and functional implications of the acid-adaptation ability of *Bifidobacterium*: A way of selecting improved probiotic strains. Int. Dairy J. 17: 1284–1290.

Schmelz, E.M., M.C. Sullards, D.L. Dillehay and A.H. Merrill, Jr. 2000. Colonic cell proliferation and aberrant crypt foci formation are inhibited by dairy glycosphingolipids in 1, 2-dimethylhydrazine-treated CF1 mice. J. Nutr. 130: 522–527.

Scholz-Ahrens, K.E. and J. Schrezenmeir. 2000. Effects of bioactive substances in milk on mineral and trace element metabolism with special reference to casein phosphopeptides. Br. J. Nutr. 84: 147–153.

Schumann, C. 2002. Medical, nutritional and technological properties of lactulose: An update. Eur. J. Nutr. 41: 17–25.

Seppo, L., T. Jauhiainen, T. Poussa and R. Korpela. 2003. A fermented milk high in bioactive peptides has a blood pressure-lowering effect in hypertensive subjects. Am. J. Clin. Nutr. 77: 326–336.

Silva, S.V. and F.X. Malcata. 2005. Caseins as source of bioactive peptides. Int. Dairy J. 15: 1–15.

Singh, H. 2006. The milk fat globule membrane—A biophysical system for food applications. Curr. Opin. Colloid Interface Sci. 11: 154–163.

Spitsberg, V.L. 2005. Bovine milk fat globule membrane as a potential nutraceutical. J. Dairy Sci. 88: 2289–2294.

Talwalkar, A., C.W. Miller, K. Kailasapathy and M.H. Nugyen. 2004. Effect of packaging conditions and dissolved oxygen on the survival probiotic bacteria in yogurt. Int. J. Food Sci. Technol. 39: 605–611.

Tavares, T., M.-A. Sevilla, M.J. Montero, R. Carrón and F.X. Malcata. 2012. Acute effect of whey peptides upon blood pressure of hypertensive rats, and relationship with their angiotensin-converting enzyme inhibitory activity. Mol. Nutr. Food Res. 56: 316–324.

Tellez, A.M., A. Corredig, R. Guri, M.W. Zanabria and V. Griffiths-Delcenserie. 2012. Bovine milk fat globule membrane affects virulence expression in *Escherichia coli* O157:H7. J. Dairy Sci. 95: 6313–6319.

Thamer, K.G. and A.L.B. Penna. 2005. Efeito do teor de soro, açúcar e de frutooligossacarídeos sobre a população de bactérias lácticas probióticas em bebidas fermentadas. Brazil. J. Pharmac. Sci. 41: 393–400.

Thamer, K.G. and A.L.B. Penna. 2006. Caracterização de bebidas lácteas funcionais fermentadas por probióticos e acrescidas de prebiótico. Ciência Tecnol. Aliment. 26: 589–592.

Tiihonen, K., T. Suomalainen, S. Tynkkynen and N. Rautonen. 2008. Effect of prebiotic supplementation on a prebiotic bacteria mixture: Comparison between a rat model and clinical trials. British J. Nutr. 99: 826–831.

Tulipano, G., O. Bulgari, S. Chessa, A. Nardone, D. Cocchi and A. Caroli. 2010. Direct effects of casein phosphopeptides on growth and differentiation of *in vitro* cultured osteoblastic cells (MC3T3- E1). Regul. Pept. 160: 168–174.

Tuohy, K.M., G.C.M. Rouzaud, W.M. Bruck and G.R. Gibson. 2005. Modulation of the human gut microflora towards improved health using prebiotics: Assessment of efficacy. Curr. Pharm. Des. 11: 75–90.

Tymczyszyn, E.E., E. Gerbino, A. Illanes and A. Gómez-Zavaglia. 2011. Galacto-oligosaccharides as protective molecules in the preservation of *Lactobacillus delbrueckii* subsp. *bulgaricus*. Cryobiol. 62: 123–129.

Vandenplas, Y. 2002. Oligosaccharides in infant formula. British J. Nutr. 87: 293–296.

van Hooijdonk, A.C., K.D. Kussendrager and J.M. Steijns. 2000. *In vivo* antimicrobial and antiviral activity of components in bovine milk and colostrum involved in non-specific defence. British J. Nutr. 84: 127–134.

Vulevic, J., A. Drakoularakou, P. Yaqoob, G. Tzortzis and G.R. Gibson. 2008. Modulation of the fecal microflora profile and immune function by a novel trans-galactooligosaccharide mixture (β-GOS) in healthy elderly volunteers. Am. J. Clin. Nutr. 88: 1438–1446.

Walter, J., C. Hertel, G.W. Tannock, C.M. Lis, K. Munro and W.P. Hammes. 2001. Detection of *Lactobacillus, Pediococcus, Leuconostoc,* and *Weissella* species in human feces by using group-specific PCR primers and denaturing gradient gel electrophoresis. Appl. Environ. Microbiol. 67: 2578–2785.

Walstra, P., J.T.M. Wouters and T.J. Geurts. 2006. Dairy Science and Technology. CRC Press, Boca Raton, FL, USA.

Westerbeek, E.A.M., J.P. van den Berg, H.N. Lafeber, W.P.F. Fetter, G. Boehm and T. van Elburg J.W.R. 2010. Neutral and acidic oligosaccharides in preterm. Am. J. Clin. Nut. 91: 679–686.

Xu, S., T.D. Boylston and B.A. Glatz. 2005. Conjugated linoleic acid content and organoleptic attributes of fermented milk products produced with probiotic bacteria. J. Agric. Food. Chem. 53: 9064–9072.

Yadav, H., S. Jain and P.R. Sinha. 2007. Production of free acids and conjugated linoleic acid in probiotic dahi containing *Lactobacillus acidophilus* and *Lactobacillus casei* during fermentation and storage. Int. Dairy J. 17: 1006–1010.

Yang, S.T. and E.M. Silva. 1995. Novel products and new technologies for use of a familiar carbohydrate, milk lactose. J. Dairy Sci. 78: 2541–2548.

Zanabria-Eyzaguirre, R. and M. Corredig. 2011. Milk lipids. Buttermilk and milk fat globule membrane fractions. pp. 691–697. *In:* J.W. Fuquay [ed.]. Encyclopedia of Dairy Sciences. 2nd edition. Academic Press, San Diego, CA, USA.

Zoellner, S.S., A.G. Cruz, J.A.F. Faria, H.M.A. Bolini, M.R.L. Moura, L.M.J. Varvalho and A.S. Sant'ana. 2009. Whey beverage with acid pulp as food carrier of probiotic bacteria. Aus. J. Dairy Technol. 64: 165–172.

CHAPTER 13

Non-Thermal Processing of Milk and Milk Products for Microbial Safety

Gulsun Akdemir Evrendilek

13.1 Introduction

Due to their perishable nature and being spoiled by microorganisms, safety and microbial quality of milk and dairy products are important issues that need to be addressed. The most common way of ensuring food safety and shelf-life stability is thermal treatment of food. Despite its high efficiency in inactivating pathogenic or spoilage microorganisms, thermal treatments may well cause color modifications, and flavor changes in foods as well as significant nutritional losses. Consumers' demand towards more nutritional foods has triggered the efforts of food industry to design alternative strategies to produce safer foods with minimal thermal treatments. Such efforts have emerged novel technologies alternative to heat treatment in food processing. Different thermal and non-thermal processing technologies such as pulsed electric fields (PEF), high pressure processing (HPP), ultrasonication, cold plasma, etc. have been tested for processing of milk and dairy products (Cruz-Romero et al. 2006, del Pozo-Insfran et al. 2007, Tahiri et al. 2006). This chapter deals with potential application fields of these technologies in milk and dairy processing.

13.2 Pulsed electric fields

Application of short burst of high voltage electric pulses in the range of 20–80 kV cm^{-1} placed between two electrodes is involved in pulsed electric fields (PEF) processing. Treatment time of PEF lasts in microseconds; thus, increase in treatment temperature during PEF processing is minimized. The applied high voltage results in an electric field that causes microbial inactivation. Depending on the system design and circuit development, the electric field may be applied in the form of exponentially decaying, square wave monopolar, square wave bipolar or instant charge reversal pulses and at ambient, sub-ambient, or slightly above ambient temperatures.

The basic components of PEF system include a high voltage pulse generator, controlling and monitoring system, treatment chambers and fluid handling system. Food product to be processed is placed in either a static or a continuous design treatment chamber, where two electrodes are connected together with a non-conductive material to avoid electrical flow from one to the other. High voltage electrical pulses generated in the system are applied to the electrodes, which then conduct the high intensity electrical pulse to the product placed between the two electrodes (Zimmermann and Benz 1980).

Microbial inactivation by PEF is provided by structural damages of cell membrane, which lead to ion leakage, metabolite losses, protein releases and increased uptake of drugs, molecular probes and DNA. Structural fatigue, because of induced membrane potential and mechanical stress, causes primary effect on microbial inactivation. Material flow after the loss of integrity of cellular membrane by the electric field, local heating and membrane stress and cell swelling or shrinking and disruption due to the unbalanced osmotic pressure between the cytosol and external medium are included in the secondary and tertiary effects, respectively (Bryant and Wolfe 1987, Chang and Reese 1990). Under normal conditions, cell membrane acts as an insulator to the cytoplasm due to its six- or eight-time lower electrical conductivity than that of cytoplasm. When a cell is exposed to electrical fields, positive and negative electrical charges are accumulated in the cell membrane forming transmembrane potential. The charges attract each other and generate compression pressure, which causes the membrane to decrease in thickness. If electrical field strength increases beyond a critical transmembrane potential, it causes pore formation in cell membrane (electroporation).

When transmembrane potential is approximately 1 V, cell lysis with loss of membrane integrity occurs (Sale and Hamilton 1968). Depending on the pulse duration time, number of pulses and PEF treatment temperature, this critical electrical potential varies (Zhang et al. 1994, 1995). Studies with different microorganisms treated by PEF have demonstrated the disruption of organelles and lack of ribosomes, leading to microbial inhibitions (Harrison et al. 1997). For example, structural damage occurred in *Staphylococcus aureus* cells after PEF treatment (64 pulses with 20, 30 and 40 kV cm^{-1}) suspended in simulated milk ultrafiltrate (SMUF) observed by electron microscopy indicated that cells exhibited rough surfaces, and the ones treated under more severe conditions showed small holes in the membrane and leakage of cellular contents (Pothakamury et al. 1997). The inactivation mechanism of thermal treatment and PEF are different, since heat treatment (i.e., 66°C for 10 min) causes a great damage of the cell organelles, and cell-wall rupture as induced by PEF treatments is not observed (Pothakumary et al. 1997).

Applied electrical field is transmitted to food samples through the ions. Thus, PEF is mostly applied to high acid foods such as fruit juices which have higher hydrogen ions content and higher electrical conductivity to transmit applied electrical field strength. Moreover, compared to neutral pH, increased acidity provides more microbial inactivation. Composition of the food samples is also important for efficiency of PEF treatment. Air and fat are not good electric conductors; therefore, food samples which have lower fat and protein contents are more prone to be successfully processed by PEF. Fat particles of foods, especially milk, seem to protect the bacteria against applied electric pulses (Grahl and Märkl 1996). The inactivation of *Escherichia coli* using PEF is more limited in skim milk than in buffer solution when exposed to similar treatment conditions of field intensity and number of pulses due to the complex composition of skim milk (i.e., lower electrical resistance and the presence of proteins in the latter) (Martin et al. 1997).

Microbial inactivation by PEF also depends on the nature and physical state of the microorganisms. The size of the cell is an important factor for inactivation by PEF. As the size of the cell gets bigger, the chance of inactivation by PEF gets higher. Therefore, yeast cells are easier to be inactivated by PEF than bacteria. Similarly, vegetative cells are less resistant to PEF than spores. In fact, much higher electric field strengths in combination with heat should be applied to inactivate spores. Moreover, growth stage of the microorganisms also affects the level of inactivation. In general, the cells in steady state are more resistant to PEF than those in lag, exponentially growth, and exponentially decreasing phases.

Inactivation of *Listeria innocua* and *Pseudomonas fluorescens* in pasteurized skim milk by PEF at 50 kV cm^{-1} electric field strength up to 200 μs treatment time resulted in 2.6 and 2.7 log cfu ml^{-1} reductions, respectively (Fernandez-Molina 2001). When raw milk inoculated with *S. aureus* and coagulase-negative *Staphylococcus* spp. was treated with PEF, 4 and 2 log cfu ml^{-1} reductions in the counts of bacteria were obtained, respectively. On the other hand, PEF processing under the same conditions did not effectively inactivate *Corynebacterium* spp. or *Xanthomonas maltophilia* (Raso et al. 1999).

PEF processing parameters also play important role on the efficacy of microbial inactivation. There is a positive relationship between applied electric field strength and microbial inactivation. Similarly, with the increase in the treatment time, rate of microbial inactivation also increases (Peleg 1995). Treatment time is calculated by the number of pulses applied and the duration of each pulse. Thus, increase in pulse width causes increase in treatment time. Shape of the electric field affects the rate of microbial inactivation. Generally bipolar pulses are more effective than monopolar pulses (Zhang et al. 1995). The square wave pulses due to the minimization of the energy absorption in foods are more effective than both exponentially decay and oscillatory pulses for inhibiting microorganisms (Knorr et al. 1994, Zhang et al. 1994).

Microbial inactivation is also related to the treatment temperature during PEF processing. Pothakamury et al. (1996) reported that inactivation rate of *E. coli* inoculated into SMUF processed by PEF in a range of temperatures from 3 to 40°C increased with the temperature. In addition, PEF can be combined with heat treatment to increase the inhibitory effect of both preservation methods. Combination of PEF with heat treatment at 50°C for the inactivation of *Listeria monocytogenes* Scott A in pasteurized whole, 2% fat, and skim milk caused 4 log cfu ml^{-1} reduction in counts, while at 25°C the reduction in the number of this bacteria was far lower (i.e., 1 to 3 log cfu ml^{-1}) (Reina et al. 1998).

13.2.1 Processing of dairy products by pulsed electric fields

Processing of milk samples by PEF is mostly carried out to determine the effect of PEF on different microorganisms that are likely present in milk. One of the earlier studies was conducted to inactivate *Salmonella* Dublin in homogenized milk by using PEF with 36.7 kV cm^{-1} and 40 pulses over a 25 min time period. Under these conditions, the target bacteria was completely inactivated in the samples stored at 7 to 9°C for eight days; however, a 3 log reduction was obtained in *E. coli* cells under the same conditions (Dunn 1996). Population of indigenous milk microflora increased to 10^7 cfu ml^{-1} in the untreated milk; whereas, the PEF-treated milk had a bacterial load of approximately 4×10^2 cfu ml^{-1} (Dunn and Pearlman 1987). The effects of PEF on the microbial load of milk and milk products are summarized in Table 13.1.

The shelf-life of the PEF-treated (40 kV cm^{-1}, 30 pulses, and 2 μs pulse width using exponential decaying pulses) raw skim milk (0.2% milk fat) was extended to two weeks at 4°C (Fernandez-Molina et al. 2000, 2005), and processing of raw milk with 2% milk fat by PEF with an electric field of 40 kV cm^{-1} provided a shelf-life of two weeks at refrigeration temperature (Qin et al. 1995). PEF processing (30.76 to 53.84 kV cm^{-1} electric field

Table 13.1 Effect of PEF treatment on the inactivation rate of selected pathogens in milk and milk products.

Product	Target microorganism	Processing conditions	References
Homogenized milk	*Salmonella* Dublin (full reduction) *Eschericia coli* (3 log reduction)	36.7 kV cm^{-1}, 40 pulses, 25 min	Dunn (1996)
Raw skim milk	*Listeria innocua* (4.3 log reduction)	30 and 40 kV cm^{-1}, 1–30 pulses, 20–72°C, <10 s. Best balance of inactivation was achieved at 55°C with 40 kV cm^{-1}	Guerro-Beltran et al. (2010)
Raw skim milk	*Listeria innocua* (2.5 log reduction)	30, 40 or 50 kV cm^{-1}	Calderon-Miranda et al. (1999)
Raw skim milk	*Listeria innocua* (2.0, 2.7 and 3.4 log reductions)	30, 40 or 50 kV cm^{-1} plus 10 IU nisin application	Calderon-Miranda et al. (1999)
UHT milk	*Geobacillus stearothermophilus* (3 log reduction)	60 kV cm^{-1}, 26–210 μs	Shin et al. (2007)
UHT milk	*Pseudomonas fluorescens, Bacillus cereus, Lactococcus lactis* (0.3–3.0 log reductions)	35 kV cm^{-1} with 64 pulses of bipolar square wave for 188 μs	Michalac et al. (2003)
UHT milk (whole)	*Pseudomonas* isolates (complete inactivation)	31 kV cm^{-1}, 20 μs, 55°C	Craven et al. (2008)
Cheese whey	*Listeria innocua,* (3.0–5.0 log reductions) *Zygosaccharomyces bailii* (7.9–8.8 log reductions)	40 kV cm^{-1}, 4937 μs PEF plus UV (with 7.7 s, 229 mJ ml^{-1} dosage)	Dave et al. (2012)

strength and 12, 24 and 30 pulse numbers) of skim milk (SM) and whole milk (WM) in combination with mild heating (20, 30 and 40°C) led to small variations in physicochemical properties of both milks after processing. While psychrophilic bacteria grew faster in both SM and WM, the growth of mesophilic bacteria was delayed in both milks after PEF processing, showing up to 6 and 7 log cfu ml^{-1} growths after 25 days of the storage at 4°C, respectively (Bermudez-Aguirre et al. 2011).

The effect of PEF on physical and biochemical properties of milk samples as well as changes in some bioactive compounds have been also investigated. PEF processing with two steps of seven pulses each and one step of six pulses with an electric field of 40 kV cm^{-1} provided no apparent changes in physical and chemical properties of milk, and no significant difference in sensory attributes between heat pasteurized and PEF-treated milk was observed (Qin et al. 1995). PEF treatment of UHT skim milk inoculated with *P. fluorescens*, *Lactococcus lactis* and *Bacillus cereus* caused no changes in total solids and protein levels, and color, pH, particle size and conductivity values (Michalac et al. 2003). Shin et al. (2007) demonstrated that PEF treatment did not affect the pH and titration acidity of milk samples.

Viscosity of PEF-treated milk decreased and coagulation properties were enhanced at high electric field levels (45–55 kV cm^{-1}) with 2.1 to 3.5 μs treatment time (Floury et al. 2006). Few studies exist on flavored milk processed by PEF with the main concern of product stability. After PEF processing at 40 kV cm^{-1} with 48 pulses (2.5 μs, 55°C), only minor changes were observed in color, Allura Red concentration, and pH. Color of the samples showed significant decreases in a*, and hue angle and Chroma values of flavored milk during storage. However, PEF affected the stability of Allura Red in milk when additional ingredients were not added to the product (Bermudez-Aguirre et al. 2010).

The effect of PEF processing on milk micronutrients such as B group vitamins, cholecalciferol, tocopherol, ascorbic acid and vitamin A showed that approximately 90% of the ascorbic acid presented in milk was retained after PEF treatment, and the other vitamins remained unaltered at electric fields between 1.8 and 2.7 kV mm^{-1}. It was also observed that the macronutrients, including fat content and protein integrity, were unaffected at field strengths as high as 80 kV cm^{-1} (Deeth et al. 2007). PEF processing of milk with 36.7 kV cm^{-1} electric field strength and 40 pulses over a 25 min time period resulted in less flavor degradation and no chemical or physical changes in the quality attributes of milk used in cheese-making (Dunn 1996).

Inactivation of alkaline phosphatase by HTST pasteurization or of lactoperoxidase by high heat treatments is the principal indicator of the efficacy of these treatments to milk (Walstra et al. 2006). Therefore, inactivation of alkaline phosphatase and other enzymes (i.e., plasmin, peroxidase, microbial proteases and lipases) by PEF treatment has been tested in many studies (Elez-Martínez and Martin-Belloso 2007, Sampedro et al. 2005). Application of 2.2 kV mm^{-1} electric field strength caused the inactivation of alkaline phosphatase up to 60% in raw milk (Castro et al. 2001). Although results about the effects of electric fields on proteins especially on enzymes are controversial, generally they include the association or dissociation of functional groups, movements of charged chains, and changes in alignment of helices (Tsong and Astunian 1986). For example, alkaline phosphatase molecules treated by PEF at 22.3 kV cm^{-1} electric field strength with 0.78 ms pulse width tended to associate and aggregate. It was reported that the polarization created by electrical charges of dipoles on the enzyme could cause the aggregate formation. The polarization leading to the aggregation of the enzyme was proposed as the mechanism of the inactivation of alkaline phosphatase by PEF (Castro et al.

2001). PEF processing at 45 kV cm^{-1} with 50 pulses resulted in reduction of plasmin activity by 90% in a SMUF (Vega-Mercado et al. 1995). Inactivation of lipase, peroxidase and alkaline phosphatase has been reported as 65, 25 and <5%, respectively, after the PEF treatment of 21.5 kV cm^{-1} electric field strength with 400 kJ l^{-1} energy (Castro et al. 2001, Grahl and Märkl 1996).

Although most of the studies have focused on PEF processing of milk, the possible applications of PEF in dairy processing have been also investigated in a few studies. For example, PEF processing of yogurt caused about 2 log cfu g^{-1} reductions in counts of *Lactobacillus brevis, Streptococcus thermophilus, Lactobacillus delbrueckii* subsp. *bulgaricus,* and *Saccharomyces cerevisiae* (Dunn and Pearlman 1987). PEF and heat processing (at 60°C for 30 s) of formulated yogurt-based drink caused no significant differences in L, a, b values, °Brix, pH and selected sensory attributes between the control and the treated samples. Microbial counts of the samples treated with the combination of heat (at 60°C) and PEF were lower than that of the control samples stored at 4 or 22°C (Evrendilek et al. 2004).

Dairy products treated with PEF have similar sensory properties to those that are heat-treated and have good consumer acceptability (Sobrino-Lopez et al. 2009, Sampedro et al. 2005). Cheddar cheese produced from PEF-treated milk had better flavor profile than that of the cheese samples made from milk pasteurized at 63°C for 30 min. Hardness and springiness values of the cheese made from PEF-treated milk increased while the other textural attributes (i.e., adhesiveness and cohesiveness) remained unchanged (Sepulveda-Ahumada et al. 2000).

The level of proteolysis of the cheese curd made from milk treated with PEF (2 µs pulse width, 2 Hz pulse frequency and up to 120 pulses) was lower than those made from raw milk but higher than that made from pasteurized milk. It has been recently reported that the rennet coagulation time of PEF-treated milk was increased by 10% (Garcia-Amezquita et al. 2013).

13.3 High pressure processing

High pressure processing (HPP), also called as high hydrostatic pressure (HHP) or ultra high pressure (UHP), involves application of elevated pressures with or without external heating to obtain microbial inactivation or to alter food attributes (Farkas and Hoover 2000). Although history of pressure application goes back to the late 1800s (Hite 1899), the industrial use of this technology—parallel to the developments of engineering—has relatively recently developed (Mertens and Knorr 1992). Pressure can be applied at ambient temperature, thus thermally induced cooked off-flavors can be eliminated. Finally, this technology can be used to process both liquid and solid foods in batch and semi-continuous equipment. During HHP processing, pressure is transmitted to food sample very rapidly and uniformly thus, pressure application is not dependent on the size and shape of the product which provides homogenous processing of irregular shaped food products (Tewari 2007, Balasubramaniam et al. 2008, Smelt 1998).

Effect of HHP on food products can be explained by Le Châtelier's Principle. According to the principle, system always acts to oppose changes in chemical equilibrium; to restore equilibrium, the system will favor a chemical pathway to reduce or eliminate the disturbance so as to restabilize at thermodynamic equilibrium. It also states that "the application of pressure shifts the system equilibrium toward the state that occupies the smallest volume" (Farkas and Hoover 2000, Pfister and Dehne 2001). The effect of HHP is also explained by isostatic principle which illustrates the transmission of pressure to food samples. The details of the working principle and basic components of HHP system are discussed in Balasubramaniam et al. (2008), Farkas and Hoover (2000), Hogan et al. (2005), US-FDA (2000) and Karim (2011).

Microbial inactivation by HPP has drawn a great attention of food industry. Thus, both vegetative and spore forms of bacteria, molds, yeasts and even viruses have been studied in both model and food systems. The processing conditions such as initial sample temperature, water temperature circulating in pressure level, pressurizing medium and holding time not only influence the level of inactivation but also affects the nutritional and sensory characteristics of foods (Zhang et al. 2011).

High-pressure treatments are generally effective in inactivating most vegetative pathogenic and spoilage microorganisms at pressures ranging between 200 and 600 MPa with temperatures at or below ambient temperature (Patterson 2005). HHP causes damage in cell membrane integrity (Pagán and Mackey 2000) as well as ribosomal destruction (Niven et al. 1999), enzyme inactivation (Degraeve et al. 1996, Simpson and Gilmour 1997), inactivation of membrane-bound transport systems (Ulmer et al. 2002), and damage to the proton efflux system (Wouters et al. 1998). Dramatic changes of the cell structure exposed to HHP such as the collapse of intracellular vacuoles at relatively mild pressures (i.e., 500 MPa), separation between the cell-wall and cytoplasmic membrane, and ribosomal destruction occur (Ritz et al. 2001, Kaletunc et al. 2004). The damages in the cell morphology cause the impairment of cell functions, slowing growth rate or causing cell death. Damages in the cell membrane caused by HHP result in the leakage of intracellular components and loss of homeostasis (Farkas and Hoover 2000). HHP treatment causes an increase in extracellular levels of adenosine triphosphate (ATP) (Smelt et al. 1994) and an increase in the uptake of propidium iodide (Ulmer et al. 2000) and ethidium bromide (Benito et al. 1999), indicating loss of membrane permeability and function.

Microbial cell death occurs depending on the level of the injuries caused by HHP (Fig. 13.1) (Masschalck et al. 2000, Smelt 1998, Tay et al. 2003). Generally, Gram-positive bacteria are more resistant to environmental stresses such as heat and pressure than Gram-negative bacteria, and cocci are more resistant than rod or spirochete-shaped bacteria (Alpas et al. 1999, Patterson 2005, Pilavtepe-Celik et al. 2008, Ritz et al. 2002, Smelt 1998). Moreover, different level of inactivation of strains of the same bacteria can be obtained under the same processing conditions. Processing of Cheddar cheese inoculated with *E. coli* O157:H7 and *L. monocytogenes* Scott A by HHP indicated that *E. coli* was more baroresistant, with decimal reduction values (D-values) of 14.5 and 3.6 min at 300 MPa and 25°C, respectively (Shao et al. 2007).

HHP application of 500–600 MPa at 25°C for 10 min required to inactivate Gram-positive microorganisms, while Gram-negative microorganisms were inactivated at relatively lower pressures under the same time and temperature conditions (Smelt 1998). It is thought that temperature including adiabatic heating during HPP can have a significant effect on microbial survival (Farkas and Hoover 2000). Moderate pressure treatments and shorter pressure holding times when combining with high temperatures can end up with greater anti-microbial impact; however, pressure application at high temperatures can lead to undesirable effects in certain cheese quality parameters. For example, pressure application at 50°C resulted in higher volume of whey losses and unacceptable textural characteristics in the cheese (Shao et al. 2007).

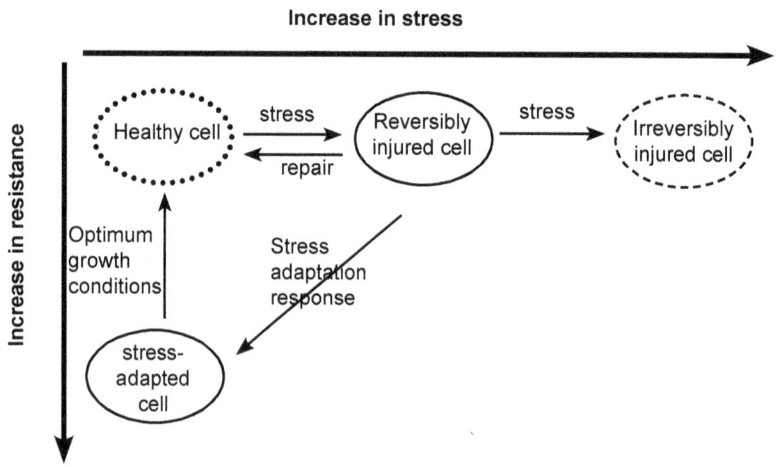

Figure 13.1 Microbial inactivation mechanism by HPP. After: Lado and Yousef (2002).

Anti-microbial effect of HPP is largely influenced by the pH, water activity or chemical composition of foods. The barotolerance of the microorganisms is getting higher as the growth medium gets richer, but microorganisms injured by pressure are generally more sensitive to low water activity. For example, carbohydrates are generally more protective than salts, and recovery of pressure treated cells is much lower when 2% salt is added to the medium (Patterson et al. 1995, Oxen and Knorr 1993). The net effect of lower water activity is not always easy to predict; however, the protective effect of different carbohydrates against high pressure is same as the protective effect of different carbohydrates on the membrane during freezing and it is in the order of fructose < glucose < sucrose < trehalose (Smelt et al. 1997).

Pressure resistance (at 483 MPa for 30 min) of *Bacillus subtilis* inoculated into milk changed as pH of milk raised to 8 (Timson and Short 1965). Pressure tolerance or resistance of bacteria can change from species to species and even may also be different in a single species grown under different conditions or in different growth media (Hoover et al. 1989). The growth phase of microorganisms also plays a role in determining their pressure sensitivity and inactivation. Cells in the lag or stationary phase of growth are generally more pressure resistant than those in the log or exponential phase; this is probably due to the fact that the cellular membrane is more robust, and stress induced genes that synthesize proteins that protect against stress conditions, can be turned on more readily in the stationary phase cells (Alpas et al. 1999, Bowman et al. 2008, McClements et al. 2001, Manas and Mackey 2004). The effect of HPP on the stationary and exponential phase cell membrane of *E. coli* reveals that the increased sensitivity to pressure of exponential phase cell membranes is linked to physical perturbations of the cell envelope, loss of osmotic responsiveness, and exclusion of intracellular proteins and RNA (Manas and Mackey 2004). Moreover, loss of viability which is related to a permanent loss of membrane integrity in exponentially phase cells and leaky cell membrane has also been reported. However cells in the stationary phase are able to reseal partially or wholly after decompression (Pagán and Mackey 2000). Reduced activity of the integral membrane protein F0F1 ATPase in *Lactobacillus plantarum* in addition to impairment of acid reflux and maintenance of intracellular pH was also reported after pressure application at 250 MPa (Wouters et al. 1998).

Inactivation of *L. monocytogenes*, *B. cereus* and *P. fluorescens* in UHT milk treated by HHP presented that exponential phase cells were significantly less resistant to pressure than stationary phase cells for all of the three species studied. Growth temperature was found to have a significant effect at two growth stages studied. Exponential cells grown at 8°C were more resistant than those grown at 30°C, but the reverse was true for the stationary phase cells. *B. cereus* stationary phase cells grown at 30°C were the most pressure resistant, and *L. monocytogenes* showed the most sub-lethal damage compared to *B. cereus* and *P. fluorescence* (McClements et al. 2001). Inactivation of *L. monocytogenes* Scott A processed by HHP in UHT whole milk revealed that elevated temperatures substantially promoted the pressure-induced inactivation of *L. monocytogenes* (Chen and Hoover 2003). Pressure inactivation (400 MPa at 20–25°C) of *L. monocytogenes* in the mid-stationary phase of growth at 4, 15, 25, 35 or 43°C, inoculated into whole UHT milk showed that both growth temperature and plating medium had a significant effect on the inactivation of stationary phase *L. monocytogenes* by HPP. It was also reported that cells of *S. aureus* ATCC 6538 and *E. coli* K-12 in the exponential phase of growth were more sensitive to HPP in Cheddar cheese slurry than cells in the stationary phase (O'Reilly et al. 2000a).

Compared to vegetative cells, endospores are far more resistant against high pressure, requiring treatment at pressures exceeding 1000 MPa and temperature of more than 80°C for full inactivation (Abee and Wouters 1999, Rastogi et al. 2007, Smelt 1998). Thus, high pressure processing is proved to inactivate bacterial spores more effectively when used in combination with heat (Ahn et al. 2007, Black et al. 2007, 2011, Heinz and Knorr 2002, Smelt 1998, Wuytack et al. 1998), and spores germinated at lower pressures are in turn more sensitive to subsequent pressure treatments (Setlow et al. 2001). *B. cereus* spores are more resistant to pressure than vegetative cells, and pressure treatment at 400 MPa for 25 min at 30°C provided about 0.45 log cfu m^{-1} reduction in spores of this bacterium. Pressure treatment at 8°C induced significantly less spore germination than 30°C (McClements et al. 2001).

Most vegetative yeast and molds are inactivated within a few minutes by 300–400 MPa at 25°C. However, heat-resistant molds such as *Byssochlamys, Neosartorya* and *Talaromyces* are generally considered to be extremely resistant to high pressure (Voldrich et al. 2004). Studies have shown that virus inactivation by pressure is due to the denaturation of capsid proteins essential for host cell attachment to initiate infection (Buckow et al. 2008, Khadre and Yousef 2002, Kingsley et al. 2002).

13.3.1 High pressure processing of dairy products

Most of the studies with milk processed by HHP have performed for the inactivation of pathogenic and spoilage microorganisms. Depending on the microbiological quality of milk, processing at 400–600 MPa was comparable to that of heat pasteurization (72.8°C, 15 s) (Buffa et al. 2000, 2001, Kolakowski et al. 1997); however, it was not comparable to sterilized milk due to the resistance of spores to high pressure. Effects of HPP on milk can be summarized as disruptive to the casein micelles and the structure of whey proteins, shifting the mineral balance in milk, inducing crystallization of milk fat, increasing the pH of milk, reducing the turbidity of milk, reducing the rennet or acid coagulation time of milk, and increasing cheese yield (Considine et al. 2007, Huppertz et al. 2002, Lopez-Fandino 2006, Trujillo et al. 2002). The disintegration of casein micelles is accompanied by increases in caseins and calcium phosphate levels in the diffusible or serum phase of milk and by decreases in both non-casein nitrogen and serum nitrogen fractions, suggesting that the whey proteins become sedimentable by centrifugation and precipitable at pH 4.6 (Law et al. 1998). High-pressure treatment markedly increases the transfer of individual caseins from colloidal to soluble phase of milk (Lopez et al. 1998), and dissociation of individual caseins in milk by high pressure is dependent on pH (Arias et al. 2000, Huppertz et al. 2002).

β-lactoglobulin is one of the pressure sensitive proteins, being easily denatured under the pressure treatment up to 500 MPa at 25°C. On the other hand, denaturation of immunoglobulins and α-lactalbumin occurs only at much higher pressures and particularly at 50°C (Schrader et al. 1997). HPP not only causes solubilization of colloidal calcium phosphate, but also resolubilizes insoluble heat-induced crystalline calcium phosphate (i.e., in UHT milk) (Felipe et al. 1997).

Some modifications in size and distribution of milk fat globules were observed on ewe's milk treated by HHP up to 500 MPa (Gervilla et al. 2001). No damage occurred on the milk fat globule membrane, being proved by the lack of increased lipolysis. Lactose in milk and milk products may isomerize in lactulose by heating and then degrade to form acids and other sugars. No changes in these compounds were observed after pressurization (100–400 MPa for 10–60 min at 25°C), suggesting that no Maillard reaction or lactose isomerization occur in milk after pressure treatment (Lopez-Fandino et al. 1996). HHP induces the crystallization of milk fat during pressure treatment of cream (Buchheim et al. 1996). No increase in the products of lipolysis was observed following treatment of milk with pressures ranging from 400 to 800 MPa (Buchheim et al. 1996).

Enzyme inactivation by high pressure is more difficult than microorganisms, since enzymes in milk are more resistant to high pressure treatment (Huppertz et al. 2004a). Studies with enzyme inactivation reveal that alkaline phosphatase, lactoperoxidase, phosphohexose-isomerase and γ–glutamyltransferase are resistant to pressures lower than 400 MPa at 25°C (Lopez-Fandino 1996, Rademacher et al. 1998, Seyderhelm et al. 1996). On the contrary, lactate dehydrogenase activity was reduced substantially at 206 MPa (Kouassi et al. 2007). Rennet coagulation time of milk treated with high pressure at >300 MPa was increased, while at lower pressures the opposite was observed (Buffa et al. 2001, Lopez-Fandino et al. 1996).

The small molecules such as vitamins, amino acids, simple sugars and flavor compounds are not affected by HPP, because contrary to thermal treatments where covalent as well as non-covalent bonds are affected, HHP at room and mild temperatures only disrupts relatively weak chemical bonds (i.e., hydrogen and ionic bonds, and hydrophobic effect) (Cheftel 1992). Therefore, HPP of milk at 400 MPa (at a rate of 2.5 MPa s^{-1} for 30 min at 25°C) resulted in no significant loss of vitamins B_1 and B_6 (Sierra et al. 2000).

Effect of HPP on cheese indicated that high pressure ranging from 200 to 600 MPa results in cheese with acceptable quality and increased yield (Drake et al. 1997, Huppertz et al. 2002, Lopez-Fandino 1996, Trujillo et al. 2002). No significant difference was noted in the flavor profile of Cheddar cheese made from pasteurized or high-pressure treated milk; however, HPP-treated milk produced a pasty, weak texture attributed to the presence of whey proteins and greater moisture content (Drake et al. 1997). With the increase in pressure, the volume of separated whey is reduced.

Application of high pressure can reduce the variability of the moisture content existing within or between cheese blocks and can create new texture in cheese (Torres-Mora et al. 1996). Effect of pressure application on cheese ripening in Gouda-type cheese revealed that salt uptake increased as the pressure and processing time increased from 100 to 400 MPa and from 0.5 to 4 hr, respectively. Disruption of *para*-casein network by HHP at 300 MPa for 5 min led to increase in the concentrations of peptides and proteins in whey (Messens et al. 1998).

Treatment of Turkish white cheese with high pressure from 50 to 600 MPa for 5 or 10 min at 25°C resulted in decreases in the counts of *L. monocytogenes*, total aerobic mesophilic bacteria, molds and yeasts, *Lactococcus* spp. and *Lactobacillus* spp. (Evrendilek et al. 2008). No changes were reported in the pH and water activity of the cheeses treated with high pressure. Gallot-Lavalle (1998) demonstrated that the counts of *L. monocytogenes* in goat's milk cheese made from raw milk processed at 450 MPa for 10 min or at 500 MPa for 5 min decreased more than 5.6 log cfu g^{-1} without significantly affecting the organoleptic characteristics of the end product. Processing of vacuum packaged and salted curds with pressure at 400 MPa for 10 min at 2°C yielded cheese with very low levels of contaminant flora and little modifications of rennet or plasmin activities (Trujillo et al. 2000). Initial number of *E. coli* in cheese samples treated at 400 MPa at 25°C for 5 min was reduced by 7 log cfu g^{-1} and it was not detected after one week at 4°C (Capellas et al. 1996). Combined treatment of nisin and high pressure was found to be the most effective way in extending the shelf-life of the cheese samples, because nisin mainly acted over the sporulated population (Ray 1992), while the combined treatment inactivated a fraction of the microbial population that is resistant to both treatments when applied separately (Hauben et al. 1996, Kalchayanand et al. 1994).

No viable cells of *E. coli* O59:H21 and O157:H7 were detected in washed-curd cheese treated at 500 MPa (De Lamo-Castellvi et al. 2005, 2006). Pressure applications at both 300 and 400 MPa to cheese (pH about 4.82) caused complete inactivation of *S.* Enteritidis CECT 4300 and *S.* Typhimurium CECT 443 with no recovery following 15 d of storage at 12°C (De Lamo-Castellvi et al. 2007). Inactivation of seven hemolytic strains belonging to serotype 1/2a of *L. monocytogenes* in Gorgonzola cheese rind by pressure treatment at 400 to 700 MPa for 1 to 15 min at 30°C ended up with strong resistance of the microorganism to pressures up to 500 MPa; however, pressure treatment at 700 MPa for 15 min led to a 5 log cfu g^{-1} reduction in *L. monocytogenes* counts (Carminati et al. 2004). HPP at 450 MPa for 10 min or 500 MPa for 5 min at 11°C caused 5.6 log cfu g^{-1} reductions of *L. monocytogenes* F13 in 14 day-old Sainte Maure de Touraine cheese (Gallot-Lavalle 1998). Pressure application of 400 MPa for 5 min yielded 3 log cfu g^{-1} reductions in starter flora (Saldo et al. 2000a). It was found that acidification capacity of lactic acid bacteria was reduced after pressure treatment at 400 MPa, even without changes in cell viability (Casal and Gomez 1999). Table 13.2 shows the effect of HHP on pathogenic microorganisms in cheese.

Previous studies have revealed that HPP is able to accelerate cheese ripening by causing alterations in enzyme structure, conformational changes in the casein matrix making it more susceptible to the action of proteases, and/or bacterial lysis enhancing the release of microbial enzymes that promote biochemical reactions (Garde et al. 2007, Martinez-Rodriguez et al. 2012, Messens et al. 1998, O'Reilly et al. 2000b, 2003, Saldo et al. 2000b, 2002, Voigt et al. 2010).

Pressure induced bacterial lysis is strain dependent and affects aminopeptidase activity (Juan et al. 2007, Malone et al. 2002). When pressure was applied at 300 MPa for 10 min at 12°C, the autolysis of cheese starter bacteria (*L. lactis* subsp. *lactis* and *L. lactis* subsp. *cremoris*) was higher in ewes' milk cheese at the beginning of ripening than that at day 15 (Juan et al. 2008).

Treatments below 150 MPa resulted in the greatest increase in the pH 4.6 SN/TN ratio in cheese; however, total free amino acids (FAA) production decreased as pressure increased from 100 to 400 MPa, and increased processing time up to 60 hr raised total FAA levels (O'Reilly et al. 2000b). Processing of 1 and 4 month-old commercial cheeses with pressures ranging from 200 to 800 MPa for 5 min at 25°C showed that pressure treatments higher than 400 MPa could be useful for controlling cheese ripening without any textural changes (Wick et al. 2004). Rynne et al. (2008) reported that processing of one day-old full-fat Cheddar cheese pressure-treated at 400 MPa for 10 min at room temperature slowed down ripening effectively.

Table 13.2 Effect of high hydrostatic pressure on the inactivation of some pathogenic bacteria in cheese.

Product	Target microorganisms	Treatment conditions	References
Washed curd cheese	*Yersinia enterocolitica* CECT 4055 (serotype O3) CECT 559 CECT 4054	Strain CECT more barotolerant at 300 MPa	De Lamo-Castello et al. (2005)
Washed curd cheese	*Listeria monocytogenes* NCTC 11994 Scott A	Strain NCTC 11994 was more sensitive to HPP at 400–500 MPa for 10 min	Lopez-Pedemonte et al. (2007a)
Swiss cheese slurry	Coliforms, yeasts, molds, presumptive coagulase-positive *Staphylococcus*, starter lactic acid bacteria	The counts of all microbial groups reduced greatly at 345–550 MPa for 10–30 min	Ding et al. (2001)
Gouda cheese	*Aeromonas hydrophila*	D-values of 32.05, 12.97 and 2.43 min at 100, 200 and 300 MPa at 50°C, respectively.	Fonberg-Broczek et al. (2005)
Washed curd cheese	*Staphylococcus aureus* CECT 4013 ATCC 13565 Staphylococcal enterotoxin A (artificially inoculated)	Rate of inactivation increased with increasing pressure from 300 to 500 MPa	Lopez-Pedemonte et al. (2007b)
Mato cheese	*Staphylococcus carnosus* 4491	No remarkable decrease in counts of bacteria at 500 MPa for 30 min at 10 or 25°C. 7 log reduction at the same pressure at 55°C for 5 min	Capellas et al. (2000)

Moderate HPP conditions (50 to 150 MPa) applied to young Cheddar cheese may be effective in accelerating proteolysis; whereas, higher HPP conditions (≥400 MPa) may help cheese manufacturers arresting the ripening process at a desired stage, thus maintaining optimum "commercial attributes" for a longer period. Pressure treatments to commercial Irish Cheddar cheese at 50 MPa for 72 hr at 25°C caused significant decrease in the level of α_{s1}-casein and increase in the accumulation of α_{s1}-I-casein in cheese during ripening (O'Reilly et al. 2003). On the other hand, pressure treatments performed between 350 and 400 MPa reduced the rate of accumulation of α_{s1}-casein degradation products significantly (O'Reilly et al. 2003). Accelerated proteolysis as a result of HPP has also been achieved in other cheese varieties such as smear and mold-ripened cheese, Garrotxa, and ewes' milk cheese, but not in Gouda, Edam, or Mozzarella cheese (Martinez-Rodriguez et al. 2012). The effects of HPP on cheese proteolysis vary depending on the cheese type and ripening conditions (Voigt et al. 2010), and the effect of HHP applied directly to cheese on the activity of proteolytic enzymes of secondary cultures is not well known (Bansal et al. 2007, 2009).

Studies related to whey proteins—especially with β-lactoglobulin—show that HPP at 200 MPa causes enhancement of proteolysis (Hayashi et al. 1987). Compared to native form of the β-lactoglobulin, pressure application at 300–900 MPa reduces the foamability of β-lactoglobulin (Pittia et al. 1996). Pressures between 200 and 900 MPa caused irreversible molecular modifications in tertiary structure, surface hydrophobicity and association state of β-lactoglobulin (Belloque et al. 2000, Yang et al. 2003). α-lactalbumin is more baroresistant than β-lactoglobulin in both milk and whey, and both proteins—when treated by pressure in whey—are more resistant than that of in milk (Huppertz et al. 2004b).

Yogurts made from high-pressure-treated milk are coagulated at a higher pH, and have similar physical properties (i.e., firmness and water-holding capacity) to yogurts made from heat-treated milk (Ferragut et al. 2000, Harte et al. 2002). Increasing pressure treatment causes an increase in the firmness of the yogurt made from high pressure treated milk and the firmness of the yogurt has the maximum value when milk is treated with HPP at 55°C compared to the milk treated at 10 or 25°C (Needs et al. 2000, Ferragut et al. 2000).

Although studies related to HPP of different dairy products such as yogurt drink (ayran) is very limited, influences of HPP on some physical properties of ayran as well as inactivation of *L. monocytogenes* and *L. innocua* in this product with or without addition of mint essential oil revealed that pressure treatment alone or in combination with mint essential oil did not cause significant changes in pH, water activity, color and whey separation. Addition of mint essential oil enhanced inactivation of both pathogenic bacteria by more than 1 log cfu ml^{-1} (Evrendilek and Balasubramaniam 2011).

13.4 Ultrasonication

Ultrasound (US) defined as the use of pressure waves with a frequency of above 20 kHz, which is undetectable by human ear, also referred to as "power ultrasound", has a potential to be used as food processing method for microbial inactivation (Butz and Tauscher 2002). The use of ultrasound to inactivate microorganisms was reported in the late 1920s (Harvey and Loomis 1929), but due to its limited lethal effect on both spoilage and pathogenic microorganisms, the use of US as a sterilization method is not commonly employed by the food industry (Cameron et al. 2009).

Low or high energy US has a potential to be used in food industry for different purposes (Piyasena et al. 2003). Lower than 1 W cm^{-2} intensities and higher than 2–3 MHz frequencies are used in low energy US, while higher than 1 W cm^{-2} intensities with frequencies between 18 and 100 kHz are used in high energy US. Generally, low energy US is used for non-invasive purposes such as determination of unwanted foreign matters in raw materials and final products, and characterization of food stuffs (Režek Jambrak et al. 2010). High energy ultrasound is employed in milk homogenization, bacteria and enzyme inactivation (i.e., chymosin) and β-galactosidase extraction (Ashokkumar and Mason 2007, Bosiljkov et al. 2011, Jeličić et al. 2010, Patel et al. 2008, Režek-Jambrak et al. 2009).

Destructive effect of US on bacterial cells stems from the formation and implosion of bubbles in a liquid, what is known as cavitation (growth and collapse of microbubbles) which can produce high localized temperatures, pressures and turbulence (Ashokkumar and Grieser 2007, Brnčić et al. 2010). Since US does not affect a large area on a food surface, it may not efficiently destroy resistant bacteria and spores (Piyasena et al. 2003, Raso et al. 1998). Combination of US with heat treatment known as thermosonication or thermoultrasonication shows better lethal effect on microorganisms. It was demonstrated that combination of ultrasonication (20 kHz and 160 W) with heat (5 to 62°C) resulted in better efficiency regarding time and energy consumption compared to either treatment individually (Ordonez et al. 1984). Similar results were obtained when US was combined with extreme pH and chlorination. Sonication, manosonication, thermosonication and manothermosonication processes have been studied for their efficiencies on microbial inactivation by Manas et al. (2000), McClements (1995), Miles et al. (1995), Ordonez et al. (1984) and Raso et al. (1998).

Several factors such as temperature, pressure, amplitude of the ultrasonic waves, exposure/contact time, volume of food being processed, and the composition of the food are important for efficacy of US processing. Temperature control is provided by dissipating excess heat evolved during US, using cold water circulated through the cooling coil that is placed inside the treatment chamber. Pressure control is monitored by a manometer that is also placed in the treatment chamber.

Microbial inactivation efficacy of US ranges depending on several factors such as cell morphology, pH and viscosity. In general, Gram-negative, rod shape bacteria are more sensitive to US treatment than Gram-positive, coccus-shaped bacteria (Hulsen 1999). However, it was also reported that there was no difference between Gram-negative (i.e., *Pseudomonas aeruginosa* and *E. coli*) and Gram-positive (i.e., *S. aureus* and *B. subtilis*) bacteria regarding US-induced inactivation kinetics (Scherba et al. 1991). According to these results, it is argued that this morphological feature does not seem to be a sole factor for grouping of the organisms based on their resistance against ultrasonic treatment, and instead it is proposed that the target of ultrasonic damage might be the inner (cytoplasmic) membrane, which consists of a lipoprotein bilayer (Villamiel and de Jong 2000). The effect of pH on microbial inactivation reveals that the best inactivation is observed at pHs between 6.8 and 7.1 for milk samples. Effectiveness of the US for microbial inactivation decreases with increasing viscosity, and increases with raising pressure up to 3000 kPa (Hulsen 1999).

13.4.1 Processing of dairy products by ultrasound

Compared to other novel technologies, studies related to US processing of foods are rather limited. However, the use of US on milk processing can have several advantages such as homogenization of fat globules, removal of gases, and enhancement of anti-oxidant activity (Mason 1998, Villamiel and de Jong 2000). Therefore, continuous-flow ultrasonic treatment can be a promising technique for milk processing.

Application of US with 20 kHz frequency for 10 min caused reduction in *E. coli* counts by 100%. It was reported that viable counts of *P. fluorescens* were reduced by 100% after 6 min and *L. monocytogenes* was reduced by 99% after 10 min. Protein or lactose contents of both raw and pasteurized milk do not change with US, but it may cause an increase in the fat concentration. Unfortunately, US does not cause inactivation of alkaline phosphatase and lactoperoxidase activities (Cameron et al. 2009). Processing of pasteurized homogenized skim milk by ultrasound with 20 kHz at 20 and 41 W under controlled temperature conditions for different time intervals up to 60 min causes no changes in viscosity, but reduces the turbidity. The sizes of the casein micelles, fat globules and soluble particles after 60 min of sonication change with changes in energy generated. Denaturation is observed in milk whey proteins, forming soluble whey protein aggregates. These aggregates further interact with casein micelles to form micellar aggregates during the first 30 min of sonication. Increase in sonication time causes the partial disruption of some whey proteins from these aggregates (Shanmugam et al. 2012).

US treatment of brain heart infusion broth, skimmilk and liquid egg inoculated with S. Typhimurium at 20 and 40°C for 30 min led to 1 and 3 log cfu ml^{-1} reductions in the counts of this bacterium, respectively (Wrigley and Llorca 1992). The counts of coliform bacteria in milk samples treated with 800 kHz for 1 min with a power intensity of 8.4 W cm^{-2} was reduced by 93%. Combination of UV irradiation with sonication increased the inactivation rate of coliforms to 99% with a possible explanation of that fat globules are broken up by the ultrasound, allowing deeper penetration of the UV, thus resulting in a more efficient process for microbial inactivation (Munkacsi and Elhami 1976).

The combined effect of ultrasonic (20 KHz, 150 W) and heat treatment on the survival of two strains of *Bacillus subtilis* in three suspending media (distilled water, glycerol and milk) revealed that when spores are suspended in water or milk were subjected to ultrasonic waves before heat treatments a little or no decrease of the heat resistance was observed.

However, when both sporicidal agents were applied simultaneously (thermo-ultrasonication), initial number of *B. subtilis*, var. *niger*-40 and *B. subtilis* ATCC 6051 decreased by 63 and 74%, respectively.

13.5 Pulsed ultraviolet light

Although it has been mainly used for surface treatment, there is a growing interest to use ultraviolet (UV) light for food preservation. Studies focused on food preservation by UV light have been mostly performed in recent years even though the earlier studies using UV light as a bactericidal agent goes back to 1828 (Demirci and Krishnamurthy 2011, Sizer and Balasubramaniam 1999). UV processing wavelength ranges from 100 to 400 nm having higher energy than infrared and lower energy than X-rays (Bintsis et al. 2000, Sastry et al. 2000).

The UV light in the range of 220 and 300 nm is considered germicidal against microorganisms such as bacteria, viruses, protozoa, molds and yeasts, and algae (Bintsis et al. 2000, Perchonok 2003, Sizer and Balasubramaniam 1999). The highest germicidal effect is obtained between 250 and 270 nm, but it may decrease as the wavelength is increased (e.g., above 300 nm, the germicidal effect is negated) (Bachmann 1975). Therefore, 254 nm wavelength (UV-C, generated by LPM lamps) is used for disinfection of surfaces, water and some food products. Bacteria suspended in air are more sensitive to UV-C light than those suspended in liquids due to the different penetration capacity of UV light through different physical media (Bintsis et al. 2000).

Energy transfer from UV light to an exposed material is realized by photons. Energy of photons in UV light range is very high and this energy can even cause ionization of molecules, whereas visible light and infrared region cause vibration and rotation of molecules, respectively. Molecules that absorb energy are elevated from the ground state to an excited state. The excited molecule can either (i) relax back to the ground state by releasing the energy as heat, (ii) relax back to the ground state by releasing energy as photons, or (iii) can induce some chemical changes (Demirci and Krishnamurthy 2011).

The effect of UV radiation on microbial inactivation may vary from species to species and, in the same species, may depend on the strain, growth media, stage of culture, number of microorganisms and other characteristics, such as type and composition of the food (Wright et al. 2000). Generally, fungi and yeasts which have bigger cell size are more resistant than bacteria during disinfection; however, high microbial levels should be taken into account when using UV-C for disinfection (Bachmann 1975).

Inactivation of microorganisms by radiation may occur due to the radiation absorbed by DNA which may stop cell growth and lead to cell death (Liltved and Landfald 2000). Physical shifting of electrons to render splitting of the DNA bonds, delay of reproduction or cell death is caused by the absorbed UV-C light (Anonymous 2002) indicating that the bactericidal effect of UV-C occurs mainly at the nucleic acid level (Wright et al. 2000). UV-C radiation causes a cross-linking between neighboring thymine and cytosine (pyrimidine nucleoside bases) in the same DNA strand. One of the most commonly occurring DNA photoproducts is the cyclobutyl pyrimidine dimers. DNA mutations might be produced due to the cross-linking effects of UV-C irradiation in the injured organism (Sastry et al. 2000). UV-A (315–400 nm) has better penetration capacity than UV-C, and it affects bacterial cells by causing membrane damages and/or generate active oxygen species or H_2O_2 on cell structure (Kramer and Ames 1987). UV-C light only penetrates a very short depth into the surface of liquids other than clear water (Sharma 1999). UV light penetration into juices is about 1 mm for absorption of 90% of the light (Sizer and Balasubramaniam 1999). Intensity of penetration decreases as the level of solids in liquids increases (Bintsis et al. 2000, Sharma 1999). It is reported that in order to ensure an adequate reduction in the counts of microorganisms in milk, it is necessary for all parts of the fluid to be exposed to at least 400 J m^{-2} of UV light at 254 nm (Anonymous 1999).

Because of the structural differences, spore inactivation mechanism by UV light is different from that of vegetative cells (Riesenman and Nicholson 2000). Resistance of spores is induced by the thick protein coat present in spores. The DNA of bacterial spore has a different conformation than those of the vegetative cell DNA, and thus, no detectable amount of thymine-containing dimers are reported in *Bacillus* spores. The main photoproduct is reported as 5-thyminyl-5,6 dihydrothymine adduct; later, it is termed as "spore photoproduct" (Setlow and Setlow 1987).

13.5.1 Processing of dairy products by pulsed UV light

Processing of milk and other dairy products by UV light aims mostly inactivation of pathogenic and spoilage bacteria. The efficacy of pulsed UV light at 5, 8, or 11 cm distance from a UV light strobe with 20, 30, or 40 ml min^{-1} flow rate up to three times by re-circulation for continuous-flow milk treatment caused 0.55 to 7.26 log cfu ml^{-1} reduction in *S. aureus* counts (Krishnamurthy et al. 2007). Processing of raw cow's milk by continuous flow coiled tube ultraviolet reactor for a residence time of 17 s with cumulative UV-C dose of 16.822 mJ cm^{-2} resulted in 2.3 log cfu ml^{-1} reduction in total microbial count with no significant difference on the odor compounds of UV light-treated, untreated and control samples. Although UV-processing caused detectible changes in the odor of milk samples, no significant difference between the malondialdehyde and other reactive substances were detected between the untreated and UV-treated milk samples right after the treatment and during the storage. It is also reported that fresh or untreated raw milk samples have lower lipid oxidation products than that of UV-treated milk (Bandla et al. 2012).

UV light was less effective for the inactivation of *Mycobacterium avium* subsp. *paratuberculosis* in milk (0.5 to 1.0 log reduction per 1000 mJ ml^{-1}) than that of suspended in Middlebrook 7H9 broth (2.5 to 3.3 log reduction per 1000 mJ ml^{-1}) (Altic et al. 2007). Processing of milk by UV light and heat treatment causes loss of vitamin C. UV light treatment causes a decrease in the concentration of vitamin B_2 after several passes of milk through the UV system, and it also causes decreases in the levels of vitamins A and E. UV light sensitivity of vitamins is in the order of C > E > A > B_2 (Guneser and Karagul Yuceer 2012).

Goat's milk samples processed for 12 consecutive times using a UV fluid processor for a cumulative exposure time of 18 s and targeted UV dose of 15.8 +/– 1.6 mJ cm^{-2} reveal a significant difference between the odor of raw goat's milk and UV irradiated milk. As the UV dose is increased, the oxidation and hydrolytic rancidity of raw milk are increased. UV-C treatment caused more than 5 log cfu ml^{-1} reduction in numbers of *L. monocytogenes* in goat's milk (Matak et al. 2005). Processing of milk by UV dose of 21.3 mJ cm^{-2}, with a flow rate of 28.8 l hr^{-1} and a tube's inner diameter of 1.5 mm caused more than 6 log cfu ml^{-1} reduction in the initial counts of bacteria. A recent study showed that application of the UV dose with 21.3 mJ cm^{-2} caused 3–4 log cfu ml^{-1} reduction in *Salmonella* spp., *Shigella* spp., *L. monocytogenes*, *Staphylococcus* spp., Enterobacteriaceae, LAB, pseudomonads and the total aerobic bacteria (Lu et al. 2011).

Processing of whey protein isolate (WPI) solutions at concentrations of 1% and 5% (w/v) circulated at flow rates ranging from 30 to 800 ml min^{-1} suggests that UV treatment increased the concentration of total and accessible thiol groups in 1% WPI solutions, while no change in 5% WPI solutions was observed. As a result of UV treatment, oxidation products of aromatic amino acids (e.g., N-formylkynurenine and dityrosine) occurred, and protein conformation was changed leading to increased susceptibility of whey protein to hydrolysis by pepsin (Kristo et al. 2012).

13.6 Cold plasma

Non-thermal plasma (NTP), also referred to as cold atmospheric plasma or cold plasma, is one of the most novel non-thermal technologies being investigated as an alternative food processing tool. It was first described by Sir William Crookes in 1879 (Crookes 1879), and electron beams were first identified by Sir J.J. Thomson in 1897 (Sarghini et al. 2008, 2011). This state of matter is described as: "Except near the electrodes, where there are *sheaths* containing very few electrons, the ionized gas contains ions and electrons in about equal numbers so that the resultant space charge is very small. We shall use the name *plasma* to describe this region containing balanced charges of ions and electrons" (Langmuir 1928). Plasma is considered to be the fourth state of matter, and defined as a quasi-neutral gas, following the more familiar states of solid, liquid and gas, and constitutes more than 99% matter of the universe. The concept of fourth state of matter occurs with the idea that phase transactions happens by progressively providing energy to the matter from the solid to the liquid up to the gas state. As more and more energy is supplied to the system, further phase transition from the gas state to the plasma occurs. It has net zero electrical charge formed by a gaseous mixture of both negatively and positively charged particles. It is more or less an electrified gas with a chemically reactive media that consists of a large number of different species such as electrons, positive and negative ions, free radicals, free electrons, gas atoms and activated neutral species (exited and radical) (Sasai et al. 2011).

It is possible to obtain plasma with extremely broad range of temperature and pressure. By coupling energy to a gaseous medium by several means such as mechanical, thermal, chemical, radiant, nuclear, or by applying a voltage, or by injecting electromagnetic waves and also by a combination of these to dissociate the gaseous component molecules into a collection of ions, electrons, charge-neutral gas molecules, and other species, plasma can be produced at low-pressure or atmospheric pressure (Sarghini et al. 2008). Plasma, based on the differences in their characteristics, is classified into two types as thermal and cold plasma (Sasai et al. 2011).

The thermal plasma can be generated under atmospheric pressure. It is described as the state of fully ionized gas characterized by a high gas temperature and an approximate equality between the gas and electron temperature ($Tg \approx Te$). Because the energy level of this plasma is very high, it is able to break any chemical bond. On the other hand, cold plasma has a low gas temperature and a high electron temperature ($Tg << Te$), and thus, it is easier to obtain cold plasma by electrical discharges under reduced pressure. Cold plasma irradiation of surface treatment has limited penetration (ca. 500–1000 Å); therefore, only the surface properties of a food can be changed without affecting the bulk properties (Georghiou et al. 2005, Gibalov and Pietsch 2004).

13.6.1 Processing of dairy products by cold plasma

Although it is used in medical, chemistry and polymer industries, limited studies have been carried out on the application of cold plasma to milk and dairy products. In one study, the effects of plasma application on the inactivation rate of *E. coli, S. aureus* and *S.* Typhimurium in whole, semi-skimmed and skimmed milk stored at 4°C for 42 days were investigated, and the reductions in the counts of *E. coli*, *S.* Typhimurium and *S. aureus* after plasma application at 20 kV were reported as 3.63, 2.00 and 2.62 log cfu ml^{-1}, respectively. No significant changes were noted in pH and color of the milk samples. No viable cells were detected after one week examination in whole milk samples, and the samples were remained stable for over six-week storage.

In order to determine the reduction in the allergenicity of isolated major milk proteins, milk was processed by ultrasound, non-thermal atmospheric plasma and UV-C light treatments. Ultrasound and plasma treatments did not show any differences in SDS-PAGE patterns of casein, β-lactoglobulin and α-lactalbumin indicating no significant change in protein concentration and/or composition. No significant difference was detected in IgE binding values of the control and the samples treated with ultrasound and cold plasma. It is reported that while the UV-C light treatment can reduce the allergenicity of major milk proteins to some extent, high intensity ultrasound and non-thermal atmospheric plasma treatments fail to generate effective results for reducing allergenicity (Tammineedi 2012).

Inactivation of *L. monocytogenes* inoculated into sliced cheese by atmospheric pressure plasma (APP), which is capable of operating at atmospheric pressure in air with 75, 100, 125 and 150 W input powers and 60, 90 and 120 s plasma exposure times revealed that the rate of microbial reduction increased with increased input power and plasma exposure time. The viable cells of *L. monocytogenes* was reduced by 1.70, 2.78 and 5.82 log cfu g^{-1} in sliced cheese after 120 s APP treatments at 75, 100 and 125 W, respectively. The exposure time required to inactivate 90% of microbial population of sliced cheese using APP (with 75, 100, 125 and 150 W) were 71.43, 62.50, 19.65 and 17.27 s, respectively. No viable cells were detected at 125 and 150 W of APP treatments in sliced cheese (Song et al. 2009).

References

Abee, T. and J.A. Wouters. 1999. Microbial stress response in minimal processing. Int. J. Food. Microbiol. 50: 65–91.

Ahn, J., V.M. Balasubramaniam and A.E. Yousef. 2007. Inactivation kinetics of selected aerobic and anaerobic bacterial spores by pressure-assisted thermal processing. Int. J. Food Microbiol. 113: 321–329.

Alpas, H., N. Kalchayanand, F. Bozoglu, A. Sikes, C.P. Dunne and B. Ray. 1999. Variation in resistance to hydrostatic pressure among strains of food-borne pathogens. Appl. Environ. Microbiol. 65: 4248–4251.

Altic, L.C., M.T. Rowe and I.R. Grant. 2007. UV light inactivation of *Mycobacterium avium* subsp. *paratuberculosis* in milk as assessed by *FASTPlaque* TB phage assay and culture. Appl. Environ. Microbiol. 73: 3728–3733.

Anonymous. 1999. UV light provides alternative to heat pasteurization of juices. Food Technol. 53: 144–144.

Anonymous. 2002. Some Microorganisms Deactivated by Ultraviolet Germicidal Light. Hauppauge, Atlantic Ultraviolet Co., New York, NY, USA.

Arias, M., R. Lopez-Fandino and A. Olano. 2000. Influence of pH on the effects of high pressure on milk proteins. Milchwiss. 55: 191–194.

Ashokkumar, M. and F. Grieser. 2007. The effect of surface active solutes on bubbles in an acoustic field. Phys. Chem. Chem. Phys. (PCCP) 9: 5631–5643.

Ashokkumar, M. and T.J. Mason. 2007. Sonochemistry Kirk-Othmer. Encyc. J. Dairy Sci. 37: 645.

Bachmann, R. 1975. Sterilization by intense ultraviolet radiation. The Brown Boveri Rev. 62: 206–209.

Balasubramaniam, V.M., D. Farkas and E.J. Turek. 2008. Preserving foods through high-pressure processing. Food Technol. 62: 32–38.

Bandla, S., R. Choudhary, D.G. Watson and J. Haddock. 2012. Impact of UV-C processing of raw cow milk treated in a continuous flow coiled tube ultraviolet reactor. Agric. Eng. Int. CIGR J. 14: 1–11.

Bansal, N., P.F. Fox and P.L.H. McSweeney. 2007. Factors affecting the retention of rennet in cheese curd. J. Agric. Food Chem. 55: 9219–9225.

Bansal, N., P.F. Fox and P.H.L. McSweeney. 2009. Comparison of the level of residual coagulant activity in different cheese varieties. J. Dairy Res. 76: 290–293.

Belloque, J., R. Lopez-Fandino and G.M. Smith. 2000. A HNMR study on the effect of high pressures on β-Lactoglobulin. J. Agric. Food Chem. 48: 3906–3912.

Benito, A., G. Ventoura, M. Casadei, T. Robinson and B. Mackey. 1999. Variation in resistance of natural isolates of *Escherichia coli* O157 to high hydrostatic pressure, mild heat and other stresses. Appl. Environ. Microbiol. 65: 1564–1569.

Bermudez-Aguirre, D., J.A. Yanez, C.P. Dunne, N.M. Davies and G.V. Barbosa-Canovas. 2010. Study of strawberry flavored milk under pulsed electric field processing. Food Res. Int. 43: 2201–2207.

Bermudez-Aguirre, D., S. Fernández, H. Esquivel, P.C. Dunne and G.V. Barbosa-Cánovas. 2011. Milk processed by pulsed electric fields: Evaluation of microbial quality, physicochemical characteristics, and selected nutrients at different storage conditions. J. Food Sci. 76: 289–299.

Bintsis, T., E. Litopoulou-Tzanetaki and R.K. Robinson. 2000. Existing and potential applications of ultraviolet light in the food industry—A critical review. J. Sci. Food Agric. 80: 637–645.

Black, E.P., P. Setlow, A.D. Hocking, C.M. Stewart, A.L. Kelly and D.G. Hoover. 2007. Response of spores to high-pressure processing. Comp. Rev. Food Sci. Food Safety 6: 103–119.

Black, E.P., C.M. Stewart and D.G. Hoover. 2011. Microbiological aspects of high pressure food processing. pp. 51–71. *In:* H.Q. Zhang, G.V. Barbosa-Cánovas, V.M. Balasubramaniam, C.P. Dunne, D.F. Farkas and J.T.C. Yuan [eds.]. Nonthermal Processing Technologies for Food. Wiley-Blackwell John Wiley & Sons Ltd., West Sussex, UK.

Bosiljkov, T., B. Tripalo, M. Brnčić, D. Ježek, S. Karlović and I. Jagušt. 2011. Influence of high intensity ultrasound with different probe diameter on the degree of homogenization (variance) and physical properties of cow milk. African J. Biotechnol. 10: 34–41.

Bowman, J.P., C.R. Bittencourt and T. Ross. 2008. Differential gene expression of *Listeria monocytogenes* during high hydrostatic pressure processing. Microbiol. 154: 462–475.

Brnčić, M., S. Karlović, S. Rimac-Brnčić, A. Penava, T. Bosiljkov, D. Ježek and B. Tripalo. 2010. Textural properties of infra-red dried apple slices as affected by high power ultrasound pre-treatment. African J. Biotechnol. 9: 6907–6915.

Bryant, G. and J. Wolfe. 1987. Electomechanical stress produced in the plasma membranes of suspended cells by applied electrical fields. J. Membrane Biol. 96: 129–139.

Buchheim, W., M. Schutt and E. Frede. 1996. High pressure effects on emulsified fats. pp. 331–336. *In:* R. Hayash and C. Balny [eds.]. High Pressure Bioscience and Biotechnology. Elsevier Science, Amsterdam, The Netherlands.

Buckow, R., S. Isbarn, D. Knorr, V. Heinz and A. Lehmacher. 2008. Predictive model for inactivation of feline calicivirus, a norovirus surrogate, by heat and high hydrostatic pressure. Appl. Environ. Microbiol. 74: 1030–1038.

Buffa, M., A.J. Trujillo, C. Royo and B. Guamis. 2000. Changes in chemical and microbiological characteristics of goat cheese made from raw, pasteurized or high-pressure-treated milk. High Pres. Res. 19: 27–32.

Buffa, M., B. Guamis, C. Royo and A.J. Trujillo. 2001. Microbiological changes throughout ripening of goat cheese made from raw, pasteurized and high-pressure-treated milk. Food Microbiol. 1: 45–51.

Butz, P. and B. Tauscher. 2002. Emerging technologies: Chemical aspects. Food Res. Int. 35: 279–284.

Calderon-Miranda, M., G. Barbosa-Canovas and B. Swanson. 1999. Inactivation of *Listeria innocua* in skim milk by pulsed electric fields and nisin. Int. J. Food Microbiol. 51: 19–30.

Cameron, M., L.D. McMaster and T.J. Britz. 2009. Impact of ultrasound on dairy spoilage microbes and milk components. Dairy Sci. Technol. 89: 83–98.

Capellas, M., M. Mor-Mur, E. Sendra, R. Pla and B. Guamis. 1996. Populations of aerobic mesophiles and inoculated *Escherichia coli* during storage of fresh goat's milk cheese treated with high pressure. J. Food Protect. 59: 582–587.

Capellas, M., M. Mor-Mur, R. Gervilla, J. Yuste and B. Guamis. 2000. Effect of high pressure combined with mild heat or nisin on inoculated bacteria and mesophiles of goats' milk fresh cheese. Food Microbiol. 17: 633–641.

Carminati, D., M. Gatti, B. Bonvini, E. Neviani and G. Muchetti. 2004. High-pressure processing of Gorgonzola cheese: Influence on *Listeria monocytogenes* inactivation and on sensory characteristics. J. Food Protect. 67: 1671–1675.

Casal, V. and R. Gomez. 1999. Effect of high pressure on the viability and enzymatic activity of mesophilic lactic acid bacteria isolated from caprine cheese. J. Dairy Sci. 82: 1092–1098.

Castro, A.J., B.G. Swanson, G.V. Barbosa-Canovas and A.K. Dunker. 2001. Pulsed electric field modification of milk alkaline phosphatase activity. pp. 83–103. *In:* G.V. Barbosa-Canovas and Q.H. Zhang [eds.]. Pulsed Electric Fields in Food Processing: Fundamental Aspects and Applications. Technomic Publishing Company Inc., Lancaster, PA, USA.

Chang, D.C. and T.S. Reese. 1990. Changes in membrane structure induced by electroporation as revealed by rapid-freezing electron microscopy. Biophys. J. 58: 1–12.

Cheftel, J.C. 1992. Effect of high hydrostatic pressure on food constituents: An overview. pp. 195–209. *In:* C. Balny, R. Hayashi, K. Heremans and P. Masson [eds.]. High Pressure and Biotechnology. Vol. 224. Coloque INSERM/John Libbey Eurotext, London, UK.

Chen, H. and D.G. Hoover. 2003. Modeling the combined effect of high hydrostatic pressure and mild heat on the inactivation kinetics of *Listeria monocytogenes* Scott A in whole milk. Innov. Food Sci. Technol. 4: 25–34.

Considine, T., H.A. Patel, S.G. Anema, H. Singh and L.K. Creamer. 2007. Interactions of milk proteins during heat and high hydrostatic pressure treatments: A review. Innov. Food Sci. Emerg. Technol. 8: 1–23.

Craven, H.M., P. Swiergon, S. Ng, J. Midgely, C. Versteeg, M.J. Coventry and J. Wan. 2008. Evaluation of pulsed electric field and minimal heat treatments for inactivation of pseudomonads and enhancement of milk shelf-life. Innov. Food Sci. Emerg. Technol. 9: 211–216.

Crookes, W. 1879. Radiant Matter, A resume of the principal lectures and papers of Prof. William Crookes on the fourth state of matter. Electric Spacecraft, USA.

Cruz-Romero, M., A.L. Kelly and J.P. Kerry. 2006. Effects of high-pressure and heat treatments on physical and biochemical characteristics of oysters (*Crassostrea gigas*). Innov. Food Sci. Emerg. Technol. 8: 30–38.

Dave, A., M. Walkling-Ribeiro, O. Rodríguez-González, M.W. Griffiths and M. Corredig. 2012. Effect of PEF and UV and their combination on selected microorganisms and physico-chemical properties in whey. J. Dairy Sci. 95(Suppl.): 168.

Deeth, H.C., N. Datta, A.I.V. Ross and X.T. Dam. 2007. Pulsed electric field technology: Effect on milk and fruit juices. pp. 241–269. *In:* G. Tewari and V.K. Juneja [eds.]. Advances in Thermal and Non-thermal Food Preservation. Blackwell Publishing. Ames, USA.

Degraeve, P., P. Delorme and P. Lemay. 1996. Pressure-induced inactivation of *E. coli* 3-galactosidase: Influence of pH and temperature. Biochim. Biophys. Acta. 1292: 61–68.

De Lamo-Castellvi, S., M. Capellas, T. Lopez-Pedemonte, M.M. Hernandez-Herrero, B. Guamis and A.X. Roig-Sagues. 2005. Behavior of *Yersinia enterocolitica* strains inoculated in model cheese treated with high hydrostatic pressure. J. Food Protect. 68: 528–533.

De Lamo-Castellvi, S., M. Capellas, A.X. Roig-Sagues, T. Lopez-Pedemonte, M.M. Hernandez-Herrero and B. Guamis. 2006. Fate of *Escherichia coli* strains inoculated in model cheese elaborated with or without starter and treated by high hydrostatic pressure. J. Food Protect. 69: 2856–2864.

De Lamo-Castellvi, S., A.X. Roig-Sagues, T. Lopez-Pedemonte, M.M. Hernandez-Herrero, B. Guamis and M. Capellas. 2007. Response of two *Salmonella enterica* strains inoculated in model cheese treated with high hydrostatic pressure. J. Dairy Sci. 90: 99–109.

del Pozo-Insfran, D., M.O. Balaban and S.T. Talcott. 2007. Inactivation of polyphenol oxidase in muscadine grape juice by dense phase-CO_2 processing. Food Res. Int. 40: 894–899.

Demirci, A. and K. Krishnamurthy. 2011. Pulsed ultraviolet light. pp. 225–261. *In:* H.Q. Zhang, G.V. Barbosa-Cánovas, V.M. Balasubramaniam, C.P. Dunne, D.F. Farkas and J.T.C. Yuan [eds.]. Nonthermal Processing Technologies for Food. Blackwell Publishing Ltd., London, UK.

Ding, Y.T., W.G. Sang, Z. Jin and J.W. Harper. 2001. High pressure treatment of Swiss cheese slurries inactivation of selected microorganisms after treatment and during accelerated ripening. J. Zhejiang Univ. 2: 204–208.

Drake, M.A., S.L. Harrison, M. Asplund, G. Barbosa-Canovas and B.G. Swanson. 1997. High pressure treatment of milk and effects of microbiological and sensory quality of Cheddar cheese. J. Food Sci. 62: 843–860.

Dunn, J.E. 1996. Pulsed light and pulsed electric field for foods and eggs. Poultry Sci. 75: 1133–1136.

Dunn, J.E. and J.S. Pearlman. 1987. Methods and apparatus for extending the shelf life of fluid food products. Patent No. US 4695472, USA.

Elez-Martinez, P. and O. Martin-Belloso. 2007. Effects of high intensity pulsed electric field processing conditions on vitamin C and antioxidant capacity of orange juice and gazpacho, a cold vegetable soup. Food Chem. 102: 201–209.

Evrendilek, G.A., H.W. Yeom, Z.T. Jin and Q.H. Zhang. 2004. Safety and quality evaluation of yogurt-based drink processed by a pilot plant PEF system. J. Food Proc. Eng. 27: 197–212.

Evrendilek, G.A., N. Koca, J.W. Harper and V.M. Balasubramanian. 2008. High-pressure processing of Turkish white cheese for microbial inactivation. J. Food Protect. 71: 102–108.

Evrendilek, G.A. and V.M. Balasubramaniam. 2011. Inactivation of *Listeria monocytogenes* and *Listeria innocua* in yogurt drink applying combination of high pressure processing and mint essential oils. Food Cont. 22: 1435–1441.

Farkas, D.F. and D.G. Hoover. 2000. High-pressure processing. J. Food Sci. (Suppl): 47–64.

Felipe, X., M. Capellas and A.R. Law. 1997. Comparison of the effects of high-pressure treatments and heat pasteurisation on the whey proteins in goat's milk. J. Agric. Food Chem. 45: 627–631.

Fernandez-Molina, J.J. 2001. Inactivation of *Listeria innocua* and *Pseudomonas fluorescens* in skim milk treated with pulsed electric fields. pp. 149–166. *In:* G.V. Barbosa-Canovas and Q.H. Zhang [eds.]. Pulsed Electric Fields in Food Processing: Fundamental Aspects and Applications. Technomic Publishing Company Inc., Lancaster, PA, USA.

Fernandez-Molina, J.J., G.V. Barbosa-Canovas and B.G. Swanson. 2000. Pasteurization of skim milk with heat and pulsed electric fields. pp. 172. *In:* IFT Annual Meeting Technical Programme, Institute of Food Technologists. Book of Abstracts. Dallas, Texas, USA.

Fernandez-Molina, J.J., S.A. Fernandez-Gutierrez, B. Altunakar, D. Bermudez-Aguirre, B.G. Swanson and G.V. Barbosa-Canovas. 2005. The combined effect of pulsed electric fields and conventional heating on the microbial quality and shelf-life of skim milk. J. Food Proc. Preserv. 29: 390–406.

Ferragut, V., V.M. Martinez, A.J. Trujillo and B. Guamis. 2000. Properties of yoghurts made from whole ewe's milk treated by high hydrostatic pressure. Milchwiss. 55: 267–269.

Floury, J., N. Grosset, N. Leconte, M. Pasco, M.N. Madec and R. Jeantet. 2006. Continuous raw skim milk processing by pulsed electric field at non-lethal temperature: Effect on microbial inactivation and functional properties. Lait 86: 43–57.

Fonberg-Broczek, M., B. Windyga, J. Szczawinski, M. Szczawinska, D. Pietrzak and G. Prestamo. 2005. High pressure processing for food safety. Acta. Biochim. Pol. 52: 721–724.

Gallot-Lavalle, T. 1998. Efficiency of high pressure treatment for destruction of *Listeria monocytogenes* in goat cheese from raw milk. Sci. Alimen. 18: 647–655.

Garcia-Amezquita, L.E., A.R. Primo-Mora, J.A. Guerrero-Beltran, G.V. Barbosa-Canovas and D.R. Sepulveda. 2013. Rennetability of cheese-making milk processed by nonthermal technologies. J. Food Proc. Eng. 36: 247–253.

Garde, S., J.L. Arques, P. Gaya, M. Medina and M. Nunez. 2007. Effect of high-pressure treatments on proteolysis and texture of ewes' raw milk La Serena cheese. Int. Dairy J. 17: 1424–1433.

Georghiou, G.E., A.P. Papadakis, R. Morrow and A.C. Metaxas. 2005. Numerical modeling of atmospheric pressure gas discharges leading to plasma production. J. Phys. D: Appl. Phys. 38: 303–328.

Gervilla, R., V. Ferragut and B. Guamis. 2001. High hydrostatic pressure effects on colour and milk-fat globule of ewe's milk. J. Food Sci. 66: 880–885.

Gibalov, V.I. and G.J. Pietsch. 2004. Dynamics of dielectric barrier discharges in coplanar arrangements. J. Phys. D: Appl. Phys. 37: 2082–2092.

Grahl, T. and H. Märkl. 1996. Killing of microorganisms by pulsed electric fields. Appl. Microbiol. Biotechnol. 45: 148–157.

Guerrero-Beltran, J.Á., D.R. Sepulveda, M.M. Gongora-Nieto, B. Swanson and G.V. Barbosa-Canovas. 2010. Milk thermization by pulsed electric fields (PEF) and electrically induced heat. J. Food Eng. 100: 56–60.

Guneser, O. and Y. Karagul-Yuceer. 2012. Effect of ultraviolet light on water- and fat-soluble vitamins in cow and goat milk. J. Dairy Sci. 95: 6230–6241.

Harrison, S.L., G.V. Barbosa-Canovas and B.G. Swanson. 1997. *Saccharomyces cerevisiae* structural changes induced by pulsed electric field treatment. Lebensmitt. Wiss. Technol. 30: 236–240.

Harte, F., M. Amonte, L. Luedecke, B.G. Swanson and G.V. Barbosa-Canovas. 2002. Yield stress and microstructure of set yogurt made from high hydrostatic pressure-treated full fat milk. J. Food Sci. 67: 2245–2250.

Harvey, E.N. and A.L. Loomis. 1929. The destruction of luminous bacteria by high frequency sound waves. J. Bacteriol. 17: 373–376.

Hauben, J.A., E.Y. Wuytack, C.C.F. Soontjens and C.W. Michiels. 1996. High-pressure transient sensitisation of *Escherichia coli* to lysozime and nisin by disruption of outer-membrane permeability. J. Food Protect. 59: 350–355.

Hayashi, R., Y. Kawamura and S. Kunugi. 1987. Introduction of high pressure to food processing: Preferential proteolysis of β-lactoglobulin in milk whey. J. Food Sci. 52: 1107–1108.

Heinz, V. and D. Knorr. 2002. Effects of high pressure on spores. pp. 77–113. *In:* M.E. Hendrickx and D. Knorr [eds.]. Ultra High Pressure Treatments of Foods. Kluwer/Plenum, New York, NY, USA.

Hite, B.H. 1899. The effect of pressure in the preservation of milk. West Virginia Agric. Exper. Station Bull. 58: 15–35.

Hogan, E., A.L. Kelly and D.-W. Sun. 2005. High pressure processing of foods: An overview. pp. 3–31. *In:* S. Da-Wen [ed.]. Emerging Technologies for Food Processing. Academic Press, London, UK.
Hoover, D.G., C. Metrick, A.M. Papineau, D.F. Farkas and D. Knorr. 1989. Biological effects of high hydrostatic pressure on food microorganisms. Food Technol. 43: 99–107.
Hulsen, U. 1999. Alternative heat treatment processes. Eur. Dairy Magazine 3: 20–24.
Huppertz, T., A.L. Kelly and P.F. Fox. 2002. Effects of high pressure on constituents and properties of milk. Int. Dairy J. 12: 561–572.
Huppertz, T., P.F. Fox and A.L. Kelly. 2004a. High pressure-induced denaturation of α-lactalbumin and β-lactoglobulin in bovine milk and whey: A possible mechanism. J. Dairy Res. 71: 489–495.
Huppertz, T., P.F. Fox and A.L. Kelly. 2004b. High pressure treatment of bovine milk: Effects of casein micelles and whey proteins. J. Dairy Res. 71: 97–106.
Jeličić, I., R. Božanić, Lj. Tratnik and K. Lisak. 2010. Possibilities of implementing nonthermal processing methods in the dairy industry. Mljekarstvo. 60: 113–126.
Juan, B., A.J. Trujillo, V. Guamis, M. Buffa and V. Ferragut. 2007. Rheological, textural and sensory characteristics of high-pressure treated semi-hard ewes' milk cheese. Int. Dairy J. 17: 248–254.
Juan, B., V. Ferragut, B. Guamis and A.J. Trujillo. 2008. The effect of high pressure treatment at 300 MPa on ripening of ewe's milk cheese. Int. Dairy J. 18: 129–138.
Kalchayanand, N., T. Sikes, C.P. Dunne and B. Ray. 1994. Hydrostatic pressure and electroporation have increased bactericidal efficiency in combination with bacteriocins. Appl. Environ. Microbiol. 60: 4174–4177.
Kaletunc, G., J. Lee, H. Alpas and F. Bozoglu. 2004. Evaluation of structural changes induced by high hydrostatic pressure in *Leuconostoc mesenteroides*. Appl. Environ. Microbiol. 70: 1116–1122.
Karim, P.A. 2011. High pressure processing as an alternative food preservation technology and its applications for fruits and vegetables. M.Sc. Thesis, Kansas State University, Manhattan Kansas, USA.
Khadre, M.A. and A.E. Yousef. 2002. Susceptibility of human rotavirus to ozone, high pressure, and electric field. J. Food Protect. 65: 1441–1446.
Kingsley, D.H., D.G. Hoover, E. Papafragkou and G.P. Richards. 2002. Inactivation of hepatitis A virus and a calcivirus by high hydrostatic pressure. J. Food Protect. 65: 1605–1609.
Kolakowski, P., A. Reps, F. Dajnowiec, J. Szczepek and S. Porowski. 1997. Effect of high pressures on the microflora of raw cow's milk. pp. 46–50. *In:* J.C. Oliveira. F.A.R. Oliveira [eds.]. Proceedings of Process Optimisation and Minimal Processing of Foods. Leuven University Press, Leuven, Belgium.
Knorr, D., M. Geulen, T. Grahl and W. Sitzman. 1994. Food application of high electric field pulses. Trends Food Sci. Technol. 5: 71–75.
Kouassi, G.K., R.C. Anantheswaran, S.J. Knabel and J.D. Floros. 2007. Effect of high-pressure processing on activity and structure of alkaline phosphatase and lactate dehydrogenase in buffer and milk. J. Agric. Food Chem. 55: 9520–9529.
Kramer, G.F. and B.N. Ames. 1987. Oxidative mechanisms of toxicity of low-intensity near-UV light in *Salmonella* Typhimurium. J. Bacteriol. 169: 2259–2266.
Krishnamurthy, K., A. Demirci and J.M. Irudayaraj. 2007. Inactivation of *Staphylococcus aureus* in milk using flow-through pulsed UV-light treatment system. J. Food Sci. 72: 233–239.
Kristo, E., A. Hazizaj and M. Corredig. 2012. Structural changes imposed on whey proteins by UV irradiation in a continuous UV light reactor. J. Agric. Food Chem. 60: 6204–6209.
Lado, B.H. and A.E. Yousef. 2002. Alternative food-preservation technologies: Efficacy and mechanisms. Microbes Infec. 4: 433–440.
Langmuir, I. 1928. Oscillations in ionized gases. Proc. Nat. Aca. Sci. US 14: 628.
Law, A.J.R., J. Leaver, X. Felipe, V. Ferragut, R. Pla and B. Guamis. 1998. Comparison of the effects of high pressure and thermal treatments on the casein micelles in goat's milk. J. Agric. Food Chem. 46: 2523–2530.

Liltved, H. and B. Landfald. 2000. Effects of high intensity light on ultraviolet-irradiated and non-irradiated fish pathogenic bacteria. Water Res. 34: 481–486.

Lopez, M.A., E. Palou, G.V. Barbosa-Canovas, J. Welti-Chanes and B.G. Swanson. 1998. Polyphenoloxidase activity and color changes during storage of high hydrostatic pressure treated avocado puree. Food Res. Int. 31: 549–556.

Lopez-Fandino, R., A.V. Carrascosa and A. Olano. 1996. The effect of high pressure on whey protein denaturation and cheese making properties of raw milk. J. Dairy Sci. 79: 929–936.

Lopez-Fandino, R. 2006. High-pressure-induced changes in milk proteins and possible applications in dairy technology. Int. Dairy J. 16: 1119–1131.

Lopez-Pedemonte, T., A. Roig-Sagues, S. De Lamo, M. Hernandez-Herrero and B. Guamis. 2007a. Reduction of counts of *Listeria monocytogenes* in cheese by means of high hydrostatic pressure. Food Microbiol. 24: 59–66.

Lopez-Pedemonte, T., A.X. Roig-Sagues, S. De Lamo, R. Gervilla and B. Guamis. 2007b. High hydrostatic pressure treatment applied to model cheeses made from cows' milk inoculated with *Staphylococcus aureus*. Food Cont. 18: 441–447.

Lu, G., C. Li and P. Liu. 2011. UV inactivation of milk-related microorganisms with a novel electrodeless lamp apparatus. Eur. Food Res. Technol. 233: 79–87.

Malone, A.S., T.H. Shellhammer and P.D. Courtney. 2002. Effects of high pressure on the viability, morphology, lysis and cell wall activity of *Lactococcus lactis* subsp. *cremoris*. Appl. Environ. Microbiol. 68: 4357–4363.

Manas, P. and B.M. Mackey. 2004. Morphological and physiological changes induced by high hydrostatic pressure in exponential and stationary-phase cells of *Escherichia coli*: Relationship with cell death. Appl. Environ. Microbiol. 70: 1545–1554.

Manas, P., R. Pagan, J. Raso, F.J. Sala and S. Condon. 2000. Inactivation of *Salmonella* Typhimurium, and *Salmonella* Senftenberg by ultrasonic waves under pressure. J. Food Protect. 63: 451–456.

Martin, O., B.L. Qin, F.J. Chang, G.V. Barbosa-Canovas and B.G. Swanson. 1997. Inactivation of *Escherichia coli* in skim milk by high intensity pulsed electric fields. J. Food Proc. Eng. 20: 317–336.

Martinez-Rodriguez, Y., C. Acosta-Muniz, G.I. Olivas, J. Guerrero-Beltran, D. Rodrigo-Aliaga and D.R. Sepulveda. 2012. High hydrostatic pressure processing of cheese. Comp. Rev. Food Sci. Food Safety 11: 399–416.

Mason, T.J. 1998. Power ultrasound in food processing—The way forward. pp. 105–126. *In:* M.J.W. Povey and T.J. Mason [eds.]. Ultrasound in Food Processing. Blackie Academic & Professional, London, UK.

Masschalck, B., R. van Houdt and C.W. Michiels. 2000. High pressure increases bactericidal activity and spectrum of lactoferrin, lactoferricin and nisin. Int. J. Food Microbiol. 64: 325–332.

Matak, K.E., J.J. Churey, R.W. Worobo, S.S. Sumner, E. Hovingh, C.R. Hackney and M.D. Pierson. 2005. Efficacy of UV light for the reduction of *Listeria monocytogenes* in goat's milk. J. Food Protect. 68: 2212–2216.

McClements, D.J. 1995. Advances in the application of ultrasound in food analysis and processing. Trends Food Sci. Technol. 6: 293–299.

McClements, J.M.J., M.F. Patterson and M. Linton. 2001. The effect of growth stage and growth temperature on high hydrostatic pressure inactivation of some psychrotrophic bacteria in milk. J. Food Protect. 64: 514–522.

Mertens, B. and D. Knorr. 1992. Developments of nonthermal processes for food preservation. Food Technol. 46: 124–133.

Messens, W., K. Dewettinck, J. van Camp and A. Huyghebaert. 1998. High pressure brining of Gouda cheese and its effect on the cheese serum. Lebensmitt. Wiss. Technol. 31: 552–558.

Michalac, S., V. Alvarez, Z.T. Jin and Q.H. Zhang. 2003. Inactivation of selected microorganisms and properties of pulsed electric field processed milk. J. Food Proc. Pres. 27: 137–151.

Miles, C.A., M.J. Morley, W.R. Hudson and B.M. Mackey. 1995. Principles of separating microorganisms from suspensions using ultrasound. J. Appl. Bacteriol. 78: 47–54.

Munkacsi, F. and M. Elhami. 1976. Effect of ultrasonic and ultraviolet irradiation on chemical and bacteriological quality of milk. Egyptian J. Dairy Sci. 4: 1–6.

Needs, E.C., R.A. Stenning, A.L. Gill, V. Ferragut and G.T. Rich. 2000. High pressure treatment of milk: Effects of casein micelle structure and on enzymatic coagulation. J. Dairy Res. 67: 31–42.

Niven, G.W., C.A. Miles and B.M. Mackey. 1999. The effects of hydrostatic pressure on ribosome conformation in *Escherichia coli*: An *in vivo* study using differential scanning calorimetry. Microbiol. 145: 419–425.

Ordonez, J.A., B. Sanz, P.E. Hernandez and P. Lopez-Lorenzo. 1984. A note on the effect of combined ultrasonic and heat treatments on the survival of thermoduric streptococci. J. Appl. Bacteriol. 54: 175–177.

O'Reilly, C.E., P.M. O'Connor, A.L. Kelly, T.P. Beresford and P.M. Murphy. 2000a. Use of hydrostatic pressure for inactivation of microbial contaminants in cheese. Appl. Environ. Microbiol. 66: 4890–4896.

O'Reilly, C.E., P.M. O'Connor, P.M. Murphy, A.L. Kelly and T.P. Beresford. 2000b. The effect of exposure to pressure of 50 MPa on Cheddar cheese ripening. Innov. Food Sci. Emerg. Technol. 1: 109–117.

O'Reilly, C.E., A.L. Kelly, J.C. Oliviera, P.M. Murphy, M.A.E. Auty and T.P. Beresford. 2003. Effect of varying high-pressure treatment conditions on acceleration of ripening of Cheddar cheese. Innov. Food Sci. Emerg. Technol. 4: 277–284.

Oxen, P. and D. Knorr. 1993. Baroprotective effects of high solute concentrations against inactivation of *Rhodotorula rubra*. Lebensmitt. Wiss. Technol. 26: 220–223.

Pagán, R. and B.M. Mackey. 2000. Relationship between membrane damage and cell death in pressure-treated *Escherichia coli* cells: Differences between exponential and stationary phase cells and variation among strains. Appl. Environ. Microbiol. 66: 2829–2834.

Patel, H.A., T. Carroll and A.L. Kelly. 2008. Nonthermal preservation technologies for dairy applications. pp. 465–482. *In:* R.C. Chandan, A. Kilara and N.P. Shah [eds.]. Dairy Processing and Quality Assurance. Wiley-Blackwell, Ames, Iowa, USA.

Patterson, M.F., M. Quinn, R. Simpson and A. Gilmour. 1995. The sensitivity of vegetative pathogens to high hydrostatic pressure treatment in phosphate buffered saline and foods. J. Food Protect. 58(5): 524–529.

Patterson, M.F. 2005. A review: Microbiology of pressure-treated foods. J. Appl. Microbiol. 98: 1400–1409.

Peleg, M. 1995. A model of microbial survival after exposure to pulsed electric fields. J. Sci. Food Agric. 67: 93–99.

Perchonok, M. 2003. Advanced food technology workshop report. Vol. 1. Houston, TX: National Aeronautics and Space Administration. Available at: http://advlifesupport.jsc.nasa.gov.

Pfister, M. and L.I. Dehne. 2001. High pressure processing-an overview of chemical changes in food. Deutsche Lebensmittel-Rundschau. 97: 257–268.

Pilavtepe-Celik, M., M.O. Balaban, H. Alpas and A.E. Yousef. 2008. Image analysis based quantification of bacterial volume change with high hydrostatic pressure. J. Food Sci. 73: 423–429.

Pittia, P., P.J. Wilde, F.A. Husband and D.C. Clark. 1996. Functional and structural properties of β-lactoglobulin as affected by high pressure treatment. J. Food Sci. 61: 1123–1128.

Piyasena, P., E. Mohareb and R.C. Mckellar. 2003. Inactivation of microbes using ultrasound: A review. Int. J. Food Microbiol. 87: 207–216.

Pothakamury, U.R., H. Vega, Q. Zhang, G.V. Barbosa-Canovas and B.G. Swanson. 1996. Effect of growth stage and processing temperature on the inactivation of *Escherichia coli* by pulsed electric fields. J. Food Protect. 59: 1167–1171.

Pothakamury, U.R., G.V. Barbosa-Canovas, B.G. Swanson and K.D. Spence. 1997. Ultrastructural changes in *Staphylococcus aureus* treated with pulsed electric fields. Food Sci. Technol. Int. 3: 113–121.

Qin, B.L., U.R. Pothakamury, H. Vega, O. Martin, G.V. Barbosa-Canovas and B.G. Swanson. 1995. Food pasteurization using high-intensity pulsed electric fields. Food Technol. 49: 55–60.

Rademacher, B., B. Pfiffer and H.G. Kessler. 1998. Inactivation of microorganisms and enzymes in pressure treated raw milk. pp. 145–151. *In:* N.S. Issacs [ed.]. High Pressure Food Science, Bioscience and Chemistry. The Royal Society of Chemistry, Cambridge, UK.

Raso, J., R. Pagan, S. Condon and F.J. Sala. 1998. Influence of temperature and pressure on the lethality of ultrasound. Appl. Environ. Microbiol. 64: 465–471.

Raso, J., M.M. Gongora, M.L. Calderon, G.V. Barbosa-Canovas and B.G. Swanson. 1999. Resistant micro-organisms to high intensity pulsed electric field pasteurization of raw skim milk. IFT Annual Meeting Technical Program. Institute of Food Technologists, Chicago, Illinois, USA. 208 p.

Rastogi, N.K., K.S.M.S. Raghavarao, V.M. Balasubramaniam, K. Niranjan and D. Knorr. 2007. Opportunities and challenges in high pressure processing of foods. Crit. Rev. Food Sci. 47: 69–112.

Ray, B. 1992. Nisin of *Lactoccocus lactis* spp. *lactis* as a food biopreservative. pp. 207–264. *In:* B. Ray and M.A. Daeschel [eds.]. Food Biopreservatives of Microbial Origin. CRC Press, Roca Baton, FL, USA.

Reina, L.D., Z.T. Jin, Q.H. Zhang and A.E. Yousef. 1998. Inactivation of *Listeria monocytogenes* in milk by pulsed electric field. J. Food Protect. 61: 1203–1206.

Režek-Jambrak, A., V. Lelas, Z. Herceg, M. Badanjak, V. Batur and M. Muža. 2009. Advantages and disadvantages of high power ultrasound application in the dairy industry. Mljekarstvo 59: 267–281.

Režek-Jambrak, A., Z. Herceg, D. Šubarić, J. Babić, M. Brnčić, S. Rimac-Brnčić, T. Bosiljkov, D. Čvek, B. Tripalo and J. Gelo. 2010. Ultrasound effect on physical properties of corn starch. Carbohydrate Polymers. 79: 91–100.

Riesenman, P.J. and W.L. Nicholson. 2000. Role of the spore coat layers in *Bacillus subtilis* spore resistance to hydrogen peroxide, artificial UV-C, UV-B, and solar UV radiation. Appl. Environ. Microbiol. 66: 620–626.

Ritz, M., J.L. Tholozan, M. Federighi and M.F. Pilet. 2001. Morphological and physiological characterization of *Listeria monocytogenes* subjected to high hydrostatic pressure. Appl. Environ. Microbiol. 76: 2240–2247.

Ritz, M., J.L. Tholozan and M. Federighi. 2002. Physiological damages of *Listeria monocytogenes* treated by high hydrostatic pressure. Int. J. Food Microbiol. 79: 47–53.

Rynne, N.M., T.P. Beresford, T.P. Guinee, E. Sheehan, C.M. Delahunty and A.L. Kelly. 2008. Effect of high-pressure treatment of 1 day-old full-fat Cheddar cheese on subsequent quality and ripening. Innov. Food Sci. Emerg. Technol. 9: 429–440.

Saldo, J., P.L.H. McSweeney, E. Sendra, A.L. Kelly and B. Guamis. 2000a. Changes in curd acidification caused by high pressure treatment. Irish J. Agric. Food Res. 39: 169–173.

Saldo, J., E. Sendra and B. Guamis. 2000b. High hydrostatic pressure for accelerating ripening of goats' milk cheese: Proteolysis and texture. J. Food Sci. 65: 636–640.

Saldo, J., P.L.H. McSweeney, E. Sendra, A.L. Kelly and B. Guamis. 2002. Proteolysis in caprine milk cheese treated by high pressure to accelerate cheese ripening. Int. Dairy J. 12: 35–44.

Sale, A.J.H. and W.A. Hamilton. 1968. Effects of high electric fields in microorganisms. III. Lysis of erythrocytes and protoplasts. Biochim. Biophys. Acta. 163: 37–43.

Sampedro, F., M. Rodrigo, A. Martinez, D. Rodrigo and G.V. Barbosa-Canovas. 2005. Quality and safety aspects of PEF application in milk and milk products. Cri. Rev. Food Sci. Nutr. 45: 25–42.

Sarghini, S., S. Paulussen and H. Terryn. 2008. Atmospheric plasma deposition of antibacterial coatings. pp. 1–5. *In:* Proceedings of the 11th International Symposium on High

Pressure Low Temperature Plasma Chemistry (Hakone XI), Oléron Island, France, 7–12 September.

Sarghini, S., S. Paulussen and H. Terryn. 2011. Plasma Process. Polymer. 8: 59.

Sasai, Y., S.-I. Kondo, Y. Yamauchi and M. Kuzuya. 2011. Cold plasma techniques for pharmaceutical and biomedical engineering. pp. 102–122. *In:* A.N. Laskovski [ed.]. Biomedical Engineering, Trends in Materials Science. In Tech Publishing, Rijeka, Croatia.

Sastry, S.K., A.K. Datta and R.W. Worobo. 2000. Ultraviolet light. J. Food Sci. Suppl. 65: 90–92.

Scherba, G., R.M. Weigel and W.D. O'Brien. 1991. Quantitative assessment of the germicidal efficacy of ultrasonic energy. Appl. Environ. Microbiol. 57: 2079–2084.

Schrader, K., W. Buchheim and C.V. Morr. 1997. High pressure effects on the colloidal calcium phosphate and the structural integrity of micellar casein in milk. Part 1. High pressure dissolution of colloidal calcium phosphate in heated milk systems. Nahrung. 41: 133–138.

Sepulveda-Ahumada, D.R., E. Ortega-Rivas and G.V. Barbosa-Canovas. 2000. Quality aspects of Cheddar cheese obtained with pasteurized milk by pulsed fields. Food Bioproducts Process. 78: 66–71.

Setlow, B. and P. Setlow. 1987. Thymine-containing dimers as well as spore photoproducts are found in ultraviolet irradiated *Bacillus subtilis* spores that lack small acid-soluble proteins. Proc. Nat. Acad. Sci. 84: 421–423.

Setlow, B., E. Melly and P. Setlow. 2001. Properties of spores of *Bacillus subtilis* blocked at an intermediate stage in spore germination. J. Bacteriol. 183: 4894–4899.

Seyderhelm, I., S. Boguslawski, G. Michaels and D. Knorr. 1996. Pressure induced inactivation of selected food enzymes. J. Food Sci. 61: 308–310.

Sharma, G. 1999. Ultraviolet light. pp. 2208–2214. *In:* R.K. Robinson, C. Batt and P. Patel [eds.]. Encyclopedia of Food Microbiology. Academic Press, San Diego, CA, USA.

Shanmugam, A., J. Chandrapala and M. Ashokkumar. 2012. The effect of ultrasound on the physical and functional properties of skim milk. Innov. Food Sci. Emerg. Technol. 16: 251–258.

Shao, Y., H.S. Ramaswamy and S. Zhu. 2007. High-pressure destruction kinetics of spoilage and pathogenic bacteria in raw milk cheese. J. Food Proc. Eng. 30: 357–374.

Shin, J.K., K.J. Jung, Y.R. Pyun and M.S. Chung. 2007. Application of pulsed electric fields with square wave pulse to milk inoculated with *E. coli, P. fluorescens* and *B. stearothermophilus*. Food Sci. Biotechnol. 116: 1082–1084.

Sierra, I., C. Vidal-Valverde and R. Lopez Fandino. 2000. Effect of high pressure on the vitamin B1 and B6 content in milk. Milchwiss. 55: 365–367.

Simpson, R.K. and A. Gilmour. 1997. The effect of high hydrostatic pressure on *Listeria monocytogenes* in phosphate-buffered saline and model food systems. J. Appl. Microbiol. 83: 181–188.

Sizer, C.E. and V.M. Balasubramaniam. 1999. New intervention processes for minimally processed juices. Food Technol. 53: 64–67.

Smelt, J.P.P.M., A.G.F. Rijke and A. Hayhurst. 1994. Possible mechanism of high pressure inactivation of microorganisms. High Pres. Res. 12: 199–203.

Smelt, J.P.P.M., A.G.F. Rijke and P.C. Wouters. 1997. Inactivation of microorganisms by high pressure. pp. 398–417. *In:* D. Reid [ed.]. The Properties of Water in Foods. ISOPOW 6 Conference. Blackie Academic & Professional, London, UK.

Smelt, J.P.P.M. 1998. Recent advances in the microbiology of high pressure processing. Trends Food Sci. Technol. 9: 152–158.

Sobrino-Lopez, A., P. Viedma-Martínez, H. Abriouel, E. Valdivia, A. Gálvez and O. Martin-Belloso. 2009. The effect of adding antimicrobial peptides to milk inoculated with *Staphylococcus aureus* and processed by high-intensity pulsed-electric field. J. Dairy Sci. 92: 2514–2523.

Song, H.P., B. Kim, J.H. Choe, S. Jung, S.Y. Moon, W. Cgoe and C. Jo. 2009. Evaluation of atmospheric pressure plasma to improve the safety of sliced cheese and ham inoculated by 3-strain cocktail *Listeria monocytogenes*. Food Microbiol. 26: 432–436.

Tahiri, I., J. Makhlouf, P. Paquin and I. Fliss. 2006. Inactivation of food spoilage bacteria and *Escherichia coli* O157:H7 in phosphate buffer and orange juice using dynamic high pressure. Food Res. Int. 39: 98–105.

Tammineedi, C.V.R.K. 2012. Effect of UV-C light, high intensity ultrasound and nonthermal atmospheric plasma treatments on the allergenity of major cow milk proteins. M.Sc. Thesis. Department of Plant, Soil and Agricultural Systems in the Graduate School Southern Illinois University Carbondale, Carbondale, IL, USA.

Tay, A., T.H. Shellhammer, A.E. Yousef and G.W. Chism. 2003. Pressure death and tailing behavior of *Listeria monocytogenes* strains having different barotolerance. J. Food Protect. 66: 2057–2061.

Tewari, G. 2007. Thermal processing of liquid foods with or without particulates. pp. 35–43. *In:* G. Tewari and V.K. Juneja [eds.]. Advances in Thermal and Non-Thermal Food Preservation. Blackwell Publishing, Ames, Iowa, USA.

Timson, W.J. and A.J. Short. 1965. Resistance of microorganisms to hydrostatic pressure. Biotechnol. Bioeng. 7: 139–159.

Torres-Mora, M.A., A. Soeldner, E.Y. Ting, A.C.O. Hawes, G.D. Aleman, G.S. Bakshi, W.R. McManus, C.L. Hansen and J.A. Torres. 1996. Early microstructure changes in Cheddar cheese and the effect of ultra high pressure curd processing. Book of Abstracts of IFT Annual Meeting, New Orleans, USA. 9 p.

Trujillo, A.J., B. Guamis and C. Carretero. 2000. A procedure for the manufacture of goat milk with controlled microflora by means of high hydrostatic pressure. Food Chem. 69: 73–79.

Trujillo, A.J., M. Capellas, J. Saldo, R. Gervilla and B. Guamis. 2002. Applications of high pressure on milk and dairy products: A review. Innov. Food Sci. Emerg. Technol. 3: 295–307.

Tsong, T.Y. and R.D. Astunian. 1986. Absorption and conversion of electric field energy by membrane bound ATPases. Bioelectrochem. Bioenerg. 15: 457–476.

US-FDA. 2000. Food and Drug Administration, Center for Food Safety and Applied Nutrition. Kinetics of microbial inactivation for alternative food processing technologies—High pressure processing. Available at: http://fda.gov/Food/ScienceResearchAreas/Safe_Practices_for FoodProcesses/ucm101456.html_documents/foodsysdoc/Final_Report_Vol_2.pdf accessed on February 5, 2013.

Ulmer, H.M., M.G. Gänzle and R.F. Vogel. 2000. Effects of high pressure on survival and metabolic activity of *Lactobacillus plantarum* TMW 1.460. Appl. Environ. Microbiol. 66: 3966–3973.

Ulmer, H.M., H. Herberhold, S. Fahsel, M.G. Gänzle, R. Winter and R.F. Vogel. 2002. Effects of pressure induced membrane phase transitions on inactivation of HorA, an ATP-dependent multidrug resistance transporter, in *Lactobacillus plantarum*. Appl. Environ. Microbiol. 68: 1088–1095.

Vega-Mercado, H., J.R. Powers, G.V. Barbosa-Canovas and B.G. Swanson. 1995. Plasmin inactivation with pulsed electric fields. J. Food Sci. 60: 1143–1146.

Villamiel, M. and P. de Jong. 2000. Inactivation of *Pseudomonas fluorescens* and *Streptococcus thermophilus* in trypticase soy broth and total bacteria in milk by continuous-flow ultrasonic treatment and conventional heating. J. Food Eng. 45: 171–179.

Voigt, D.D., F. Chevalier, M.C. Qian and A.L. Kelly. 2010. Effect of high-pressure treatment on microbiology, proteolysis, lipolysis and levels of flavor compounds in mature blue-veined cheese. Innov. Food Sci. Emerg. Technol. 11: 68–77.

Voldrich, M., J. Dobias, L. Ticha, M. Cerovsky and J. Kratka. 2004. Resistance of vegetative cells and ascospores of heat resistant mould *Talaromyces avellaneus* to the high pressure treatment in apple juice. J. Food Eng. 61: 541–543.

Walstra, P., J.T.M. Wouters and T.J. Geurts. 2006. Dairy Science and Technology. Second edition. CRC Press, Boca Raton, FL, USA.

Wick, C., U. Nienaber, O. Anggraeni, T.H. Shellhammer and P.D. Courtney. 2004. Texture, proteolysis and viable lactic acid bacteria in commercial Cheddar cheeses treated with high pressure. J. Dairy Res. 71: 107–115.

Wouters, P.C., E. Glaasker and J.P.P.M. Smelt. 1998. Effects of high pressure on inactivation kinetics and events related to proton efflux in *Lactobacillus plantarum*. Appl. Environ. Microbiol. 64: 509–514.

Wright, J.R., S.S. Sumner, C.R. Hackney, M.D. Pierson and B.W. Zoecklein. 2000. Efficacy of ultraviolet light for reducing *Escherichia coli* O157:H7 in unpasteurized apple cider. J. Food Protect. 63: 563–567.

Wrigley, D.M. and N.G. Llorca. 1992. Decrease of *Salmonella* Typhimurium in skim milk and egg by heat and ultrasonic wave treatment. J. Food Protect. 55: 678–680.

Wuytack, E.Y., S. Boven and C.W. Michiels. 1998. Comparative study of pressure-induced germination of *Bacillus subtilis* spores at low and high pressure. Environ. Microbiol. 64: 3220–3224.

Yang, J., J.R. Powers, S. Clark, A.K. Dunker and B.G. Swanson. 2003. Ligand and flavor binding functional properties of *β*-lactoglobulin in the molten globule state induced by high pressure. J. Food Sci. 68: 444–452.

Zhang, Q., A. Monsalve-Gonzalez, B.L. Qin, G.V. Barbosa-Canovas and B.G. Swanson. 1994. Inactivation of *Saccharomyces cerevisiae* in apple juice by square wave and exponential-decay pulsed electric field. J. Food Process Eng. 17: 469–478.

Zhang, Q., G.V. Barbosa-Canovas and B.G. Swanson. 1995. Engineering aspects of pulsed electric field pasteurization. J. Food Eng. 25: 261–281.

Zhang, H.Q., G.V. Barbosa-Cánovas, V.M. Balasubramaniam, C.P. Dunne, D.F. Farkas and J.T.C. Yuan. 2011. Nonthermal Processing Technologies for Food. Blackwell Publishing Ltd. John Wiley & Sons Ltd., Cambridge, UK.

Zimmermann, U. and R. Benz. 1980. Dependence of the electrical breakdown voltage on the charging time in Valonia utricularis. J. Membr. Biol. 53: 33–43.

CHAPTER 14

Microbiological Safety Systems for Dairy Processing

Theo Varzakas

14.1 Introduction

The life cycle of milk and milk products commences with the production of fresh milk in dairy farms. Milk is then processed to produce pasteurized and homogenized market milk, butter, cheese, yogurt, custard and dairy desserts, long-life (UHT), condensed, evaporated or powdered milk products. Dairy products are packaged into consumer portions and distributed to retail outlets. For fresh dairy products, refrigerated storage is required to maintain eating appeal and prevent microbiological spoilage.

Dairy processing occurs world-wide; however, the structure of the dairy industry varies from one country to another. In less developed countries, milk is generally sold directly to the public, but in major milk producing countries raw milk is mostly sold on a wholesale basis. In Ireland and Australia, for example, many of the large-scale processors are owned by the farmers as co-operatives, while in the United States individual contracts are agreed between farmers and processors.

A number of microorganisms or agents can lead to illnesses via food systems, but the leading causes of the most frequently reported food-borne illnesses in milk and dairy products are bacteria or the toxins that some bacteria produce (FSIS 1990). Of particular concerns in food service are the followings: *Salmonella, Campylobacter jejuni, Clostridium perfringens, Staphylococcus aureus, Clostridium botulinum, Escherichia coli, Listeria monocytogenes*, and *Yersinia enterocolitica* (FSIS 1990).

Numerous microorganisms, including bacteria, yeasts and molds, constitute the complex ecosystem present in milk and fermented dairy products. Bacteria in the unknown ecosystems were assigned an identity by comparison with a comprehensive bacterial reference database of ~150 species that included useful dairy microorganisms (lactic acid bacteria), spoilage bacteria (e.g., *Pseudomonas* and Enterobacteriaceae), and pathogenic bacteria (e.g., *L. monocytogenes* and *S. aureus*) (Ogier et al. 2004).

Hence, the necessity of monitoring and controlling health-threatening microorganisms in milk and milk products arises and this will be made clear in the following sections.

14.2 Risk assessment-HACCP

Risk assessment is defined as "a process of evaluation including the identification of the attendant uncertainties, of the likelihood and severity of an adverse effect(s)/event(s) occurring to man or the environment following exposure under defined conditions to a risk source(s)" (EC 2000). Risk assessment consists of hazard identification, hazard characterization, exposure assessment and risk characterization (Notermans et al. 1996).

Risk assessment of food products has been interwoven with risk analysis system and Hazard Analysis Critical Control Point (HACCP). The HACCP system refers to physical, chemical and microbiological hazards occurring in raw material/process over the food production (Mortimore and Wallace 1995). However, some hazards or risks may escape from the HACCP system because this risk covers a wider range than that of the food product process. Such an example is the risk of a characteristic that does not have an obvious relationship with physical, chemical and microbiological hazards. This risk mainly refers to the acceptance of the product by the public, particularly negatively predisposed towards genetically-modified organisms (GMOs). Its assessment could be part of the recently introduced ISO 22000 but could not be included in the processes of the HACCP system. A synoptical HACCP monitoring system for dairy products is given in Table 14.1 and hazard analysis of raw materials used in the manufacture of dairy products is presented in Table 14.2.

The production of safe food is based on the use of good quality raw materials, and the application of Good Manufacturing Practices (GMP) and the HACCP system. In addition, risk analysis is becoming the new cornerstone in producing acceptable and safe food. According to the agreements of the World Trade Organization (WTO), especially the agreement on the application of sanitary and phytosanitary measures (SPS), the setting of control criteria should have a scientific basis. For this purpose, quantitative risk analysis is considered to be a logical approach that can provide the necessary insight into the process of setting such criteria.

Elements of quantitative risk analysis can also be introduced into the HACCP system, for example, in setting criteria at critical control points

Table 14.1 Synoptical HACCP monitoring system (Adapted from Arvanitoyannis et al. 2009).

Product	Processing stage	Description of possible dangers	Controls	Critical limits	Personnel-in-charge	Corrective action	Records
ALL PRODUCTS	**Milk receipt**	Milk with high microbial load	Determination of pH	6.4–6.6	Receipts supervisor	Return of non-conforming lot	Non-conformance sheet, corrective and preventive action
		Milk contaminated with pathogenic microorganisms	Control of receipt temperature and cleaning of transportation vehicle	T<7°C			
		Transportation under non-hygienic conditions					
		Antibiotics	Antibiotics kits	According to current legislation		Recommendations to suppliers	Non-conforming book Incoming materials sheet
		Adulterated milk (with water or cheaper milk)	Specific gravity	1.029–1.045 kg l^{-1} at 15°C		Recommendations to suppliers	
CHEESE	**Curd formation** (Feta-type, semi-hard, hard-cheeses)	Weakness of fast growth of beneficial microorganisms in starter cultures, prevailing of harmful microorganisms	Weight quantity of rennet		Production supervisor	Increase or decrease temperature of cheese making and milk	Production book
			Monitoring of temperature of cheese boiler	According to milk condition and type of cheese approx. 32–33°C		Increase or decrease quantity of rennet based on experience	
		Rennet contaminated with undesirable microorganisms	Monitoring of curding time and checking coagulating power of the enzyme	Rennetting according to instructions of use		Replace fresh rennet with the old one	

			Control of milk temperature	Entrance temperature of 32°C in the boiler			
FETA-TYPE CHEESES	**Brining**	Weak brine Weakness of anti-microbial action	Salt content	pH when the cheese curd gets into brine of approximately 6.2 16 Bé	Production supervisor	Addition of extra salt	Production book
	Dry salting (Ripening in cheese tables and in open containers)	Failure to increase salt coefficient Drop of pH and moisture removal Immature stop of fermentation	pH monitoring Control of room temperature	At the end of salting (one day cheese) pH<5.2 At the end of prematuration (before entrance to the fridge) pH<4.8 Minimum 16°C	Production supervisor	Extension of ripening outside the fridge Adaptation of expiry date/fast distribution and consumption within two months after the product is introduced to the markets Air conditioning	Production book
SEMI HARD CHEESES	**Cutting/ Immersion/ Fermentation**	Failure of death of microorganisms	Control of temperature of hot water, time	Minimum 70°C /15 min	Production supervisor	Adjustment of temperature, time extension	Production book
	Salting/Ripening	Failure to remove moisture	Control and adjustment of temperature and	15–16°C RH 80–90%	Production supervisor	Air conditioning/ adjustment of temperature	Production book
		Growth of undesirable microorganisms	moisture in the maturation chamber				

Table 14.1 contd....

Table 14.1 contd.

Product	Processing stage	Description of possible dangers	Controls	Critical limits	Personnel-in-charge	Corrective action	Records
HARD CHEESES	**Pressing/ change of paddy/ reversing**	Failure to remove moisture, prevailing of good microorganisms	Control of room temperature conditions	15–16°C end of pressing pH<4.8	Production supervisor	Air conditioning/ adjustment of temperature	Production book
	Brining	Weak brine/weakness of anti-microbial action	Salt content	Minimum 16 Bé	Production supervisor	Addition of extra salt	Production book
	Ripening	Failure to remove moisture Growth of undesirable microorganisms	Control and adjustment of temperature and moisture in the maturation chamber	Graviera: 2 weeks 12–14°C 4 weeks 18°C 6 weeks 12–14°C RH 85%	Production supervisor	Air conditioning/ adjustment of temperature	Production book
				Kefalotyri: 3 months/12–16°C/ RH 85%			
WHEY CHEESES	**Receipt of whey/ Filtration**	Stay of cheese crumbs	Macroscopic control of raw materials Control of pH	No visible piece pH>6.3	Production supervisor	Addition of NaOH Rejection, Re-filtration	Production book
	Pasteurization	Survival of pathogenic microorganisms, weakness to reduce the initial microbial load satisfactorily Burning during thermal processing	Temperature and time control	Increase from 30 to 75°C in 30 min and from 75 to 90°C in 15 min Total heating time is 45 min	Production supervisor	Extension of pasteurization/ adjustment of temperature Stirring during thermal processing	Production book

	Packaging	Vacuum failure, hermetic sealing, growth of microorganisms	Control of parameters of device, sealing control	Hermetic sealing No non-conformance	Production supervisor	Repetition	Production book
BUTTER	**Pasteurization**	Survival of pathogenic microorganisms, weakness to reduce initial microbial load satisfactorily	Control of temperature and time	63°C/30 min	Production supervisor	Extension of pasteurization/adjustment of temperature	Production book
ALL PRODUCTS	**Cooling**	Growth of undesirable microorganisms Release before the end of ripening	Control of temperature of refrigeration chambers Control of date of entrance/exit Periodical checks of the final product in an external lab	0–4°C RH 85–95% Depending on product and legislation	Production supervisor	Adjustment of temperature/humidity Repairing of malfunction Rejection 1st distribution/adaptation of shelf life	Temperature recording meters Laboratory control results
	Distribution	Growth of undesirable microorganisms	Temperature control of the cooling chambers in the vehicles	0–4°C	Driver	Adjustment of temperature/repairing of malfunction Reject	Recording meters

Table 14.2 Hazard analysis of raw materials.

Raw materials	Biological hazard	Chemical hazard	Physical hazard
Milk	Presence of pathogenic microorganisms: *Listeria monocytogenes, Yersinia enterocolitica, Campylobacter jejuni, Bacillus cereus, Salmonella* spp., *Staphylococcus aureus, Escherichia coli* (O157:H7), *Shigella* spp., *Mycobacterium* spp., *Brucella* spp., *Bacillus anthracis, Streptococcus* spp., *Clostridium botulinum, Clostridium perfringens* Viruses Increased temperature during receipt of milk (>6°C). This means danger for microbial growth and multiplication of pathogenic microorganisms	Antibiotic residues Detergents remnants Presence of mycotoxins Increase in pesticide residues, dioxins, polychlorinated biphenyls (PCBs)	Foreign matter
Salt	Increased moisture, contamination	Heavy metals	Foreign matter
Enzymes/ microorganisms	Contamination, genetically modified organisms		

(CCPs) (Notermans et al. 1996). Notermans et al. (1998) outlined a risk assessment approach to food safety evaluation, based on testing a particular type of food against particular type of risk (for example, the risk assessment of *Bacillus cereus* in pasteurized milk). The results obtained are related to possible adverse effects on the health of consumers. The proposed system provides information on the exposure of consumers to microbial pathogens when the food is consumed. It reflects the successful application of good manufacturing practices and HACCP principles on the producer, and the effect of consumer handling of the product on the exposure rate. The information obtained on factors affecting exposure to microbial hazards and their impact on consumers would allow risk management and communication to be carried out effectively.

Henson and Holt (1999) explored incentives for the adoption of food safety controls by businesses in the UK dairy sector. Four key factors were found to have motivated the adoption of HACCP and these were internal efficiency, commercial pressure, external requirements and good practice.

Consumer concern about livestock production methodologies has been increasing over the last decades due to various outbreaks of food-borne zoonoses and animal diseases. Quality assurance programs in different production chains have been installed by industry to counteract the problems occurring. The primary producers, like dairy farms, are not formally comprised in such programs. Yet, quality control at dairy farm level goes beyond the quality control of the milk product alone (Noordhuizen and Metz 2005). For better safeguarding food safety and public health, as well as animal health and welfare, the whole production process on the dairy farm should be addressed.

A well-known welfare report regards the Tierges and Heits Index (TGI-Animal Health Index), applied in Germany and Austria, mostly in organic farms. The TGI addresses categories like movement possibilities, opportunities for social contacts, floor design of housing facilities, climatic conditions in the barns, and intensity of care by the farmer. Disadvantage of an index is that good categories may cover up for deficient categories. The TGI has features which are comparable to those in good farming practice codes; hence, TGI might be called a good welfare practice code. The emphasis in welfare monitoring in general is on deviant animals as well as on risky environmental conditions of the farm (von Borell 2000).

In the case of cattle welfare, the focus should be on those areas which contribute significantly to the occurrence of welfare disorders. Examples are housing (space per cow, floor design for locomotion, cubicle design for resting and lying, maintenance standards, space for social interaction), barn climate (humidity, temperature, ventilation, draughts), feed and water availability, ration composition and quality of feedstuffs. Therefore, any inspection focusing on welfare issues should address both the animals and the risk conditions in the cow's environment. In addition, several countries have started with the stepwise implementation of, either voluntary or compulsory, quality control programs on dairy farms. Monitoring of cows (prevalence, incidence) and farm conditions (risk factors) is part of Dairy Chain Quality Program (KKM) in the Netherlands. The information gathered is also used for on-farm consultancies by the veterinary practitioners in herd health programs. It is expected that the KKM-program will ultimately result in a HACCP-based quality management program incorporating good farming practice codes with risk identification, risk management and prevention playing a significant role.

Noordhuizen and Metz (2005) concluded that the application of HACCP-compatible programs on the dairy farms will be conducted in the near future based on developments within the dairy sector and at EU political level. HACCP and nowadays ISO 22000 are compatible programs in use within the dairy sector in EU. This application helps in identifying and managing the quality hazards and risks occurring in the production process on dairy farms, and in providing the consumer with more certainty about the quality of products of animal origin.

14.3 Hygiene monitoring and pre-requisite programs

In a typical dairy operation, Relative Light Units (RLU) data, reflecting the ATP bioluminescence technique for rapid hygiene testing, are collected over a period of time from a Control Point (CP) of milk filling machine and analyzed in retrospect to assess the hygiene status of various CPs in any HACCP system. The measurement in RLU is used to give a Pass/Fail status to the CPs tested (Hayes et al. 1997). The analysis showed that the Cusum and Individuals Charts established a proper trend analysis of the RLU data. The findings showed the use of Statistical Process Control (SPC) in the monitoring and prevention of CPs from failing the hygiene test. Furthermore, they stressed that if SPC was adopted, the total number of failed CPs would decrease. This would lead to more effective hygiene management and more efficient production.

Moreover, Bactoscan analysis and ATP bioluminescence technique could be applied as emergency brakes by food companies as explained by te Giffel et al. (2001). They reported that the main problem is the sensitivity of the techniques. In the future, process control systems can be developed by integrating microbiological results and predictive models into process control software. Information technologies (IT), neural networks and fuzzy logic could help in this direction. "Fuzzy" cognitive maps convert stimuli to responses attributive, so that they translate mathematically the appropriate actions for sanitary hygienic control.

Fuzzy logic spares the determination of sample sizes based on the pre-suppositions of the statistical inference. The Fuzzy logic is based on the Postulates of the Theory of Possibility, where there is not a necessity to assist the axioms of probability, such as no negativity and unity (Braga et al. 1995). The employment of the "Fuzzy Cognitive Maps" is feasible and can be a facilitating alternative technique, mainly in samples of reduced dimensions. The Fuzzy method has been applied in quality control, when the parameters used do not allow rigid limits. This is the peculiar case of the controls in the hot and cold chains in collective feeding. The Fuzzy maps are easy and agile instruments to evaluate the conditions of temperature of the food supply and to determine the quality of the productive process.

Attitudes towards food hygiene management strategies in their companies were measured by a mail survey designed and distributed to 87 Finnish food manufacturing companies in order to be distributed to 870 employees representing both workers and managers by Hielm et al. (2006). Answers were stratified according to four job categories and four industry sectors (meat, dairy, fish and bakery). All 30 companies that responded to the survey had a functioning own-checking plan (OCP), while other quality management programs were less prevalent. When asked what had caused most difficulties in devising the OCP/HACCP plan, the most common answers were choosing the critical control points, committing the firm's entire workforce and organizing the documentation of monitoring results. According to the respondents, the biggest benefits of the OCP/HACCP plan were product safety and quality. In Finland, as in the other Nordic countries, the preventive risk management strategy in the food industry is based on good hygiene practices (GHPs or hygiene pre-requisites) and HACCP plans when appropriate. This approach became mandatory for Finnish food operators, caterers and retailers in 1995.

Due to difficulties in implementing HACCP, the hygiene pre-requisites have until now formed a substantial part of the own-checking system in Finland. Panisello and Quantick (2001) have categorized requirements for successful HACCP-implementation into four "segments": management commitment, education and training, availability of resources and external pressures. Shortcomings are explained by three "technical barriers" including (1) prior to, (2) during the process of and (3) after HACCP implementation. The technical barriers seem a far more formidable hurdle to overcome, as these encompass all those practices, attitudes and perceptions that negatively affect the understanding and proper and effective implementation of the HACCP principles (Panisello and Quantick 2001). Behavioral studies from the United Kingdom (Taylor and Taylor 2004), Italy (Angelillo et al. 2001), the United States (Henroid and Sneed 2004), Poland (Konecka-Matyjek et al. 2005), and the Philippines (Azanza and Zamora-Luna 2005) assert that such barriers are of a universal nature.

Although OCP/HACCP is in widespread use in the EU, few surveys have been published on the attitudes towards the systems among food company employees. In the UK, Panisello et al. (1999) surveyed food companies to establish the level of, and barriers to, HACCP-implementation. While HACCP was implemented in 73% of the responding companies, the authors recognized several barriers to the penetration of HACCP, especially in SMEs: lack of knowledge, expertise and adequate resources. Similarly, a Spanish study regarding perceived barriers to the implementation of HACCP, focused on attitudes of company quality managers and external consultants (Vela and Fernandez 2003). Too little emphasis was placed on the attitudes of the company laborers. It is well known that risk management strategies work only if they are internalized by all company employees, and that a successful implementation of HACCP demands commitment by the whole personnel (Mortimore 2001, Panisello and Quantick 2001).

The prevalence of biofilms in dairy processing is an important reservoir of both spoilage and pathogenic microflora which can lead to spoilage of finished product and transmission of diseases. It is recommended that each plant should be evaluated for the prevalence of biofilms. Effective sanitation program should then be devised based on *in vitro* studies that could be invariably repeated under *in situ* conditions in order to control the biofilms prevalent in dairy processing areas.

Biofilm status of different segments of pasteurization lines of commercial plant (CP) and an experimental dairy plant (EDP) was evaluated by Sharma and Anand (2002). Biochemical differentiation of organisms in biofilms revealed the predominance of genus *Bacillus* (37% and 44%, respectively) in both plants. The other microflora of CP included *Lactobacillus, Streptococcus, Lactococcus* and *Staphylococcus,* while microflora of EDP additionally had *Micrococcus* spp. The Gram-negative genera in the constitutive microflora of biofilms were mainly *Shigella, E. coli, Enterobacter aerogenes, Citrobacter, Flavobacterium* and *Proteus* in CP, while EDP additionally had *Klebsiella* spp. A sanitizer, iodophore, at a concentration of 10 ppm with a contact time of 20 min was found to be most effective to control mixed species biofilms of CP under *in vitro* and *in situ* conditions. Therefore, evaluation of biofilms status and development of an effective sanitation plan should be part of the HACCP plan in conjunction with ISO 9000 specifications for the dairy processing industry.

Roberto et al. (2006) presented results of a marketing study associated with the implementation and maintenance of HACCP in a pasteurized milk plant. The Good Manufacturing Practices/Standard Operating Procedures/ Sanitation Standard Operating Procedures (GMP/SOP/SSOP) pre-requisites were evaluated in the plant. Two HACCP plans were proposed: the first plan was developed under the actual operating conditions, without previous compliance of GMP/SSOP pre-requisites, and a second plan in compliance with GMP/SSOP. The cost estimation for implementation and maintenance of HACCP, with or without previous adoption of the pre-requisite programs, was performed and a comparative analysis of the estimated values was carried out. The results suggested that a previous compliance of GMP/SSOP pre-requisites was essential for developing an effective HACCP plan with low number of critical control points, leading to lower costs and investments for implementation and maintenance of HACCP. With the compliance of the pre-requisite programs (GMP/SSOP) it was possible to reduce these costs approximately by 24.2% emphasizing the importance of a solid pre-requisite program to improve economic viability for HACCP implementation.

Packaging materials are often considered a CCP in HACCP systems of dairy plants. Methods for the determination of the microbial contamination rate of plastic cups, must reliably detect single molds, yeasts or coliforms. Even if packaging is sterilized before dispensing, a low contamination rate in the delivered packaging materials might occur due to inefficient GMPs,

Good Hygienic Practices (GHPs), SSOPs or due to cross contamination with the food material.

The effectiveness of common sterilization techniques such as UV sterilization can be reduced by dust. The application of sterilization by heat or steam is limited by possible thermal deformation of plastic cups. Thus, a sterilization method for plastic cups needs to be employed to ensure excellent hygienic conditions. Electron beam irradiation is a simple and easily applicable method to decontaminate plastics without changing the mechanical characteristics. The method has been used for the sterilization of medical disposable products such as Petri dishes or catheters for many years. It is based on the acceleration of electrons in an electric field and their focusing into a beam. Their interaction with the product and deposition of their energy in the form of ionizing radiation result in sterile products. Employing of simple measures in the production and during transport of packaging materials, such as dust-prevention or tight sealing in polyethylene bags, heavily reduced microbial contamination rates of packaging material. To reduce contamination rates further, electron beam irradiation at 4–5 kGy was applied to plastic cups sealed in polyethylene bags. This treatment resulted in sterile polystyrene and polypropylene cups without influencing mechanical characteristics of the packaging material.

Tacker et al. (2002) compared a specially adapted coating method, impedance method, direct inoculation and membrane filter technique to determine contamination with yeasts, molds, coliforms and total bacterial counts using the appropriate agar in each case. The coating method was recommended for determining yeasts, molds and coliforms as it allowed the localization of the microorganisms as well as the determination of single microorganisms. For total bacterial count, a direct inoculation technique was proposed.

14.4 Production of raw and pasteurized milk-potential hazards

Pasteurized milk is the largest selling milk in most industrialized countries because the consumption of raw milk carries the risk of infection by milk-borne pathogens, especially *Salmonella* (Small and Sharp 1979) and *Campylobacter* (Heeschen 1996, Potter et al. 1983, Summer 1996). The International Dairy Federation (IDF) has defined pasteurization as, "a process applied to a product with the object of minimizing possible health hazards arising from pathogenic microorganisms associated with milk, by heat treatment, which is consistent with minimal chemical, physical and sensory changes in the product" (Varnam and Sutherland 1996, Mossel 1981). However, in some countries, farms are still allowed to sell raw milk for household use. In the UK and several other countries, bottled raw milk can be directly delivered to customers provided that milk-producing cows have been attested free from tuberculosis and brucellosis (Mossel et al. 1995).

Bovine milk-derived ingredients (non-fat dried milk-NFDM), milk protein concentrate (MPC), and whey protein concentrate (WPC), from cows hyper-immunized with a variety of antigens have been available experimentally or commercially for several decades. Although the safety of milk is rarely questioned, an assessment of ingredients derived from the milk of cows hyper-immunized with a proprietary bacterin (S100) consisting of heat-killed cultures of 26 bacterial pathogens, originally isolated from humans, obtained from the American Type Culture Collection (ATCC), was made to determine that these ingredients share the human food safety profile traditionally ascribed to regular milk (Gingerich and McPhillips 2005). The critical elements in this determination are: (1) quantitative determination of the difference between S100 and regular dairy ingredients, (2) comparison of exposures resulting from proposed uses to background exposures already in the American diet, and (3) corroborative results of controlled clinical trials comparing safety outcomes from consumption of S100 with those of conventional dairy ingredients. Estimated daily intake projections showed that use of S100 ingredients at maximum proposed levels resulted in exposures to active (undenatured) immunoglobulin G (IgG) below background in the present American diet in infants but above background in children and adults, whose intake of conventional dairy products is markedly lower. Safety of this consumption level is corroborated by clinical results showing no difference in safety outcomes between S100 ingredients, consumed at exaggerated levels, and conventional dairy products, in a variety of adult populations. There is no evidence that demonstrates a hazard to the public when S100 ingredients are used at levels that might reasonably be expected from the proposed applications.

Raw milk is an excellent medium for the growth of microorganisms which can be derived from the udder, the environment, the milk handling equipment, and the personnel (Mossel et al. 1995). *E. coli*, *S. aureus*, *Corynebacterium bovis*, *Streptococcus agalactiae*, *S. dysagalactiae* and *S. uberis* may cause under certain circumstances mastitis, leading to significant economic losses (Hahn 1996, Barkema et al. 1998, Elbers et al. 1998). In winter months, feed and bedding are the main sources of thermoduric spoilage organisms, while milk handling equipment is the major source of Gram-negative, psychrotrophic spoilage bacteria. Employees suffering clinical symptoms of infection and feces may contaminate milk with *Campylobacter* spp. and *Salmonella* spp.

Milk should only be accepted at the dairy plant when obtained from animals which are not suffering from tuberculosis and brucellosis (Romero et al. 1995); are free from contagious diseases (Heeschen 1996, Troutt et al. 1995); are not suffering from clinical mastitis (Mossel et al. 1995); have not been treated with antibiotics unless milk has been obtained after expiration of the retention period following veterinary treatment (Troutt et al. 1995); are subjected to proper supervision and support from relevant authorities (Tschumi 1997) and does not suffer from infections or tissue damages of the udder (Burgess et al. 1994).

Pathogens that have been involved in food-borne outbreaks associated with the consumption of milk include *L. monocytogenes*, *Salmonella* spp., *Campylobacter* spp., *S. aureus*, *B. cereus* and *C. botulinum*. *E. coli* O157:H7 has become a serious threat to the dairy industry with several outbreaks reported in developed countries ranging from mild diarrhea to potentially fatal hemolytic uremic syndrome (HUS), hemorrhagic colitis and thrombotic thrombocytopenic purpura (Coia et al. 2001). Readers are recommended to refer to Chapter 1 for more information about pathogenic microorganisms associated with raw milk and milk products.

The toxins produced by *Clostridium botulinum* are among some of the most potent, naturally occurring toxic substances. If spores of *C. botulinum* are present in food and conditions are not inhibitory to germination and growth, the organism can proliferate and produce toxin. The organism is widespread in soil and aquatic sediments and also in the gastrointestinal tracts of ruminants, fish and birds (Bell and Kyriakides 2000). The toxins produced by *C. botulinum* are neurotoxins which attack the nervous system of humans. They block the release of acetylcholine, a neurotransmitter at the peripheral nerve ends. Transmission of nerve impulses at the neuromuscular junction is prevented and no muscle stimulation occurs, resulting in flaccid paralysis. The nature and the properties of botulinum neurotoxins and their mode of action are discussed in Hauschild (1989). Toxic doses for humans of all botulinum toxin types, however, are estimated to be very low, i.e., at levels of <1 µg for toxin types A and B and approximately 10 µg for toxin types E and F (Shone 1987). Three categories of botulism are recognized in humans; in addition to food-borne botulism, there are also wound and infant botulism. Botulism may be easily misdiagnosed because symptoms resemble other illnesses (Bell and Kyriakides 2000).

Finally, probiotics, commonly defined as viable microorganisms (bacteria or yeasts) that exhibit a beneficial effect on the health of the host when they are ingested, are used in foods, especially in fermented dairy products, but also in pharmaceutical preparations. The development of new probiotic strains aims at more active beneficial organisms. In the case of novel microorganisms and modified organisms the question of their safety and the risk to benefit ratio have to be assessed. Lactic acid bacteria (LAB) in foods have a long history of safe use. Members of the genera *Lactococcus* and *Lactobacillus* are most commonly given generally-recognized-as-safe (GRAS) status whilst members of the genera *Streptococcus* and *Enterococcus* and some other genera of LAB contain some opportunistic pathogens.

Lactic acid bacteria are intrinsically resistant to many antibiotics. In many cases resistances are not, however, transmissible, and the species are also sensitive to many clinically used antibiotics even in the case of a lactic acid bacteria-associated opportunistic infection. Therefore no particular safety concern is associated with intrinsic type of resistance. Plasmid-associated antibiotic resistance, which occasionally occurs, is another matter because of the possibility of the resistance spreading to other, more harmful species and genera. The transmissible enterococcal resistance against glycopeptide antibiotics (i.e., vancomycin and teicoplanin) is particularly noteworthy as reported by Salminen et al. (1998), as vancomycin is one of the last effective antibiotics left in the treatment of certain multidrug-resistant pathogens.

New species and more specific strains of probiotic bacteria are continuously identified. Prior to incorporating new strains into products their efficacy should be carefully assessed, and a case by case evaluation as to whether they share the safety status of traditional food-grade organisms should be made. These microorganisms can produce a wide variety of antagonistic primary and secondary metabolites including organic acids, diacetyl, CO_2 and even antibiotics such as reuterocyclin produced by *Lactobacillus reuteri*. Moreover, members of the group can also produce a wide range of bacteriocins, some of which have activity against food pathogens such as *L. monocytogenes* and *C. botulinum*. Indeed, nisin has been used as an effective bio-preservative in some dairy products for decades, while a number of more recently discovered bacteriocins, such as lacticin 3147, demonstrate increasing potential in a number of food applications. Both of these lactococcal bacteriocins belong to the lantibiotic family of post-translationally modified bacteriocins that contain lanthionine, β-methyllanthionine and dehydrated amino acids. The exploitation of such naturally produced antagonists holds tremendous potential for extension of shelf-life and improvement of safety of a variety of foods (see also Chapter 4) (Ross et al. 2002).

Exposure to mycotoxins through food is widely recognized as human health hazard (Bhat and Vasanthi 1999). Of all the mycotoxins, Aflatoxin B1 (AFB1) is considered to be the most toxic/carcinogenic compound (IARC 1993). It is biotransformed by hepatic microsomal cytochrome P450 to Aflatoxin M1 (AFM1) which possesses 10 times lower carcinogenic potential with respect to the parent molecule (Cullen et al. 1987) and has been listed as Class 2B carcinogen (Smith et al. 1994). AFM1 has been shown to be excreted in milk following exposure to AFB1 contaminated feed (Heathcoate and Hibbert 1978). Since milk is a major commodity for introducing aflatoxins in the human diet, the occurrence of AFM1 in this product is of concern (Stoloff 1980). Evidence of hazardous human exposure to AFM1 through dairy products has been shown by several investigators (Galvano et al. 1996).

Regulatory limits for AFM1 throughout the world are greatly influenced by economic considerations and may vary from one country to another (Stoloff et al. 1991). The European Community and Codex Alimentarius prescribe that the maximum level of AFM1 in liquid milk and dried or processed milk products should not exceed 50 ng kg^{-1} (Codex Alimentarius Commission 2001). This limit has been established in compliance with the ALARA (as low as reasonable achievable) principle.

Bovine viral-diarrhea (BVD), enzootic bovine leucosis (EBL), Johne's disease (JD), and neosporosis are common infectious diseases on dairy farms in Canada and elsewhere. A survey carried out by van Leeuwen et al. (2001) showed that 37.6, 20.8, 2.6 and 20.3% of a random sample of dairy cattle in the Maritimes provinces of Canada were seropositive to the agents that cause BVD (in cattle not vaccinated for BVD), EBL, JD and neosporosis, respectively.

Kan and Meijer (2007) reported that toxic substances such as dioxins, mycotoxins, heavy metals, pesticides, veterinary drugs and polycyclic aromatic hydrocarbons are also present in ingredients for animal feed. Adequate risk management depends on knowledge of absorption, metabolism, carry-over and toxicological profile of these substances and on practical measures to reduce especially the latter two. Generally, toxic substances are metabolized before or after absorption through the intestinal tract. Depending on their physico-chemical characteristics, some substances such as veterinary drugs and feed additives are metabolized into naturally occurring and generally harmless constituents. Other substances like dioxins are persistent and remain in the animal and in animal products. Heavy metals are not metabolized at all. Some metals are irreversibly bound to body tissues, e.g., lead to bone or cadmium to kidneys. Plant-associated toxins can enter the human food supply as endogenous components of the food that humans eat, as contaminants in foods such as grain products and honey, or as contaminant residues in animal-derived food such as meat, dairy products and eggs. Animals are exposed to these toxins via normal grazing and browsing, or being supplied with otherwise healthy feed contaminated by the toxin-producing plants or plant-associated microorganisms.

Exposure prevention is by far the preferred risk management tool. Use of adsorbents has been tested extensively both for organochlorine compounds and mycotoxins. Heavy metal and drug contamination of feeds are also controlled at the feed mill, by selection of feed ingredients and through proper manufacturing practices. Mycotoxin control is quite difficult to be executed as weather conditions play a pivotal role in fungal growth and mycotoxin formation. Absorbents added to the feed may sometimes alleviate the problem. Withdrawal times should be followed if prescribed legally and may in some instances provide a solution. In the case of milk and eggs, produced on a daily basis, the animal products are most likely to show increased residue levels during prolonged exposure (Kan and Meijer 2007).

Reduction in contamination levels, due to dilution as a result of growth, will also help to obtain a product in compliance with legal residue limits. On-farm HACCP programs for monitoring antibiotic residues are costly to implement as reported by Gardner (1997). The estimated annual cost of testing to detect beta-lactam antibiotic residues in tanker truck milk in the US is between $8 and $35 million. Extension of testing to other drugs, chemicals, and pathogens in bulk tank milk stretch the dairy industry resources. The lack of field-validated tests for most of the chemical and infectious agents of concern makes it difficult to ensure that stated goals of HACCP programs are consistently achieved. More basic and field researches are needed to develop and validate these tests, which can only be achieved through a substantial infusion of new research funding. Validation of tests is essential for selection of the most appropriate testing strategies, to estimate predictive values and for appropriate test interpretation. The determination of the numbers of samples to be tested in HACCP monitoring programs depends on the specific purpose of the test and the likely prevalence of the agent or residue at the CCPs. The tree diagram for CCP detection is given in Fig. 14.1. Figure 14.2 also shows the recognition of pre-requisite programs as described in ISO 22000.

14.5 Model HACCP programs for dairy products

14.5.1 Manufacture of pasteurized milk

The processes taking place at a typical milk manufacturing plant include receipt and filtration/clarification of the raw milk; separation of all or part of the milk fat (for standardization of market milk, production of cream and butter and other fat-based products, and production of milk powders); pasteurization; homogenization (if required); deodorization (if required); further product-specific processing; packaging and storage, including cold storage for perishable products; distribution of final products. Figure 14.3 shows the flow diagram outlining the basic steps in the production of pasteurized milk.

14.5.2 Manufacture of milk powder

Milk used for making milk powder, whether it is whole or skimmed milk, is not pasteurized before use. The milk is preheated in tubular heat exchangers before being dried. The preheating temperature depends on the season (which affects the stability of the protein in the milk) and on the characteristics desired for the powdered product. The preheated milk

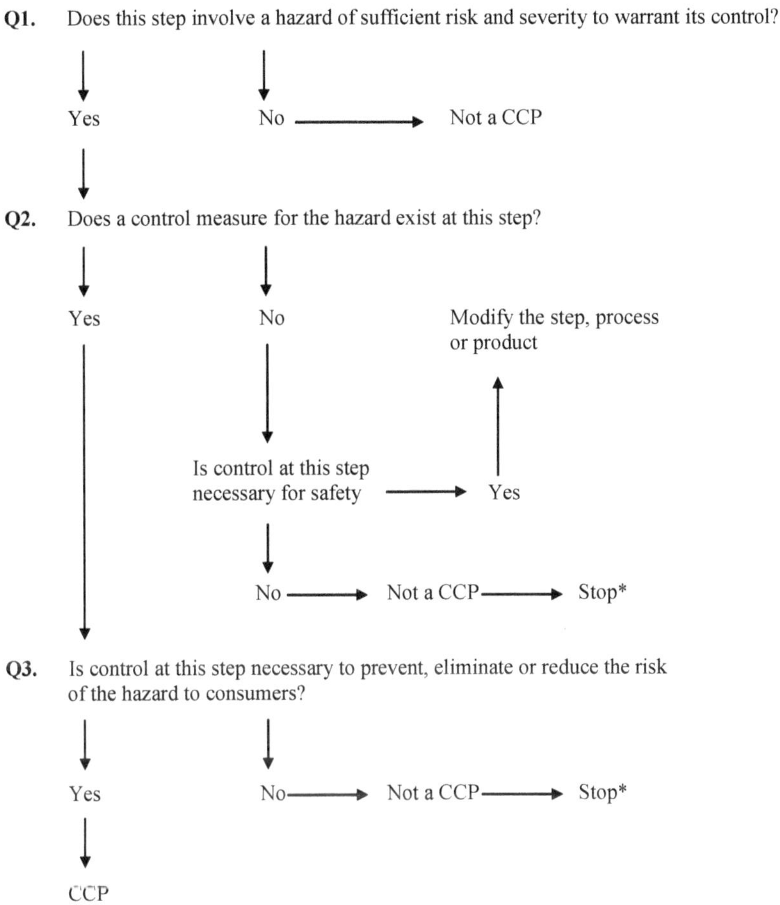

Figure 14.1 Tree diagram for CCP detection.

is fed to an evaporator to increase the concentration of total solids. The solids concentration that can be reached depends on the efficiency of the equipment and the amount of heat that can be applied without unduly degrading the milk protein. The milk concentrate is then pumped to the atomizer of a drying chamber. In the drying chamber the milk is dispersed as a fine fog-like mist into a rapidly moving hot air stream, which causes the individual mist droplets to instantly evaporate. Milk powder falls to the bottom of the chamber, from where it is removed. Finer milk powder particles are carried out of the chamber along with the hot air stream and collected in cyclone separators. Milk powders are normally packed and distributed in bulk containers or in 25 kg paper packaging systems.

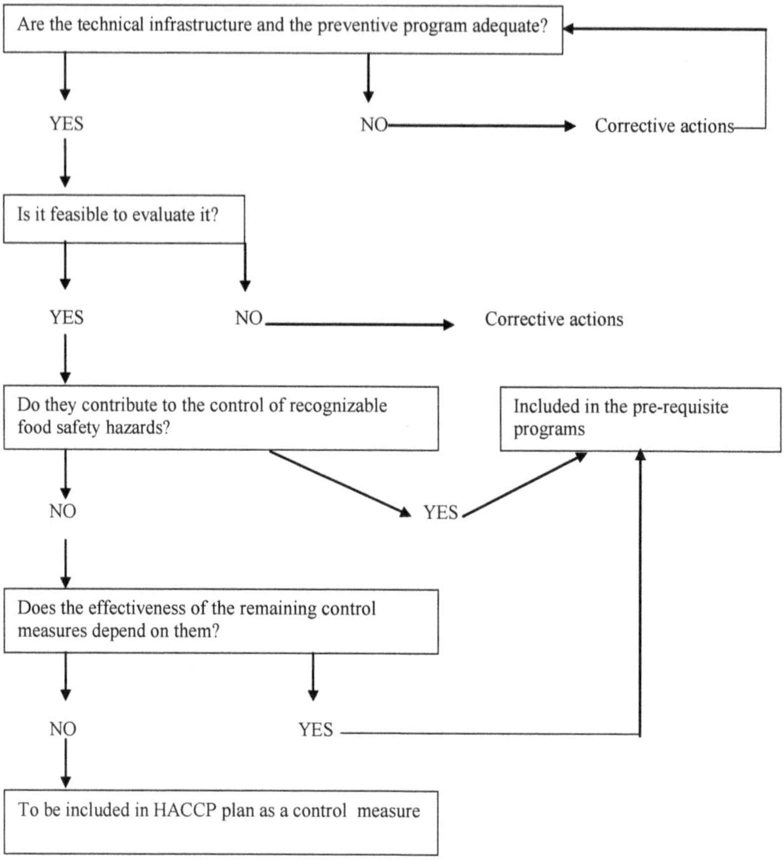

Figure 14.2 Recognition of pre-requisite programs as described in ISO 22000.

14.5.3 Manufacture of butter

Figure 14.4 illustrates the flow diagram of butter-making. The initial steps (filtration/clarification, separation and pasteurization of the milk) are the same as described in milk manufacturing (see Fig. 14.3). Milk destined for butter-making must not be homogenized, because the cream must remain in a separate phase. After separation, cream to be used for butter-making is heat-treated and cooled under conditions that facilitate good whipping and churning. It may then be ripened with a culture that increases the content of diacetyl, the major compound responsible for the flavor of butter. Alternatively, culture inoculation may take place during churning. Butter of which flavor is enhanced using this process is termed lactic, ripened or cultured butter. This process is very common in continental European

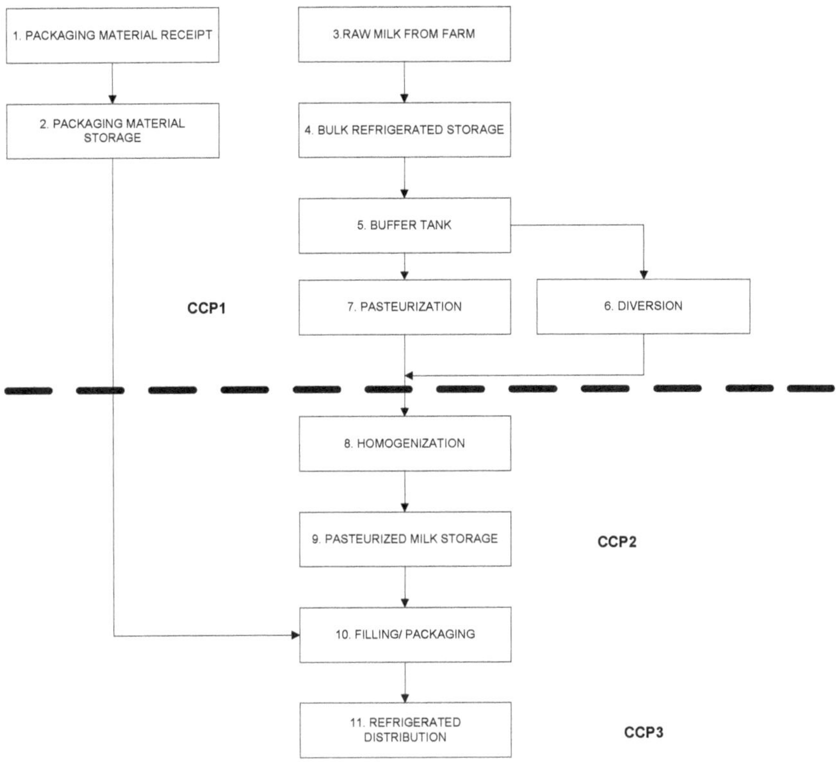

Figure 14.3 Flow diagram of pasteurized milk.

countries. Although the product is claimed to have a superior flavor, the storage life is limited. Butter made without the addition of a culture is called sweet cream butter. Most butter made in the English-speaking world is of this nature. Both cultured and sweet cream butter can be produced with or without the addition of salt. The presence of salt affects both the flavor and the keeping quality of the end product.

14.5.4 Manufacture of cheese

Virtually all cheeses are made by coagulating milk protein (casein) in a manner that traps milk solids and milk fat into a curd matrix. This curd matrix is then consolidated to express the liquid fraction, cheese whey (Figs. 14.5 and 14.6). Cheese whey contains those milk solids which are not held in the curd mass, in particular most of the milk sugar (lactose) and a number of soluble proteins.

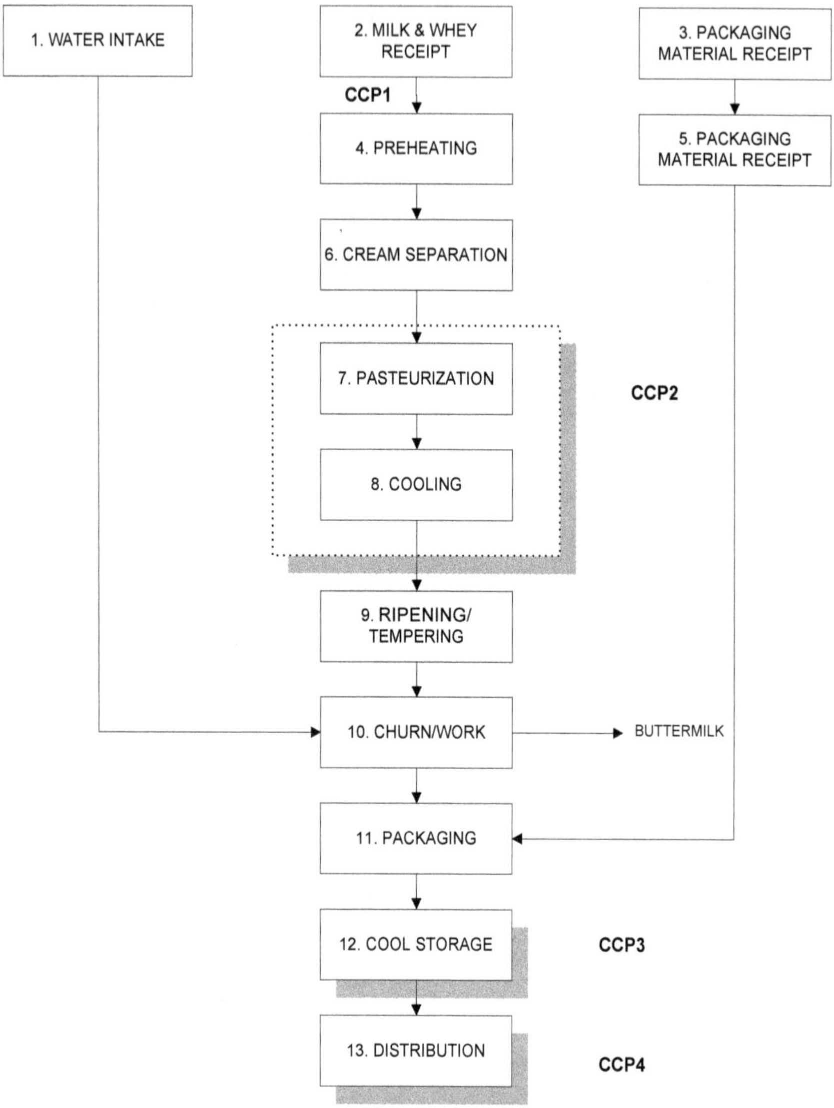

Figure 14.4 Flow diagram of butter manufacturing.

Manufacture of Feta cheese

Feta is a traditional white soft Greek cheese and representative of the cheeses that are ripened and kept in brine (Anifantakis 1991a, Abou-Donia 1991, Tamime and Kirkegaard 1991). The distinguishing characteristics of

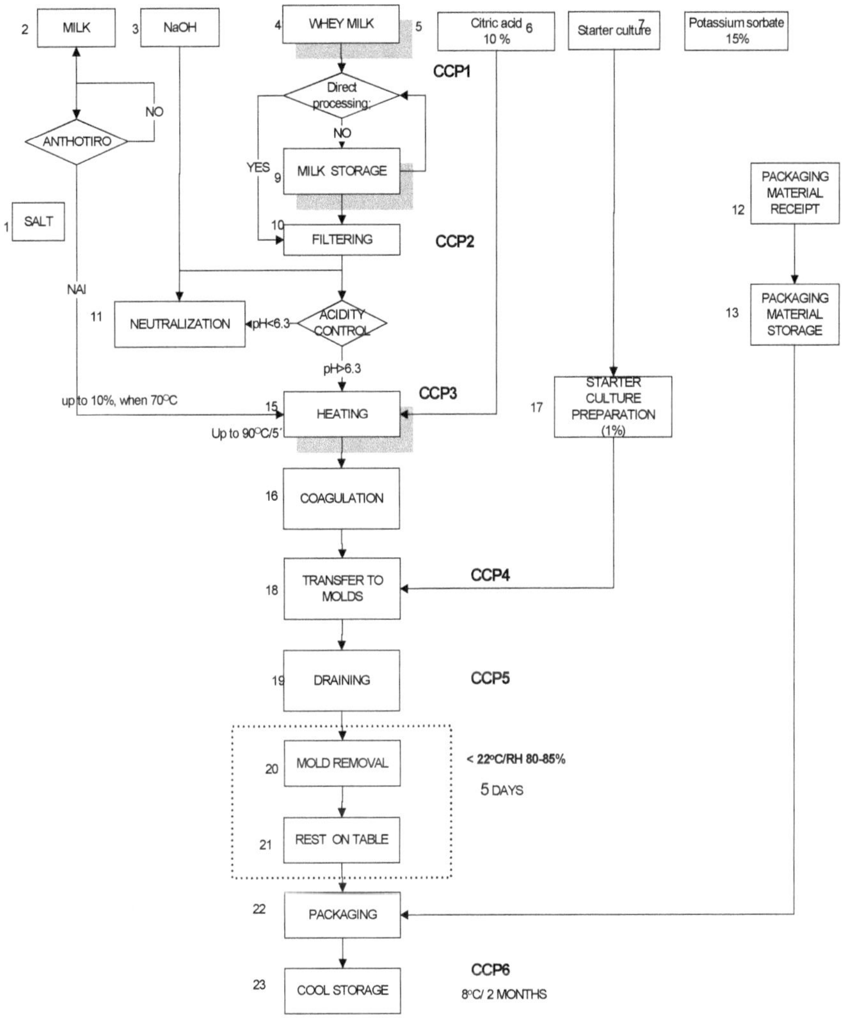

Figure 14.5 Flow diagram of whey cheese manufacturing.

Feta are the creamy and rich flavor, the soft texture with some irregular small mechanical openings, and the white color and the rectangular shape (Zerfiridis 1994, Tsotsanis 1996). It is produced from sheep's milk or mixed sheep's and goat's milk in a ratio up to 7:3, respectively (Greek Codex of Foods and Drinks 1998).

Moisture content and minimum fat-in-dry matter level of Feta cheese should be 52–56% and 43%, respectively (Abd El-Salam et al. 1993, Vastardis and Anifantakis 1992). The lactic cultures used for Feta cheese are *Lactobacillus delbrueckii* subsp. *bulgaricus–Streptococcus thermophilus* (1:1)

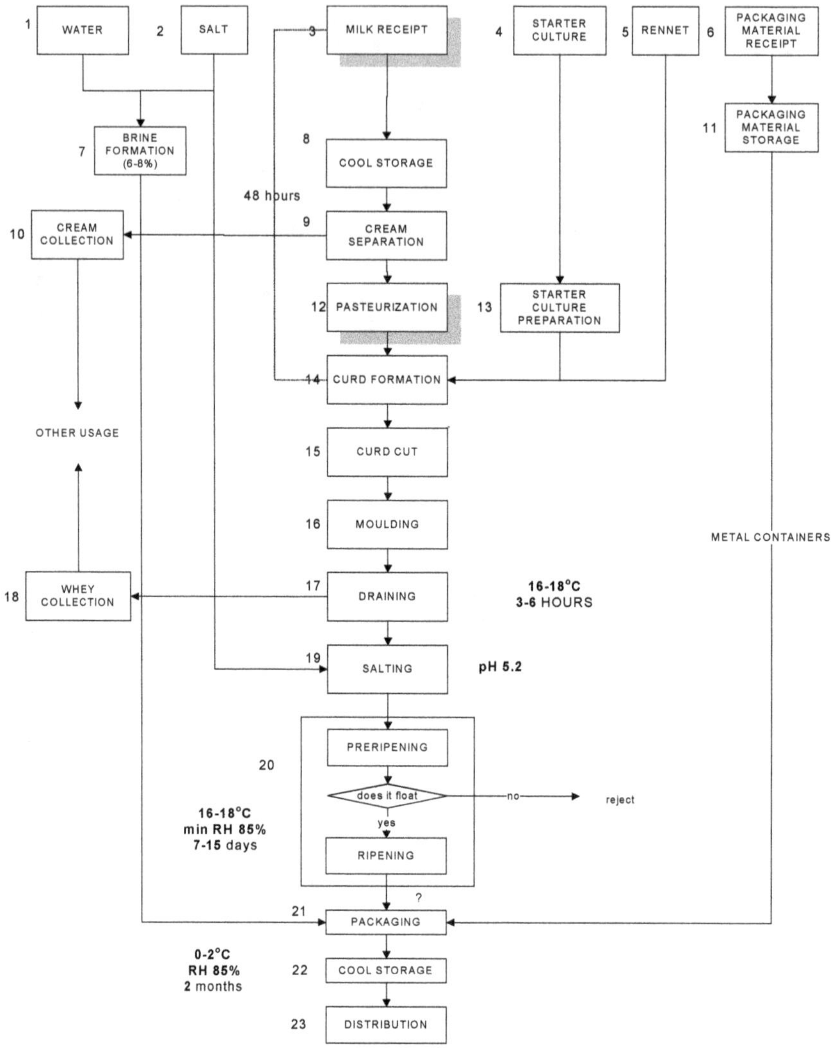

Figure 14.6 Flow diagram of Feta cheese manufacturing.

(Pappas and Zerfiridis 1989), *Lb. delbrueckii* subsp. *bulgaricus–Lactococcus lactis* (3:1), and *Lactobacillus casei–L. lactis* (1:1) (Abd El-Salam et al. 1993). Starter culture should be inoculated at a level of 1% and inoculated milk should be left at 32–34°C for 15–30 min (milk ripening). The activity of the lactic culture should be verified by monitoring the acidity development in milk. Critical Control Points (CCP) for the production of Feta cheese are shown in Fig. 14.5.

Whey cheeses (with specific examples of Myzithra, Anthotyros and Manouri)

Myzithra, Anthotyros and Manouri are whey cheeses traditionally produced in Greece from whey derived from cheeses made from mixtures of ewe's and goat's milk. Whey from Feta cheese is mostly used for preparation of these whey cheeses. It is about fresh cheeses with high moisture and low acidity (pH approx. 6.0), that do not undergo ripening and due to heating of whey cheese they get deprived of the natural lactic acid flora, hence the pathogenic microorganisms do not appear.

They have a short shelf-life and tests carried out at the end of production date are specific for a certain lot and hence implementation of corrective actions or recall is difficult to be carried out. Heating is achieved at 80–95°C for 15–20 min. At this stage the coagulation of whey proteins occurs. When the temperature exceeds 60°C, milk, cream, salt or other additives are added to the whey. The thermal processing removes pathogenic microorganisms from the product without destroying the thermo-resistant spores of sporogenic bacteria and their toxins. For Myzithra that gets dried in semi-basement places special measures are required (e.g., placement of the appropriate screen) to avoid contamination from the environment. Transfer to molds of whey cheeses is the most critical and dangerous processing stage in terms of safety because if contamination occurs at this stage this cannot be removed by another step of the processing. Refrigeration temperature (4°C) should be controlled effectively by recording charts to avoid multiplication of undesirable microorganisms. After the end of processing, whey cheese is refrigerated and distributed under refrigeration. CCPs of whey cheese production line are illustrated in Fig. 14.6 and description of dangers, critical limits and corrective actions are presented in Table 14.1.

Semi-hard cheeses

Kasseri is a semi-hard cheese with maximum moisture level of 40% and minimum fat-in-dry-matter level of 40% (Greek Codex of Foods and Drinks 1998). It is a typical Balkan pasta-filata cheese, which is traditionally made from sheep's milk or from a mixture of sheep's and up to 20% of goat's milk (Greek Code of Food and Drinks 1998). Kasseri cheese belongs to the product category of "protected name of origin" in the EU (Regulation 1017/1996/EU). The major advantage of this cheese is that it can be produced from milk with high acidity (Zerfiridis 1997).

The typical form of Kasseri cheese is flat, cylindrical and its typical size is: diameter, 25–30 cm; height, 7–10 cm; and weight, 7–8 kg. Its texture is firm and few holes, if any, are uniformly distributed. The cheese has a white-yellow color and a rind of the same color. The addition of goat's milk leads to harder cheeses. The HACCP monitoring system for semi-hard cheeses is presented in Table 14.1.

Hard cheeses

Kefalotiri is a traditional Greek cheese made from sheep's or goat's milk or a mixture of them (Anifantakis 1991b). Kefalotiri is of "protected name of origin" in the EU (Regulation 1017/1996/EU). The name refers to the "head-shaped" appearance of the cheese. The main characteristics of Kefalotiri cheese are firm texture, salty taste and strong flavor. It is a hard-rind cheese of white-yellow color. Small gas holes and bigger slit holes exist in the mass of cheese (Greek Codex of Foods and Drinks 1998). Kefalotiri has a flat cylindrical shape, diameter of 32–34 cm, height of 12–14 cm and weight of 8–10 kg. The chemical constituents of the final product are: moisture 36%, fat 28%, fat-in-dry-matter 45%, pH 5.6 and salt content 4%.

The thermophilic starters used consist of *S. thermophilus* and *Lb. delbrueckii* subsp. *bulgaricus* (2:1) (Robinson 1995), or *S. thermophilus* and *Lb. casei* (8:2) (Pappas and Zerfiridis 1989), or *S. thermophilus, Lactococcus lactis* subsp. *lactis* var. *diacetylactis* and *S. durans* (4:4:2) (Tzanetaki 1993). Coagulation of milk is achieved by adding rennet of animal or microbial origin at 32°C within 30 min. The curd is dipped in the whey at 38–42°C. When the pH drops to 5.1–5.2, the curd is cut into slices and kneaded in hot water (i.e., 80°C) for 15 min. The cheese is filled into the molds at 55–60°C, and then pressed and salted in brine. After brining the pressed cheese is dry-salted at 15°C and at 80% relative humidity. The cheese is then ripened for three months.

14.5.5 Manufacture of ice cream

Blending

The liquid and dry ingredients are weighed and blended together by means of rapid agitation at high speed blenders. It is of key importance that the dry ingredients disperse fully in the mix; therefore blending tanks should be fitted with highly efficient turbine agitators. Dispersion can also be aided by introducing solid ingredients into the liquid in the pipe feeding the blending tank. The flow diagram of ice cream manufacture is shown with identified CCPs in Fig. 14.7.

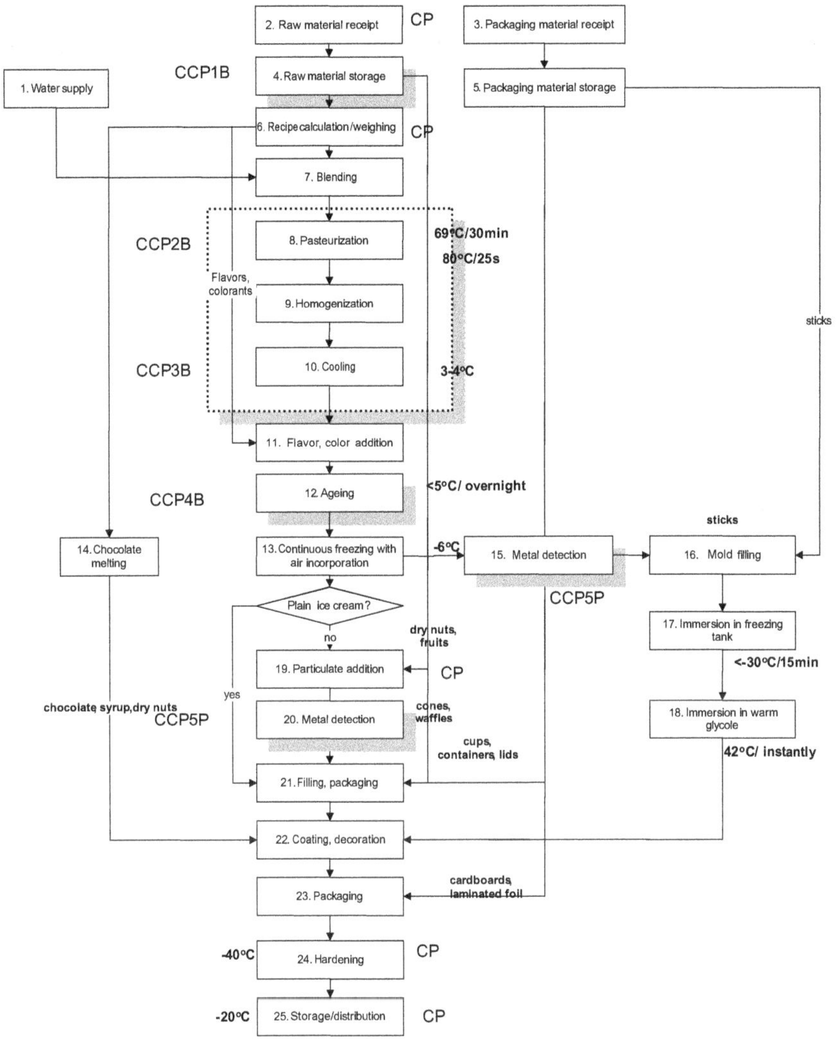

Figure 14.7 Flow diagram of ice cream manufacturing.

Pasteurization

The mix should be heated to a level sufficient to destroy vegetative pathogens and especially *L. monocytogenes*. Moreover, pasteurization also reduces the number of spoilage organisms such as psychrotrophs and helps to hydrate some of the components (proteins, stabilizers). In the pasteurization, batch pasteurizer or continuous (HTST) method is used. Low temperature-long time (LTLT) pasteurization, which takes place in a steam or water jacketed vat, results in more whey protein denaturation, giving a better body to the ice cream. The product is heated in the vat to at least 69°C and held for 30 min to satisfy legal requirements for pasteurization. Various time-temperature combinations can be used for heat treatment of the mix. Holding time is controlled by automatic timers and valves. Plate heat exchangers are commonly used for HTST processing, although tubular heaters may be preferred where space is limited. Preheating at 30 to 40°C is necessary for the components to dissolve easily. The HTST system is equipped with a heating section, a cooling section, and a regeneration section. Cooling sections of ice cream mix HTST presses are usually larger than milk HTST presses. Due to the preheating of the mix, regeneration is lost and mix entering the cooling section is still quite warm. Fouling is a serious potential problem during HTST pasteurization but can be minimized by taking rigorous precautions against the incorporation of excess air into the mix.

Homogenization

The reduction of the size of fat globules is required during ice cream manufacture to prevent churning and to improve whipping properties and air incorporation by allowing proteins to absorb onto the surface of fat globules. This is accomplished by homogenization, a process that reduces the size of the fat globules found in milk or cream to less than 1 μm. Two-stage homogenization (15 MPa in the first and 4 MPa in the second stage) is usually preferred for ice cream mix. Homogenization of the mix should take place at the pasteurizing temperature. The high temperature produces more efficient breaking up of the fat globules at any given pressure. Clumping or clustering of the fat is reduced thereby producing a thinner, more rapidly whipped mix with improved melting-down. Overall, the higher the fat and total solids in the mix, the lower the pressure should be.

Ageing

The mix is then aged for at least four hours and usually overnight. This allows time for the fat to cool down and crystallize and for the proteins and polysaccharides to fully hydrate. Ageing improves whipping qualities of mix, and body and texture of ice cream. Ageing is performed in insulated or refrigerated storage tanks, silos, etc. Mix temperature should be maintained as low as possible without freezing, at or below 5°C. The longer the ageing time, the better the results under average plant conditions. A "green" or "un-aged" mix is usually quickly detected at the freezer.

Freezing and hardening

Traditionally, ice cream freezing is a two-stage process. In the first stage the temperature is reduced under agitation, air being incorporated to give an aerated product. The second stage, which is much slower, involves no air incorporation and takes place under quiescent conditions in a hardening room or tunnel. The process is not complete and even at very low temperatures some water remains unfrozen. Freezing takes place in a "barrel" freezer—a scraped-surface, tubular heat exchanger, which is jacketed with a boiling refrigerant such as ammonia or Freon. Mix is pumped through this freezer and is drawn off the other end within 30 s (or 10 to 15 min in the case of batch freezers) with about 50% of its water frozen. There are rotating blades inside the barrel that keep the ice scraped off the surface of the freezer and also dashers inside the machine which help to whip the mix and incorporate air. The crystallization stage is of major importance as the texture of ice cream is largely determined by the size of the ice crystals. Fast freezing rates decrease the size of crystals, making them undetectable in the mouth.

Flavorings, coloring and finely chopped fruits and nuts may be added to the mix directly before freezing, whereas larger pieces must be added as the ice cream leaves the freezer. After the particulates have been added, ice cream is packaged and placed into a blast freezer at –30 to –40°C where most of the remaining water is frozen. Below –25°C, ice cream is stable for indefinite periods without danger of ice crystal growth; however, above this temperature, ice crystal growth is possible and the rate of crystal growth is dependent upon the temperature of storage, limiting the shelf-life. Hardening is the static (still, quiescent) freezing of the packaged products in blast freezers. Freezing rate must still be rapid, so freezing techniques involve low temperature (–40°C) with either enhanced convection (freezing tunnels with forced air fans) or enhanced conduction (plate freezers).

14.5.6 Manufacture of yogurt

Heating at 85–95°C/5–30 min

Apart from the technological necessities (i.e., heat-induced denaturation of serum proteins), heating of yogurt milk is essential to inactivate microorganisms and indigenous milk enzymes that might cause changes in certain ingredients of milk. Time and temperature of heating should be carefully controlled. If heating is carried out at an open tank, the danger of cross-contamination from the environment (foreign bodies, insects) should be checked. Moreover, cleaning of the tank should be controlled.

Cooling at 40–43°C-starter culture addition-incubation

Yogurt is characterized by the formation of a lipoprotein matrix (curd) formed by slow acidification achieved by *Lb. delbrueckii* subsp. *bulgaricus* and *S. thermophilus*. Following heat treatment, milk is cooled to 42–43°C and the yogurt starters are added. Fermenting milk is filled into cups and left for incubation at 42–43°C which lasts within 3–4 hr. During incubation, yogurt cups should be left uncovered to prevent re-introduction of water evaporated into the yogurt. Hence, special attention should be given to the cleaning and maintaining of the incubation chambers to avoid environmental contamination (e.g., mold growth) since this stage is carried out after thermal processing. Yogurt might not coagulate easily due to the inhibition of the lactic acid bacteria in the presence of antimicrobial substances in milk such as antibiotic residues, cleaning and disinfectants residues.

Refrigeration (≤4°C)

After the coagulation is complete, yogurt is cooled to ≤4°C to stop the bacterial growth. At this stage, temperature should be controlled and contamination risk should be evaluated. Container lids should be clean before use. The flow diagram of yogurt production is given with CCPs in Fig. 14.8.

Desserts based on yogurt (cream or rice milk)

The quality of auxiliary raw materials (rice, sugar, cinnamon) should be controlled and they are stored in safe and clean places. Evaluation of suppliers of these raw materials should be carried out frequently.

14.6 Conclusion

In the present chapter, microbiological safety systems for dairy processing based on HACCP have been addressed. Potential dangers of raw milk have been described. Model HACCP programs for raw milk, butter, Feta cheese, ice cream, whey cheeses, semi-hard and hard-cheeses and yogurt manufacturing have been incorporated including risk analysis and risk assessment. These tools have proved to be effective in the management of a dairy processing plant to avoid contamination and implementing good hygienic and safety measures. Management has to take action in the effective implementation of these systems and programs to avoid recalls and food crisis world-wide. Moreover, consumers have to be educated and in alert to read labeling.

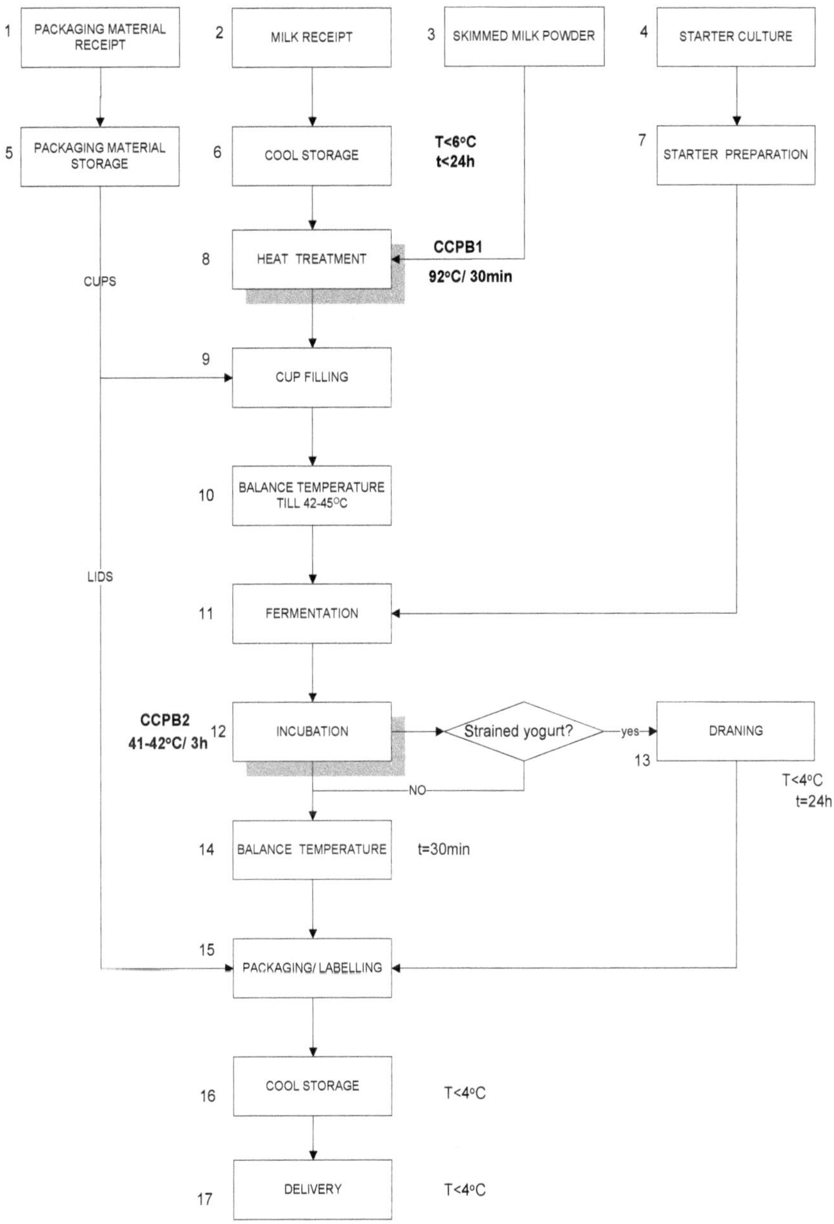

Figure 14.8 Flow diagram of yogurt manufacturing.

References

Abd El-Salam, M.E., E. Alichanidis and G.K. Zerfiridis. 1993. Domiati and Feta type cheeses. pp. 301–335. *In:* P.F. Fox [ed.]. Cheese: Chemistry, Physics and Microbiology, 2nd edition. Vol. 2. Chapman and Hall, London, UK.

Abou-Donia, S.A. 1991. Manufacture of Egyptian, soft, pickled cheeses. pp. 160–280. *In:* R.K. Robinson and A.Y. Tamime [eds.]. Feta and Related Cheeses, Ellis Horwood, London, UK.

Angelillo, I.F., N.M. Viggiani, R.M. Greco and D. Rito. 2001. HACCP and food hygiene in hospitals: Knowledge, attitudes, and practices of food services staff in Calabria, Italy. Collaborative group. Infect. Cont. Hospital Epidemiol. 22: 363–369.

Anifantakis, E.M. 1991a. Traditional Feta cheese. pp. 49–69. *In:* R.K. Robinson and A.Y. Tamime [eds.]. Feta and Related Cheeses, Ellis Horwood, London, UK.

Anifantakis, E.M. 1991b. Greek cheeses. National Dairy Committee of Greece, Athens, Greece.

Arvanitoyannis, I.S., T. Varzakas and M. Koukaliaroglou-van Houwelingen. 2009. Implementing HACCP and ISO22000 for foods of animal origin-dairy foods. pp. 91–180. *In:* I.S. Arvanitoyannis. [ed.]. Handbook of HACCP and ISO22000: Application to foods of animal origin. Wiley-Blackwell, Oxford, UK.

Azanza, M.P.V. and M.B.V. Zamora-Luna. 2005. Barriers of HACCP team members to guideline adherence. Food Cont. 16: 15–22.

Barkema, H.M., Y.H. Schukken, T.G.M. Lam, M.L. Beiboer, H. Wilmine, G. Benediktus and A. Brand. 1998. Incidence of clinical mastitis in dairy herds grouped in three categories by bulkmilk somatic cell counts. J. Dairy Sci. 81: 411–419.

Bell, C. and A. Kyriakides. 2000. *Clostridium botulinum*: A Practical Approach to the Organism and its Control in Foods. Practical Food Microbiology Series. Blackwell Publishing, Oxford, UK.

Bhat, R.V. and S. Vasanthi. 1999. Mycotoxin contamination of foods and feeds. Overview, occurrence and impact on food availability trade exposure of farm animals and related economic losses. Presented paper in the III joint FAO/WHO/UNEP international conference on mycotoxins. Tunis Tunisia, MYC-CONF/99/4a.

Braga, M.J.F., J.M. Barreto and M.A.S. Machado. 1995. Concepts of mathematical modeling and theory of possibility. Artes and Rabiskus, Rio de Janeiro, Brazil.

Burgess, K., C. Heggum, S. Walker and M. van Schothorst. 1994. Recommendations for the hygienic manufacture of milk and milk-based products. Bulletin of the IDF 292: 12–19.

Codex Alimentarius Commission. 2001. Comments submitted on the draft maximum level for aflatoxin M1 in milk. Codex Committee on Food Additives and Contaminants. 33rd session, Hague, The Netherlands.

Coia, J.E., Y. Johnston, N.J. Steers and M.F. Hanson. 2001. A survey of the prevalence of *Escherichia coli* in raw meats, raw cows' milk and raw-milk cheeses in south-east Scotland. Int. J. Food Microbiol. 66: 63–69.

Cullen, J.M., B.H. Ruebner, L.S. Hsieh, D.M. Hyde and D.P.H. Hsieh. 1987. Carcinogenicity of dietary Aflatoxin M1 in male Fischer rats composed to Aflatoxin B1. Cancer Res. 47: 1913–1917.

EC. 2000. Report of the scientific steering committee's working on harmonisation of risk assessment procedures in the scientific committees advising the European Commission in the area of human and environmental health—26–27 October. http://europa.eu.int/comm/food/fs/sc/ssc/out83_en.pdf, Appendices: http://europa.eu.int/comm/food/fs/sc/ssc/out84_en.pdf.

Elbers, A.R.W., J.D. Miltenburg, D. Lange, A.P.P. de Crauwels, H.W. Barkema and Y.H. Shukken. 1998. Risk factors for clinical mastitis in a random sample of dairy herds from the southern part of the Netherlands. J. Dairy Sci. 81: 420–426.

FSIS (Food Safety and Inspection Service). 1990. FSIS Facts. Bacteria that cause foodborne illness (FSIS-40). United States Department of Agriculture, Washington, D.C., USA.

Galvano, F., V. Galofaro and G. Galvano. 1996. Occurrence and stability of Aflatoxin M1 in milk and milk products: A worldwide review. J. Food Protect. 59: 1079–1090.

Gardner, I.A. 1997. Testing to fulfill HACCP (Hazard Analysis Critical Control Points) requirements: Principles and examples. J. Dairy Sci. 80: 3453–3457.

Gingerich, D.A. and C.A. McPhillips. 2005. Analytical approach to determination of safety of milk ingredients from hyperimmunised cows. Regul. Toxic. Pharma. 41: 102–112.

Greek Codex of Foods and Drinks. 1998. Milk, eggs and their products. Ministry of Economics, Athens, Greece.

Hahn, G. 1996. Pathogenic bacteria in raw milk—situation and significance. pp. 67–83. *In:* Proceedings of Symposium on Bacteriological Quality of Raw Milk. IDF Special Issue 9601, IDF, Brussels, Belgium.

Hauschild, A.H.W. 1989. *Clostridium botulinum*. pp. 111–189. *In:* M.P. Doyle [ed.]. Foodborne Bacterial Pathogens. Marcel Dekker Inc., New York, NY, USA.

Hayes, G.D., A.J. Scallan and J.H.F. Wong. 1997. Applying statistical process control to monitor and evaluate the hazard analysis critical control point hygiene data. Food Cont. 8: 173–176.

Heathcoate, J.G. and J.R. Hibbert. 1978. Aflatoxins: Chemical and Biological Aspects. Elsevier Applied Science, Amsterdam, The Netherlands.

Heeschen, W.H. 1996. Bacteriological quality of raw milk: Legal requirements and payment systems—Situation in the EU and IDF member countries. pp. 1–18. *In:* Proceeding of Symposium on Bacteriological Quality of Raw Milk. IDF Special Issue 9601, IDF, Brussels, Belgium.

Henroid, D. and J. Sneed. 2004. Readiness to implement hazard analysis and critical control point (HACCP) systems in Iowa schools. J. Am. Diet. Assoc. 104: 180–185.

Henson, S. and G. Holt. 1999. Exploring incentives for the adoption of food safety controls: HACCP implementation in the UK dairy sector. Rev. Agric. Eco. 22: 407–420.

Hielm, S., P. Tuominen, K. Aarnisalo, L. Raaska and R. Maijala. 2006. Attitudes towards own-checking and HACCP plans among Finnish food industry employees. Food Cont. 17: 402–407.

IARC. 1993. Some naturally occurring substances: Food items and constituents, heterocyclic aromatic amines and mycotoxins. IARC, Lyon, France 56: 245–395.

Kan, C.A. and G.A.L. Meijer. 2007. The risk of contamination of food with toxic substances present in animal feed. Animal Feed Sci. Technol. 133: 84–108.

Konecka-Matyjek, E., H. Turlejska, U. Pelzner and L. Szponar. 2005. Actual situation in the area of implementing quality assurance systems GMP, GHP and HACCP in Polish food production and processing plants. Food Cont. 16: 1–9.

Mortimore, S. 2001. How to make HACCP really work in practice. Food Cont. 12: 209–215.

Mortimore, S. and C. Wallace. 1995. HACCP: A practical approach. Chapman and Hall, London, UK.

Mossel, D.A.A. 1981. Coliform test for cheese and other foods. Lancet 2: 1425–1426.

Mossel, D.A.A., J.E.L. Corry, C.B. Struijk and R.M. Baird. 1995. Essentials of the Microbiology of Foods: A Textbook for Advanced Studies. John Wiley and Sons, Chichester, New York, NY, USA.

Noordhuizen, J.P.T.M. and J.H.M. Metz. 2005. Quality control on dairy farms with emphasis on public health, food safety, animal health and welfare. Live. Prod. Sci. 94: 51–59.

Notermans, S., G.C. Mead and J.L. Jouve. 1996. Food products and consumer protection: A conceptual approach and a glossary of terms. Int. J. Food Microbiol. 30: 175–185.

Notermans, S., M.J. Nauta, J. Jansen, J.L. Jouve and G.C. Mead. 1998. A risk assessment approach to evaluating food safety based on product surveillance. Food Cont. 9: 217–223.

Ogier, J.C., V. Lafarge, V. Girard, A. Rault, V. Maladen, A. Gruss, J.Y. Leveau and A. Delacroix-Buchet. 2004. Molecular fingerprinting of dairy microbial ecosystems by use of temporal temperature and denaturing gradient gel electrophoresis. Appl. Environ. Microbiol. 70: 5628–5643.

Panisello, P.J. and P.C. Quantick. 2001. Technical barriers to hazard analysis critical control point (HACCP). Food Cont. 12: 165–173.

Panisello, P.J., P.C. Quantick and M.J. Knowles. 1999. Towards the implementation of HACCP: Results of a UK regional survey. Food Cont. 10: 87–98.

Pappas, C.P. and G.K. Zerfiridis. 1989. Seminars in dairy technology. Greek National Dairy Committee, Athens. pp. 77–125.

Potter, M.E., M.J. Blaser and R.K. Sikes. 1983. Human *Campylobacter* infection associated with certified raw milk. Am. J. Epidemiol. 117: 475–483.

Roberto, C.D., S.C.C. Brandao and C.A.B. da Silva. 2006. Costs and investments of implementing and maintaining HACCP in a pasteurized milk plant. Food Cont. 17: 599–603.

Robinson, R.K. 1995. A colour guide to: Cheese and fermented milks. pp. 70. Vol. 59–60. Chapman and Hall, London, UK.

Romero, C., M. Pardo, M.J. Grillo, R. Diaz, J.M. Blasco and I. Lopez-Goni. 1995. Evaluation of PCR and indirect enzyme-linked immunosorbent assay on milk samples for diagnosis of brucellosis in dairy cattle. J. Clin. Microbiol. 33: 3198–3200.

Ross, R.P., S. Morgan and C. Hill. 2002. Preservation and fermentation: Past, present and future. Int. J. Food Microbiol. 79: 3–16.

Salminen, S., A. von Wright, L. Morelli, P. Marteau, D. Brassart, W.M. de Vos, R. Fonden, M. Saxelin, K. Collins, G. Mogensen, S.E. Birkeland and T. Mattila-Sandholm. 1998. Demonstration of safety of probiotics—A review. Int. J. Food Microbiol. 44: 93–106.

Sharma, M. and S.K. Anand. 2002. Biofilms evaluation as an essential component of HACCP for food/dairy processing—A case. Food Cont. 13: 469–477.

Shone, C.C. 1987. Understanding toxin action: *Clostridium botulinum* neurotoxins, their structures and modes of action. pp. 11–57. *In:* D.H. Watson [ed.]. Natural Toxicants in Foods. Ellis Horwood Ltd., Chichester, UK.

Small, R.G. and J.M.C. Sharp. 1979. A milk-borne outbreak of *Salmonella Dublin*. J. Hygiene 82: 95–100.

Smith, J.E., C.W. Lewis, J.G. Anderson and G.L. Solomons. 1994. Mycotoxins in human nutrition and health. European Commission Directorate General XII. Brussels Report No. EUR 16048.

Stoloff, L. 1980. Aflatoxin M1 in perspective. J. Food Protect. 43: 226–230.

Stoloff, L., H.P. van Egmond and D.L. Park. 1991. Rationales for establishment of limits and regulations for mycotoxins. Food Add. Cont. 8: 213–222.

Summer, J. 1996. Farm production influences on milk hygiene quality. pp. 94–102. *In:* Proceeding of Symposium on Bacteriological Quality of Raw Milk. IDF Special Issue 9601. IDF, Brussels, Belgium.

Tacker, M., C. Hametner and B. Wepner. 2002. Determination of microbial contamination of plastic cups for dairy products and utilization of electron beam treatment for sterilization. Food Add. Cont. 19(Suppl. 1): 178–184.

Tamime, A.Y. and J. Kirkegaard. 1991. Manufacture of Feta cheese-industrial. pp. 70–143. *In:* R.K. Robinson and A.Y. Tamime [eds.]. Feta and Related Cheeses. Ellis Horwood, London, UK.

Taylor, E. and J.Z. Taylor. 2004. Using qualitative psychology to investigate HACCP implementation barriers. Int. J. Environ. Health Res. 14: 53–63.

te Giffel, M.C., M.C. Meeuwisse and P. de Jong. 2001. Control of milk processing based on rapid detection of microorganisms. Food Cont. 12: 305–309.

Troutt, H.F., J. Gillespie and B.I. Osburn. 1995. Implementation of HACCP program on farms and ranches. pp. 36–57. *In:* A.M. Pearson and T.R. Dutson [eds.]. HACCP in Meat, Poultry and Fish Processing. Blackie Academic and Professional, London, UK.

Tschumi, M. 1997. Quality controls for pasteurized milk. Schweizer. Milch. 123: 3–4.

Tsotsanis, M. 1996. Problems of Feta cheese. Eur. Food Law Rev. 7: 339–349.

Tzanetaki, E. 1993. Milk microbiology. Notes of ATH. Thessaloniki, Greece. pp. 280–286.

Varnam, A.H. and J.P. Sutherland. 1996. Milk and Milk Products: Technology, Chemistry and Microbiology. Chapman and Hall, London, UK.

Vastardis, I.G. and E.M. Anifantakis. 1992. Comparative study of physicochemical characteristics of imported cheeses in brine placed on the Greek market. Greek J. Dairy Sci. Technol. 1: 7–20.

van Leeuwen, J.A., G. Keefe, R. Tremblay, C. Power and J.J. Wichtel. 2001. Seroprevalence of infection with *Mycobacterium avium* subspecies *paratuberculosis*, bovine leukemia virus, and bovine viral-diarrhoea virus in the Maritimes, Canada dairy cattle. Can. Vet. J. 42: 193–198.

Vela, A.R. and J.M. Fernandez. 2003. Barriers for the developing and implementation of HACCP plans: Results from a Spanish regional survey. Food Cont. 14: 333–337.

von Borell, E. 2000. Assessment of pig housing based on the HACCP concept-critical control points for welfare, health and management. pp. 320–368. *In:* H. Blokhuis and D. Wechsler [eds.]. Improving Health and Welfare in Animal Production. EAAP. Vol. 102. Wageningen Pers Publications, Wageningen, The Netherlands.

Zerfiridis, G.K. 1994. Technology of Milk Products: Cheese Manufacture. Giahoudi-Giapouli publications, Thessaloniki, Greece.

Zerfiridis, G.K. 1997. Technology of milk products. pp. 165–169. *In:* Cheese-making. Vol. 155–163. 3rd edition. Giahoudi-Giapouli publications, Thessaloniki, Greece.

CHAPTER 15

Strategies for Rapid Detection of Milk-borne Pathogens[#]

Keith A. Lampel

15.1 Introduction

The Centers for Disease Control and Prevention (CDC) recently published a report on outbreaks in the United States from 1993–2006 due to the ingestion of contaminated dairy products derived from either non-pasteurized or pasteurized milk (Langer et al. 2012). The pathogens identified as the etiological agents in non-pasteurized dairy products were *Campylobacter* spp., *Salmonella* spp., Shiga-toxin producing *Escherichia coli* (STEC), *Brucella* spp., *Listeria* spp. and *Shigella* spp. For pasteurized dairy products, the leading causative agent was norovirus followed by *Salmonella* spp., *Campylobacter* spp., *Staphylococcus aureus, Clostridium perfringens, Bacillus cereus, Listeria* spp. and *Shigella* spp. Other microbial pathogens that have been associated with milk-related outbreaks are enterohemorrhagic *Escherichia coli,* beta hemolytic *Streptococcus, Yersinia enterocolitica* and *Coxiella burnetti.* Depending on the source of milk, other pathogenic microbes have been found which includes the parasitic protozoa *Toxoplasma gondii, E. coli* (enterotoxigenic *E. coli,* ETEC; enteroinvasive *E. coli,* EIEC; enteropathogenic *E. coli,* EPEC), *Mycobacterium* spp. and, in addition, microbial products such as Staphylococcal enterotoxins and fungal aflatoxins.

[#] **Disclaimer**: This book chapter reflects the views of the author and should not be construed to represent FDA's views or policies nor is there any endorsement given to any manufacturer mentioned.

Public health issues pertaining to pasteurized and raw milk should be considered as separate concerns. For the former, contamination of the milk usually occurs with either inadequate pasteurization or post-pasteurization. As for the latter, several points along the processing of milk have been identified as potential hazard points. Routine testing for the presence of pathogens in pasteurized milk is usually not recommended (ICMFS 2005).

Testing for the presence of microbial pathogens or their respective toxins can be performed beyond the actual final product, milk. These on-farm sites include bulk tank milk, in-line milk filters, manure, and feces. The primary target pathogens are STEC, notably *E. coli* O157:H7, *Salmonella* spp., *Campylobacter* spp. and *L. monocytogenes* since these microbes have been responsible for several noted outbreaks. A broad perspective of the prevalence and other factors that affect the presence of microbial pathogens in milk was compiled by Oliver et al. (2005, 2009). Other pathogens or their metabolic products of concern include *Helicobacter pylori* from bovine and non-bovine milk, viruses, fungi (and aflatoxins), and parasites. For each of these microbes or protein toxins, methods are designed to either isolate or detect the presence of that agent. These include basic bacteriological regimens to current, state of the art technology, most based on serological reactions or molecular biology. These technologies, and the matrix, whether food or environmental, could be considered the driving forces for method development and application.

15.2 Methods for detection of milk-borne pathogens

On a broad level, the technology that underlies methods available presently is based on culture, molecular biology, immunology and microscopy. Each methodology aims to fit a particular niche as to a particular application, whether it is for detection/isolation for presence/absence testing or for isolation of the pathogen. In addition to targeting microbial pathogen cells, laboratory analytical protocols also determine the presence of toxins produced by several microorganisms, such as fungal or bacterial toxins.

As with most diagnostic applications, sample preparation is critical to the overall success of the assay, regardless of technology applied. Milk and milk products present a formidable obstacle to yielding a quality and appropriate quantity of analyte, whether it is a microbial cell or spore, nucleic acid or protein. As with any food matrix, dairy foods present their own inherent source of inhibitors, such as fat or lipid content, calcium concentration, presence of indigenous enzymes, e.g., proteinases, and levels of protein (Rossen 1992). These additional factors, i.e., potential inhibitors, may pose a detrimental influence on the overall integrity of an accurate and sensitive assay.

Although the literature is replete with newer technologies, different versions of prior-published methods, or combinations of different-technology protocols, the major key to yielding the correct diagnosis is not always in the method employed but in sample preparation. Several manufacturers provide kits or protocols that allow the investigator to extract and remove components of the food matrix that could interfere with the assay or ability to culture the microbe of interest.

Overall, and in a simplistic approach, the goal of any assay should be rapid, ease of use, accurate, specific for intended target(s), and with a minimum of false-negative and false-positive results. Each aspect can have its own ramifications as to allow contaminated dairy products into commerce. Published or commercially available methods for the isolation and/or detection of microbes or their metabolic products should provide well-documented sensitivity and specificity attributes, supporting validated data, and indicate which food matrices were tested. False-negative results can have a profound, negative health impact to the consumer whereas false-positive results present further economic obstacles to the producer or manufacturer.

15.2.1 Nucleic acid-based assays

DNA-based assays

The polymerase chain reaction (PCR) is undoubtedly the most commonly applied molecular-based method to screen foods for the presence of microbial pathogens. There have been many different renditions of the initial concept of PCR as put forth by K. Mullis (Saiki et al. 1985). Since the advent of conventional PCR, the overarching goal has been to shorten the time of amplification and to make this process in 'real-time', that is to capture the *in vitro* increase in targeted amplicons as it is happening. For most PCR assays, the target molecule can either be DNA or RNA, with qualitative or quantitative analyses. One of the critical aspects of molecular-based detection systems, particularly PCR, is the sequence characteristics of the primers and/or DNA probes used. As one of the critical facets that drive the specificity of these molecular-based assays, their overall success is dependent upon the degree of homology with the target nucleotide sequence. As such, these short DNAs drive the specificity, efficacy, and sensitivity of the overall amplification/detection process. Further details of PCR and any other DNA-based technology can be found in other reviews (Gorski and Csordas 2010, Hoorfar 2011, Mandal et al. 2011, Naum and Lampel 2011).

Rapid, sensitive methods, such as PCR, offer a reliable and effective screening system to dairy-related food commodities and environmental samples on dairy farms. These are adequate for determining absence or presence of a specific pathogen; however, the critical information that is missing relates to the viability status of the pathogen, i.e., live versus dead cells. Real-time PCR has become a standard analytical tool in many laboratories providing multiple targets in one assay. However, conventional and real-time PCR-based assays have limitations due to the number of targets that can be amplified in one reaction. Multiplex real-time PCR assays have been developed that can target several specific DNAs, but the total number of specific amplifications is limited to the number of fluorescent probes used. A few commercially available instruments have a maximum number of channels to read the fluorescent peaks. Furthermore, if five channels are available, at least one is dedicated to an internal control reaction and possibly another channel is used for positive controls. However, many of these instruments can analyze from 96 wells or higher and, therefore, the permutations exist to detect all targeted pathogens. If routine analyses are performed, laboratories can automate nearly the entire system reducing time of analysis, labor costs, and hands-on operations. As with many molecular-based detection methods, sample preparation may be the limiting factor in many aspects, one of which is generating false-negative results due to either the presence of inhibitors or the number of target molecules is below the level of sensitivity of that assay.

An example of the application of PCR-based assays as a means to decrease the time of analysis of milk was included in a study of the prevalence of *L. monocytogenes* on Irish farms (Fox et al. 2009). In this study, samples were initially enriched in broth, plated, and presumptive positive, individual colonies were identified as *L. monocytogenes* by PCR targeting the *hly* gene of the pathogen. Of the 298 samples tested from 16 farms, 19% (57/298) were found to be positive for *L. monocytogenes*.

Consumption of raw milk and raw-milk dairy products poses a serious health risk due to the presence of microbial pathogens that may be present (Oliver et al. 2009). Common target bacterial pathogens that are incorporated in PCR-based assays include *L. monocytogenes*, *Salmonella* spp. and *E. coli* O157:H7. In a recent study, a multiplex PCR-based assay was compared to culture-based isolation methods targeting the aforementioned pathogens in different environmental samples, bulk milk tank, in-line filters, manure and feces. In this comparative study, the multiplex PCR assay outperformed the culture based method in determining the presence of these pathogens in each sample (Amagliani et al. 2012). A report on the prevalence of the aforementioned pathogens in bulk tank milk and in-line filters from dairies within the United States were based on PCR data for *Salmonella* spp. and toxigenic strains of *E. coli*; *Listeria* were isolated by culture methods (van Kessel et al. 2011).

There are different variations of PCR that can be applied to the detection of dairy microbial pathogens in milk and milk products. One alternative real-time PCR approach integrated the use of molecular beacons, hair-pin shaped, short oligonucleotides (probes) that have attached on one end a fluorophore and on the opposite end a quenching molecule. When these oligonucleotides are not hybridized to a complementary sequence of an amplicon, the quenching dye prevents fluorescence, commonly referred to as fluorescence resonance energy transfer (FRET). However, when hybridized to a complementary sequence, the change in conformation of the probe separates the fluorophore from the quenching dye allowing fluorescence. A quantitative, multiplex real-time PCR assay was applied to detect *E. coli* O157:H7 and *L. monocytogenes* in several dairy products, including raw milk (Singh et al. 2009). The labeled probes were designed to hybridize to the *rfb* gene and the *hly* gene, respectively.

Another example of a multiplex PCR assay, but in combination with a magnetic separation step, was reported by Omiccioli et al. (2009a). The PCR assay targeted three pathogens, *E. coli* O157:H7, *L. monocytogenes* and *Salmonella* spp. As for the overall assay, nanoparticles (beads) were added to lysed bacterial cells to bind double stranded DNA and PCR inhibitors were washed from the bound DNA prior to elution for PCR. This is a non-specific means to capture all double-stranded DNA that does not affect directly the specificity of the overall assay; that aspect was delivered by the selection of primers that targeted only the three pathogens. As an added precautionary step, an internal control was integrated into the multiplex assay to assess the integrity of the PCR. In another report by this group, a real-time PCR platform was designed to target the same pathogens but with an enrichment step added which increased the sensitivity of their detection assay by one log as compared to the conventional PCR assay (Omiccioli et al. 2009b).

Other methods have been developed to provide greater latitude in the number of microbes that can be targeted in one assay, and in one way, circumvent some of the minor detractions of PCR-based assays. Multiple targeted systems, such as microarrays or other detection technology, such as Bio-Plex (Bio-Rad, Inc., CA, USA) are designed, as in the latter method, so that theoretically up to 100 amplicons or other molecules (proteins) can be detected in one platform. The extended number of targets not only provides a more in-depth interpretation of data collected, but gives an analytical laboratory more flexibility. With conventional and real time PCR, additional reactions are most likely required to accurately state whether a specific pathogen is present. An example would be a lack of qualified and complete identification of the pathogen if the PCR assay was designed only to be genus specific; additional information needed would be species identification and at times, serotype.

As mentioned above, sample preparation is a critical step in molecular based assays, specifically those that use nucleic acid as the target molecule. In a recent report, a comparison of different extraction methods using raw milk and raw milk cheese as the matrix was presented (Quigley et al. 2012). Not only is the matrix a concern, but whether the target organism is Gram-positive or Gram-negative becomes a factor. Studies such as this can be used as a guide for consideration but the ultimate selection rests with the laboratory personnel performing the assays.

Enterotoxigenic *S. aureus* poses a complex picture in regard to its presence in milk and dairy products. This pathogen produces several enterotoxins, some have a greater propensity to be associated with human illness than others. PCR-based assays have been developed that target human-relevant *S. aureus* enterotoxins genes, including *sea, sec, sed, seg, sej* and *sel*, or now considered the enterotoxin gene cluster (*egc*) (Fusco et al. 2011). Although the genes may be present in staphylococci in milk, this is not affirmation that the toxins are present. Detection of the staphylococcal toxin is necessary to ensure any contaminated food is not capable of causing food-borne intoxication.

Mycobacterium avium subsp. *paratuberculosis* (MAP), the causative agent of chronic granulomatous enteritis (paratuberculosis), also known as Johne's disease, has been found in raw milk. A real-time, quantitative PCR assay, with an internal amplification control, was developed to detect this pathogen in milk. DNA preparation was accomplished using a commercially-available kit after a pre-processing step that included two centrifugations. In examining 345 milk samples from one farm 111 samples (32.5%) were shown to be positive by this assay for MAP. However, no MAP isolates were recovered from anyone of these milk samples (Slana et al. 2008). This brings to the forefront the continued debate to the decision process on whether a food sample is considered contaminated if PCR data indicate a positive result yet no viable cells are isolated. Again, PCR and other DNA-based methods may be a rapid, sensitive and specific means to indicate the presence of a particular pathogen(s), yet for conclusive or confirmative analysis, culture methods may be indicated.

The combined use of immune-magnetic beads or microspheres with PCR-based assays have been demonstrated as an effective method for *Salmonella, Campylobacter, E. coli, Listeria* and enterotoxin B (SEB) from *S. aureus* (Kim et al. 2010). The technology used in this study is based on the Luminex system (Bio-Rad, CA, USA), a rapid, multiplex method, but encountered diminishing return with respect to sensitivity in particular when milk was used as the food matrix.

Alternative PCR assays to conventional and real-time platforms include randomly amplified polymorphic DNAs (RAPD-PCR), fluorescence *in situ* hybridization (FISH), loop-mediated isothermal conditions (LAMP). One of the disadvantages for some PCR assays is that these assays are performed in highly sophisticated instruments (thermal cyclers) that regulate the cycle time and temperatures. Isothermal amplifications offer alternative amplifications parameters which include reactions at one temperature and preclude the use of thermal cyclers. One LAMP-based assay was recently developed for the detection of *L. monocytogenes* that indicates the presence of the pathogen in raw milk samples in approximately 30 minutes, including DNA preparation, and had a reported sensitivity of 8 to 10 cells per reaction (Wang et al. 2011). A LAMP-based assay was developed to target *Brucella* spp. (*B. abortus, B. melitensis, B. suis, B. canis, B. neotomae* and *B. ovis*) found in milk and other fluids. Overall, the specificity of this LAMP assay was shown to be specific for the intended targets and was 10 times as more sensitive than the nested PCR-based assay (Lin et al. 2011).

In addition to detecting human pathogens, a major concern to dairy farms is the prevalence of etiological agents causing bovine mastitis. Primary pathogens include *S. aureus, Streptococcus* species (*S. agalactiae, S. bovis, S. dysagalactiae* and *S. uberis*) and *Mycoplasma* spp. with *Corynebacterium bovis* common also. A biochip, rapid assay was designed to detect these seven bovine pathogens from milk which involves DNA extraction, PCR amplification of specific DNA target regions, hybridization of the amplicons to the chip and capturing a colorimetric product to indicate a positive reaction (Lee et al. 2008). Another published method used a PCR-based assay that targeted the 16S rDNA and 23S rDNA of *S. agalactiae* (Group B *Streptococcus*), *S. aureus* and *E. coli*. This assay was designed to be used to detect these pathogens in dairy products (Chotár et al. 2006). Another study developed an amplification reaction combined with a microarray system to detect microbial pathogens known to cause bovine, ovine and caprine mastitis and food-borne illnesses (Cremonesi et al. 2009). These microbes included *S. aureus*, several *Streptococcus* spp., *Mycoplasma* spp., *Bacillus* spp., *Salmonella* spp., *Campylobacter* spp. and *E. coli*. Other PCR-based assays have found additional etiological agents of mastitis in milk including *Klebsiella oxytoca* and *K. pneumonia*, and *S. dysgalactiae* (Cressier and Bissonnette 2011).

PCR-based amplifications can be used to detect the presence of multiple numbers of microbial organisms, whether pathogenic or not, in an array based platform. One method used this technology to amplify specific regions in hypervariable or variable regions in the 16S rDNA and 23S rDNA, respectively, as well as specific virulence genes. The targeted microbes were non-pathogenic (spoilage organisms) lactic acid bacteria (LAB) as well as the pathogens *C. jejuni, E. coli, S.* Typhimurium, *L. monocytogenes* and *Y. enterocolitica*. Amplified PCR products were hybridized against oligonucleotides probes bound to the array surface (Giannino et al. 2009).

15.2.2 Immunological-based methods

The underlying success of immunological-based assays, regardless of format, relies on the interaction of an antibody with an antigen. The quality of antibody used in the assay as well as the amount of target antigen present in the sample can be predictive values of any detection based assay. In addition, the means to capture the antibody-antigen complex requires a detection method, sometimes quantitative, that typically uses a secondary antibody that is conjugated to an enzyme or fluorophore molecule. Antibody-based assays utilize different formats and types of antibodies. As for the latter, polyclonal, monoclonal and recombinant antibodies are available, and each has different attributes that can be exploited for specific applications. Polyclonal antibodies have been incorporated into detection methods targeting many milk-related pathogens and toxins and cost less than monoclonal antibodies. A positive feature of monoclonal antibodies is their specificity for one antigenic structure; therefore, they provide a degree of certainty of the resulting reaction as being positive.

As indicated above, a common format of immunological-based assays includes a primary antibody attached to a solid surface which captures specific target molecules (antigens) present in the sample being analyzed. The secondary enzyme, conjugated as described above, attaches to the captured antigen and generates a colored enzymatic product. Other formats, such as the enzyme-linked immunosorbent assays (ELISA), use similar underlying antibody-antigen interactions, but generate a fluorescent signal that is subsequently captured on a plate reader. Therefore, these assays, in addition to a sensitive and specific detection method, can also provide a quantitative result. Enzyme-linked fluorescence assays are available from several commercial sources that target common milk-borne pathogens, such as *E. coli* O157:H7, *Salmonella* and *L. monocytogenes*.

FISH methods utilize fluorescent probes that bind (hybridize) to specific areas within the DNA or mRNA molecule. The use of FISH detected the presence of *H. pylori* in raw bovine milk. The specific probe hybridized to a specific region of the pathogen's 16S rDNA (Rahimi and Kheirabadi 2012). Whether the presence of *H. pylori* in milk is recognized as a commonly used means of transmission to humans has not been established.

In some applications, a combination of immunomagnetic capture and PCR are used to increase the level of sensitivity of the entire assay. Mercanoglu et al. (2009) reported the use of immunomagnetic separation after selected enrichment times to increase the sensitivity of the PCR assay to a level of 1–10 cfu ml^{-1}. There are commercially available enrichment/ capture instruments such as the Pathotrix (Life Technologies, CA, USA) that can be applicable for analysis of milk pathogens.

15.2.3 Biosensors

Biosensors are composed of two basic units, a bioreceptor and the transducer, that are the basis for the detection of a number of different targets, such as microbial cells (e.g., pathogens), and toxins. Bioreceptors usually are, but not limited to, antibodies, enzymes or nucleic acids. These biochemical recognition elements recognize specific target molecules in a sample and the subsequent biochemical reactions are converted to electrical signals by a physical transducer that are displayed and recorded by the instrument. A review of this process is provided by Velusamy et al. (2010).

Biosensors offer several advantages for their application to food safety, particularly with dairy products. They have a wide range of analytes that can be targeted including major dairy related pathogens such as *E. coli* O157:H7, *S. aureus*, *Salmonella* and *L. monocytogenes*, and microbial toxins, e.g., staphylococcal enterotoxins and mycotoxins. For toxins, the level of sensitivity can range in the nanogram per milliliter level and for bacteria, <100 cfu ml^{-1}. An example of the application of a dairy-related pathogen detection was described by Shriver-Lake et al. (2007) in which the level of detection of *S.* Typhimurium in milk was 5×10^6 cfu ml^{-1}. In another study, *E. coli* O157:H7 was spiked into milk and the level of detection was 5.3×10^2 cfu ml^{-1} without enrichment and the total time of sample preparation and analysis was given as 3 hr (Chen et al. 2008). Another noted potential of biosensors for diagnosis of food samples is the development of assays to capture simultaneously multiple targets. Biosensors are small and portable, features that lend itself for any point in the dairy food industry. A more comprehensive review of the potential of biosensors for the dairy industry was recently presented by Rasooly and Herold (2011) and McGrath et al. (2012).

15.3 Conclusion

Enrichment remains a critical step to yield sufficient quantities of target pathogen to ensure that the limit of sensitivity of any assay, respective of analyte, is met. In some cases, such as viruses, parasites and toxins, this additional step is not an option. However, alternative means, such as capturing the non-amplifiable target or low, stressed bacterial pathogens via immune-magnetic beads, are available for some, such as bacteriophages. Real-time testing seems to be a driving force for method development that meets the needs of the industry; rapid, portable, user-friendly, multiplex capability, accurate and sensitive are also desirable features. The utility of rapid methods in sample analysis is depicted in Fig. 15.1. The examination of foods prior to or immediately after sample preparation would expedite the overall analysis time and reduce the costs for labor and laboratory reagents

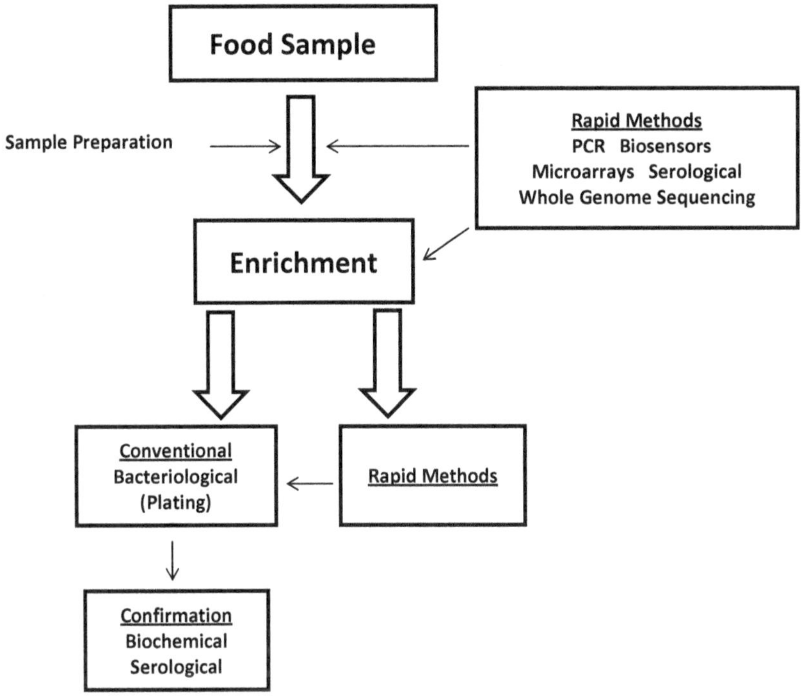

Figure 15.1 A general flow diagram depicting laboratory analysis of foods.

Food samples can be processed by several different regimens including conventional application of enrichment in broth medium followed by bacteriological protocols, i.e., plating enrichment cultures on solid agar media. Presumptive positive colonies are typically confirmed by biochemical tests and/or serological reactions. Rapid methods can be applied at different steps including after sample preparation, but prior to enrichment, after growth in enrichment broth and plating on agar media. Rapid methods can reduce the time of analysis considerably with a high degree of specificity and accuracy.

(no need for enrichment). The application of rapid methods in the analysis of post-enrichment cultures can be used to identify pathogens from broth medium or colonies, the latter after growth on solid agar medium after appropriate incubation conditions.

In addition to rapid method detection as one of its primary applications, there are uses of molecular-based methods that can decrease the time of analysis to identify microbial pathogens. Multilocus sequence typing (MLST) involves an initial amplification step targeting specific housekeeping genes followed by sequence analysis of the amplicons generated. MLST is used to compare the sequence of the amplified DNA to a database to determine or confirm the strain identity. These allelic profiles have the discriminatory power to identify clones within pathogen populations, and can be a useful aid in linking food-borne illnesses in regard to clinical, environmental and food samples. An alternative method is pulsed-field gel electrophoresis (PFGE) in which genomic DNA is digested with select restriction enzymes to generate a specific pattern on gel. This has become a powerful tool on an international level to determine if an etiological agent of an outbreak is related to isolates from clinical, environmental and food samples. In the United States, federal and state health laboratories submit PFGE profiles to the CDC (http://www.cdc.gov/pulsenet/) which houses databases for several bacterial pathogens. This network provides a rapid means to track the extent of an outbreak and alerts the appropriate authorities to respond. Another alternative typing method is PCR-restriction fragment length polymorphism (PCR-RFLP).

Other methods available for rapid detection of milk-borne pathogens include reverse-transcriptase PCR and nucleic acid sequence-based amplification (NASBA), which targets mRNA of bacterial cells, RAPD-PCR, and microarrays. The latter process uses multiple well format and potentially can have 92 to 364 wells used for amplification and detection protocols. Over the past several years, microfluidics and other systems, such as biosensors (lab-on-a-chip) technology, have progressed where there are instruments based on these protocols that have reached the market. In the near future, small, efficient and portable instruments could be available and used either on-site on a dairy farm or in a near-by laboratory to monitor the microbial quality of milk prior to distribution, either before or after pasteurization.

The next generation technology, the "omics", may become a future application for the detection of microbial pathogens and toxins present in milk. Genomics, proteomics and metabolomics have the potential to identify genomes or specific microbial proteins or other metabolic products and match the unknown generated data to an expansive database. Whole genome sequencing community has seen the cost of sequencing plummet significantly and may, in the near future, become economically feasible to be established in analytical laboratories. The potential for rapid diagnosis exists but this technology is not just limited to this application. Of keen interest in the research field is the study of host-pathogen interactions. Direct analysis of milk for dairy-related pathogens has its own merit to ensure that product is safe to consume. However, the potential use of the next generation technology, such as whole genome sequencing, extends beyond direct analysis of milk. For instance, determining the underlying genetic and pathogenic potential of microbes that cause mastitis, that may also cause human illness, can also provide a means to reduce the risk and incidences of this disease among dairy herds including cows, goats and sheep. Such a concept was put forth from a study reported by Ward et al. (2009). This group analyzed and presented genomic data that putatively identify genetic loci that enhanced *S. uberis* to be an opportunistic pathogen and colonize the bovine's udder. Extension of this type of research can have a significant impact on the dairy industry. For example, if virulence/genetic determinants responsible of niche-adaptation of dairy-related pathogens can be identified, perhaps prophylactic means can be developed and introduced to reduce colonization of dairy herds to reduce colonization of these animals. Therefore, a two-fold benefit can be realized; reduction of infections among dairy animals and, perhaps lessening the spread of pathogens from these zoonotic sources to humans.

As technology advances the application of rapid methods to ensure pathogen-free milk can safely enter the food market has several hurdles to address. In addition to cost, ease of use, and portability, the concerns of sample preparation to yield the appropriate amount of template target molecule that meets the minimum sensitivity of the assay as well as its quality, i.e., free of inhibitors, remain. The tools presented in this work reflect the potential of this technology for its use in food safety. The importance of a safe milk supply reaches different levels of concern, from direct consumption, such as raw milk, to industry-produced dairy products, e.g., raw-milk derived cheeses. Rapid detection assays can have a pivotal impact on providing a short time turn-around in determining the presence of pathogenic microbes. Since the advent of molecular based techniques, the integration of such methods has provided the milk industry and regulatory agencies with an impressive arsenal to assure the consumer of its purchase and consumption of a safe milk product.

References

Amagliani, G., A. Petruzzelli, E. Omiccioli, F. Tonucci, M. Magnani and G. Brandi. 2012. Microbiological surveillance of a bovine raw milk farm through multiplex real-time PCR. Foodborne Path. Dis. 9: 406–411.

Chen, S.H., V.C. Wu, Y.C. Chuang and C.S. Lin. 2008. Using oligonucleotide-functionalized Au nanoparticles to rapidly detect foodborne pathogens on a piezoelectric biosensor. J. Microbiol. Methods. 73: 7–17.

Chotár, M., B. Vidová and A. Godány. 2006. Development of specific and rapid detection of bacterial pathogens in milk and dairy products by PCR. Folia Microbiol. 51: 639–646.

Cremonesi, P., G. Pisoni, M. Severgnini, C. Consolandi, P. Moroni, M. Raschetti and B. Castiglioni. 2009. Pathogen detection in milk samples by ligation detection reaction-mediated universal array method. J. Dairy Sci. 92: 3027–3039.

Cressier, B. and N. Bissonnette. 2011. Assessment of an extraction protocol to detect the major mastitis-causing pathogens in bovine milk. J. Dairy Sci. 94: 2171–2184.

Fox, E., T. O'Mahony, M. Clancy, R. Dempsey, M. O'Brien and K. Jordan. 2009. *Listeria monocytogenes* in the Irish dairy farm environment. J. Food. Protect. 72: 1450–1456.

Fusco, V., G.M. Quero, M. Morea, G. Blaiotta and A. Viscont. 2011. Rapid and reliable identification of *Staphylococcus aureus* harbouring the enterotoxin gene cluster (*egc*) and quantitative detection in raw milk by real time PCR. Int. J. Food Microbiol. 144: 528–537.

Giannino, M.L., M. Aliprandi, M. Feligini, L. Vanoni, M. Brasca and F. Fracchetti. 2009. A DNA array based assay for the characterization of microbial community in raw milk. J. Microbiol. Methods 78: 181–188.

Gorski, L. and A. Csordas. 2010. Molecular detection: Principles and methods. pp. 1–22. *In:* D. Liu [ed.]. Molecular Detection of Foodborne Pathogens. CRC Press, Boca Raton, Florida, USA.

Hoorfar, J. 2011. Rapid detection, characterization, and enumeration of foodborne pathogens. APMIS 119(Suppl. 133): 1–24.

ICMSF. 2005. Milk and dairy products. pp. 643–715. *In:* Micro-organisms in Foods. 6: Microbial Ecology of Food Commodities, 2nd edition. Kluwer Academic/Plenum Publishers, New York, USA.

Kim, J.S., C.R. Taitt, F.S. Ligler and G.P. Anderson. 2010. Multiplexed magnetic microsphere immunoassays for detection of pathogens in foods. Sens. Inst. Food Qual. Saf. 4: 73–81.

Langer, A.J., T. Ayers, J. Grass, M. Lynch, F.J. Angulo and B.E. Mahon. 2012. Nonpasteurized dairy products, disease outbreaks, and state laws—United States, 1993–2006. Emerg. Infect. Dis. 18: 385–391.

Lee, K.H., J.W. Lee, S.W. Wang, L.Y. Liu, M.F. Lee, S.T. Chuang, Y.M. Shy, C.L. Chang, M.C. Wu and C.H. Chi. 2008. Development of a novel biochip for rapid multiplex detection of seven mastitis-causing pathogens in bovine milk samples. J. Vet. Diag. Invest. 20: 463–471.

Lin, G.-Z., F.-Y. Zheng, J.-Z. Zhou, X.-W. Gong, G.-H. Wang, X.-A. Cao and C.-Q. Qi. 2011. Loop-mediated isothermal amplification assay targeting the *omp*25 gene for rapid detection of *Brucella* spp. Mol. Cell. Probes 25: 126–129.

Mandal, P.K., A.K. Biswas, K. Choi and U.K. Pal. 2011. Methods for rapid detection of foodborne pathogens: An overview. Am. J. Food Technol. 6: 87–102.

McGrath, T.F., C.T. Elliott and T.L. Fodey. 2012. Biosensors for the analysis of microbiological and chemical contaminants in food. Anal. Bioanal. Chem. 403: 75–92.

Mercanoglu, T.B., U. Ben and S.A. Aytac. 2009. Rapid detection of *Salmonella* in milk by combined immunomagnetic separation-polymerase chain reaction assay. J. Dairy Sci. 92: 2382–2388.

Naum, M. and K.A. Lampel. 2011. DNA-based assays. pp. 221–225. *In:* J.W. Fuquay, P.F. Fox and P. McSweeney [eds.]. Encyclopedia of Dairy Science, 2nd edition. Elsevier Ltd., Oxford, UK.

Oliver, S.P., B.M. Jayarao and R.A. Almeida. 2005. Foodborne pathogens in milk and the dairy farm environment: Food safety and public health implications. Foodborne Path. Dis. 2: 115–129.

Oliver, S.P., K.J. Boor, S.C. Murphy and S.E. Murinda. 2009. Food safety hazards associated with consumption of raw milk. Foodborne Path. Dis. 6: 793–806.

Omiccioli, E., G. Amagliana, G. Brandi, I.J. Bruce and M. Magnani. 2009a. Simultaneous direct detection of *Salmonella* spp., *Listeria monocytogenes* and *Escherichia coli* O157 in milk samples by magnetic extraction and multiplex PCR. J. Rapid Meth. 17: 195–213.

Omiccioli, E., G. Amagliana, G. Brandi and M. Magnani. 2009b. A new platform for real-time PCR detection of *Salmonella* spp., *Listeria monocytogenes* and *Escherichia coli* O157 in milk. Food Microbiol. 26: 615–622.

Quigley, L., O. O'Sullivan, T.P. Beresford, R.P. Ross, G.F. Fitzgerald and P.D. Cotter. 2012. A comparison of methods used to extract bacterial DNA from raw milk and raw milk cheese. J. Appl. Microbiol. 113: 96–105.

Rahimi, E. and E.K. Kheirabadi. 2012. Detection of *Helicobacter pylori* in bovine, buffalo, camel, ovine, and caprine milk in Iran. Foodborne Path. Dis. 9: 453–456.

Rasooly, A. and K. Herold. 2011. Analytical methods—biosensors. pp. 235–247. *In:* J.W. Fuquay, P.F. Fox and P. McSweeney [eds.]. Encyclopedia of Dairy Science, 2nd edition. Elsevier Ltd., Oxford, UK.

Rossen, L., P. Norskov, K. Holmstrom and O.F. Rasmussen. 1992. Inhibition of PCR by components of food samples, microbial diagnostic assays and DNA-extraction solutions. Int. J. Food Microbiol. 17: 37–45.

Saiki, R.K., S. Scharf, F. Faloona, K.B. Mullis, G.T. Horn, H.A. Erlich and N. Arnheim. 1985. Enzymatic amplification of β-globin genomic sequences and restriction site analysis for diagnosis of sickle cell anemia. Science 230: 1350–1354.
Shriver-Lake, L.C., J.S. Erickson, K.E. Sapsford, M.M. Ngundi, K.M. Shaffer, N.V. Kulagina, J.E. Hu, S.A. GrayIII, J.P. Golden, F.S. Ligler and C.R. Taitt. 2007. Blind laboratory trials for multiple pathogens in spiked food matrices. Anal. Lett. 40: 3219–3231.
Singh, J.V., K. Batish and S. Grover. 2009. A molecular beacon-based duplex real-time polymerase chain reaction assay for simultaneous detection of *Escherichia coli* O157:H7 and *Listeria monocytogenes* in milk and milk products. Food. Path. Dis. 6: 1195–1199.
Slana, I., P. Kralik, A. Kralova and I. Pavlik. 2008. On-farm spread of *Mycobacterium avium* subsp. *paratuberculosis* in raw milk studied by IS900 and F57 competitive real time quantitative PCR and culture examination. Int. J. Food Microbiol. 128: 250–257.
van Kessel, J.A., J.S. Karns, J.E. Lombard and C.A. Kopral. 2011. Prevalence of *Salmonella enterica, Listeria monocytogenes* and *Escherichia coli* virulence factors in bulk tank milk and in-line filters from U.S. dairies. J. Food Protect. 74: 759–768.
Velusamy, V., K. Arshak, O. Korostynska, K. Oliwa and C. Adley. 2010. An overview of foodborne pathogen detection: In the perspective of biosensors. Biotechnol. Advances 28: 232–254.
Wang, D., G. Zhang, C. Lu, R. Deng, A. Zhi, J. Guo, D. Zhao and Z. Xu. 2011. Rapid detection of *Listeria monocytogenes* in raw milk with loop-mediated isothermal amplification and chemosensor. J. Food Sci. 76: M611–M615.
Ward, P.N., M.T.G. Holden, J.A. Leigh, N. Lennard, A. Bignell, A. Barron, L. Clark, M.A. Quail, J. Woodward, B.G. Barrell, S.A. Egan, T.R. Field, D. Maskell, M. Kehoe, C.G. Dowson, N. Chanter, A.M. Whatmore, S.D. Bentley and J. Parkhill. 2009. Evidence for niche adaptation in the genome of the bovine pathogen *Streptococcus uberis.* BMC Genomics. 10: 54–70.

CHAPTER 16

Current Regulations in Microbiological Control of Milk and Dairy Products

Theo Varzakas, Ilya Vladimirovich Nikolaev* and *Olga Vladimirovna Koroleva*

16.1 Introduction

Raw milk is inherently dangerous and may contain pathogens including Enterotoxigenic *Staphylococcus aureus, Campylobacter jejuni, Salmonella* species, *Escherichia coli* (*E. coli* 0157H:7, Enterohemorrhagic *E. coli*—EHEC, Enterotoxigenic *E. coli*—ETEC), *Listeria monocytogenes, Mycobacterium tuberculosis, M. bovis, Brucella* species (*B. abortus* being mainly associated with cows, and *B. melitensis* being mainly associated with goats), *Coxiella burnetii, Yersinia enterocolitica* and others. There are variations in incidence rates between countries and even within regions of countries for the presence of pathogens in raw milk. There are also variations in incidence rates reported for the three main commercial milks (bovine, ovine and caprine).

Apart from pathogenic bacteria, it is also important to consider the yeast population of milk, which can impart important flavors on dairy products (de Freitas et al. 2009). Cocolin et al. (2002) carried out analyses which focused on the yeast population in raw cow's milk using a combination of culturing and denaturing gradient gel electrophoresis (DGGE) fingerprinting. The classical culture-based methods identified six *Candida* species (which accounted for 56.2% of the population), two *Kluyveromyces* species, *Saccharomyces cerevisiae, Pichia guilliermondii, Trichosporon mucoides*

*Corresponding author

and *Cryptococcus curvatus* as being the main yeast constituents. When the samples were analyzed by DGGE additional yeasts such as *Galactomyces* spp., *Candida kefyr*, *Candida humilis* and *Saccharomyces bayanus/pastorianus* were detected. Callon et al. (2007) revealed the presence of a diverse population of bacteria and yeast in the milk as was also reported by Quigley et al. (2011). In addition to species commonly encountered in milk, some species which are atypical of goat's milk or had previously only been associated with cheeses, including a number of corynebacteria and brachybacteria, were identified thereby highlighting the sensitivity of molecular-based tests. Another unexpected finding was the detection of several halophilic species which are atypical of milk, including *Jeotgalicoccus psychrophilus*, *Salinicoccus* spp., *Dietzia maris*, *Exiguobacterium* spp., *Ornithinicoccus* spp. and *Hahella chejuensis*. This indicates that employing sensitive methodologies to monitor microbiological quality of foods and thus to ensure food safety is as important as taking safety precautions during food processing.

Although dairy industry in developed countries has already established food safety systems to produce safer and extended shelf-life products, their production processes are under continuous inspection by the legal authorities. Food business operators (FBOs) must provide the competent authorities with all the assistance needed in carrying out the control, notably as regards access to premises and the presentation of documentation or records. The official controls include audits of good hygiene practices and HACCP (Hazard Analysis and Critical Control Points) principles, as well as specific controls whose requirements are determined by sector (milk and dairy products).

This chapter deals with the current EU, US, Japanese and Russian legislations regarding food safety with specific references to raw milk, processed milk and dairy products.

16.2 European Union regulations for dairy products

From 1 January 2006 new food hygiene legislation came into force throughout the European Union. The legislation affects all food businesses, including caterers, primary producers (such as farmers), manufacturers, distributors and retailers. For the dairy industry, the legislation has replaced the requirements of the Dairy Hygiene Directive 92/46/EEC. This directive contained many detailed prescriptive requirements, some of which were not consistent across the directives applying to other foods. The new legislation achieves consistency by applying to all foods, but with additional requirements applying to foods of animal origin such as dairy products. Although the result is a simplification of the requirements, unfortunately the task of achieving this has turned out to be more complicated than initially envisaged. The directives are being replaced by two main regulations together with subsidiary regulations and guidance documents. Underpinning the new hygiene package is Regulation 178/2002. This lays down the general principles and requirements of food law, and also establishes the European Food Safety Authority. The main regulations in the new package are:

- *Regulation 852/2004 on the hygiene of foodstuffs.* This regulation defines the food safety objectives to be achieved, leaving the food operators responsible for adopting the safety measures to be implemented in order to guarantee food safety. This applies to all foods, and begins at primary production and continues through processing, distribution and retail. Finally, Regulation 852/2004 encourages national and community guides to the legislation.
- *Regulation 853/2004 on specific hygiene rules.* This regulation lays down specific hygiene rules for food of animal origin in order to guarantee a high level of food safety and public health. This gives additional requirements for foods of animal origin and composite products such as pizzas are excluded from the scope while applying the additional requirements to dairy and other foods of animal origin.
- *Regulation 854/2004 laying down specific rules for the organization of official controls on products of animal origin intended for human consumption.*
- *Regulation 882/2004 on official controls performed to ensure the verification of compliance with feed and food law, animal health and welfare rules.*
- There are also regulations giving transition measures, implementing measures and microbiological criteria. As well as this, the European Commission is issuing guidance on the interpretation of certain aspects of the legislation.

The new legislation requires FBOs (except primary producers) to put in place, implement and maintain a permanent procedure, or procedures, based on the HACCP principles. The legislation is structured in this way so that it can be applied flexibly in all food businesses regardless of their type or size.

16.2.1 Regulation 852/2004

Regulation 852/2004 includes that, as far as possible, FBOs are to ensure that primary products are protected against contamination, having regard to any processing that primary products will subsequently undergo. FBOs producing primary products of animal origin are to take appropriate measure(s) to ensure the cleanliness of animals taken for slaughter and, where necessary, production animals. They also must keep clean and, where necessary after cleaning, to disinfect equipment, containers, crates, vehicles and vessels in an appropriate manner.

FBOs must ensure that staff handling foodstuffs are in good health and undergo training on health risks. FBOs carrying out any stage of production, processing and distribution of food after above stages must comply with the general hygiene requirements laid down in Annex II of Regulation 852/2004 and any specific requirements provided for in Regulation 853/2004.

The key requirements of Regulation 852/2004 are:

- Primary responsibility for food safety rests with the FBOs,
- Food safety throughout the food chain starts with primary production,
- Maintenance of the cold chain is important for food that cannot be stored safely at ambient temperatures, particularly frozen food,
- HACCP principles, together with the application of good hygiene practice, should reinforce FBOs' responsibility,
- Guides to good practice are to be encouraged,
- Microbiological criteria and temperature control requirements will be established based on a scientific risk assessment, and
- Imported foods must be of the same hygiene standard or an equivalent standard as food produced in the European Community.

Primary producers will need to follow good practice and manage their operations as set out in Annex I of Regulation 852/2004. Primary producers must ensure that hazards are acceptably controlled and respect other existing legislation but primary producers are not required to apply HACCP-based procedures. Primary producers will have to be registered with competent authorities, although existing forms of registration may be used for this purpose.

Hazard analysis and critical control points (HACCP)

Regulation 852/2004 requires FBOs to develop and apply procedures based on the following HACCP principles:

- Identifying all hazards,
- Identifying critical control points,
- Establishing critical limits at critical control points,
- Establishing and implementing effective monitoring procedures at critical control points,
- Establishing corrective actions when monitoring indicates that a critical control point is not under control,
- Establishing procedures to verify that the above measures are working effectively, and
- Establishing documents and records commensurate with the nature and size of the food business.

Readers may refer to Chapter 14 for more detail about implementation of HACCP programs in production of processed milk and milk products.

General requirements for heat treatment

1. Any heat treatment process must (i) raise every part of the product treated to a given temperature for a given period of time; and (ii) prevent the product from becoming contaminated during the process.
2. FBOs must check regularly the main relevant parameters (particularly temperature, pressure, sealing and microbiology), and the use of automatic devices.
3. The process used should conform to an internationally recognized standard (for example, pasteurization, ultra-high temperature or sterilization).

Specific requirements of Annex I of Regulation 852/2004

Specific requirements include the following:

- To protect primary products against contamination,
- To ensure the cleanliness of animals as far as possible,
- To keep clean any facilities used to store and handle feed,
- To ensure that staff handling foodstuffs are in good health and undergo training on health risks,
- To store and handle waste and hazardous substances so as to prevent contamination,
- To prevent the introduction and spread of contagious diseases transmissible to humans through food, including taking precautionary measures when introducing new animals and reporting suspected outbreaks of such diseases to the competent authority,
- To take account of the results of any relevant analyses carried out on samples taken from animals or other samples that have importance to human health,
- To use feed additives and veterinary medicinal products correctly, as required by relevant legislation, and
- To keep and retain records relating to measures to control hazards in an appropriate manner and for an appropriate period. These records should include (i) the nature and the origin of feed fed to animals, (ii) veterinary medicinal products or other treatments administered to the animals, dates of administration and withdrawal periods, (iii) the occurrence of diseases that may affect the safety of the products, (iv) the results of any analyses carried out on samples taken from animals

or other samples taken for diagnostic purposes that have importance for human health, and (v) any relevant reports on checks carried out on animals or products of animal origin.

FBOs are obliged to take appropriate remedial action when informed of problems identified during official controls.

16.2.2 Regulation 853/2004

This regulation lays down specific hygiene requirements that must be implemented by food businesses handling food of animal origin. It requires that FBOs place products of animal origin on the market only if they have been prepared and handled exclusively in establishments that meet the relevant requirements of Regulation 852/2004, the requirements of Annexes II and III of Regulation 853/2004 and other relevant requirements of food law; and that the competent authority has registered or, where required, 'approved' the food business. Annex III Section IX of Regulation 853/2004 as amended sets out the specific requirements that apply to raw milk, colostrum, dairy products and colostrum-based products. This Annex is divided into a number of chapters. These are:

Chapter I—Raw milk and colostrum-primary production

- Animal health requirements for raw milk and colostrum production
- Hygiene on milk and colostrum production holdings
- Requirements for premises and equipment
- Hygiene during milking, collection and transport
- Staff hygiene
- Criteria for raw milk and colostrum

Chapter II—Requirements concerning dairy and colostrum-based products

- Temperature requirements
- Requirements for heat treatment
- Criteria for raw cow's milk

Chapter III—Wrapping and packaging
Chapter IV—Labeling
Chapter V—Identification marking

Specific requirements of Regulation 853/2004

There are specific additional dairy requirements in Regulation 853/2004 applying to primary production. In the main, these are similar or more flexible than the requirements of Directive 92/46/EEC but the following small differences are worth noting:

- Raw milk from any animal showing individually a positive reaction for tuberculosis or brucellosis must not be used for human consumption.
- The inspection requirements for milk from each animal have been modified to permit automatic milking. The new wording requires that milk from each animal be checked for organoleptic or physico-chemical abnormalities by the milker or by a method achieving similar results and that milk presenting such abnormalities is not used for human consumption.
- Milk from animals showing clinical signs of udder disease must not be used for human consumption except in accordance with the instructions of a veterinarian.
- Teat dips or sprays are to be used only after authorization or registration.
- FBOs must initiate procedures to ensure that raw milk is not placed on the market if it contains antibiotic residues above permitted limits.
- When raw milk fails to comply with the standards for antibiotic residues, plate count or somatic cell count, the FBOs must inform the competent authority and take measures to correct the situation.

Regulation 853/2004 does not include additional heat-treatment requirements for dairy products but the Commission's Regulation laying down implementing measures (Regulation 2074/2005) has reintroduced definitions for pasteurization and UHT as follows:

1. Pasteurization is achieved by a treatment involving (i) a high temperature for a short time (at least 72°C for 15 s); (ii) a low temperature for a long time (at least 63°C for 30 min); or (iii) any other combination of time and temperature conditions to obtain an equivalent effect, such that the products show, where applicable, a negative reaction to an alkaline phosphatase test immediately after such treatment.
2. Ultra high temperature (UHT) treatment is achieved by a treatment (i) involving a continuous flow of heat at a high temperature for a short time (not less than 135°C in combination with a suitable holding time) such that there are no viable microorganisms or spores capable of growing in the treated product when kept in an aseptic closed container at ambient temperature; and (ii) sufficient to ensure that the products remain microbiologically stable after incubating for 15 days at 30°C in closed containers, or for seven days at 55°C in closed containers, or after any other method demonstrating that the appropriate heat treatment has been applied (Arvanitoyannis et al. 2009) .

16.2.3 Other relevant regulations

Regulation 605/2010 lays down animal and public health and veterinary certification conditions for the introduction of raw milk and dairy products into the European Union intended for human consumption. Milk and dairy products which are imported into the European Union must meet basic animal and public health conditions. Compliance with these conditions should guarantee that milk and dairy products fulfill the requirements of European legislation. In addition, compliance with these conditions aims to maintain the public health situation within the European Union.

Regulation 1664/2006 (amending Regulation 2074/2005) implements measures for certain products of animal origin intended for human consumption and repealing certain implementing measures. Foodstuffs should not contain microorganisms or their toxins or metabolites in quantities that present an unacceptable risk for human health. The safety of foodstuffs is mainly ensured by a preventive approach, such as implementation of good hygiene practice and application of procedures based on HACCP principles.

Regulation 2073/2005 contains microbiological criteria for milk and dairy products. Testing against the criteria set down in the Regulation should be undertaken when validating or verifying the correct functioning of systems in place. In Table 16.1 microbiological criteria for milk and dairy foods are presented. According to Article 4 of Regulation 852/2004, FBOs are to comply with microbiological criteria. This should include testing against the values set for the criteria through the taking of samples, the conduct of analyses and the implementation of corrective actions, in accordance with food law and the instructions given by the competent authority. It is therefore appropriate to lay down implementing measures concerning the analytical methods, including, where necessary, the measurement of uncertainty, the sampling plan, the microbiological limits, and the number of analytical units that should comply with these limits. Furthermore, it is appropriate to lay down implementing measures concerning the foodstuff to which the criterion applies the points of the food chain where the criterion applies, as well as the actions to be taken when the criterion is not met. The measures to be taken by the FBOs in order to ensure compliance with criteria defining the acceptability of a process may include, among other things, controls of raw materials, hygiene, temperature and shelf-life of the product.

The Scientific Committee on Veterinary Measures relating to Public Health SCVPH issued an opinion on verotoxigenic *E. coli* (VTEC) in foodstuffs. This opinion concluded that applying an end-product microbiological standard for VTEC O157 is unlikely to deliver meaningful reductions in the associated risk for the consumers. However, microbiological

Table 16.1 Microbiological criteria for milk and dairy products.

Food	Microorganisms	Sampling plan	Limits	Analytical reference method	Stage where criterion applies
Cheeses, butter and cream made from raw milk or milk that has undergone a lower heat treatment than pasteurization	*Salmonella*	$n = 5, c = 0$	Absence in 25 g	EN/ISO 6579	Products placed on the market during their shelf-life
Milk powder and whey powder	*Salmonella*	$n = 5, c = 0$	Absence in 25 g	EN/ISO 6579	Products placed on the market during their shelf-life
Ice cream excluding products where the manufacturing process or the composition of the product will eliminate the salmonella risk	*Salmonella*	$n = 5, c = 0$	Absence in 25 g	EN/ISO 6579	Products placed on the market during their shelf-life
Cheeses, milk powder and whey powder, as referred to in the coagulase-positive staphylococci criteria in Chapter 2.2 of the Annex	Staphylococcal enterotoxins	$n = 5, c = 0$	Not detected in 25 g	European screening method of the CRL for milk	Products placed on the market during their shelf-life
Dried infant formulae and dried dietary foods for special medical purposes intended for infants below six months of age, as referred to in the Enterobacteriaceae criterion in Chapter 2.2 of the Annex	*Enterobacter sakazakii*, *Salmonella*	$n = 30, c = 0$	Absence in 10 g Absence in 25 g	ISO/DTS 22964, EN/ ISO 6579	Products placed on the market during their shelf-life

n: number of units comprising the sample; ***c***: number of sample units giving values over m or between m and M guidelines aimed at reducing the fecal contamination along the food chain can contribute to a reduction in public health risks, including VTEC. The SCVPH identified the following food categories where VTEC represents a hazard to public health: raw milk and raw milk products and other products as mentioned in Regulation 2073/2005.

The Scientific Panel on Biological Hazards (BIOHAZ Panel) of the European Food Safety Authority (EFSA) issued an opinion on the microbiological risks in infant formulae and follow-on formulae on 9 September 2004. It concluded that *Salmonella* and *Cronobacter sakazakii* are the microorganisms of greatest concern in infant formulae, formulae for special medical purposes and follow-on formulae. The presence of these pathogens constitutes a considerable risk if conditions after reconstitution permit multiplication. *Enterobacteriaceae*, which are often present in above-mentioned foods, could be used as an indicator for risk. Monitoring and testing of *Enterobacteriaceae* was recommended in both the manufacturing environment and the finished product by the EFSA (Regulation 2073/2005).

Regulation 178/2002 lays down the pursuit of a high level of protection of human life and health is one of the fundamental objectives of food law. The Regulation also describes other common principles and definitions for national and Community food law. It places primary responsibility to produce safe food on the FBOs. The Regulation defines a food business as "any undertaking, whether for profit or not and whether public or private, carrying out any of the activities related to any stage of production, processing and distribution of food" and FBO as "the natural or legal persons responsible for ensuring that the requirements of food law are met within the food business under their control".

Regulation 273/2008 lays down certain reference methods for the chemical, physical and microbiological analysis and sensory evaluation of milk and milk products to be used under the arrangements provided for in the common organization of the market in milk and milk products established by Regulation No 1255/1999 and the rules for applying those methods.

Regulation 1234/2007 (further amended by Regulation 247/2008 and 361/2008, respectively) contains the rules of establishing a common organization of agricultural markets and specific provisions for certain agricultural products. Since 1st July 2008, the rules set out in Regulation 1234/2007 replaced those of the Regulation 2597/97/EC as amended. They apply in lieu of the provisions of the Dairy Produce Acts 1924–1984, the Creamery Acts 1928–1934 and the Milk and Dairies Acts 1935–1956. Accordingly, only the provisions of these Acts which do not touch on matters covered by the Regulations (mainly administrative and commercial aspects) are now enforced. It is intended in due course to replace all outdated legislation with measures dovetailed with the Regulations.

16.2.4 Registration and approval of establishments

Establishments handling products of animal origin must be registered and, where necessary, approved by the competent authority in their Member State. This does not apply to establishments engaged only in primary production, transport or storage of products not requiring temperature-controlled storage conditions, or retail operations not subject to the Regulation 882/2004.

In accordance with Regulation 882/2004 on official controls performed to ensure the verification of compliance with feed and food law and with the provisions relating to animal health and animal welfare, Member States must keep up-to-date lists of approved establishments, which are given an approval number with additional codes indicating the type of products of animal origin manufactured. In addition to the common control requirements, specific official controls shall include:

(i) *Inspection of holdings.* Animals must undergo regular veterinary inspections to ensure compliance with the health requirements for raw milk production (health status of the animals, use of veterinary medicinal products).
(ii) *Control of raw milk upon collection.* The competent authority shall organize control schemes in order to ensure compliance with the standards that apply to raw milk. When the raw milk fails to meet mandatory food safety criteria, the competent authority may suspend the delivering of the milk in question and ask the farmer to take the necessary measures.

16.2.5 Importation of products of animal origin from non-EU member countries

The Commission, assisted by the Standing Committee on the Food Chain, draws up lists of third countries or parts of third countries from which the importation of products of animal origin is authorized. A third country may only be listed if the country in question provides the appropriate guarantees and after a Community inspection has been carried out in the country. Furthermore, the Commission draws up a list of establishments from which products of animal origin may be imported or dispatched. An establishment may only be listed if the competent authority in the third country of origin guarantees that the said establishment complies with the relevant Community requirements. Regular Community inspections are carried out in order to check these guarantees.

Furthermore, Regulation (EC) 854/2004 provides that an establishment may only be included in these lists if the competent authority in the originating third country guarantees that:

- the aforementioned establishment, as well as any other establishment handling raw materials of animal origin used in the production of the products of animal origin concerned, comply with the relevant European requirements, specifically with those of Regulation 853/2004 or those which have been defined as equivalent to these requirements following the decision to include this third country in the relevant list, in accordance with Article 11;
- an official inspection service in this country carries out monitoring of the establishments and makes available to the Commission, where necessary, all relevant information on the establishments providing raw materials; and
- this service has the power to prevent establishments from exporting to the European Union in the event that they do not comply with the requirements detailed above.

When drawing up these lists, particular account is taken of:

- the existing legislation of the third country, and the organization and powers of the competent authority and the inspection services;
- where applicable, the situation regarding animal health, zoonoses and plant health, as well as the procedures for notifying the Commission and the competent international bodies of animal or plant diseases which occur;
- experience gained in relation to marketing vis-à-vis the third country and its collaboration in exchanging information, particularly on health risks;
- the results of EU inspections/audits carried out in the country concerned; and
- the existence, in the third country concerned, of legislation on animal nutrition and programs for monitoring zoonoses and residues.

16.2.6 Labeling

In addition to the complying with requirements of Directive 2000/13/EC and its amendments on the labeling, advertising and presentation of foodstuffs, Chapter IV of Annex III, Section IX of Regulation 853/2004 requires that in the case of products made with raw milk whose manufacturing process does not include any heat treatment or any physical or chemical treatment the labeling must clearly show the words 'made with raw milk'. 'Raw milk' is defined as milk produced by the secretion of the mammary gland of farmed animals that has not been heated to more than 40°C or undergone any treatment that has an equivalent effect. The labeling information concerning products made from raw milk applies up to the point of sale and may be provided with packaging, document, notice, label, ring or collar accompanying or referring to the concerned products.

16.2.7 Preserved milk

Council Directive 2001/114/EC relates to certain partly or wholly dehydrated preserved milk for human consumption. This Directive has been amended by Council Directive 2007/61/EC. According to current Directive, partly dehydrated milk is defined as "the liquid product, whether or not sweetened, obtained by the partial removal of water from milk, from wholly or partly skimmed milk or from a mixture of these products, which may have an admixture of cream or of wholly dehydrated milk or both, the addition of wholly dehydrated milk not to exceed, in the finished products, 25% of total milk solids". 'Totally dehydrated milk' means the solid product, where the water content does not exceed 5% by weight of the finished product, obtained by the removal of water from milk, from wholly or partly skimmed milk, from cream or from a mixture of these products. The general labeling provisions of Directive 2000/13/EC apply in addition to the specific requirements set out in Directive 2001/114/EC as amended. The labeling must state the percentage of milk fat (except in the case of condensed skimmed milk, sweetened condensed skimmed milk and dried skimmed milk or skimmed-milk powder) and the percentage of fat-free dried milk extract in the case of the partly dehydrated milk. These particulars must appear near the trade name.

In the case of totally dehydrated milk, the label must state the recommendations as to the method of dilution or reconstitution, including details of the fat content of the product thus diluted or reconstituted. The labeling of totally dehydrated milk must state that the product is 'not intended as a food for infants under 12 months'.

The preserved milk samples are analyzed in accordance with the rules specified in Council Directive 87/524/EEC.

16.3 US-FDA regulations of milk and dairy products

Guidance for Industry: Dairy Farms, Bulk Milk Transporters, Bulk Milk Transfer Stations and Fluid Milk Processors: Food Security Preventive Measures Guidance (2007) is designed as an aid to operators of dairy farms, bulk milk transportation operations, bulk milk transfer stations and fluid milk processing facilities. It identifies the kinds of preventive measures. Operators of these establishments may take to minimize the risk that fluid milk under their control will be subject to tampering or other malicious, criminal, or terrorist actions.

16.3.1 Raw milk

Overall rules of raw milk delivery procedure indicated in the Guidance are outlined below:

- Accepting only those incoming tanker loads of raw milk for which all openings are either locked or sealed, with a serially numbered seal, from the time the tanker was last washed until the load is delivered. Exception may be provided for incoming loads for which a thorough investigation demonstrates that there is a verified, reasonable explanation for a deviation. Seals or locks need not be in place during those times that the tanker was under the direct, visual supervision of the driver.
- Using only known, reputable transportation companies.
- Establishing delivery schedules for raw milk, not accepting unexplained, unscheduled deliveries or drivers, and investigating delayed or missed shipments.
- Supervising off-loading of incoming milk.
- Reconciling the amount received with the amount listed on the shipping documents.
- Verifying that operators of bulk milk transfer stations that supply raw milk adhere to the preventive measures listed in this guidance.

State and local regulatory agencies are responsible for the enforcement of sanitation requirements on dairy farms, in processing plants, dry milk plants, receiving and transfer stations (establishments where raw milk or cream is received for further transport). The Public Health Service has no legal jurisdiction in the enforcement of milk sanitation standards except on interstate carriers and milk and milk products shipped in interstate commerce. United States Food and Drug Administration's (FDA) primary function under the Federal/State Milk Safety Cooperative Program (IMS Grade A products) is to provide technical assistance to the states in the implementation and enforcement of their milk regulations. This assistance is provided through the district and regional milk specialists and the Center for Food Safety and Nutrition's (CFSAN), Milk Safety Branch.

To assist states and municipalities in initiating and maintaining effective programs for prevention of milk-borne diseases, the Public Health Service developed a model regulation in 1924 known as the Standard

Milk Ordinance for voluntary adoption by state and local milk control agencies. This model milk ordinance, now titled "Grade A Pasteurized Milk Ordinance" (PMO), also provides administrative and technical details as to satisfactory compliance with the ordinance. The ordinance is incorporated by reference in federal specifications, for procurement of milk and milk products served on interstate carriers, and is recognized by public health agencies and the milk industry as a national standard for milk sanitation.

FDA is responsible for direct inspectional coverage of all non-grade A products shipped in interstate commerce which include: milk and milk products, butter, dried milk products, cheese and cheese products, ice cream and related products (frozen desserts), filled milk and imitation milk products.

In addition to applicable Code Food Regulations (CFR) references, the PMO can be used as an inspectional guide to cover specific operations in the dairy industry, including pasteurization equipment, packaging, quality control and record keeping requirements. Although the PMO does not have the force of regulations, it provides procedures and standards of general applicability that are acceptable to FDA. Deviations from the PMO can be listed as objectionable conditions if the procedures and standards in use at a firm are not deemed to be equivalent to those in the PMO.

van Kessel et al. (2004) reported on the prevalence of *Salmonella* spp. and *L. monocytogenes* in bulk tanks on U.S. dairies. They reported a 2.6% incidence rate for *Salmonella* spp. and a 6.5% incidence rate for *L. monocytogenes*. They commented that although the prevalence of these organisms was low, inappropriate handling of raw milk could result in bacterial growth and substantially increase the potential risk to consumers of raw milk and raw milk products. These incidence rates were reported even with very low standard plate counts (total bacterial counts) at <5000 cfu ml^{-1}, being reported for the vast majority of samples analyzed for the pathogens. This is important to note because it is clear illustration of the fact that a simple standard plate count is not an indication of the safety of milk. A low standard plate count clearly does not mean that milk will be pathogen-free (Sheehan 2009).

Many of the above-mentioned microorganisms can cause very serious, sometimes life-altering and sometimes even fatal disease conditions in humans. With pregnant women, *L. monocytogenes*-caused illness can result in miscarriage, fetal death, or illness or death of a newborn infant. Enterohemorrhagic *E. coli* (EHEC) infection has been linked to hemolytic uremic syndrome (HUS), a condition that can cause kidney failure and death. If infected with EHEC, young children are particularly susceptible to contracting HUS as has recently happened in the US.

Werner et al. (1979) reported on the incidence of *Salmonella* Dublin infections in California between 1971 and 1975 and said that the mean annual incidence of *Salmonella* Dublin infections in California increased five-fold. Keene et al. (1997) reported on a prolonged outbreak of *E. coli* O157:H7 which was caused by the consumption of raw milk sold at Oregon grocery stores. Outbreaks began in 1992 and continued until June of 1994. When the dairy that was the source of the raw milk was identified, it was discovered that four of the 132 animals in the herd were initially positive for *E. coli* O157:H7. Despite public warnings, new labeling requirements and increased monitoring of the culprit dairy, illnesses continued until June 1994, when retail sales were finally stopped. Proctor and Davis (2000) reported on *E. coli* O157:H7 infections in Wisconsin between 1992 and 1999. During that timeframe, there were 1333 cases, even though the disease only became reportable in Wisconsin in April 2000. The highest age-specific mean annual incidence, at 13.2 cases per 100,000 population occurred in children aged 3–5 years old. Among case patient identifiable exposures, consumption of raw milk/milk products was among the top three causes most frequently noted.

An outbreak of food-borne illness involving *E. coli* O157:H7 also occurred in California in 2006. This outbreak was determined by California to likely be caused by a dairy owned by a raw milk advocate. Schmid et al. (1987) reported on *Campylobacter jejuni* infections in Dubuque, Iowa over a twelve-month period.

A number of food-borne disease outbreaks have been associated with *L. monocytogenes*. An outbreak in Massachusetts revealed that this organism was present in whole and 2% milk which had undergone pasteurization (Fleming et al. 1985). A second outbreak in Los Angeles in 1985 pointed to the possibility of both improper manufacturing practices and post-processing contamination as sources of *L. monocytogenes* in a Mexican style soft-cheese (Linnan et al. 1988). In an attempt to determine the sources of *Listeria* contamination, the Food and Drug Administration (FDA) embarked on a study of dairy processing plants across the United States. The results of this study, known as the Dairy Safety Initiatives Program, indicated that pathogens such as *Listeria* spp., *Yersinia enterocolitica, Salmonella* spp. and *Aeromonas hydrophila* were present in finished products produced by some of the plants tested. In total, 2.9% of the finished products were contaminated with *Listeria* species. The second year status report issued by the FDA (U.S. Food and Drug Administration 1987) indicated that post-pasteurization contamination was a significant factor in those plants which manufactured products containing pathogens.

Stringent cleaning and sanitizing regimes have been implemented in many dairy processing facilities in order to reduce environmental contamination caused by *L. monocytogenes*. The efficacy of various sanitizing agents on eliminating *Listeria* has been investigated by numerous researchers. It has been reported that the use of chlorine-based sanitizer at

100 ppm, iodine-based sanitizer at 25–45 ppm, acid anionic-based sanitizer at 200 ppm and quaternary ammonium-type sanitizer at 100–200 ppm were all effective in eliminating *L. monocytogenes* (Ryser and Marth 1991).

Pritchard et al. (1995) compared the incidence of *Listeria* contamination of processing equipment with that of the general dairy processing environment. A total of 378 sponge samples obtained from 21 dairy plants were analyzed for *Listeria* using three different enrichment media. Eighty (80) of 378 sites (21.2%) were identified as *Listeria*-positive. *Listeria innocua* was isolated from 59 of the 80 (73.8%) positive samples, *L. monocytogenes* was identified in 35 (43.8%) of the positive samples, and *L. seeligeti* was isolated from five (6.3%) of the *Listeria*-positive samples. Positive equipment samples were obtained from six of the 21 (28.6%) plants and 19 of the 21 (90.5%) plants had positive environmental sites. Seventeen of the 215 (7.9%) samples from equipment were positive for *Listeria* species. Eleven of these sites, including three holding tanks, two table tops, three conveyor/chain systems, one pasta filata wheel, one pint milk filler and one brine pre-filter machine, were positive for *L. monocytogenes*.Nineteen of the 21 (90.5%) plants had positive environmental sites. Sixty-three of the 163 (41.1%) samples from environmental sites were *Listeria*-positive and 24 were positive for *L. monocytogenes*. Two-tailed student t-test analysis of the mean frequencies indicated that the level of contamination was significantly higher in 'environmental' (49.7%) as opposed to 'equipment' samples (7.0%).

FDA's opinion regarding raw milk consumption is entirely consistent with that of the American Medical Association, which holds as policy the position that "all milk sold for human consumption should be required to be pasteurized" (H-150.980, Milk and Human Health).

The Centers for Disease Control and Prevention (CDC's) Morbidity and Mortality Weekly Report (MMWR) for the week of March 2, 2007 reported that from 1998 to May 2005, 45 outbreaks of food-borne illness implicated unpasteurized milk, or cheese made from unpasteurized milk. Those outbreaks accounted for 1007 illnesses, 104 hospitalizations, and two deaths. The CDC also noted that between 1973 and 1992, 87% of the raw milk outbreaks occurred in those states which allowed for raw milk sales to consumers while consumption of raw milk was estimated to have been less than 1% of the total milk sold in those states (Sheehan 2009).

16.3.2 Grade "A" milk and milk products

Grade "A" milk and milk products include:

1. All milk and milk products with a standard of identity provided for in 21 CFR Part 131, excluding 21 CFR 131.120 Sweetened Condensed Milk.
2. Cottage cheese (21 CFR 133.128) and dry curd cottage cheese (21 CFR

131.129). 3. Whey and whey products as defined in 21 CFR 184.1979, 184.1979a, 184.1979b, 184.1979c, and Section 1, Definition SS of this *Ordinance*. 4. Modified versions of these foods listed above in Items 1 and 2, pursuant to 21 CFR 130.10—requirements for foods named by use of a nutrient content claim and a standardized term. 5. Milk and milk products as defined in Items 1, 2, 3 and 4 above, packaged in combination with food(s) not included in this definition that are appropriately labeled with a statement of identity to describe the food(s) in final packaged form, e.g., "cottage cheese with pineapple" and "fat free milk with plant sterols". 6. Products not included in Items 1–5 are Grade "A" milk products which have a minimum of 2.0% milk protein and a minimum of sixty-five percent by weight milk, milk product or a combination of milk products (PMO 2011).

16.3.3 Food safety and raw milk

Pasteurization of milk was adopted decades ago as a basic public health measure to kill dangerous bacteria and largely eliminate the risk of getting sick from one of the most important staples of the American diet. In 1987, FDA issued a regulation prohibiting the interstate sale of raw milk. In recent years, however, a small number of Americans (less than one percent) have rejected pasteurization in favor of raw (or unpasteurized) milk, citing a range of taste, nutritional and health benefits they believe are associated with raw milk consumption, as well as a general preference for unprocessed food. Today, 20 states explicitly prohibit intrastate raw milk sales in some form and 30 allow it.

While the perceived nutritional and health benefits of raw milk consumption have not been scientifically substantiated, the health risks are clear. Since 1987, there have been 143 reported outbreaks of illness—some involving miscarriages, still births, kidney failure and deaths—associated with consumption of raw milk and raw milk products that were contaminated with pathogenic bacteria. Because *E. coli* can spread from one child to another, the risk is not just to the one that drank the milk.

As a science-based, public health regulatory agency, FDA strongly supports the application of effective measures, such as pasteurization, to protect the safety of the food supply and maintain public confidence in such important, healthy staples of the diet as milk. However, in light of concerns that have been raised about potential FDA actions, this agency does not regulate the intrastate sale or distribution of raw milk. Whether to permit the sale and distribution of raw milk within a state is for the state to decide.

With respect to the interstate sale and distribution of raw milk, FDA has never taken, nor does it intend to take, enforcement action against an individual who purchased and transported raw milk across state lines solely for his or her own personal consumption.

Consumers who purchase raw milk are urged to understand the health risks involved. While raw milk puts all consumers at risk, the elderly, immune-compromised people, children and pregnant women are especially vulnerable to the hazards of raw milk consumption. FDA's consumer education continues to focus on helping consumers understand the risk to these populations.

The FDA's position on raw milk is in concert with the Centers for Disease Control (CDC) and the American Academy of Pediatricians. The FDA Food Safety Modernization Act (FSMA) came into effect on January 4, 2011. FSMA is the greatest expansion of FDA's food regulatory authority since the enactment in 1938 of the Federal Food, Drug, and Cosmetic Act.

FSMA enlarges FDA's power to regulate the safety procedures, domestically and internationally, for the producing, manufacturing, packaging, storing, and transporting food. Also, FSMA provides FDA with new powers to regulate food imports by increasing FDA's ability to inspect foreign and domestic facilities, increasing FDA's enforcement powers (including broadening FDA's administrative detention powers and FDA's ability to obtain information), and granting FDA the authority to issue mandatory recalls. Food manufacturers, importers and distributors must begin preparing now to comply with the new requirements.

In 2008, the intake of milk and infant formula which have been highly adulterated by melamine resulted in a major outbreak of renal disease (urolithiasis) and the associated deaths of numerous infants in China (Zhu et al. 2009, Zhang et al. 2009, Dobson et al. 2008). A safety limit of melamine ingestion has been officially set by the US FDA at 2.5 ppm for adults' food, and at 1 ppm for infant formula (Zhu et al. 2009). However, the melamine concentration in several adulterated milk products reaches 3300 ppm, posing extreme danger to consumers (Zhu et al. 2009). The control of the amount of melamine in milk is of utmost importance in food safety due to the high intake of this product, especially by infants.

Regulating food safety in the United States is complex. This complexity is due largely to the historical division of food safety responsibility among different Federal agencies and to evolving public attitudes towards the safety of food and concern about the changing nature of food-borne illnesses (Arvanitoyannis et al. 2006).

16.3.4 Microbiological control of milk and milk products in the United States

It is verified that all farm bulk tanker trucks unloading raw milk at any dairy processing plant are screened for beta-lactam drug residues prior to processing the raw milk. Quality control test results are compared with finished product standards with regard to Standard Plate Count, coliforms, drug residues, phosphatase, vitamin testing and temperature.

Regarding milk and whey products, quality control specifications for raw materials including bacterial load, antibiotics, pesticides, butterfat, sediment, etc., are determined (American Dry Milk Institute 2012, DMO 2012, CFR Sections 1/10; 101; 108; 110; 113; 130; 131; 133; 173.310; and 178.1010).

Sampling, test procedures and results of incoming raw milk, pasteurized milk, base powder, and other ingredients are evaluated. Residual phosphatase activity is the test commonly used to indicate the effectiveness of pasteurization. False-positive phosphatase results may be found when pasteurized milk is held at 50°F (10°C)–93°F (33.8°C), pasteurization temperatures exceed 163°F (72.7°C) and salt such as magnesium chloride is added after pasteurization. Scope of salmonella testing of water supply, and the air supply, at critical processing points, in the plant environment, and in finished products is ascertained.

Regarding bacteriological examination of cheese during inspections in-line and finished product, samples are collected to document suspected or observed bacteriological problems. If inspectional evidence indicates slow acid formation, a 1/4 pound sample is collected aseptically from each vat of cured cheese available (with a maximum of five vats) immediately prior to removal of the cheese from the vat for hooping and pressing. If possible, sample is collected before cheese curd is salted. *Staphylococci* die rapidly after cheese has been pressed and for this reason sample only freshly made cheese. Cottage cheese, baker's cheese, cream cheese, Neufchatel cheese or other unripened cheese samples would be collected if staphylococcal or other bacterial contamination is suspected.

CSR 80-2.060—the examination of milk and milk products

CSR 80-2.060 specifies sampling frequency and required chemical and bacteriological tests to be conducted both on raw and pasteurized Grade A dairy products. This rule corresponds with Part II, Section 6 of the Grade A Pasteurized Milk Ordinance with Administrative Procedures-2009 Recommendations of the United States Public Health Service/Food and Drug Administration (PMO). During any consecutive six months, at least four samples of raw milk for pasteurization, taken from each producer and delivered in accordance with 2 CSR 80-2.070 (Section 7 of the PMO) of these rules, shall be obtained by the hauler under the direction of the regulatory agency or shall be taken from each producer by the regulatory agency.

In addition, during any consecutive six (6) months, at least four (4) samples of pasteurized milk and at least four (4) samples of each milk product defined in these rules shall be taken from every milk plant. Samples of milk and milk products shall be taken while in possession of the producer or distributor at any time prior to final delivery. Samples of milk and milk products from dairy retail stores, food service establishments, grocery stores, and other places where milk and milk products are sold shall be examined periodically as determined by the regulatory agency; and the results of the examination shall be used to determine compliance with 2 CSR 80-2.020, 2 CSR 80-2.040, and 2 CSR 80-2.100.

Required bacterial counts, somatic cell counts, and cooling temperature checks shall be performed on raw milk for pasteurization. In addition, antibiotic tests on each producer's milk or on commingled raw milk shall be conducted at least four times during any consecutive six months. When commingled milk is tested, all producers shall be represented in the sample. All individual sources of milk shall be tested when test results on the commingled milk are positive. Required bacterial counts, antibiotic tests, coliform determinations, phosphatase, and cooling temperature checks shall be performed on pasteurized milk and milk products. Whenever two of the last four consecutive bacterial counts, somatic cell counts, coliform determinations, or cooling temperatures, taken on separate days, exceed the limit of the standard for the milk, milk products, or both, the regulatory agency shall send a written notice to the person concerned. This notice shall be in effect so long as two of the last four consecutive samples exceed the limit of the standard. An additional sample shall be taken within twenty-one days of the sending of the notice, but not before the lapse of three days. Immediate suspension of the permit in accordance with 2 CSR 80-2.030, court action, or both, shall be instituted whenever the standard is violated by three of the last five bacterial counts, coliform determinations, cooling temperatures or somatic cell counts.

Whenever a phosphatase test is positive, the cause shall be determined. When the cause is improper pasteurization, it shall be corrected and any milk or milk product involved shall not be offered for sale. Whenever antibiotic or pesticide residue test is positive, immediate permit suspension in accordance with 2 CSR 80-2.030 (Section 3 of the PMO), court action, or both shall be instituted. Samples shall be analyzed at an official or appropriate officially designated laboratory. All sampling procedures and required laboratory examinations shall be in substantial compliance with the current edition of *Standard Methods for the Examination of Dairy Products of the American Public Health Association*, and the current edition of *Official Methods of Analysis of the Association of Official Analytical Chemists*.

These procedures, including the certification of sample collectors and examinations shall be evaluated in accordance with 2005 *Evaluation of Milk Laboratories, Recommendations of the U.S. Public Health Service/Food and Drug Administration*. Examinations and tests to detect adulterants, including pesticides, shall be conducted as the regulatory agency requires. Assays of milk and milk products to which vitamin(s) A, D, or both have been added, shall be made at least annually in a laboratory acceptable to the regulatory agency (Code of State Regulations CSR 80.2, 2011).

16.3.5 Ice cream

Ice cream in the US has a legal definition found in the Code of Federal Regulations (CFR 2003a), which specifies solids, fat and air contents. For example, vanilla ice cream must contain a minimum of 10% milk fat by weight and a minimum of 20% milk solids. Reduced calorie ice creams also exist in the States and they must meet specific nutrient claims regarding reduced fat (CFR 2003b). Minimum contents are normally stated in the federal requirements; however, commercial ice creams vary in body, flavor, melt and texture. According to the International Dairy Foods Association (IDFA 2002), 61% of all frozen desserts manufactured in the US belong to ice cream category whereas 26% are non-fat and low fat ice creams. Frozen yogurt has a 5%, water ices 4% and sherbets 3% completing the list of the frozen desserts. In the US batch pasteurization requires a minimum temperature of 68.3°C for 30 min. High-temperature short-time (HTST) pasteurization requires a minimum of 79.4°C for 25 seconds (CFR 2003a). This relationship of time and temperature is a function of the microbial load, fat content and microbial inactivation rate.

16.3.6 Yogurt

Yogurt must come from milk from FDA supervised Grade A dairy farms and manufacturing plants according to the Pasteurized Milk Ordinance regulations (US Department of Health and Human Services 2011). For microbiological control, refrigerated bulk milk tanks should cool to 10°C in one hour and less than 5°C in two hours. Unnecessary agitation should be avoided to prevent lipolytic deterioration of milk flavor. Yogurt is made from a mix standardized from whole, partially defatted milk, condensed skimmed milk, cream and non-fat dry milk. Usually milk-solids non-fat (SNF) are supplemented with non-fat dry milk. The US dairy industry goes up to 12% SNF or non-fat milk solids to make a thick-custard like product. Full-fat yogurt contains up to 3.25% milk fat whereas low fat yogurt is made from a mix containing 1.11% milk fat (Chandan 1997, 2004). These levels correspond to the FDA requirement for nutritional labeling of different types of yogurt. Moreover, ingredients containing mastitis milk and rancid milk should be avoided. Finally, milk partially fermented by contaminating organisms and milk containing antibiotic and sanitizing chemical residues cannot be used in yogurt manufacture. Most of the yogurt sold in the USA is manufactured and fermented with *Streptococcus thermophilus* and *Lactobacillus delbrueckii* subsp. *bulgaricus*. Direct microscopic control of the starter will show the presence of budding cells or mold mycelium filaments. Moreover, plating the starter on acidified potato dextrose agar confirms the results.

Gassiness in yogurt is a result of excessive carbon dioxide and hydrogen production attributed to defects in starters or contamination with sporeforming *Bacillus* species, coliforms or yeasts. Most spoilage flora in yogurt are yeasts and molds tolerant to low pH and capable of growing at refrigeration temperatures. Yeast growth during shelf-life of yogurt is a major problem. Potassium sorbate controls the growth of yeasts and molds. Hypochlorites and iodophors are effective sanitizers for fungal control on food-contact surfaces.

Efficiency of equipment and environmental sanitation can be verified by enumeration techniques involving exposure of poured plates to the atmosphere in the plant or making a smear of the contact surfaces of the equipment, followed by plating.

16.4 Japanese dairy regulation

In Japan, the revision of Food Sanitation Law (Law No. 233, 1947) came into force in 1998. By this amendment, a voluntary HACCP and GHP based approval system was introduced to milk and milk products, meat products, and then expanded to surimi-based products, low acid canned food, and soft drink producing establishments (FAO/WHO 2006, JETRO 2006). Since microbiological criteria for final product and standards for processing and storage are articulated in the Ministerial Ordinance on Milk and Milk Products Concerning Compositional Standards (MHLW Ordinance No. 52, 1951) (JETRO 2008), the number of food-borne outbreaks associated with milk consumption is very limited in Japan. In 2010, 156 approvals have been given by the Ministry of Health, Labor and Welfare (MHLW) of Japan to multiple milk processing establishments, while 497 milk processing establishments were registered. It is estimated that 60% of milk plants which process more than two tons of raw milk gained approval (Japan Dairy Association 2005). Sampers et al. (2012) investigated how the implementation of the voluntary HACCP and GHP based approval system affected the Food Safety Management Systems (FSMS) implemented in Japanese milk processing plants by an independent analysis and gained an insight into the actual performance of current FSMS in the Japanese milk industry. FSMS-diagnostic instrument was applied to assess the level of the core control and assurance activities in the FSMS and to judge the risk level of the context wherein the companies operate. The data were collected in 13 dairy companies (mostly located around Tokyo area) and involved in-depth interviews performed by the National Institute of Public Health with responsible quality assurance persons of respective companies. The results revealed that the microbial food safety output was higher for companies with national HACCP approval. They have more advanced FSMS in combination with a less risky context. All Japanese companies scored high on technology-dependent activities (i.e., preventive measures and intervention processes), but less in managerial activities as monitoring and typical quality assurance activities as validation and verification of the FSMS. Japan has a detailed vertical legislation, leading to a "hazard-based" and "legislation-based" FSMS compared to a "science- or risk-based" FSMS common in Europe (Sampers et al. 2012).

All companies scored 3 (high risk) for preventive control activities as 'cooling facilities' and 'raw material control', as the Milk Ordinance articulated a storage requirement of final product and microbiological and other criteria for raw milk (JETRO 2008). For the monitoring activities 'Standards and tolerances design', 'Measuring equipment to monitor process/product status' and 'Sampling design and measuring plan', moderate to high risk levels were found (2–3) as in Japan the methodology of analysis is prescribed by the legislation (Forsythe and Hayes 1998).

All companies shall implement regulatory requirements, and the harmonization of test methods with internationally recognized methods in Japan is behind compared to other industrial countries; this is one of the reasons why test methods used in these milk plants were not internationally standardized methods (level 1), e.g., ISO method (Sampers et al. 2012).

Most companies in Japan do not perform scientific validation studies (e.g., microbiological testing, challenge test) and just follow scientific literatures and/or industry guidelines values for the Critical Limit (i.e., 120–130°C for 2 s). Validation and verification procedures seem to be missing from the implementation of a food safety management system from Japanese dairy companies. Hence, food safety assurance is questionable.

The Japanese companies implement the legal requirements, which strictly prescribe control measures and intervention steps for ensuring safety of milk and milk products. This is indeed reflected in the high scores for the corresponding control activities for all companies. However, the lower scores for other control activities as CCP determination, microbiological analyses, sampling, maintenance and calibration, indicate that they do not yet tailor these activities for company's specific characteristics. This is certainly lacking for the companies without national HACCP approval which are mainly smaller companies (Sampers et al. 2012).

16.5 Dairy and milk-containing products: microbiological safety assurance in the Russian Federation

The basic difference in general food and feed systems existing in European Union (EU) and Russian Federation (RF) is that within the EU the "from the farm to the fork" approach is followed for ensuring food safety; however, in RF the safety of the end product is controlled. Moreover there are different classifications of raw materials and different safety parameters to be determined. Generally, the Russian microbiological safety requirements can be described as stricter and more detailed as those in the EU.

16.5.1 Classification of dairy and milk-containing products in RF

Prior to discussion of the Russian microbiological safety norms and regulations, it is essential to overview the classification of dairy and milk-containing products (Fig. 16.1) (Dunchenko et al. 2007, GOST R 52738-2007). All products containing in the dry matter at least 25% of milk-originating compounds (milk proteins and milk fat) are classified as dairy and milk containing products (Dunchenko et al. 2007, SanPiN 2.3.2.1078-01). The class of dairy and milk containing foodstuffs is divided into sub-classes and types of products. For each type of product there are corresponding microbiological safety requirements indicated by the national standard and other normative and technical regulations.

Dairy and milk containing products are further sub-divided into three sub-classes: (a) dairy products, (b) milk containing products with substitution of milk fat and milk proteins with non-milk ingredients from 2 to 49% and (c) from 51 to 90%. Further classification criteria specifying the type of the product include fat content (low fat, classic, fatty), consistency (liquid, paste-like, dried) and regimen of heat treatment. Also a specific group according to GOST R 52738-2007 constitutes traditional fermented dairy products, including curd (Tvorog), sour-cream (Smetana), fermented milk beverages (Kefir, Acidophilin, sour clotted milk Prostokvasha), beverages from fermented backed milk (Ryazhenka, Varenetz). Additionally a class of foodstuffs for particular nutritional uses also includes dairy products (e.g., milk formulae, specific milk processing products) that can be divided into: foods for infants and young children, foods for pre-school and school-children nutrition, special products for dietary nutrition of infants and young children and foods for pregnant and nursing women (SanPiN 2.3.2.1078-01, Federal Law No. 88-FZ 2008).

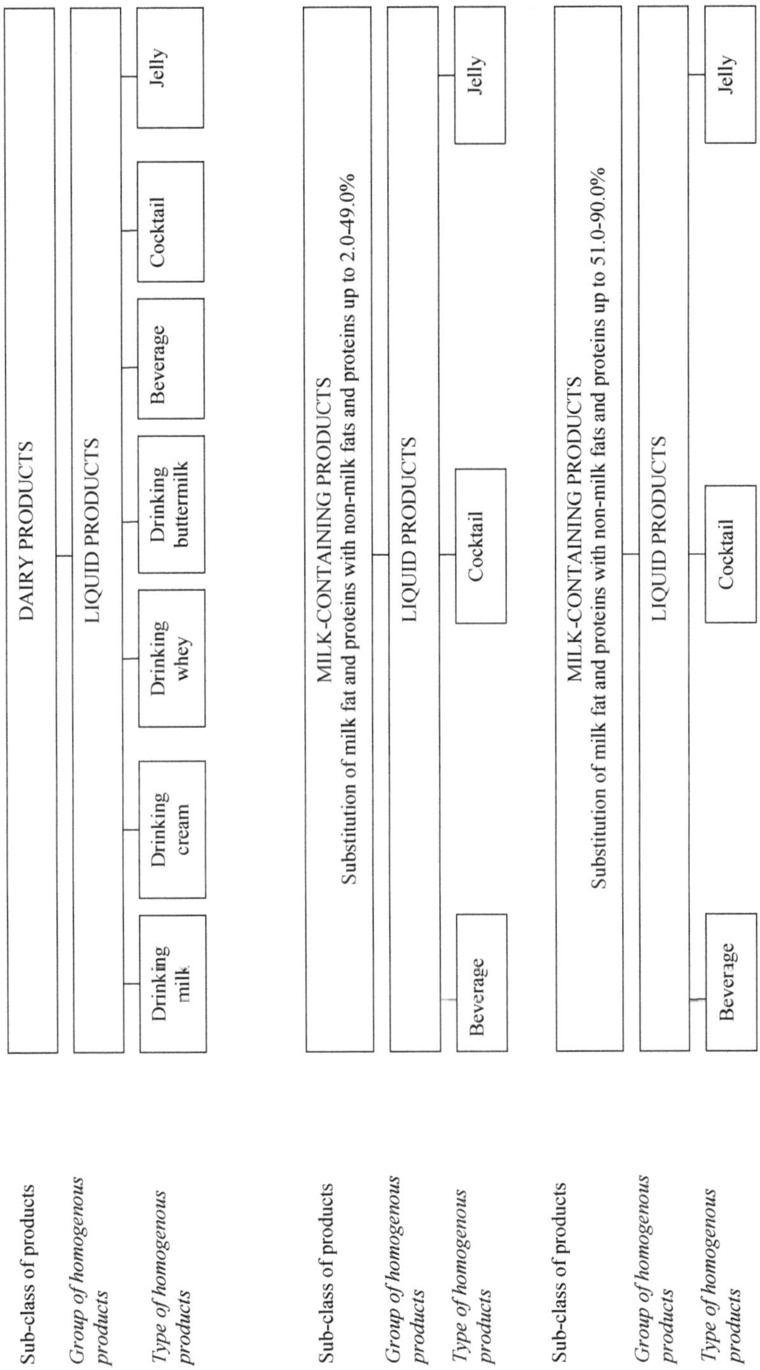

Figure 16.1 Russian classification of dairy products.

16.5.2 Normative and technical regulations specifying microbiological safety parameters of dairy and milk-containing products in RF

In RF there are different regulating documents and standards for dairy and milk-containing products concerning their microbiological safety criteria and the corresponding control methods. The most significant RF documents concerning the safety aspects are:

(a) Federal Law No. 29-FZ of January 2, 2000. Quality and safety of food products;
(b) Federal Law No. 88-FZ of June 12, 2008. Technical regulations on milk and milk processing products; and
(c) SanPiN 2.3.2.1078-01—Hygienic requirements for safety and nutrition value of foodstuff.

According to Section 1 of the RF Federal Law No. 29-FZ:

Safety of food products—a state of reasonable confidence that food products under the normal conditions of their use are not harmful and do not constitute any danger to the health of present and future generations.

Indicators of dairy and milk containing products' safety include both chemical and microbiological safety parameters. The system of the Russian legislation in the field of food production somewhat reminds Codex Alimentarius, systems of the European Union and the USA. It contains both *horizontal* normative documents, and *vertical*. Its unique, but principal difference is a full closeness of procedure of development of documents and, following from this, an arbitrariness of departments in an establishment of obligatory norms.

Russian horizontal documents concerning milk products manufacturing cover: (a) information for consumers, (b) sanitary requirements to manufactures, (c) requirements to milk products including sanitary rules, (d) requirements to the materials contacting with milk products, and (e) list of forbidden and resolved additives of the Ministry of Health of Russia. Current Russian vertical documents concerning milk products manufacturing cover: (a) terminological clarifications and classification of the products, and (b) requirements for manufacturing of milk-containing and dairy products. The legislation on microbiological safety criteria for dairy and milk-containing products in RF gets through the Federal Laws, Sanitary and Epidemiologic Rules and Regulations (SanPiN), national state standards (GOST), methodological instructive regulations (MIR) and technical.

National State Standards of Russia (GOST) are documents that define the characteristics, manufacturing methods and packing marks of a certain dairy and milk-containing product. Yet a parallel system of technical specifications (TU) has been introduced in order to facilitate development of novel products. Instead of applying the GOST standards, any manufacturer of dairy and milk-containing products may as well choose to produce them according to those technical specifications that they may define independently. Nevertheless, these specifications have to be sufficient in order to meet well the safety requirements for dairy and milk-containing products specified by SanPiN 2.3.2.1078-01, Federal Law No. 88 2008.

TU—the type of authorization document—show the conditions, requirements and regulations placed on certain product, or production process. At its core, technical specifications are a substitute for National State Standard and point out the technical documentation of the company, which specifies requirements for quality control, manufacturing production, its performance and reception of the finished commodity production. However, the claims on microbiological safety parameters of the product in the technical specifications should not have any contradictions with continuing GOST for certain types of the products. Technical specifications are necessary to prevent disputes between producers and consumers. Based on the technical conditions, the company can certify the certain dairy and milk-containing products.

In the RF certification activity is realized in certification systems. The Governmental Decree No: 982 dated December 1, 2009 cancelled mandatory certification of dairy and milk-containing products with exception of specialized food products for infants and children. For other dairy and milk-containing products categories compliance with microbiological safety criteria is testified in the form of Declaration of Conformity.

Each type of dairy and milk-containing products is characterized by certain list of microbiological safety parameters. Microbiological safety indicators common for all categories of dairy and milk-containing products except of raw whole milk are: (a) taint count of mesophilic aerobic and facultative anaerobic microorganisms, (b) absence of coliforms in certain weight of the product, and (c) absence of pathogenic microorganisms including *Salmonella* spp. in certain weight of the product. Thermally treated (e.g., sterilized, pasteurized) dairy and milk-containing products should also comply with the requirements on industrial sterility.

For dairy and milk-containing products with shelf-life >72 hr microbiological safety indicators usually include enumeration of molds and yeasts. The highest number of microbiological safety indicators is specified for milk formulae for infant nutrition. In addition to the parameters mentioned above their microbiological safety parameters include counts of *B. cereus,* acidophilic bacteria and bifidobacteria.

Globalization process accelerates creation of regional supranational economic unions raising probability of surviving. The Customs Union between Russia, Kazakhstan and Belarus came into existence on January 1, 2010. The Decision of the Customs Union Commission of 18 June 2010

No: 317 approves the list of common commodities, subjected to veterinary surveillance, including milk and dairy products, and unified Sanitary-Epidemiological Requirements for group of milk and dairy products and microbiological indexes for baby-food dairy products. For the most part, the microbiological criteria specified in novel unified Sanitary-Epidemiological Requirements are similar to the Russian SanPiN 2.3.2.1078-01. However, some parts of these requirements may differ and reflect sanitary regulations of Belarus, Kazakhstan and/or international standards.

16.5.3 Methods used for microbiological safety examination of dairy and milk-containing products in RF

Nowadays in RF the old standards still remain valid, and the test laboratories are obliged to apply only domestic normative documents. The state does not devote funds for research and development, approbation and standardization of novel and improved methods, modification of existing methods; the private sector is not interested in such developments. As a result even in 2013 the laboratories forced to carry out the tests applying the methods developed more than 20 years ago.

The demonstrative example is the situation with the analysis of microbiological safety indicators of foodstuffs including dairy and milk-containing products. All officially approved methods are based on conventional microbiological techniques, i.e., using of selective media and method of serial dilutions for enumeration of microorganisms. Less time-consuming polymerase-chain reaction (PCR) techniques are not currently introduced into the practice. Brief overview of methodology for analysis of microbiological safety indicators of dairy and milk-containing products is listed below.

Taint count of mesophilic aerobic and facultative anaerobic microorganisms (CMAFAnM) is determined according to GOST 53430-2009 by enumeration of the colonies grown on the solid agarized medium at 30°C. Presence of coliforms is determined according to the same national standard by their ability to ferment lactulose during incubation at 37°C for 24 hr. Gas formation testifies the presence of coliforms in the aliquot of the product. Count of spores of mesophilic and thermophilic aerobic and facultative anaerobic microorganisms is performed by enumerating the colonies grown on the selective solid medium during 72 hr of incubation at 30 and 55°C respectively. Counts of yeasts and molds are determined according to GOST 10444.12-88 by enumeration of the colonies with certain micro- and macro-morphology grown on the selective agarized media during incubation at 24°C for 120 hr. There are two protocols (standard and protocol with primary enrichment) for determination of *S. aureus* in the aliquot of the product. Presence of *S. aureus* in the aliquot of the product is determined through its ability to coagulate rabbit's plasma. Detection of *Salmonella* spp. is the only microbiological safety indicator of dairy and milk-containing products for which the methodology of the analytical procedures (GOST R 52814-2007) is currently harmonized with the international standards (ISO 6579:2002). The assay is based on the following sequential steps: cultivation on liquid non-selective media, cultivation on liquid selective media, cultivation on at least two solid media for differential diagnostics, screening of biochemical and serological characteristics of curtain microorganisms.

16.6 Current debate in probiotic dairy foods

Probiotics, according to FAO/WHO definition, are "living microorganisms that confer health benefits to a host when administered in adequate amounts" (FAO/WHO 2002). Probiotic bacteria predominantly belong to the genera *Lactobacillus* and *Bifidobacteria* (Kekkonen et al. 2008), along with some strains of *Enterococcus* and *Saccharomyces* (Donohue 2006). Bacteria with a potentially probiotic activity are currently considered and evaluated under EU regulation 1924/2006 on health claims on food (Verhagen et al. 2010).

Numerous studies have indicated beneficial health effects of probiotics (Bousvaros et al. 2005, Kekkonen et al. 2008). In addition, they are generally regarded as safe because of their long history of use as well as being part of the normal commensal microflora (Donohue 2006). They are rarely associated with disease, and are thought to have low pathogenicity (Bernardeau et al. 2008). However, a small portion of the population may be at risk of adverse effects following probiotic administration (Besselink et al. 2008, Sanders et al. 2010). Immune-compromised patients have, for example, been found to be susceptible to opportunistic *Lactobacillus* spp. infections (Schlegel et al. 1998), although it has not been determined if these *Lactobacillus* strains are consistent with those found in food.

The genetic stability of the bacterial strain over time, deleterious metabolic activities and the potential for pathogenicity and toxicity need to be assessed, depending on the genus and species of the microbe being used (Magnusson et al. 2012). Some applications of the microbes pertain to stimulation of the immune system. However, over-stimulation of the immune system may lead to adverse consequences, and immunological effects must be considered in terms of potential adversity, especially in infants with underdeveloped immune function (Sanders et al. 2010). An area that is a cause for concern is that the microbes meant to be probiotic are known to harbor antibiotic resistant genes and have been suggested to be capable of transferring antibiotic resistance to pathogens that can be transmitted to humans through the food chain via the commensal flora (Courvalin 2006).

There are several approaches for taking advantage of functional foods in the US. For probiotics, prebiotics, and especially dairy foods with these components, there are at least four avenues that companies and commodity organizations are or could be pursuing: taking steps to obtain health claim approval from FDA; highlighting health-promoting ingredients in products and pursuing structure/function claims; advertising; and supporting the development and dissemination of educational materials.

The most involved process, but possibly one with a large pay-back, would be to submit a health claim petition to FDA. The general requirements for a petition are detailed in the US Code of Federal Regulations (CFR 1997). Alternatively, commodity-based organizations could take the lead in submitting health claims petitions that might benefit a category of products.

A second issue is the scientific standard that FDA has set for approval of a health claim. The standard is significant scientific agreement, and many agree that FDA has done well in adhering to this standard (Dilley 1996). However, a recent development raises the possibility that health claims might be challenged even after they are approved.

Manufacturers of cultures-added dairy foods in the US and world-wide already capture consumer interest by prominently displaying that their products contain certain bacteria. In the US, the National Yogurt Association (NYA) is involved in certifying that yogurts meet NYA criteria for live and active cultures, and their seal of approval is designed to give consumers some measure of satisfaction. Pursuit of structure/function claims might be useful.

The high standards of the Federal Trade Commission (FTC) provide another reason to pursue avenues that would be acceptable to FDA, such as gaining approval for a specific health claim (Berner and O'Donell 1998).

16.7 Conclusion

EU, US, Russian and Japanese regulations on microbiological control of milk and dairy products have been compared. Examples have been given from the monitoring system. Outbreaks from the presence of pathogenic microorganisms in dairy foods have been reported in all regions described. Microbiological criteria have been set everywhere or are in the process of being established.

Regarding the Russian system for the assurance of microbiological safety of dairy and milk-containing products, it is on the way of uncompleted law reform, modernization and harmonization with other national and international standards. Moreover, the State authorities are mainly responsible for examination of microbiological safety of dairy and milk-containing products. However, official practical implementation of novel cost-efficient and less time-consuming techniques for analysis of pathogenic and contaminating microorganisms in dairy and milk-containing products will provide the basis for transferring of a large part of certain liability from the State to the manufactures, and introducing a rapid alert system for all milk processors in RF.

References

Arvanitoyannis, I.S., P. Tserkezou and T. Varzakas. 2006. An update of US food safety, food technology, GM food and water protection and management legislation. Int. J. Food Sci. Technol. 41(Suppl. 1): 130–159.

Arvanitoyannis, I.S., T. Varzakas and M. Koukaliaroglou-van Houwelingen. 2009. Implementing HACCP and ISO22000 for foods of animal origin-dairy foods. pp. 91–180. *In:* I.S. Arvanitoyannis [ed.]. Handbook of HACCP and ISO22000: Application to Foods of Animal Origin. Wiley-Blackwell, Oxford, UK.

Bernardeau, M., J.P. Vernoux, S. Henri-Dubernet and M. Gueguen. 2008. Safety assessment of dairy microorganisms: the *Lactobacillus* genus. Int. J. Food Microbiol. 126: 278–285.

Berner, L.A. and J.A. O'Donnell. 1998. Functional foods and health claims legislation: Applications to dairy foods. Int. Dairy J. 8: 355–362.

Besselink, M.G., H.C. van Santvoort, E. Buskens, M.A. Boermeester, H. van Goor, H.M. Timmerman, V.B. Nieuwenhuijs, T.L. Bollen, B. van Ramshorst, B.J. Witteman, C. Rosman, R.J. Ploeg, M.A. Brink, A.F. Schaapherder, C.H. Dejong, P.J. Wahab, C.J. van Laarhoven, E. van der Harst, C.H. van Eijck, M.A. Cuesta, L.M. Akkermans and H.G. Gooszen. 2008. Probiotic prophylaxis in patients with predicted severe acute pancreatitis: A randomised, double-blind, placebo-controlled trial. Ned. Tijdschr. Geneeskd. 152: 685–696.

Bousvaros, A., S. Guandalini, R.N. Baldassano, C. Botelho, J. Evans, G.D. Ferry, B. Goldin, L. Hartigan, S. Kugathasan, J. Levy, K.F. Murray, M. Oliva-Hemker, J.R. Rosh, V. Tolia, A. Zholudev, J.A. Vanderhoof and P.L. Hibberd. 2005. A randomized, double-blind trial of Lactobacillus GG versus placebo in addition to standard maintenance therapy for children with Crohn's disease. Inflam. Bowel Dis. 11: 833–839.

Callon, C., F. Duthoit, C. Delbes, M. Ferrand, Y. le Frileux, R. de Cremoux and M.C. Montel. 2007. Stability of microbial communities in goat milk during a lactation year: Molecular approaches. Syst. Appl. Microbiol. 30: 547–560.

CFR. 1997. Code of Federal Regulations. Subpart E—Specific Requirements for Health Claims. 21 CFR 101.70-101.81. CFR Sections 1/10; 101; 108; 110; 113; 130; 131; 133; 173.310; and 178.1010. Department of Health and Human Services, Washington, D.C.

CFR. 2003a. Code of Federal Regulations. Title 21. Part 135. Frozen desserts. Food and Drug Administration. Department of Health and Human Services, Washington, D.C.

CFR. 2003b. Code of Federal Regulations. Title 21. Part 101. Food labelling. Food and Drug Administration. Department of Health and Human Services, Washington, D.C.

Chandan, R.C. 1997. Dairy-based Ingredients. Eagan Press. St. Paul, Minnesota, USA.

Chandan, R.C. 2004. Dairy: Yogurt. pp. 297–318. *In:* S.J. Smith and Y.H. Hui [eds.]. Food Processing Principles and Applications. Blackwell Publishing, Oxford, UK.

Cocolin, L., D. Aggio, M. Manzano, C. Cantoni and G. Comi. 2002. An application of PCR-DGGE analysis to profile the yeast populations in raw milk. Int. Dairy J. 12: 407–411.

Courvalin, P. 2006. Antibiotic resistance: The pros and cons of probiotics. Dig. Liver Dis. 38(Suppl. 2): 261–265.

CSR. 2011. Code of State Regulations. Rules of Department of Agriculture Division 80—State Milk Board. Chapter 2—Grade A Pasteurized Milk Regulations. Department of Health and Human Services, Washington, D.C.

de Freitas, I., N. Pinon, J.L. Maubois, S. Lortal and A. Thierry. 2009. The addition of a cocktail of yeast species to Cantalet cheese changes bacterial survival and enhances aroma compound formation. Int. J. Food Microbiol. 129: 37–42.

Dilley, A.P. 1996. Keystone dialogue health claim recommendations. Food Technol. 50: 124–127.

Dobson, R.L.M., S. Motlagh, M. Quijano, R.T. Cambron, T.R. Baker, A.M. Pullen, B.T. Regg, A.S. Bigalow-Kern, T. Vennard, A. Fix, R. Reimschuessel, G. Overmann, Y. Shan and G.P. Daston. 2008. Identification and characterization of toxicity of contaminants in pet food leading to an outbreak of renal toxicity in cats and dogs. Toxic. Sci. 106: 251–262.

Donohue, D.C. 2006. Safety of probiotics. Asia Pac. J. Clin. Nutr. 15: 563–569.

Dunchenko, N.I., I.A. Khramtzov, I.A. Makeeva, N.B. Smirnova, L.V. Gavrilova, L.V. Golubeva, L.V. Kalinina and V.M. Pozdnyakovsky. 2007. Expertise of Milk and Milk-containing Proteins. Quality and safety. Novosibirsk: Publishing House of Siberian University, Krasnoyarsk, Russia.

FAO/WHO. 2002. Guidelines for the Evaluation of Probiotics in Food. Report of a Joint Working Group on Drafting Guidelines for the Evaluation of Probiotics in Food. pp. 1–11.

FAO/WHO. 2006. FAO/WHO guidance to governments on the application of HACCP in small and/or less-developed food businesses. FAO-Food and Nutrition paper, No. 86. pp. 53–55. http://www.who.int/foodsafety/publications/fs_management/HACCP_SLDB.pdf.

Fleming, D.W., S.L. Cochi, K.L. MacDonald, J. Brondum, P.S. Hayes, B.D. Plikaytis, M.B. Holmes, A. Audurier, C.V. Broome and A.L. Reingold. 1985. Pasteurized milk as a vehicle of infection in an outbreak Listeriosis. North Eng. J. Med. 312: 404–407.

Forsythe, S.J. and P.R. Hayes. 1998. Food Hygiene Microbiology and HACCP. 3rd edition. An Aspen Publication, Maryland, USA.

Guidance for Industry: Dairy Farms, Bulk Milk Transporters, Bulk Milk Transfer Stations and Fluid Milk Processors: Food Security Preventive Measures Guidance. 2007. U.S. Department of Health and Human Services, Food and Drug Administration, Center for Food Safety and Applied Nutrition.

IDFA. 2002. The latest scoop. International Dairy Foods Association, Washington, D.C., USA.

Japan Dairy Association. Basic knowledge on dairy farming and processing. http://www.j-milk.jp/expertise/research/8d863s000003hl1s-att/8d863s000003hl5h.pdf.

JETRO (Japan External Trade Operation). 2006. Food sanitation law in Japan. http://www.jetro.go.jp/en/reports/regulations/pdf/food-e.pdf.

JETRO (Japan External Trade Operation). 2008. Ministerial ordinance on milk and milk products concerning compositional standards, etc. (MHLW Ordinance No. 52, 1951: Latest revision No. 132 October 30, 2007). http://www.jetro.go.jp/en/reports/regulations/pdf/foodadd2010apr-e.pdf.

Keene, W.E., K. Hedberg, D.E. Herriott, R.W. McKay, T.J. Barrett and D.W. Fleming. 1997. A prolonged outbreak of *E. coli* 0157:H7 infections caused by commercially distributed raw milk. J. Infect. Dis. 176: 815–818.

Kekkonen, R.A., N. Lummela, H. Karjalainen, S. Latvala, S. Tynkkynen, S. Jarvenpaa, H. Kautiainen, I. Julkunen, H. Vapaatalo and R. Korpela. 2008. Probiotic intervention has strain-specific anti-inflammatory effects in healthy adults. World J. Gastroent. 14: 2029–2036.

Linnan, M.J., L. Mascola, X.D. Lou, V. Goulet, S. May, C. Salminen, D.W. Hird, M.L. Yonekura, P. Hayes, R. Weaver, A. Audurier, B.D. Plikaytis, S.L. Fannin, A. Kleks and C.V. Broome. 1988. Epidemic listeriosis associate with Mexican-style cheese. North Engl. J. Med. 319: 823–828.

Magnusson, S.H., H. Gunnlaugsdottir, H. van Loveren, F. Holm, N. Kalogeras, O. Leino, J.M. Luteijn, G. Odekerken, M.V. Pohjola, M.J. Tijhuis, J.T. Tuomisto, O. Ueland, B.C. White and H. Verhagen. 2012. State of the art in benefit–risk analysis: Food microbiology. Food Chem. Toxic. 50: 33–39.

PMO. 2011. Grade "A" Pasteurized Milk Ordinance, including provisions from the Grade "A" Condensed and Dry Milk Products and Condensed and Dry Whey—Supplement I to the Grade "A" Pasteurized Milk Ordinance. US Department of Health and Human Services, Public Health Service/Food and Drug Administration.

Pritchard, T.J., K.J. Flanders and C.W. Donnelly. 1995. Comparison of the incidence of *Listeria* on equipment versus environmental sites within dairy processing plants. Int. J. Food Microbiol. 26: 375–384.

Proctor, M.E. and J.P. Davis. 2000. *Escherichia coli* 0157:H7 infections in Wisconsin, 1992–1999. WMJ: Official Publication of the State Medical Society of Wisconsin. WMJ. Aug. 99: 32–37.

Quigley, L., O. O'Sullivan, T.P. Beresford, R.P. Ross, G.F. Fitzgerald and P.D. Cotter. 2011. Molecular approaches to analysing the microbial composition of raw milk and raw milk cheese. Int. J. Food Microbiol. 150: 81–94.

Ryser, E.T. and E.H. Marth. 1991. *Listeria,* Listeriosis and Food Safety. 1st edition. Marcel Dekker Inc., New York, NY, USA.

Sampers, I., H. Toyofuku, P.A. Luning, M. Uyttendaele and L. Jacxsens. 2012. Semi-quantitative study to evaluate the performance of a HACCP-based food safety management system in Japanese milk processing plants. Food Cont. 23: 227–233.

Sanders, M.E., L. Akkermans, D. Haller, C. Hammerman, J. Heimbach, G. Huys, D. Levy, D. Mack, P. Phothirath, A. Constable, G. Solano-Aguilar and E. Vaughan. 2010. Assessment of probiotic safety for human use. Gut Microbes 1: 1–22.

Schlegel, L., S. Lemerle and P. Geslin. 1998. *Lactobacillus* species as opportunistic pathogens in immunocompromised patients. Eur. J. Clin. Microbiol. Infect. Dis. 17: 887–888.

Schmid, G.P., R.E. Schaefer, B.D. Plikaytis, J.R. Schaefer, J.H. Bryner, L.A. Wintermeyer and A.F. Kaufmann. 1987. A one-year study of endemic campylobacteriosis in a midwestern city: Association with consumption of raw milk. J. Infect. Dis. 156: 218–222.

Sheehan, J.F. 2009. Testimony of Office of Food Safety. Center for Food Safety and Applied Nutrition, U.S. Food and Drug Administration.

U.S. Food and Drug Administration. 1987. FDA Dairy Safety Initiatives. Second year status report. Milk Safety Branch, Center for Food Safety and Applied Nutrition, Washington, DC, USA.

US Department of Health and Human Services (US DHHS). 2011. Grade A Pasteurized Milk Ordinance. Revision. Food and Drug Administration, Washington, D.C.

van Kessel, J.S., J.S. Karns, L. Gorski, B.J. McCluskey and M.L. Perdue. 2004. Prevalence of *Salmonellae, Listeria monocytogenes* and fecal coliforms in bulk tank milk on US dairies. J. Dairy Sci. 87: 2822–2830.

Verhagen, H., E. Vos, S. Francl, M. Heinonen and H. van Loveren. 2010. Status of nutrition and health claims in Europe. Arch. Biochem. Biophys. 501: 6–15.

Werner, S.B., G.L. Humphrey and I. Kamei. 1979. Association between raw milk and human *Salmonella* dublin infection. Br. Med. J. 28:2(6184): 238–241.

Zhang, L., L.L. Wu, Y.P. Wang, A.M. Liu, C.C. Zou and Z.Y. Zhao. 2009. Melamine contaminated milk products induced urinary tract calculi in children. World J. Pediatr. 5: 31–35.

Zhu, L., G. Gamez, H. Chen, K. Chingin and R. Zenobi. 2009. Rapid detection of melamine in untreated milk and wheat gluten by ultrasound-assisted extractive electrospray ionization mass spectrometry (EESI-MS). Chem. Commun. 5: 559–561.

Internet sources

http://www.fda.gov/Food/FoodSafety/Product-SpecificInformation/MilkSafety/ucm277854.htm.

http://www.fda.gov/NewsEvents/Newsroom/PressAnnouncements/2008/ucm116960.htm.

Index

A

ACE-inhibitory action 193
acesulfame 223
acetaldehyde 43, 50, 88, 168, 171, 184–189, 194, 198, 228, 237, 290
acetic acid bacteria 87, 194, 195
acetoin 49, 50, 158, 184, 198, 249
acetone 184
Achromobacter 3, 4, 8, 9, 24, 247, 262
Acinetobacter 2, 4, 6, 24, 105, 117, 139, 247, 254
activation energy 6, 8
adjunct cultures 45, 46, 118, 120, 127, 137, 148, 236
Aerococcus 88
Aeromonas 3, 4, 7, 9, 18, 117, 247, 264, 336, 419
aflatoxins 370, 390, 391
age gelation 106
Alcaligenes 2–4, 8, 9, 24, 102, 247, 264
amino acids 15, 28, 44, 50–52, 55, 56, 59, 75, 78, 79, 87, 120, 121, 137, 140, 143, 145, 147, 148, 151, 152, 154, 155, 169, 171, 172, 177, 185–187, 189, 190, 198, 201, 222, 227, 294, 295, 298, 303, 304, 309, 310, 334, 335, 342, 370
aminopeptidases 51, 147, 190
anti-bacterial peptides 88
anti-fungal peptides 88
anti-microbial compounds 87, 88
anti-oxidants 225, 228, 231, 309, 310
antiport system 47, 175
anti-tumoral effect 184
Arrhenius plot 8
artisanal starter cultures 45
aspartame 223
Aspergillus spp. 24
ATP bioluminescence technique 364

B

Bacillus spp. 7, 9, 11, 13–15, 24, 29, 87, 102, 103, 105, 116, 274, 283, 284, 396
bacteriocins 54–56, 60, 69, 88–90, 99, 184, 194, 227, 235, 236, 370
bacteriophages 79, 88, 171, 398
β-galactosidase 47, 60, 75, 171, 173, 175, 184, 296, 297, 338
bifidobacteria 24, 41, 48, 49, 54, 222–227, 229–232, 234, 237, 292, 294, 296–298, 431, 433
Bifidobacterium longum 215, 234
bifidogenic effect 297, 298
bioactive components 214
biofilms 7, 13, 16, 18, 20, 25, 169, 197, 264, 365, 366
biomass 174, 192
biopreservation 86–91
biosensors 398–400
bitter taste 4, 118, 273, 276
bitty cream 3, 9, 11, 15, 255
blended spread 248
blue-veined cheeses 122, 124
botulinum toxins 369
brined cheeses 116–119, 148, 151
brucellosis 18, 22, 25, 368, 410
butter 3, 16, 40, 245, 247–259, 299, 356, 361, 372, 374–376, 384, 412, 418
buttermilk 40, 248, 252, 299, 376, 429

C

carbohydrate metabolism 70, 170, 175, 194
carbonyl compounds 184, 198
Carnobacterium 88
casein micelle 4, 138, 182, 332, 333, 339
caseinophosphopeptides 289, 307
caseins 4, 7–9, 44, 51, 52, 57, 59, 96, 114, 125, 135–138, 140, 143–148, 158, 171, 182, 183, 189, 190, 198, 199, 232, 260, 262,

289, 298, 299, 304–310, 312, 332–336, 339, 344, 375
cell-wall associated proteinases 147
Cheddar cheese 16, 45, 53, 129, 146, 148, 151, 152, 234, 254, 303, 306, 307, 328, 330, 332, 334–336
cheese 3, 9, 10, 16, 19–21, 40, 41, 44–46, 49, 52, 53, 56, 57, 75, 76, 79, 80, 86, 87, 98, 113–130, 134–159, 169, 176, 215, 218–220, 222, 224, 227, 229, 233–236, 254, 289, 291–293, 300, 301, 303, 306–308, 326–328, 330, 332–336, 344, 356, 358–360, 375–380, 384, 395, 401, 405, 412, 418–421, 423
churning 248–250, 252, 253, 299, 374, 382
chymosin 135, 136, 143–147, 338
citrate 41, 44, 49–51, 57, 60, 61, 120, 128, 134, 135, 156, 158, 159, 237, 249, 272
citrate metabolism 41, 49–51, 60, 61, 128, 158, 159, 249
Clustered Regularly Interspaced Short Palindromic Repeats (CRISPR) 79
cold plasma 322, 342–344
coliforms 2, 3, 15, 22, 24–27, 118, 245, 254–256, 264, 336, 339, 366, 367, 423, 426, 431, 432
condensed milk 3, 273, 276–279, 281, 285, 420
conjugated linoleic acid (CLA) 193, 289, 292, 294, 301, 312
contaminant 1, 2, 23, 102, 105, 107, 113, 114, 124, 126, 127, 196, 259–261, 334, 371
contamination 7, 10, 13, 14, 18–20, 22–24, 58, 62, 101–109, 117, 127, 246, 247, 252, 253, 256, 257, 260, 262–264, 275–277, 282–284, 362, 366, 367, 371, 372, 379, 383, 384, 391, 406, 408, 413, 419, 420, 423, 426
cooked flavor 96, 100, 101, 246, 275
Corynebacterium 1, 2, 5, 9, 10, 18, 24, 102, 107, 115, 138, 140, 247, 324, 368, 396
Coxiella burnetii 18, 22, 98, 404
cream 3, 9, 11, 13, 15, 40, 103, 104, 128, 167, 220, 235–237, 245–265, 276, 280, 285, 291, 333, 372, 374–376, 378–384, 412, 416–418, 423, 425, 428, 429
cross contamination 367, 383
cross-flow microfiltration 99
cultured butter 40, 250, 259, 374

D

dairy 1–437
Dairy Chain Quality Program (KKM) 363
Dairy Hygiene Directive 405
dairy quality 13, 31
dairy spread 248, 249, 255
denaturation 96, 100, 221, 280, 310, 332, 333, 339, 382, 383
denaturing gradient gel electrophoresis (DGGE) 198, 404, 405
diacetyl 44, 49, 50, 61, 88, 128, 158, 168, 184, 198, 237, 249, 290, 370, 374
direct injection 105
direct-to-vat set (DVI) 40, 58, 218–221, 225, 232
DNA-based assay 392
DNA:DNA hybridization 41
docosapentaenoic acid (DPA) 294
Domiati cheese 116, 117, 119, 148
double cream 103, 255
dried milk 274, 277, 279–284, 368, 416, 418
Dutch-type cheese 121, 139, 158, 159

E

eicosapentaenoic acid (EPA) 294
electric field 322–328, 367
electron beam irradiation 367
ELISA 397
Embden Meyerhof Parnas (EMP) pathway 42, 46, 47, 53, 175, 186
emulsion phase 248
encapsulation 217, 225, 231, 312
Enterococcus 3, 10, 16, 68, 88, 90, 102, 105, 114, 117, 120, 123, 126, 137, 148, 195, 254, 255, 262, 282, 369, 433
enterotoxins 21, 103, 107, 258, 259, 283, 336, 390, 395, 398, 412
enzyme inactivation 310, 329, 333, 338
Eschericia coli 108, 127, 176, 194, 258, 264, 326
Eschericia coli O157:H7 127, 258, 264
exopolysaccharides 46, 53, 75, 76, 168, 169, 171, 178, 193, 216, 217, 229
extended shelf-life (ESL) 96, 97, 99, 100, 104, 105, 405
extracellular enzymes 3, 7, 8, 30, 102, 137, 275, 282
evaporated milk 3, 11, 271–273, 275, 276, 285

F

fat crystallization 333
fat globule 8, 9, 114, 138, 153, 182, 249, 260, 289, 299, 300, 333, 339, 382
fermentation 9, 39, 41, 42, 44, 46–49, 57, 60–62, 68, 70, 76, 77, 79, 87–90, 145,

156, 167, 168, 172–177, 185, 187–189, 191–193, 198–200, 217, 219, 222, 223, 225–232, 234, 236, 248, 250, 262, 291, 294, 296, 297, 359, 385
fermented foods 39, 69, 87, 89, 91
fermented milks 129, 167, 215, 226, 230–232, 237, 289, 291, 296, 298, 305
Feta cheese 117, 152, 233, 376–379, 384
fluorescence *in situ* hybridization (FISH) 395, 397
food business operators (FBOs) 405–411, 413
food safety 87, 91, 96, 98, 106, 260, 275, 290, 322, 362, 363, 368, 374, 398, 401, 405–407, 413, 414, 417, 421, 422, 426–428
formic acid 44, 48, 68, 172, 173, 224
fouling 382
free fatty acids 127, 139, 140, 152–154, 190, 223, 233, 254
frequency 23, 328, 337, 339, 423
frozen yogurt 215, 236, 237, 261, 262, 425
fructooligosaccharides (FOS) 80, 81, 223, 295, 296
functional foods 86, 214, 217, 288, 289, 296, 310, 311, 434
fuzzy logic 364

G

galactooligosaccharides (GOS) 223, 296–298
galactose utilization 168
gastrointestinal tract (GIT) 16, 17, 70, 75, 80, 129, 216–218, 290, 292, 294, 296, 297, 312, 369
generation time 1, 5, 7, 14, 29, 102, 104, 233
glucose metabolism 171
glycine 171, 185, 186, 188
good hygienic practices (GHP) 365, 367, 426
good manufacturing practices (GMP) 25, 127, 357, 362, 366
grade A milk 418, 423
Gram-negative 2, 4, 6, 7, 9, 16–18, 24, 25, 27, 28, 30, 31, 89, 90, 102, 104, 105, 128, 249, 254, 260, 262, 330, 338, 366, 395
Gram-positive 5, 7, 9, 10, 18, 21, 22, 25, 27–31, 41, 68, 76, 87, 89, 90, 102, 104, 105, 158, 168, 170, 256, 260, 264, 330, 338, 395
GRAS 56, 61, 89, 290, 302, 369

H

HACCP 357, 358, 362–366, 372, 374, 380, 384, 405–408, 411, 426, 427
Halloumi cheese 142
heat-stable enterotoxins 107
heat-stable spores 106, 107
heat treatment 3, 8, 9, 11, 22, 26, 30, 31, 96–101, 104, 105, 107, 114, 116, 117, 128, 146, 168, 169, 173, 182, 200, 233, 234, 245, 246, 251, 260, 272, 276, 277, 279, 282, 322, 324, 325, 327, 338, 339, 342, 367, 382, 384, 385, 408–410, 412, 415, 428
heterofermentation 88
heteropolysaccharides 53, 178, 179
hexose monophosphate (HMP) pathway 186
high pressure processing 322, 328, 332
high probiotic density (HPD) 215, 232, 233, 237
homofermenters 42, 44, 68
homogenization 22, 138, 223, 224, 246, 271–275, 277, 278, 280, 328, 339, 372, 375, 381, 382
horizontal gene transfer (HGT) 70, 75, 170
hydrogen peroxide 26, 56, 57, 88, 192, 194, 227
hydrolytic rancidity 251, 254, 342
hygiene 17, 22, 23, 25, 45, 139, 256, 257, 275, 283, 284, 309, 364, 365, 405–407, 409, 411

I

ice cream 220, 236, 237, 245, 257–265, 276, 291, 380–384, 412, 418, 425
immunological-based methods 397
immunomodulatory effect 184
incubation 45, 57, 169, 174, 178, 187, 226, 228, 229, 231, 257, 384, 385, 399, 432
indigenous microflora 1, 290
inulin 49, 174, 223, 262, 295, 296
irradiation 263, 339, 341, 343, 367

J

Johne's disease 18, 107, 371, 395

K

kefir 127, 167, 192–195, 197–200, 231
kefiran 193, 194
koumiss 167, 199–201

L

LAB genome 69, 76
labeling 384, 409, 415, 416, 419, 425
lactase 222, 228, 261
lactic acid 2, 3, 10, 16, 25, 39–49, 51–53, 56–61, 68, 75, 86–90, 113–115, 117, 120–124, 126, 134, 136, 143, 146–148, 156, 157, 168–170, 172–178, 181, 182, 184, 185, 192, 194, 197–201, 222, 227, 248–250, 255, 264, 290, 294, 297, 301, 305, 335, 336, 357, 369, 370, 379, 384, 396
lactic acid bacteria (LAB) 2, 3, 10, 16, 25, 39–43, 45, 46, 48, 49, 51–53, 56–61, 68–71, 75–82, 86–92, 113–115, 117, 120, 121, 123, 126, 134, 136, 137, 139, 143, 147, 148, 156, 157, 168–170, 175, 176, 181, 182, 184, 185, 192, 194, 197, 198, 222, 248–250, 255, 264, 290, 291, 297, 300, 301, 305–307, 335, 336, 342, 357, 369, 370, 384, 396
Lactobacillus acidophilus 46, 129, 215, 298, 300
Lactobacillus casei 45, 114, 117, 378
Lactobacillus delbrueckii subsp. *bulgaricus* 44, 168, 221, 225, 261, 298, 328, 377, 426
Lactobacillus johnsonii 222
Lactobacillus kefiranofaciens 193, 194, 197
Lactobacillus plantarum 114, 117, 331
Lactobacillus rhamnosus 114, 218, 222
Lactobacillus spp. 3, 10, 53, 87, 102, 115, 125, 137, 139, 194, 200, 261, 291, 293, 334, 433
Lactococcus 2, 3, 16, 25, 41, 44, 46, 47, 49, 60, 68, 70, 88, 113, 120, 126, 137, 139, 147, 150, 158, 169, 233, 235, 249, 255, 293, 305, 326, 334, 366, 369, 378, 380
Lactococcus lactis biovar. *diacetylactis* 44, 249
Lactococcus lactis subsp. *cremoris* 44, 139, 305
Lactococcus lactis subsp. *lactis* 249, 380
lactoferrin 26, 28, 57, 298, 304
lactoperoxidase 26, 56, 57, 100, 192, 327, 333, 339
lactose-fermenting yeasts 200, 245
lactose permease 42, 175
lactulose 100, 223, 296, 297, 333, 432
Leloir pathway 42, 43, 47, 175, 176, 181
Leuconostoc 3, 16, 25, 40, 41, 44, 47, 49, 60, 68, 70, 74, 77, 88, 114, 117, 123, 136, 147, 158, 195–197, 249, 264
Leuconostoc mesenteroides subsp. *cremoris* 44, 158, 249
lipolysis 6, 31, 120, 123, 127, 134, 138, 139, 140, 152, 153, 156, 190, 198, 223, 224, 233, 333
Listeria monocytogenes 4, 7, 18, 20, 90, 98, 108, 118, 194, 253, 258, 264, 284, 300, 325, 336, 356, 362, 404
loop-mediated isothermal conditions (LAMP) 395, 396
L-threonine 187
lysozyme 11, 26, 27, 57, 223

M

Maillard reaction 101, 275, 333
maltodextrins 223
mastitis 18, 20, 22, 23, 54, 368, 396, 400, 425
membrane techniques 174
mesophilic starters 49, 231
microbial inactivation 96, 114, 323–325, 328–330, 337–340, 425
Micrococcus 1–5, 9, 10, 102, 105, 115, 116, 123, 137, 138, 247, 254, 255, 262, 274, 282, 366
microentrapment 231, 262
microfiltration 31, 99, 105, 127
microflora 1, 2, 4–7, 31, 53, 99, 101, 104, 105, 113, 116, 119, 120, 123, 130, 137, 152, 192–195, 198–201, 246, 247, 250, 251, 275, 289, 296–298, 308, 325, 365, 366, 433
micro-sieving 99
milk-borne pathogens 98, 367, 391, 397, 400
milk fat globule membrane (MFGM) 8, 9, 138, 289, 299–302, 312, 333
milk powder 10, 97, 280–285, 291, 372, 373, 385, 412, 416
mold-ripened cheeses 121, 122, 137, 140, 146, 149, 153, 157
molds 3, 16–18, 39, 86, 87, 89, 90, 104, 113–115, 119, 121–124, 129, 137, 138, 140, 146, 149, 151–153, 157, 167, 194, 245, 252, 254–256, 260, 262, 279, 329, 332, 334, 336, 340, 356, 366, 367, 377–381, 384, 426, 431, 432
multiplex-PCR 259, 393, 394
Mycobacterium avium subsp. *paratuberculosis* (MAP) 21, 22, 98, 107, 342, 395
Mycobacterium bovis 18, 98
Mycobacterium tuberculosis 18, 98, 107, 404
mycotoxins 91, 119, 262, 362, 370, 371, 398

N

natamycin 223, 230
natural killer cell activity 69

neural networks 364
nisin 56, 76, 88–90, 99, 184, 223, 227, 230, 235, 236, 326, 334, 370
non-starter lactic acid bacteria (NSLAB) 45, 46, 113, 114, 117, 126, 134, 137, 148, 156, 158, 301, 305
non-thermal technologies 342
nutraceuticals 218, 288, 289, 302

O

Oenococcus 68, 88
off-flavors 4, 9, 16, 102, 224, 237, 280, 328
omega-3 fatty acids 294
omics 82, 400
outbreaks 19–21, 90, 96, 107–109, 247, 253, 256–258, 260, 263, 265, 274, 282–284, 363, 369, 390, 391, 408, 419–421, 426, 434
over-acidification 176, 228–230
own checking plan (OCP) 365
oxidative stability 246
oxygen sensitivity 231

P

parasites 90, 391, 398
pasta-filata cheeses 124, 126, 379
pasteurized milk 3, 7, 11, 13, 14, 26, 96, 99, 101–109, 117, 118, 121, 124, 127, 128, 139, 253, 256, 328, 339, 362, 366, 367, 372, 375, 390, 391, 418, 420, 421, 423–425
pathogenic bacteria 59, 119, 263, 296, 336, 337, 357, 404, 421
PCR 198, 200, 259, 264, 265, 392–397, 399, 400, 432
Pediococcus spp. 87, 117, 264
peptidases 8, 51, 75, 78, 120, 138, 140, 143, 147, 149, 190, 305
peptides 51, 52, 55, 59, 88–92, 120, 137, 140, 143, 146–148, 172, 190, 193, 198, 201, 216, 227, 289, 296, 298, 302–308, 310–312, 334
peroxidase test 99
peroxidases 227, 230, 231
phospho-β-galactosidase 47, 60, 75, 175
phosphoenolpyruvate (PEP) 42, 43, 47, 169, 175, 191
phospholipases 7–9
phosphotransferase system (PTS) 42, 47, 75, 78, 81, 169, 175, 183
phytosterols 214
plasmid 49, 51, 52, 54, 55, 59, 60, 73, 75, 76, 169, 170, 181, 370
plasmin 4, 106, 128, 140, 143, 146–148, 327, 328, 334
plasminogen 4, 146
plasminogen activators 4, 146
plate heat exchanger 97, 98, 101, 169, 382
post-fermentation 176
prebiotics 53, 223, 231, 233, 262, 289, 295, 296, 434
pregastric esterase 139
probioactives 216, 217
probiotics 46, 53, 69, 80, 82, 89, 91, 92, 114, 123, 129, 130, 167, 178, 201, 214–233, 235–237, 261–263, 289–295, 297, 298, 300, 305–307, 311, 312, 369, 370, 433, 434
processed liquid milk 95, 96, 99, 101, 102, 104, 106
protozoa 18, 28, 265, 340, 390
proteolysis 6, 9, 31, 44, 119, 120, 122, 123, 128, 134–137, 140, 141, 143–146, 148, 151, 153, 156, 189, 192, 198, 224, 228, 306, 307, 328, 336, 337
proteolytic activity 6, 8, 118, 122, 129, 136, 171, 189, 190, 226
Pseudomonas 2–9, 18, 24, 27, 29, 30, 102, 103, 117, 128, 139, 245–247, 252, 254–256, 262, 264, 274, 282, 300, 324, 326, 338, 357
psychrotrophic bacteria 2–7, 96, 102–104, 106, 117, 224, 275, 279, 282
psychrotrophs 2–9, 29–31, 97, 102, 117, 254, 381
public health 95, 99, 106, 119, 363, 391, 406, 411, 413, 417, 418, 421, 423, 424, 427
pulse shape 325
pulsed electric fields 322, 323, 325
pulsed UV light 341
pyruvate 42, 43, 46–50, 176, 185, 186

Q

Qualified Presumption of Safety (QPR) 89, 290
quality control 184, 218, 251, 363, 364, 418, 423, 431
quantitative risk analysis 357

R

raffinose 78, 80, 172, 223
rancidity 102, 251, 254, 310, 342
RAPD PCR 395
real-time PCR 393, 394
redox level 224, 225, 227, 228, 231

redox potential 114–116, 120, 138, 152, 298
Relative Light Units (RLU) 364
reuterin 88, 90
reverse osmosis (RO) 174, 272
Rhizopus spp. 255
ripening 39, 41, 113–115, 117–121, 123–130, 134–138, 140, 143–153, 156, 157, 235, 236, 249, 292, 300, 301, 306, 308, 334–336, 359–361, 376, 378, 379
risk assessment 357, 362, 384, 407

S

Saccharomyces spp. 87, 137, 197, 200, 262
Salmonella 17, 18, 26, 27, 30, 90, 106–108, 119, 128, 188, 194, 253, 258, 259, 262–265, 274, 283, 294, 300, 325, 326, 342, 356, 362, 367–369, 390, 391, 393–398, 404, 412, 413, 418, 419, 423, 431, 433
Salmonella Typhimurium 108, 188, 194, 258, 283
sanitation 10, 58, 95, 102, 103, 114, 128, 256, 257, 278, 283, 366, 417, 418, 426
shelf-life 1, 4, 5, 9, 28, 87, 90, 96, 99–104, 116, 128, 129, 246, 271, 277, 279, 280, 322, 325, 334, 370, 379, 383, 405, 411, 412, 425, 431
single cream 245, 248
slime formation 245
somatic cells 5, 15, 23, 25, 28, 106, 246, 410, 424
spoilage 1–3, 5–7, 10, 11, 13, 15, 16, 24, 56, 89, 91, 95, 102, 105, 106, 114, 115, 117, 119, 124, 127–129, 138, 246, 249, 251–256, 273, 275, 276, 279, 283, 322, 329, 332, 337, 341, 356, 357, 365, 368, 381, 396, 426
spoilage microorganisms 2, 3, 102, 124, 128, 251, 254, 283, 322, 329, 332
sporeforming bacteria 2, 3, 10–13, 96, 103–105, 251, 275, 276
spore inactivation 341
Staphylococcus spp. 6, 20, 21, 23, 24, 105, 115, 138, 324, 342
starter cultures 16, 39–42, 44–46, 49–51, 53–62, 80, 113, 114, 117–126, 128–130, 136, 146, 148, 182, 194, 199, 215, 219, 222, 224, 225, 227, 230–232, 249, 306, 358, 377, 378, 384, 385
Statistical Process Control (SPC) 26, 364
sterilization 13, 97, 99, 101, 105, 106, 246, 271, 272, 275–277, 337, 367, 408
Streptococcus spp. 5, 24, 102, 201, 362, 396
Streptococcus thermophilus 3, 44, 70, 113, 136, 168, 195, 226, 261, 282, 297, 298, 328, 377, 426
surface ripening bacteria 39
sweet butter 248
sweet curdling 3, 9, 11, 15, 246, 255
Swiss-type cheeses 115, 169, 235, 236
symbiosis 130, 171, 173, 174

T

Tagatose pathway 42, 47
texture 16, 39, 46, 53, 59, 69, 87, 113, 114, 117, 118, 120, 122, 124, 125, 127, 128, 134, 136, 143, 156–158, 178, 182, 192, 193, 217, 224, 292, 334, 377, 380, 382, 383, 425
thermization 27, 30, 96–98, 245
thermoduric bacteria 2, 4, 9, 10
thiocyanate 26, 57
triacylglycerol (TAG) 152, 153, 190, 193
Turkish white cheese 116, 293, 334
Tyndallization 98, 251

U

UHT 4, 8, 11, 13, 14, 96–101, 105, 106, 245, 246, 255, 272, 273, 275, 278, 280, 326, 331–333, 356, 410
ultrafiltration (UF) 62, 142, 174, 199, 233, 280, 281, 302, 311
ultra pasteurized milk 96
ultrasonication 322, 337, 338, 340
urea metabolism 173
urokinase 146
UV light 58, 260, 340–342

V

Vagococcus 88
viruses 18, 26, 27, 29, 79, 90, 91, 329, 340, 362, 391, 398

W

water activity (a_w) 86, 114, 127, 130, 153, 175, 220, 250, 252, 277, 279, 331, 334, 337
Weisella 88
whey cheeses 128, 293, 306, 360, 377, 379, 384
whey proteins 8, 51, 96, 189, 199, 275, 289, 297–299, 302, 303, 312, 333, 334, 337, 339, 342, 368, 379, 382

whipped cream 245, 247
whipping cream 104, 245, 247, 249

X

Xanthomonas maltophilia 324
X-propyl-dipeptidylamino-peptidase (X-pro-DPAP) 190, 191

Y

Yarrowia lipolytica 17, 138, 140, 151, 255
yeasts 2, 3, 9, 11, 13, 16, 17, 23, 24, 39, 86, 87, 89, 90, 104, 113–115, 117–119, 124, 127, 129, 137, 138, 140, 150–152, 157, 158, 167, 192, 194–201, 223, 232, 245, 252, 254, 255, 260–262, 279, 294, 324, 329, 332, 334, 336, 340, 356, 366, 367, 369, 404, 405, 426, 431, 432
Yersinia enterocolitica 4, 7, 18, 20, 108, 109, 118, 336, 356, 362, 390, 404, 419
yogurt 3, 16, 19, 44, 46, 53, 79, 86, 87, 97, 129, 167–169, 171–178, 180–192, 199, 214–216, 218–222, 224–237, 261, 262, 276, 289, 291, 298, 305, 306, 328, 337, 356, 383–385, 425, 426, 434

Z

Zygosaccharomyces spp. 262
Zygosaccharomyces florentinus 196

Color Plate Section

Chapter 7

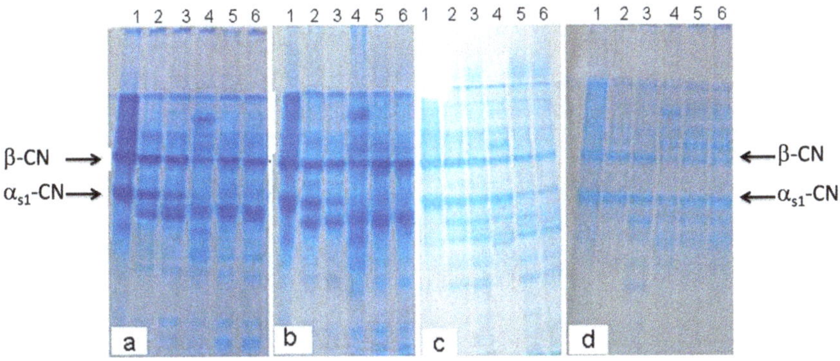

Figure 7.4 Urea-polyacrylamide gel electrophoretograms of Kasar cheeses calf rennet (a), chymosin (b), *Rhizomucor miehei* protease (c) and *Cryphonectria parasitica* protease (d). Lane 1: Na-caseinate; lanes 2–6: Kasar cheeses after 1, 15, 30, 60 or 90 days of ripening period, respectively. From Yasar and Guzeler (2011). With permission of Wiley & Sons Inc.

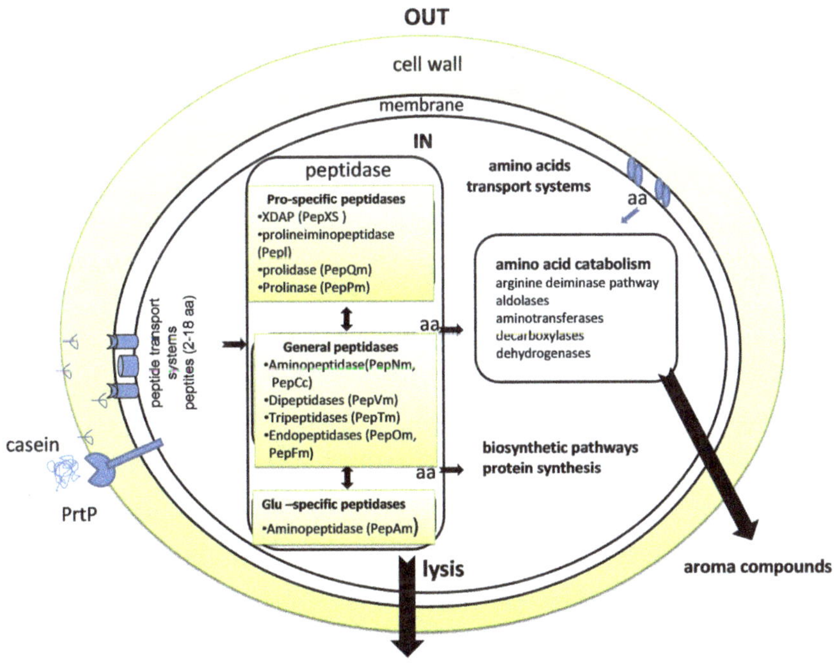

Figure 7.6 Catabolism of nitrogenous compounds in cheese during ripening and enzyme system of lactic acid bacteria. From Parente and Cogan (2004). With permission of Elsevier B.V.

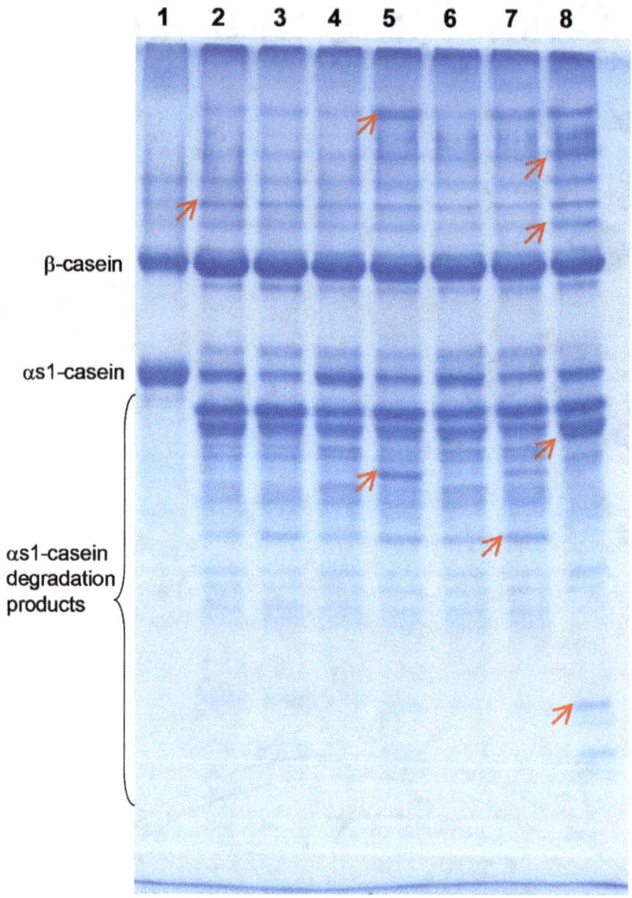

Figure 7.7 Urea-polyacrylamide gel electrophoresis electrophoretograms of Turkish White-brined cheese manufactured using defined lactococcal starters after 90 days of ripening. 1: Sodium caseinate, 2: *L. lactis* subsp. *lactis* UC317, 3: *L. lactis* subsp. *lactis* NCDO763, 4: *L. lactis* subsp. *cremoris* HP, 5: *L. lactis* subsp. *cremoris* SK11, 6: *L. lactis* subsp. *lactis* UC317 plus *L. lactis* subsp. *cremoris* HP, 7: *L. lactis* subsp. *lactis* NCDO763 plus *L. lactis* subsp. *cremoris* SK11, and 8: Starter-free. From Hayaloglu and Özer (2011). With permission of Sidas Publishing.

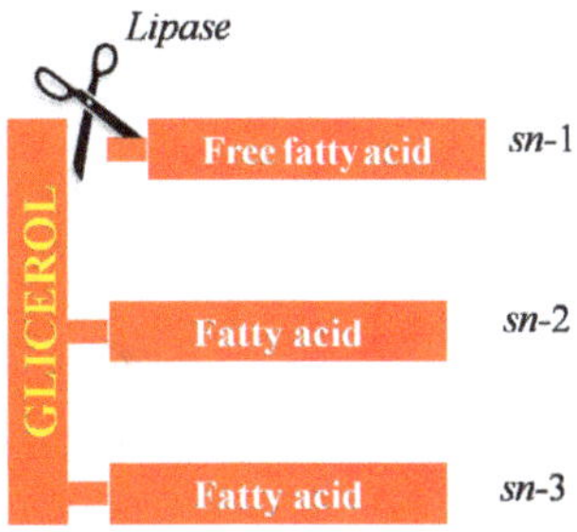

Figure 7.10 Lipase action on a triacylglycerol. From Hayaloglu and Özer (2011). With permission of Sidas Publishing.

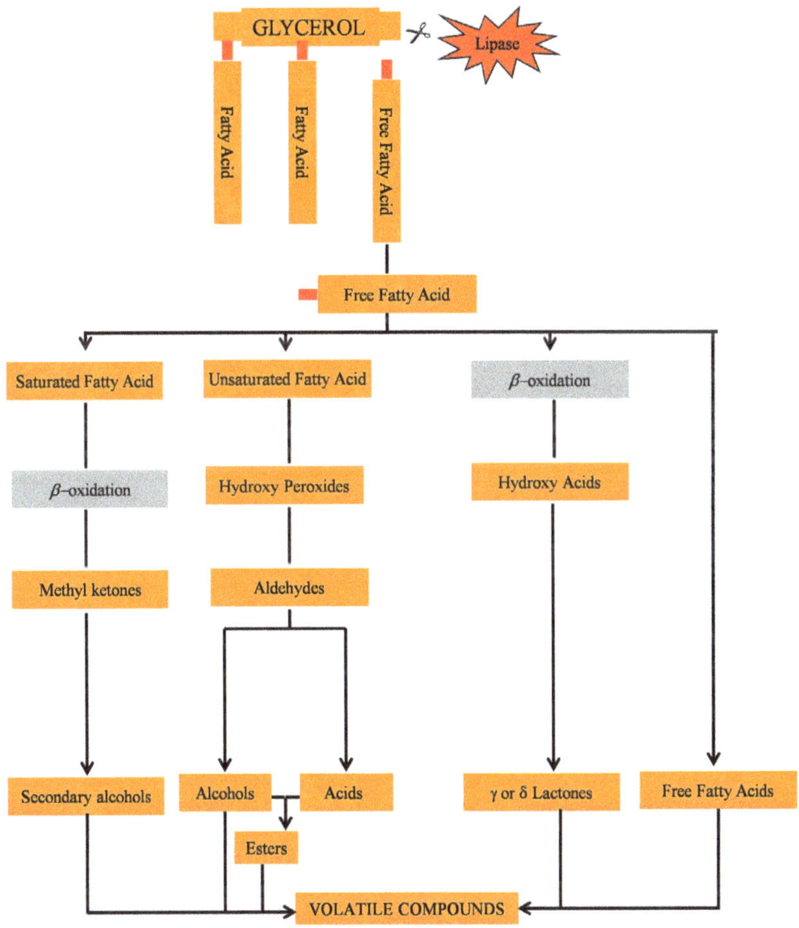

Figure 7.11 Fatty acid metabolism in cheese and volatiles originated from the fatty acids. From Molimard and Spinnler (1996), Hayaloglu and Özer (2011). With permission of Sidas Publishing.

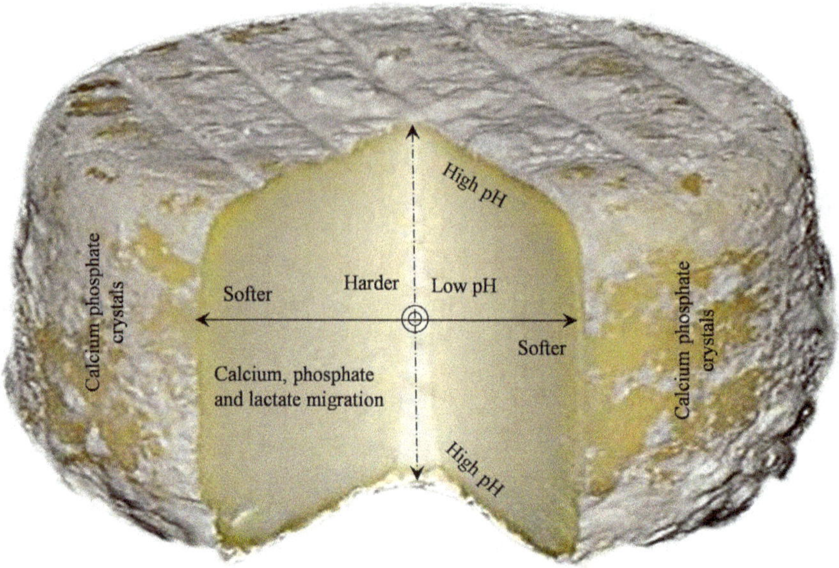

Figure 7.13 A representative demonstration of pH gradient, lactate metabolism and texture changes in a Camembert-type cheese.

For Product Safety Concerns and Information please contact our EU representative GPSR@taylorandfrancis.com
Taylor & Francis Verlag GmbH, Kaufingerstraße 24, 80331 München, Germany

www.ingramcontent.com/pod-product-compliance
Lightning Source LLC
LaVergne TN
LVHW011735250926
848950LV00003B/308
* 9 7 8 1 4 8 2 2 3 5 0 2 9 *